Archives of Toxicology, Supplement 18

Springer

Berlin
Heidelberg
New York
Barcelona
Budapest
Hong Kong
London
Milan
Paris
Santa Clara
Singapore
Tokyo

Toxicology-
From Cells to Man

Proceedings of the
1995 EUROTOX Congress
Meeting Held in Prague, Czech Republic,
August 27-30, 1995

Edited by the Publication Commitee
J. P. Seiler, O. Kroftová and V. Eybl

With 65 Figures and 41 Tables

Springer

Jürg P.Seiler

Interkantonale Kontrollstelle für Heilmittel
Erlachstrasse 8
CH-3000 Bern 9

Olga Kroftová

Charles University of Prague
Department of Pharmacology
3rd Medical School
Ruská 87

Vladislav Eybl

Charles University, Faculty of Medicine
Department of Pharmacology
Karlovarská 48
CZ-301 66 Pilsen

Library of Congress Cataloging-in-Publication Data.

EUROTOX Congress (1995 : Prague, Czech Republic)
Toxicology - from cells to man : proceedings of the 1995 EUROTOX Congress, meeting held in Prague, Czech Republic, August 27-30, 1995 / edited by the Publication Commitee, J.P. Seiler, O. Kroftová, and V. Eybl. p. cm. -- (Archives of toxicology. Supplement ; 18) Includes bibliographical reference and index. 1.

ISBN-13: 978-3-642-64696-6 e-ISBN-13: 978-3-642-61105-6
DOI: 10.1007/978-3-642-61105-6

Toxicology-Congress. I. Seiler, J.P. (Jürg P.), 1942- II. Kroftová, O. (Olga) III. Eybl, Vladislav. IV. Title. V Series. RA 1190. E8 Suppl vol. 18 [RA 1191] 615.9 s--dc20 [615.9]

ISBN-13: 978-3-642-64696-6 Springer-Verlag Berlin Heidelberg New York

Softcover reprint of the hardcover 1st edition 1996

Cover design: Springer-Verlag, Design & Production
Typesetting: Camera ready by authors
SPIN: 10526359 27/3136 – 5 4 3 2 1 0 Printed on acid free paper

Contents

Intercellular Communication

Renal Toxicology

Metal Antidotes - New Developments

Free Radicals

Toward Risk Assessment of Heterocyclic Amines in Food

The Young Scientist Poster Award

The Gerhard Zbinden Memorial Lecture

(Chair: H.M. Bolt)

EUROTOX has institued this lecture to honor the memory of Gerhard Zbinden (1924 - 1993), EUROTOX Honorary Member and recipient of the EUROTOX Merit Award. The Gerhard Zbinden Memorial Lecture aims at recognising scientific excellence in the area of drug and chemical safety. The lecture is held at the annual EUROTOX Congress by a scientist chosen for his outstanding research contributions to the science of toxicology. The lecture is sponsored by the Chemical Industries Basel (KGF: Ciba-Geigy AG, F. Hoffmann-La Roche AG, Lonza AG, Sandoz AG), Switzerland.

Alteration of Cell Signalling in Chemical Toxicity

Pierluigi Nicotera
Chair of Molecular Toxicology, Faculty of Biology, University of Konstanz, P.O. Box 5560 X911, D 78434, Konstanz, Germany

Introduction

Function of complex multicellular organisms demands that individual cells integrate in a communication network. Chemical signalling - elicited by hormones, neurotransmitters, cytokines and intercellular traffic through communicating junctions - achieves this purpose. It is therefore not surprising that alterations in cell signal transduction can have serious implications for organ function. Several toxic agents interfere with normal signal transduction. This may lead to alterations in cell plasticity and differentiation or to the activation or suppression of the cell death program, apoptosis. In turn, affected cells within a population may fail to deliver appropriate signals to neighbouring cells causing a progressive loss of trophic stimulation. Finally, this may cause further damage and ensue in long-lasting pathological effects.

Over the past several years many laboratories including our own have investigated the effect of chemical agents on cell signal transduction. One apparent conclusion of these investigations has been that adverse effects on signal transduction systems invariably precede the most classical signs of cytotoxicity such as loss of viability. Often the levels of a given toxic chemical required to cause significant alterations in cell signalling are several folds lower than those required to produce cell lysis. It has also become clear that increasing the dose of a given toxicant will result in the progressive recruitment of multiple cytotoxic mechanisms leading to rapid cell necrosis. More recently, most studies on physiological cell death, apoptosis, have emphasized the concept that interference with signalling systems can be a very efficient way to stimulate or repress cells' own death program. Based on these developments, we may then design a dose-response relationship where low doses of a given toxicant would affect cell signalling, promote alterations in cell differentiation or proliferation, and activate or suppress apotosis. As the level of exposure increases, the recruitment of multiple mechanisms would rapidly lead to cell death either by apoptosis or necrosis (Fig. 1). Within a tissue, both types of cell death may be present concomitantly and often chained in a concausal relationship. For example, when the ability of scavenger cells to engulf apoptotic bodies is overwhelmed, secondary necrosis can follow (Leist et al., 1995). On the other hand, an initial necrotic death may result in a local deprivation of trophic factors leading to apoptosis of neighbouring cells. Similarly, necrosis in a large part of the cell population can elicit inflammatory responses, production of cytokines that would in turn trigger apoptosis in the survivors. It is also noteworthy that the individual response is strongly dependent from the cell differentiation state. Thus, as described below a similar signalling alteration may sort different effects in for example differentiated versus immature neurons. In the following sections, I will briefly review some recent findings obtained in our and other laboratories and their implications for the model described above.

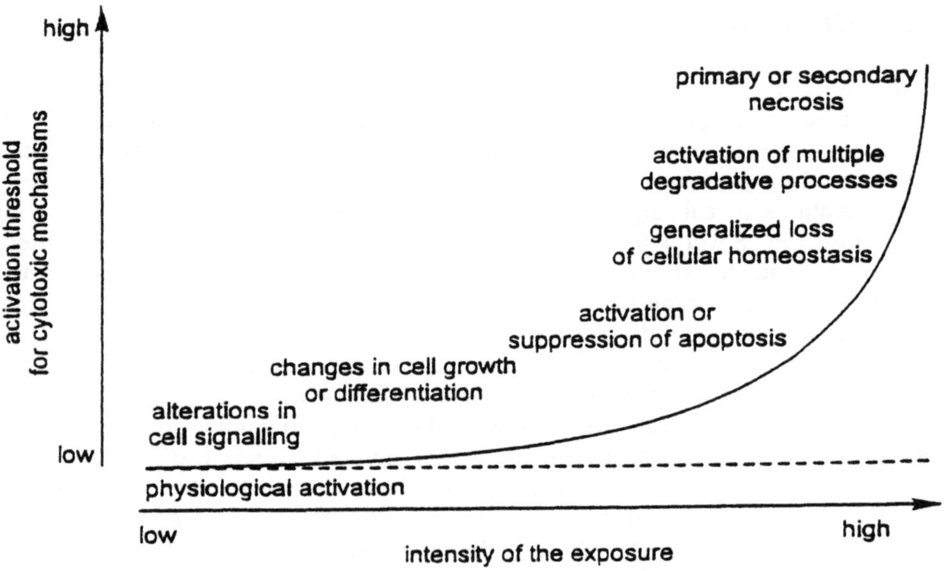

Figure 1: Cell responses to toxicants as function of the exposure intensity: different mechanisms and end-points in the dose-response relationship

Apoptosis and Necrosis: two distinct modes of cell death

Apoptosis and necrosis are two distinct forms of cell death, which have different defining morphologic and molecular features and implications for the surrounding tissue. Apoptosis is an active process of cell destruction characterized by cell shrinkage, chromatin aggregation with extensive genomic fragmentation and nuclear pyknosis (Kerr et al., 1972; Wyllie et al., 1980). *In vivo*, phagocytes normally sequester antigenically modified apoptotic cells, preventing inflammation and damage to the surrounding tissue (Duvall et al., 1985; Savill et al., 1993). Necrosis is instead characterized by passive cell swelling, intense mitochondrial damage with rapid energy loss and generalized disruption of internal homeostasis. This swiftly leads to membrane lysis and release of intracellular constituents. The latter evoke an inflammatory reaction with local cellular infiltration, vascular damage, edema, and injury to the surrounding tissue (Swartz, 1993).

Apoptosis plays an important role in both physiological and pathological conditions. In the development of the embryo, overproduction of cells ultimately requires that surplus elements die as part of the process that balances growth and differentiation with death. In the central and peripheral nervous systems, apoptosis occurs extensively during this phase (Deckwerth and Johnson, 1993a). However, disregulation of apoptosis may also underlie the aetiology of several diseases. For example, suppression of apoptosis may be involved in the proliferation of preneoplastic or neoplastic cells and in the development of autoimmune diseases (Thompson, 1995). Increased apoptosis may be instead involved in neurodegenerative diseases (Kure et al., 1991; MacManus et al., 1993; Raff et al., 1993), during retroviral infection (HIV) (Ameisen and Capron, 1991), and perhaps in the development of diabetes (Juntti-Berggren et al., 1993; Ankarcrona et al., 1994).

Signalling in apoptosis

Although the signals leading to apoptosis are probably multiple, recent work suggests that activation of a protease, or family of proteases, related to the nematode gene ced-3 and human interleukin-1β converting enzyme (ICE) (Yuan et al., 1993) may be a common converging point. Activation or depression of these protease families may be triggered by intracellular messengers such as calcium ions, cAMP or by radicals including the nitric oxide and superoxide radicals.

The biochemical machinery responsible for these events seems present in all mammalian cells, except blastomeres. Some of these proteolytic systems may be Ca^{2+} dependent (Zhivotovsky et al., 1994) and direct evidence that Ca^{2+} increases can mediate apoptotic death has been provided by several laboratories including ours (Juntti-Berggren et al., 1993; Nicotera et al., 1994 a and b and ref. therein). Thus, elevation of the intracellular Ca^{2+} level appears to represent a common trigger for apoptosis in cells of diverse tissue origins, and proteins that mediate the effects of Ca^{2+} are ubiquitously involved.

The overwhelming accumulation of reactive oxygen species resulting in the perturbation of the cellular prooxidant-antioxidant balance (oxidative stress) has also been implicated as a relevant mechanism of cytotoxicity in many biological systems. Both apoptosis and necrosis can be induced by oxidative stress. For example, low prooxidant concentrations induce apoptosis in RINm5F pancreatic β cells, whereas higher concentrations promote necrosis (Dypbukt et al., 1994). The involvement of radicals in the pathogenesis of neurodegenerative disorders is suggested by a number of observations. Behl et al. (1994) have recently reported that the amyloid β peptide kills cells by H_2O_2 -mediated mechanisms, while another study has shown that suppression of Cu, Zn-SOD activity induces apoptosis in PC12 cells (Troy & Shelanski, 1994). Analogously, antioxidants may protect cells from apoptosis (see Buttke & Sandstrom, 1994 for recent review). For example, apoptotic cell death induced by tumor necrosis factor α (TNF-α) can be prevented by antioxidants such as N-acetylcysteine (NAC) and thioredoxin (Chang et al., 1992; Matsuda et al., 1991) or by overexpression of mitochondrial manganese-superoxide dismutase (Mn-SOD) (Hirose et al., 1993). There are several possible roles for oxidation in apoptosis. Protein oxidation may be essential to influence gene expression required for the signals leading to apoptosis. Oxidative events for example are involved in changes in transcription factor DNA-binding ability (Abate et al., 1990; Meyer et al., 1993). However, it is also possible that oxidative events can cause cytoskeletal alterations, initiate cell shrinkage or promote changes in high- order chromatin structure.

Apoptosis and necrosis elicited by low and high doses of toxicants in primary neurons.

Recently, much attention has focused on the role of Ca^{2+} signalling alterations and mild oxidative stress in the development of neurotoxic disorders (Coyle and Puttfarcken, 1993). Mild oxidation can cause receptor alterations, affect Ca^{2+} channel conductance, and ultimately interfere with growth factor stimulation (Rossi et al., 1993). In contrast, marginally higher pro-oxidant levels can cause death (Dypbukt et al., 1994). Amplification of agonist-stimulated Ca^{2+} pulses in cells previously exposed to mild oxidation can potentiate physiological processes such as cell secretory activity or differentiation (Viviani et al., 1995b; Rossi et al., 1993). Again, when the duration and extent of the Ca^{2+} increase exceed the cell regulatory capacity death may ensue (Jiang et al., 1994).

Apparently the Ca^{2+} overload elicited by excess stimulation of glutamate and a concomitant oxidative stress are involved in neuronal cell death following ischemic injury. Both Ca^{2+} overload and oxidative stress have also been implicated as effectors of cell death in other acute and chronic neurologic diseases (Choi, 1988; Coyle, 1993). The cellular mechanism responsible for cell death after activation of glutamate receptor subtypes - of which the N-methyl-D-aspartate (NMDA) receptor is one of the most studied - are not yet clarified. In acute neurologic diseases, such as stroke and head trauma, excitotoxicity may be related to excessive glutamate release and/or a lack of clearance from synaptic clefts, which results in excessive stimulation of NMDA-receptors (Meldrum and Gartwaite, 1990). In chronic neurodegenerative diseases, however, a slow excitotoxic process is more likely to occur and lead to neuronal death (Beal et al., 1993).

The interaction of glutamate with excitatory amino acids (EAA) receptor subtypes initiates a cascade of events involving excessive Ca^{2+} entry and activation of several enzymes including phospholipases and nitric oxide synthase, NOS (Choi, 1988; Dawson et al., 1991; Manev et al., 1989; Trout et al., 1993). Phospholipase A2 activation leads to the generation of arachidonic acid and other metabolites as well as to the formation of oxygen free radicals. This can lead to an oxidative stress, which - together with the activation of Ca^{2+}-dependent catabolic processes (Nicotera et al., 1992) - would result in either necrosis or apoptosis. In addition, it has been shown that the $NO^{\cdot}$ - concomitantly generated by activation of NOS - can react with O_2 forming peroxynitrite (Beckman et al., 1990), a cytotoxic agent (Lipton et al., 1993) present during inflammation, sepsis and ischemia/reperfusion. In collaboration with Dr. Lipton's laboratory, we have recently investigated whether apoptosis would be induced by $NO^{\cdot}$ and related radicals in neuronal cells. We exposed cortical cultures to high or low concentrations of the glutamate agonist, NMDA or $NO^{\cdot}$-donors such as 3-morpholinosydnonimide (SIN-1), S-nitrosocysteine (SNOC) and peroxynitrite ($OONO^{-}$) (Bonfoco et al., 1995). We found that exposure of cortical cultures to relatively short durations or low concentrations of NMDA, SNOC or SIN-1, induced delayed neuronal death characterized by apoptotic features. In contrast, intense exposure to high concentrations of NMDA or peroxynitrite induced necrotic cell death. Superoxide dismutase (SOD) and catalase could attenuate neuronal cell death most likely by reducing the formation of peroxynitrite. These findings suggest that the intensity of the original insult may decide the resulting pathway to either necrotic or apoptotic cell death. The nature of the original insult and the decision to enter the necrotic versus the apoptotic pathway might have therapeutic implications in terms of the possible effectiveness of SOD/catalase or NMDA-antagonists.

We also investigated the mechanisms deciding whether exposure to glutamate would result in apoptosis or necrosis (Ankarcrona et al., 1995). Our findings proved that glutamate can induce either early necrosis or delayed apoptosis in cultures of cerebellar granule cells. During and shortly after exposure to glutamate, a subpopulation of neurons died by necrosis. In these cells, mitochondrial membrane potential collapsed, nuclei swelled, and intracellular debris were scattered in the incubation medium. Neurons surviving the early necrotic phase recovered mitochondrial potential and energy levels. Later, they underwent apoptosis, as shown by the formation of apoptotic nuclei and chromatin degradation into high and low molecular weight fragments. These results suggest that mitochondrial function is a critical factor that decides the mode of neuronal death in excitotoxicity.

Alterations of cell signalling alter differentiation and promote apoptosis in neural cells.

Cellular responses to endogenous or environmental toxicants are also quite dependent on the degree of differentiation. Thus, cell death may occur in dividing cells because of unbalanced mitogenic stimulation, whereas the same condition may not cause cytotoxicity in differentiated cells (Franklin and

Johnson, 1992; Hartley et al., 1993). Recent studies have shown that metals such as mercury and tin can affect neurotransmitter production, alter neural differentiation and plasticity, and ultimately activate apoptosis in neural cell lines (Rossi et al, 1993, 1995, Viviani et al., 1995 a). In particular, exposure of PC12 cells to inorganic mercury causes alterations of L-type Ca^{2+} channel properties, which results in potentiation of Ca^{2+} signals elicited by depolarization or agonist stimulation. Mercury potentiated intracellular Ca^{2+} responses to hormone stimulation or to KCl-induced depolarization and enhanced cell differentiation induced by NGF (Rossi et al., 1993). Conversely, Hg^{2+} concentrations that potentiate PC12 cell differentiation caused the activation of apoptosis in post-mitotic cerebellar granule cells (CGC). The observation that Hg^{2+} addition caused a sustained $[Ca^{2+}]_i$ increase after stimulation with the glutamate receptor agonists NMDA, suggested that Hg^{2+} in combination with endogenously produced glutamate (Gallo et al., 1982) elicited Ca^{2+} increases high and sustained enough to promote apoptosis (Rossi et al., 1995).

The observation that Hg^{2+} caused apoptosis in CGC cells is more relevant if one considers that Hg^{2+} concentrations that were not toxic for CGC, became lethal when associated with low NMDA concentrations. Interestingly, morphological studies in organotypic cultures of rat cerebellum have recently shown that not-toxic mercury concentrations can lower the threshold for glutamate neurotoxicity (Matyja and Albrecht, 1993). These findings and those reported here may even be more relevant in view of the demonstration that low Hg^{2+} concentrations can inhibit glutamate uptake in astrocyte cultures (Brookes, 1992).

Nevertheless, one obvious difference between PC12 cells and CGC is the stage of development and it is well known that the implications of enhanced Ca^{2+} signals in mature and immature neurons are quite different. While in immature neural cells, increased $[Ca^{2+}]_i$ protects cells lacking growth stimulation (Koike et al., 1989), Ca^{2+} overload in differentiated neurons causes altered neurotransmitter release and cell death (Hartley et al., 1993). Ca^{2+} responses to high K^+ are amplified with cell differentiation (Connors et al., 1987) and the types of ion channels expressed in mature cells are higher than those found in neuroblasts or in undifferentiated cells (Kalman et al., 1990; Janigro et al., 1989). While high Ca^{2+} levels may be required to maintain differentiated responses in adult neural cells, further Ca^{2+} increases would result in cell killing. According to this hypothesis, human neuroblastoma cells, SH-SY5Y, are more sensitive to Hg^{2+} toxicity after NGF-differentiation , despite the Ca^{2+} signalling alterations observed in both undifferentiated and differentiated clones are similar (A.D. Rossi and P. Nicotera, unpublished observations).

Conclusions

Apparently several environmental pollutants and chemical toxicants can exert their effects on cell signalling at very low concentrations. The effects observed both *in vivo* and *in vitro* include alterations in gap junction proteins, intracellular messenger production, channel conductance and gene expression. Local effects of such alterations include modification of cell differentiation and ultimately cell deletion. In addition, increased production of mediators during inflammation produced by chemical or environmental toxicants may mediate local or distal effect via cytokine production, using thereby cell signalling molecules to deliver cytotoxic effects. Moreover, the sensitivity of signal-transduction elements to chemicals makes them an ideal target for therapeutical approaches and potential indicators for early adverse effects of chemical injury.

References

Abate C, Patel N, Rauscher III, FJ and Curran T (1990). Redox regulation of Fos and Jun DNA-binding activity in vitro. Science 249:1157-1161.

Ameisen, JC and Capron, A (1991). Cells dysfuntion and depletion in AIDS: the programmed cell death hypothesis. Immunol.Today 12:102-105.

Ankarcrona M, Dypbukt J M, Bonfoco E, Zhivotovsky B, Orrenius S, Lipton SA, and Nicotera P (1995). Glutamate-induced neuronal death: a succession of necrosis or apoptosis depending on mitochondrial function. Neuron in press

Beal MF, Hyman B T and Koroshetz W (1993). Do defects in mitochondrial energy metabolism underlie the pathologie of neurodegenerative diseases? Trends Neurosci. 16:125-131.

Beckman JS, Beckman, TW, Chen J, Marshall P A and Freeman BA (1990). Apparent hydroxyl radical production by peroxynitrite: implications for endothelial injury from nitric oxide and superoxide. Proc. Natl. Acad. Sci. USA 87:1620-1624.

Behl C, Davis JB, Lesley R, and Schubert, D (1994). Hydrogen peroxide mediates amyloid - protein toxicity. Cell 77:817-827.

Bonfoco E, Krainc D, Ankarcrona M, Nicotera P and Lipton SA. (1995). Apoptosis and Necrosis: Two distinct events induced respectively by mild and intense insults with NMDA or nitric oxide/superoxide in cortical cell cultures. Proc. Natl. Acad. Sci. USA

Brookes N (1992) *In vitro* evidence for the role of glutamate in the CNS toxicity of mercury. Toxicology 76, 245-256.

Buttke TM and Sandstrom PA(1994). Oxidative stress as a mediator of apoptosis. Immunol. Today 15:7-10.

Chang D J, Ringold GM and Heller RA (1992). Cell killing and induction of manganous superoxide dismutase by tumor necrosis factor-alpha is mediated by lipoxygenase metabolites of arachidonic acid. Biochem. Biophys. Res. Commun. 188:538-546.

Choi DW (1988). Glutamate neurotoxicity and diseases of the nervous system. Neuron 1:623-634.

Coyle JT and Puttfarcken P. (1993). Oxidative stress, glutamate, and neurodegenerative disorders. Science 262:689-695.

Connor JA, Tseng H-Y and Hockberger PE (1987) Depolarization- and transmitter-induced changes in intracellular Ca^{2+} of rat cerebellar granule cells in explant cultures. J. Neurosci. 7, 1384-1400.

Dawson VL, Dawson TM, London ED, Bredt DS and Snyder S H (1991). Nitric oxide mediates glutamate neurotoxicity in primary cortical cultures. Proc. Natl. Acad. Sci. USA 88:6368-6371.

Deckwerth TL and Johnson Jr EM (1993). Neurotrophic factor deprivation-induced death. Ann. N. Y. Acad. Sci. 679:121-131

Duvall E, Wyllie AH and Morris RG (1985). Macrophage recognition of cells undergoing programmed cell death (apoptosis). Immunology 56:351-358.

Dypbukt J M, Ankarcrona M, Burkitt M, Sjöholm □, Ström K, Orrenius S and Nicotera, P. (1994). Different prooxidant levels stimulate growth, trigger apoptosis, or produce necrosis of insulin- secreting RINm5F cells: The role of intracellular polyamines. J. Biol. Chem. 269:30553-30560.

Hartley DM, Kurth MC, Bjerkness L, Weiss JH and Choi DW (1993) Glutamate receptor-induced $^{45}Ca^{2+}$ accumulation in cortical cell culture correlates with subsequent neuronal degeneration. J. Neurosci. 13, 1993-2000.

Hirose K, Longo DL, Oppenheim, JJ and Matsushima K (1993). Overexpression of mitochondrial manganese superoxide dismutase promotes the survival of tumor cells exposed to interleukin-1, tumor necrosis factor, selected anticancer drugs, and ionizing radiation. FASEB J. 7:361-368.

Jiang S-A, Chow SC, Nicotera P and Orrenius S. (1994) Intracellular Ca^{2+} signals activate apoptosis in thymocytes: studies using the Ca^{2+}-ATPase inhibitor thapsigargin. Exp.Cell Res. 212, 84-92.

Johnson JW and Ascher P (1987) Glycine potentiates the NMDA response in cultured mouse brain neurons. Nature 325, 529-531.

Janigro D, Maccaferri G and Meldolesi J (1989) Calcium channels in undifferentiated PC12 pheochromocytoma cells. FEBS Lett. 255, 398-400.

Juntti-Berggren L, Larsson O, Rorsman P, Ämmälä C, Bokvist K, Wåhlander K, Nicotera P, Dypbukt JM, Orrenius S, Hallberg A and Berggren P (1993). Increased activity of L-Type Ca^{2+} channels exposed to serum from patients with Type I diabetes. Science 261:86-90.

Kalman D, Wong B, Horvai AE, Cline MJ and O-Lague PH (1990) Nerve growth factor acts through cAMP-dependent protein kinase to increase the number of sodium channels in PC12 cells. Neuron 2, 355-366.

Kerr JFR, Wyllie AH and Currie A R (1972). Apoptosis: a basic biological phenomenon with wide ranging implications in tissue kinetics. Br. J. Cancer 26:239-257.

Kure S, Tominaga T, Yoshimoto T, Tada K and Narisawa K. (1991). Glutamate triggers internucleosomal DNA cleavage in neuronal cells. Biochem. Biophys. Res. Commun. 179:39-45.

Leist MF, Gantner F, Bohlinger I, Germann PG, Tiegs G, and Wendel A (1995). TNF-induced murine hepatic apoptosis as a pathomechanism of septic liver failure. Am. J. Pathol. 166, 1-15.

Lipton SA, Choi Y-B, Pan Z-H, Lei SZ, Chen, H-S V, Sucher N J, Loscalzo J, Singel DJ and Stamler JS. (1993). A redox-based mechanism for the neuroprotective and neurodestructive effects of nitric oxide and related nitroso-compounds. Nature 364:626-632.

MacManus J P, Buchan, AM, Hill IE, Rasquinha, I and Preston E. (1993). Global ischemia can cause DNA fragmentation indicative of apoptosis in rat brain. Neuroscience Lett. 164:89-92

Manev H, Favaron M, Guidotti A and Costa E 1989. Delayed increase of Ca^{2+} influx elicited by glutamate: role in neuronal death. Mol. Pharmacol. 36:106-112.

Matsuda M, Masutani H, Nakamura H, Miyajima S, Yamauchi A, Yonehara S, Uchida A, Irimajiri K, Horiuchi A and Yodoi J (1991). Protective activity of adult T cell leukemia-derived factor (ADF) against tumor necrosis factor dependent cytotoxicity on U937 cells. J. Immunol. 147:3837-3841.

Matyja E and Albrecht J (1993) Ultrasctructural evidence that mercury chloride lowers the threshold for glutamate neurotoxicity in an organotypic culture of rat cerebellum. Neurosci. Lett. 158: 155- 158.

Meldrum B and Garthwaite J (1990). Excitatory amino acid neurotoxicity and neurodegenerative diseases. Trends Pharmacol. Sci. 11, 379-387.(Abstract).

Meyer M, Schreck R and Baeuerle PA (1993). H2O2 and antioxidants have opposite effects on activation of NF-KB and AP-1 in intact cells: AP-1 as secondary antioxidant-responsive factor. EMBO J. 12:2005-2015.

Nicotera P, Zhivotovsky B, Bellomo G, and Orrenius S (1994a). Ion signalling in apoptosis. In: Apoptosis, eds. R. T. Schimke and E. Mihich, pp. 97-115. Plenum Press, New York.

Nicotera P, Zhivotovsky B, and Orrenius S (1994b). Nuclear calcium transport and the role of calcium in apoptosis. Cell Calcium 1994, 279-288.

Raff MC, Barres BA, Burne JF, Coles HS, Ishizaki Y, and Jacobson MD. (1993). Programmed cell death and the control of cell survival: Lessons from the nervous system. Science 262:695-700.

Rossi AD, Larsson O, Manzo L, Orrenius S, Vahter M., Berggren P-O and Nicotera P (1993) Modification of Ca^{2+} signalling by inorganic mercury in PC12 cell. FASEB J. 7, 1507-1514.

Rossi AD, Viviani B, Zhivotovsky B, Manzo L, Orrenius S, Vahter M and Nicotera M. Inorganic mercury modifies Ca^{2+} signalling, triggers apoptosis and potentiates NMDA toxicity in neural cells.(1995) (submitted for publication)

Savill JS, Fadok V, Henson P. and Haslett C (1993). Phagocyte recognition of cells undergoing apoptosis. Immunol. Today 14:131-136.

Schwartz, L M, Smith SW, Jones MEE. and Osborne BA. (1993). Do all programmed cell deaths occur via apoptosis? Proc. Natl. Acad. Sci. USA 90:980-984.

Thompson C B. (1995). Apoptosis in the pathogenesis and treatment of disease. Science 267:1456-1462.

Troy CM and Shelanski ML (1994). Down-regulation of copper/zinc superoxide dismutase causes apoptotic death in PC12 neuronal cells. Proc. Natl. Acad. Sci. USA 91: 6384-6387.

Viviani B, Rossi AD, Chow SC and Nicotera P (1995a) Organotin compounds induce calcium overload and apoptosis, in PC12 cells. Neurotoxicology. 16:19-26, 1995

Viviani B, Rossi AD, Chow SC and Nicotera P (1995b) Triethyltin interferes with Ca^{2+} signaling and potentiates norepinephrine release in PC12 cells (submitted for publication)

Wyllie AH, Kerr J FR and Currie, AR(1980) Cell death: The significance of apoptosis. Int. Rev. Cytol. 68:251-306.

Yuan J, Shaham S, Ledoux S, Ellis HM and Horvitz HR (1993). The c. elegans cell death gene ced-3 encodes a protein similar to mammalian interleukin-1β-converting enzyme. Cell 75:641-652.

Zhivotovsky B, Wade D, Gahm A, Orrenius S, and Nicotera P (1994). Formation of 50 KBP chromatin fragments in isolated liver nuclei is mediated by protease and endonuclease activation. FEBS Lett. 351:150-154.

Human Immunotoxicology and the Environment

(Special Symposium, organized by the Immunotoxicity and Chemical Allergy Speciality Section of EUROTOX)

Halogenated Organic Compounds and the Human Immune System: Preliminary Report on a Study in Hobby Fishermen

M. Løvik,[1] H.R. Johansen,[1] P.I. Gaarder,[1] G. Becher,[1] I.S. Aaberge,[1] W. Gdynia,[2] and J. Alexander.[1]
[1]National Institute of Public Health, Geitmyrsvn. 75, 0462 Oslo, Norway
[2]Herre Medical Center, 3965 Herre, Norway.

Introduction

Polychlorinated dibenzo-*p*-dioxins (PCDDs), polychlorinated dibenzofurans (PCDFs), and polychlorinated biphenyls (PCBs) are liphophilic compounds markedly resistant to biological degradation. They persist for a long time in nature and can accumulate within the food chain (Fletcher and McKay 1993). Humans, for example, fishermen consuming contaminated fish and other seafood from polluted lakes and seas, may therefore be exposed to substantially elevated levels of these compounds (Humphrey 1987, Svensson et al. 1991).

A relatively large amount of data from experimental animals is available on the effects from PCDDs, PCDFs, and PCBs on the immune system (Vos and Luster 1989), but most of the results are from acute or subchronic exposures with doses often much higher than those encountered in nature. There are only few reports on immunotoxic effects from long-time, low grade exposure to halogenated organic compounds in wild animals, exposed more in the way the general human population may be exposed (Ross et al. 1995, Lahvis et al. 1995, Swart et al. 1994). Further, the considerable species differences in toxicity of the compounds in question (Kimbrough 1995) make extrapolation to man difficult. When it comes to human data, there are some reports on accidental exposure of short duration to high doses of PCBs, PCDDs, and PCDFs (Bekesi et al. 1979, Chang et al. 1982, Elo et al. 1985, Kuratsune 1989, Lu and Wu 1985). A few reports can be found on possible toxic effects on the human immune system from exposure of long duration to low doses, the exposure situation most relevant for the general population (Evans et al. 1988, Knutsen et al. 1987, Lawton et al. 1985, Stehr et al. 1986).

The Frierfjord in South-Eastern Norway has for many years received discharge of organochlorine compounds from a magnesium production plant located at the inner part of the fjord. Contamination at various distances from the factory has been monitored in sediment and marine organisms (Knutzen and Oehme 1989, Johansen et al. 1993). Although emissions have declined much, levels of organochlorine compounds in fish and shellfish still exceed those in diffusely polluted areas by a factor of 2 - 40. There are restrictions on commercial fishing and recommendations concerning limited consumption of fish and shellfish. However, the area is popular for recreational purposes, and local hobby fishermen are known to catch and consume considerable amounts of fish and shellfish from the fjord. The crab *Cancer pagurus*, caught mainly during summer and autumn, has been found to contain high levels of organochlorine compounds showing the characteristic pattern of the emissions from the magnesium factory (Oehme et al. 1990).

The aim of our study was to determine if any correlations could be demonstrated between alterations in the immune system of crab-consuming hobby fishermen and blood levels of PCBs, PCDDs, and PCDFs. In this preliminary report, we describe the study. Further, some tendencies with regard to correlations between blood levels of PCBs/PCDDs/PCDFs and clinical chemistry/hematology/immune parameters apparent at the present stage of data analysis are indicated.

Methods

Twenty-four male hobby fishermen consuming locally caught crabs were recruited through local newspapers and radio. Ten male referents were drawn randomly from the Register of Population. All participants answered a questionnaire designed to obtain information about general health status, smoking habits, possible occupational and environmental exposure to PCDDs, PCDFs and PCBs, and fish and crab consumption. Venous blood was drawn from each participant after overnight fasting, and transported at controlled temperature to our laboratory within a few hours. EDTA was used for anticoagulation of part of the blood. The various immunological cell-based tests were performed the same afternoon, whereas clinical chemistry and hematology was performed by the Central Laboratory, Ullevål Hospital the next morning. The laboratory techniques will be described in detail elsewhere. Whole blood for toxicological analyses and serum samples were frozen and stored at -20 °C until analysed in our laboratories. The study was performed in May 1993, with a small blood sample being drawn in June 1993 to determine the response to vaccination. In February 1995, the study was repeated with a new questionnaire and blood specimens for clinical chemistry, hematology, and a somewhat reduced panel of immunological tests. Determination of PCBs, TCDDs, and TCDFs was performed only in 1993, and all clinical chemistry, hematology, and immunological test results from 1993 as well as from 1995 are analysed in relation to the 1993 toxicant levels.

In this preliminary data analysis, the non-parametric Spearman's rank sum correlation test was used throughout to determine the correlations between the various clinical chemistry, hematology, and immune parameters and the blood levels of PCBs, PCDDs, and PCDFs.

Results

Fish and crab consumption: Parallel to increased crab consumption, there was a slight, not statistically significant increase in fish intake. Data on blood levels of various fatty acids, relevant for the discussion of effects on the immune system, will be available but have not yet been incorporated into the study.

PCB, PCDD, and PCDF: No correlation was found between age (median 46 years, range 40-54 years) and blood levels of PCBs, PCDDs, or PCDFs. A large variation in blood levels of many PCDDs and PCDFs was found (Johansen et al., submitted for publication), especially for the dibenzofurans which are characteristic for the pollution in the fjord and which are also found in the crabs (Oehme et al. 1990). The PCDD/PCDF related TEQs (toxic equivalents) (Safe 1992) in the referent persons in our study was comparable with levels found in the general population in Germany (Päpke et al. 1994), whereas the mean blood level in those hobby fishermen in our group that consumed more than 50 crabs per year was about five times higher. There were statistically significant correlations between the number of crabs consumed per year and blood levels of several PCDDs/PCDFs. In contrast, no statistically significant associations were found between total PCB TEQs and crab intake, and the increase in the levels of several PCB congeners among the high-intake group (>50 crabs/year) compared to the levels in the referents was relatively small. Thus, the hobby fishermen in our study had moderately elevated levels of PCBs, of the same magnitude as other groups of fishermen not consuming crabs from the Frierfjord, and the source of the PCBs probably was not the magnesium factory.

Clinical chemistry: A statistically significant negative correlation between PCDD/PCDF levels and creatinine levels was observed. With bilirubin, a similar tendency was seen. In contrast, there was a tendency to a positive correlation between PCB levels and serum levels of the liver enzymes ALAT and γ GT.

Hematology: There was a positive correlation between some red blood cell parameters (hemoglobin, packed red blood cell volume, and red blood cell numbers) and PCB levels. In contrast, negative correlations were found between certain white blood cell parameters and the levels of PCDDs/PCDFs.

Lymphocyte subpopulations determined by flow cytometry: There was a tendency to a negative correlation between the percentage of cells positive for some B lymphocyte markers and blood levels of PCDDs/PCDFs.

Natural killer (NK) cell numbers and function: By flow cytometry, no correlation between the percentage of $CD56^{+}$ cells and PCB/PCDD/PCDF levels was observed. The same applied to NK cell function as determined by the chromium release assay.

Response to mitogen stimulation: The response to the mitogens Con A and PHA appeared to be positively correlated to blood levels of PCDDs/PCDFs.

Serum immunoglobulin levels: A negative correlation between serum IgM levels and blood levels of TCDDs/TCDFs was observed. In contrast, the corresponding correlation tended to be positive for serum IgE.

Autoantibodies: Among a number of autoantibodies examined, a negative correlation between blood PCB levels and anti-nuclear antibodies (ANA) was observed.

Response to vaccination with tetanus and diphteria toxoid: The secondary response to a booster dose with tetanus and diphteria toxoid was determined. The analysis of this response is pending awaiting confirmation of pre-booster vaccination status for some of the participants.

Conclusions

The revision of the data is not completed, nor is the data analysis final. Some additional laboratory analyses will be performed on frozen material. This will bring in some new parameters and give additional information on some existing ones. However, at this stage of analysis, it appears that in our group of hobby fishermen consuming crabs from a polluted fjord, there are positive or negative correlations between blood levels of PCDDs/PCDFs as well as PCBs and several immune system parameters, in addition to parameters of clinical chemistry and hematology. The results are encouraging with regard to the use of epidemiological studies of populations with long-term, low grade exposure as a tool to determine the immunotoxic effects of halogenated organic compounds.

References

Bekesi JG, Anderson HA, Roboz JP, Roboz J, Fischbein A, Selikoff IJ and Holland JF (1979) Immunologic dysfunction among PBB-exposed Michigan dairy farmers. Ann NY Acad Sci 320:717-728

Chang K-J, Hsieh K-H, Tang S-Y, Tung T-C and Lee T-P (1982) Immunologic evaluation of patients with polychlorinated biphenyl poisoning: evaluation of delayed-type skin hypersensitive response and its relation to clinical studies. J Toxicol Environ Health 9:217-223

Elo O, Janhunen H and Rantanen J (1985) Recent PCB accidents in Finland. Environ Health Perspect 60:315-319

Evans RG, Webb KB, Knutsen AP, Roodman ST, Roberts DW, Bagby JR, Garrett WA and Andrews JS (1988) A medical follow-up of the health effects of long-term exposure to 2,3,7,8-tetrachlorodibenzo-p-dioxin. Arch Environ Health 43:273-278

Fletcher CL and McKay WA (1993). Polychlorinated dibenzo-p-dioxins (PCDDs) and dibenzofurans in the acquatic environment - a literature review. Chemosphere 26:1041-1069

Humphrey HEB (1987) The human population - an ultimate receptor for aquatic contaminants. Hydrobiologia 149:75-80

Johansen HR, Rossland OJ and Becher G (1993) Congener specific determination of PCBs in crabs from a polluted fjord region. Chemosphere 27:1245-1252

Kimbrough RD (1995) Polychlorinated biphenyls (PCBs) and human health: an update. Crit Rev Toxicol 25:133-163

Knutsen AP, Roodman ST, Evans RG, Mueller KR, Webb KB, Green PS, Hoffman RE and Schramm WF (1987) Immune studies in dioxin-exposed Missouri residents: Quail Run. Bull Environ Contam Toxicol 39:481-489

Knutzen J and Oehme M (1989) Polychlorinated dibenzofuran (PCDF) and dibenzo-p-dioxin (PCDD) levels in organisms and sediments from the Frierfjord, Southern Norway. Chemosphere 19:1897-1909

Kuratsune M (1989) Yusho, with reference to Yu-Cheng. In: Halogenated biphenyls, terphenyls, dibenzodioxins and related products. Kimbrough RD and Jensen MA (eds) Elsevier Science Publishers, Amsterdam, pp 295-322

Lahvis GP, Wells RS, Kuehl DW, Stewart JL, Rhinehart HL and Via CS (1995) Decreased lymphocyte response in free-ranging bottlenose dolphins (*Tursiops truncatus*) are associated with increased concentrations of PCBs and DDT in peripheral blood. Environmental Health Perspect 103 (Suppl 4):67-72

Lawton RW, Ross MR, Feingold J and Brown JF (1985) Effects of PCB exposure on biochemical and hematological findings in capacitor workers. Environ Health Perspect 60:165-184

Lu Y-C and Wu Y-C (1985) Clinical findings and immunological abnormalities in Yu-Cheng patients. Environ Health Perspect 59:17-29

Oehme M, Batonova A and Knutzen J (1990) Effect of polychlorinated dibenzofuran and dibenzo-p-dioxin contamination of a coastal region using isomer profiles in crabs. Environ Sci Technol 24:1836-1841

Päpke O, Ball M, Lis ZA and Scheunert K (1994) PCDD/PCDF in humans, a 1993-update of background data. Chemosphere 29:2355-2369

Ross PS, De Swart RL, Reijnders PJH, Van Loveren H, Vos JG and Osterhaus ADME (1995) Contaminant-related suppression of delayed-type hypersensitivity and antibody responses in harbor seals fed herring from the Baltic Sea. Environ Health Perspect 5:149-162

Safe S (1992) Development, validation, and limitation of toxic equivalency factors. Chemosphere 25:61-64

Stehr PA, Stein G, Falk H, Sampson E, Smith SJ, Steinberg K, Webb K, Ayres S, Schramm W, Donnell HD and Gedney WB (1986) A pilot epidemiologic study of possible health effects associated with 2,3,7,8-tetrachlorodibenzo-p-dioxin contaminations in Missouri. Arch Environ Health 41:16-22

Svensson B-G, Nilsson A, Hansson A, Rappe C, Åkeson B and Skerfving S (1991) Exposure to dioxins an dibenzofurans through the consumption of fish. New Engl J Med 324:8-12

Swart RL, Ross PS, Vedder LJ, Timmermann HH, Helsterkamp S, Van Loveren H, Vos JG, Reijnders PJH and Osterhaus ADME (1994) Impairment of immune function in harbor seals (*Phoca vitulina*) feeding on fish from polluted waters. Ambio 23:155-159

Vos JG and Luster MI (1989) Immune alterations. In: Halogenated biphenyls, terphenyls, dibenzodioxins and related products. Kimbrough RD and Jensen MA (eds) Elsevier Science Publishers, Amsterdam, pp 295-322

Risk Assessment for the Harmful Effects of UVB Radiation on the Immunological Resistance to Infectious Diseases

Henk Van Loveren, Wim Goettsch, Wout Slob, and Johan Garssen
National Institute of Public Health and Environmental Protection, PO BOX 1, 3720 BA Bilthoven, The Netherlands

Abstract

Risk assessment comprises four steps: hazard identification, dose-response assessment, exposure assessment, and risk characterization. According this scheme, we have analysed the effects of UVB radiation on basal immune functions in rats and man, and the immunological resistance to infectious diseases in rats. Non-threshold mathemical methods were used in order to estimate the risk for the human population after increased exposure to UVB radiation. These data demonstrate that UVB radiation, at doses relevant to outdoors exposure, may affect the immunological resistance to infectious diseases in human individuals. This study may also provide a basis for a strategy to assess the risk of adverse effects of exposure to immunotoxic agents.

Introduction

Experimental data on the effects of immunotoxicants on resistance to infectious diseases, easily obtained in laboratory animals, cannot be acquired using humans for ethical reasons. A possible alternative is extrapolation of animal data to humans with respect to such effects. These extrapolated data may be utilized as a basis for risk assessment of immunotoxicological compounds.

Risk assessment is a process of analyzing relevant biological, dose-response and exposure data for a particular agent, in an attempt to establish qualitative and quantitative estimates of adverse effects on human health (Scala 1991). As defined by the National Academy of Science (1983), risk assessment comprises four steps: hazard identification, dose-response assessment, exposure assessment, and risk characterization. This general framework covers the process of assessing risk of cancer as well as non-cancer endpoints, including decreased resistance to infections.

In this report we describe our first attempt to perform an assessment of the risk of effects of UVB radiation on the immunological resistance to infectious diseases.

Hazard Identification

Immunological studies showing the effects of UVB exposure on the immune system of the rat have been described. It was shown that suberythemal doses (0.25 and 0.50 of the minimal erythemal dose {MED}) of UVB radiation daily for 7 consecutive days could induce severe suppression of the natural killer (NK) cell activity and mixed lymphocyte responsiveness (MLR) of splenocytes in rats (Goettsch et al. 1994a, Garssen et al. 1994), both immune responses considered to correlate with resistance to infections. In a Listeria monocytogenes host resistance model in the rat a similar UVB exposure protocol (i.e. 0.5 MED daily for 7 consecutive days) resulted in an 8-fold increase in the number of bacteria in the spleen, 4 days after infection (Goettsch 1995). Delayed type hypersensitivity responses, a parameter for T cell mediated immunity, to Listeria was inhibited by these UVB exposure protocols, indicating that UVB exposure induces suppression of the specific immune response to Listeria, leading to delayed clearance of bacteria from the spleen.

In vitro, in situ and in vivo UVB exposure impaired several basal immune functions, such as T cell function (MLR responses), alloreactive capacity of (epi)dermal cells (MSLR) and the presence of MHC-II positive cells in the (epi)dermis of mice, rats and humans (Goettsch et al. 1994b).

From these studies it was concluded that low doses of UVB radiation, relevant to the outdoor situation are able to impair the immune system. Thus, UVB radiation is a potential hazard to immunological functions, and hence to the immunological resistance to infectious diseases in humans. Below, we have indicated how we have characterized the risk of immunosuppression by ultraviolet irradiation of the human population. Calculations of the various relationships that were used to quantitatively assess the risk were all based on experiments described by Goettsch (1995) and Goettsch et al. (1994b), unless stated otherwise.

The approach as described herein may be applicable for other immunotoxic agents.

Dose-Response Assessment

Linear and non-linear regression methods were applied to evaluate the relation between dose and effect and to calculate a dose that results in a 50% reduction of the response (effective dose 50%; ED50).

We wanted to extrapolate the effects of UVB irradiation on resistance to Listeria monocytogenes as was established in rats to the human situation. First, the results of experiments to establish these effects indicated a dose-dependent decrease of the proliferative response of lymphoid spleen cells to the bacterium. Using log-transformed data an $ED50_{rat,Lis,LST}$ was calculated to be 6.8 kJ/m^2 (cumulative dose from the UVB source FS40 lamps). This figure was used as the dose that decreased resistance in to Listeria in the rat.

We then composed a factor to account for interspecies variation (IEV) (between different species i.e. rats and humans), which is one of the keys to extrapolate data from experimental animals to man. Since it is known that UVB suppresses predominantly cellular immune responses, we took the mixed skin lymphocyte response te compare the effects of UVB on the immune system of rats and man. So, ED50's of the effect of *in situ* exposure to UVB radiation on the mixed skin lymphocyte responses in rats and humans were calculated. Analysis of these effects expressed on a log scale indicated linear dose-response curves; IEV was calculated on this basis by dividing the $ED50_{MSLR,human}$ (4176 J/m^2) by the $ED50_{MSLR,rat}$ (1085 J/m^2) leading to a factor of 3.85.

A third key that is required for the estimation of the risk in man is the intraspecies variation that can be encountered. The human subjects studied all had skin type 2. The ED50 of the the mixed skin lymphocyte response of in situ irradiation was used to compose this factor for intraspecies variation (IAV). The average ED50 value of the $MSLR_{human,\textit{in situ}}$ was related to the most sensitive subject to provide the intraspecies variation (IAV), and was estimated at a factor 0.5.

The dose of UVB, emitted from FS40 lamps, that would have suppressed the proliferative response of T cells to Listeria from individuals with skin type 2, infected with Listeria ($ED50_{human,Lis,LST}$) can than be estimated as:

$ED50_{Lis,LST}$ * IEV * IAV = 6.8 * 3.85 * 0.5 = 13.1 kJ/m^2

Exposure Assessment and Risk Characterization

In this first attempt of a risk assessment for the effects of UVB on the immunological resistance to infectious diseases, it was presumed that only cumulative doses UVB radiation were relevant for the effects tested. Adaptation processes were not taken into account in our worst-case approach, because there were not enough data to correct for these processes, so these effects may be overestimated.

The question, which had to be raised, was the relevance of an $ED50_{human,Lis,LST}$ of 13.1 kJ/m^2 UVB radiation (280-315 nm, FS40) for the actual exposure of people outdoors. For that reason, it was necessary to translate the dose of UVB, as mentioned in the dose-response assessment, into biologically relevant doses. However for the calculation of a biologically effective dose concerning suppression of resistance to Listeria, a wavelength action spectrum has not been established. For this reason, a wavelength spectrum for contact hypersensitivity (CHS) in mice was used, as determined by de Fabo and Noonan (1990). The spectrum of the FS40 that we used was multiplied by the action spectrum for CHS. Thus, for every wavelength a biological effective irradiance (BEI_{imm}) was calculated. The total BEI_{imm} form the FS40 lamps for the wavelengths between 280 and 315 nm was 0.257 W/m^2. The integrated UVB irradiance of the FS40 curve, as measured by the fotosensitive Optronics measurement system, was 1.929

W/m^2. From this it followed, that the BEI_{imm} emitted by the FS40 lamps was only 13.3% of the total irradiance that was emitted. Hence, it was calculated that the biological effective ED50 value ($ED50_{bed,Lis,LST}$) for suppression of T-cell responses to Listeria was 13.1 * 0.133 = 1.74 kJ/m^2 (FS40). Dividing this value by the BEIimm (emitted by sunlight, as calculated by De Fabo and Noonan 1990), gave a first prediction on the duration of sun exposure (at a certain latitude, clear skies, noon) necessary to induce 50% suppression of specific T-cell response to Listeria. For instance it was predicted that 104 minutes round noon, under clear skies in July at 400 S would lead to 50% suppression of the specific T-cell responses to Listeria. However assuming a 20% of decrease of ozone, and a resulting increase of UVB irradiation, it was calculated that 94 minutes was enough to induce the same immunosuppression. As another example, in January at a latitude of 200 S, this immunosuppression might be induced after 88 minutes (Table 1).

Table 1 Predicted effects of ozone decreases on the biologically effective irradiance for immune suppression (at noon under clear skies).

Latitude	Ozone (dobson units)	% Decrease in ozone	Biologically effective irradiance (W/m^2)	Calculated time (min.) for 50% immuno-suppression
60°N January	368.5	0	0.0087	3349
	350.1	5	0.0091	3201
	331.7	10	0.0095	3066
	294.8	20	0.0103	2828
60°S July	333.8	0	0.174	167
	317.1	5	0.179	163
	300.4	10	0.184	158
	267.0	20	0.196	149
40°N January	335.6	0	0.073	399
	318.8	5	0.075	389
	302.0	10	0.078	373
	268.5	20	0.083	351
40°S July	307.9	0	0.278	104
	292.5	5	0.285	102
	277.1	10	0.292	100
	246.3	20	0.310	94
20°N January	262.6	0	0.332	88
	249.5	5	0.341	85
	236.3	10	0.350	84
	210.1	20	0.370	78
20°S July	262.1	0	0.196	149
	249.0	5	0.201	145
	235.9	10	0.206	142
	209.7	20	0.218	134

Discussion

Hazard estimation of the UVB-induced effects on the resistance to infectious diseases, indicated that UVB radiation impaired cellular immune responses, leading to decreased resistance to different infectious agents in the rat.

An important step from hazard identification to the ultimate characterization of risk is exposure assessment. Exposure assessment of UVB radiation must take into account the action spectrum used. To estimate the biologically effective dose to which humans are exposed outdoors, the spectrum of the sun was multiplied by the action spectrum of contact hypersensitivity as determined by De Fabo and Noonan (1983, 1990).It appears reasonable to suggest that an action spectrum for the UVB-induced effects on the immunological resistance to infectious diseases is similar to the action spectrum of UVB-induced immunosuppression of CHS, because the effects on the resistance to infectious agents are often mediated by effects on specific T-cell responses, as is the case with resistance to *Listeria monocytogenes*. In the CHS reaction, also effects on specific T-cell responses are measured, and it is reasonable to postulate that these action spectra are comparable. It is, however, worthwhile to work out an action spectrum for the effects of UVB on the resistance to an infectious agent in rats.

Earlier observations from our group (Goettsch et al.1994, Goettsch 1995) have indicated that the immunological effects observed after UVB exposure daily for 14 consecutive days to 750 J/m^2 were not comparable to effects observed after daily UVB exposure for 7 days to 1500 J/m^2, in spite of equal total UVB doses, and a cumulative dose of UVB radiation may therefore not provide enough information on the UVB-induced effects on the immune system; exposure intensity and exposure duration, including adaptation processes, may play a role in the effects measured. The UVB dose which induced suppression in the Listeria model described in this report, was presented as a cumulative dose. However, in our experiments, the rats obtained this dose over a period of seven days, without prior exposure. Thus it might be appropriate to divide the $BED_{human,Lis,LST}$ by 7 and present that value as the daily UVB exposure for 7 consecutive days, that could induce 50% immunosuppression of the specific T-cell response to Listeria in individuals that had no prior exposure. Thus, the total period necessary to induce 50% suppression of Listeria specific T cell response in July at 40° N, would be approximately 15 minutes daily for seven consecutive days in persons that were not exposed earlier.

Risk assessment of carcinogenic potential of UVB radiation is relatively easy as compared with risk assessment of UVB-induced suppression of the immunological resistance to infectious diseases, because the adversity of the endpoint, i.e. arising neoplasms, is beyond dispute. Mathematical techniques were used to translate an action spectrum for UVB-induced carcinogenesis by different groups (De Gruijl et al. 1993, 1994; Slaper et al. 1991) to provide a risk assessment for neoplasms. In viral infections, the activity of NK cells is important for the clearance of virus from the host. However, it was demonstrated that the effects of immunotoxic compounds on basal immune parameters, such as NK activity, did not always correlate to the effects of the same compounds on the resistance to viral infections

(Selgrade et al. 1992). In such a case, the use of the approach for the extrapolation of immunotoxic compounds as was used in the present report is debatable, because differences in susceptibility between rat and humans, measured in a NK assay, may be not representative for the actual differences in susceptibility between these species for the resistance to virus infections. Consequently, a more complex model should be applied. Also, decreases in specific T-cell responses to Listeria are difficult to quantify in terms of increases in incidence and/or severity of infectious diseases in a population. Dose-response assessment of the effects of UVB radiation on the resistance to infectious agents and basal immune functions in rats and humans, showed that there was a dose-dependent relationship between UVB exposure and several effects tested. Comparison of the effects of UVB exposure on basal immune functions and the immunological resistance to infectious diseases in rats indicated that there were no major differences in the sensitivity of those assays for UVB radiation. This result is important for the subsequent extrapolation of the effects of UVB exposure on immunological resistance in rats to humans. Yet, also for UVB exposure it should be mentioned, that comparison of the effects of UVB exposure on basal immune functions for interspecies differences depend on the assay used. For instance, the MSLR assay used for the calculation of the IEV does only account for effects of UVB radiation on the cells of Langerhans in the skin. Since UVB radiation is able to impair immune responses by different pathways, the effects measured on MSLR may not fully correlate with the effects of UVB exposure on immunological resistance to infectious diseases. Therefore the IEV that we have calculated for the present risk assessment may not be fully representative for all effects of UVB radiation on the immune system and additional data are helpful. The use of antigen-specific immunological assays, such as contact hypersensitivity and antigen-specific proliferative responses of lymphocytes in combination with host-resistance, may provide additional data on the IEV. For instance, effects of UVB exposure on specific immune responses to Herpes simplex (HSV-1) can be studied in HSV-1 positive psoriasis patients and human volunteers and be compared to similar studies in the rat. Moreover, an increased infection load, leading to clinical effects, can be defined more definitely as an adverse effect than immune responses to the infectious agent, and may be used as an endpoint for risk assessment. Particularly in HSV-1 infections, suppression of the immune response is related to pathological skin lesions (Bratcher et al. 1993; Muira et al. 1994; Norval et al. 1987). Infection studies with Mycobacterium leprae or Leishmania might give further information on how UVB-induced specific immunosuppression is quantified into pathological effects, useful for risk assessment.

Another problem that one encounters when evaluating the risk of immunosuppression by UVB irradiation is the analysis of the IAV. Studies by Vermeer et al. (1991) and by Kurimoto and Streinlein (1992) provide evidence that differences in susceptibility for the effects of UVB radiation on the CHS probably depends on several genetic factors. In mice it was demonstrated that UVB-induced immunosuppression is strain dependent, and could be attributed to several genetic factors (Cooper et al. 1985; Taylor et al. 1993). The analysis in our present study was based on a relatively small group of individuals, all with skin type 2. It may be suggested that additional studies with human volunteers have to be accomplished with large groups of individuals, in order to have a more appropriate estimation of the IAV. Another approach would be to use data from especially susceptible human individuals, such as skin cancer patients or renal transplant patients receiving immunosuppressive therapy, in order to analyze the dose-response relationship between UVB radiation and specific immunosuppression.

In conclusion, UVB exposure impairs the immunological resistance to infectious diseases in rats, and these data appear to be relevant for the human situation. The exact quantitative relations between immunosuppressive effects of UVB radiation and an increased incidence of infectious diseases in the human population needs a cautious approach and follow-up studies are essential. It is necessary to work out an action spectrum for the effect of UVB exposure on the infectious load in relation to the suppression of immunological functions. Secondly, it is necessary to perform susceptibility studies with sizeable groups of rats and human volunteers to compare the UVB-induced immunosuppression in CHS protocols, which will supply an IEV and IAV for the UVB effects on specific T-cell responses. Thirdly, clinical studies on the effects of UVB exposure on HSV-1 and HPV infections in humans may give better insight into the effects of UVB radiation on immunological resistance to infectious diseases in humans and the relation to possible pathological effects. Finally large scale epidemiological (cohort) studies may provide the definitive data whether UVB exposure affects the incidence of infectious diseases in humans. They may also validate the calculated quantitative relationships between UVB exposure and the incidence of infectious diseases in the human population, as obtained from experimental data.

References

Bratcher DF, Harrison CJ, Bourne N, Stanberry LR, Bernstein DI (1993) Effect of indomethacin on ultraviolet radiation-induced recurrent herpes simplex virus disease in guinea pigs. J Gen Virol 74: 1951-1954

Cooper KD, Oberhelman L, Hamilton TA, Baadsgaard O, Terhune M, LeVee G, Anderson H, Koren H (1985) UV exposure reduces immunization rates and promotes tolerance to epicutaneous antigens in humans: Relationship to dose, CD1a-DR+ epidermal macrophage induction, and Langerhans cell depletion. J Immunol 134: 129-137

De Fabo EC, Noonan FP (1983) Mechanism of immune suppression by ultraviolet radiation in vivo. I. Evidence for the existence of a unique photoreceptor in skin and its role in photoimmunology. J Exp Med 157: 84-98

De Fabo EC, Noonan FP (1990) Biologically effective doses of sunlight for immune suppression at various lattitudes and their relationship to changes in stratospheric ozone. Photochem Photobiol 52: 811-817

De Gruijl FR, Van der Leun JC (1993) Influence of ozone depletion on the incidence of skin cancer: quantitative prediction. In: Young AR, Bjorn LO, Moan J (eds) Environmental Photobiology. Plenum Press, New York, pp 89-112

De Gruijl FR, Van Der Leun JC (1994) Estimate of wavelength dependency of ultraviolet carcinogenesis in humans and its relevance to risk assessment of a stratospheric ozone depletion. Health Phys 67: 319-325

Garssen J, Goettsch W, De Gruijl FR, Van Loveren H (1994) Immunologic effects of UVB: Recent Progress. In: Proceedings of The 13th UOEH International Symposium, pp 159-169

Goettsch W, Garssen J, De Gruijl FR, Van Loveren H (1994) Effects of UVB on the immune system and the consequences for the resistance against infectious diseases. In: Jung EG, Holick MF (eds) Biological effects of light. Walter De Gruyter, Berlin, pp 637-641

Goettsch W, Garssen J, Hurks HMH, De Gruijl FR, Van Loveren H (1994) Effects of in vitro exposure to ultraviolet radiation on the functional activity of lymphocytes, with emphasis on susceptibility of different species. Photochem Photobiol 60: 273-281

Goettsch W (1995) Effects of ultraviolet-B radiation on the resistance to infectious diseases. Thesis, Utrecht University

Kurimoto I, Streinlein JW (1992) Cis-urocanic acid suppression of contact hypersensitivity induction is mediated via tumor necrosis factor alpha. J Invest Dermatol 99: 69S-70S

Muira S, Kulka M, Smith CC, Imufuku S, Burnett JW, Auralian L (1994) Cutaneous ultraviolet radiation inhibits Herpes simplex virus-induced lymphoproliferation in latently infected subjects. Clin Immunol 72: 62-69

National Academy of Science Report: Risk Assessment in the Federal Government: Manageing the Process (1983). National Academy Press, Washington DC

Norval M, Howie SE, Ross JA, Maingay JP (1987) A murine model of Herpes simplex virus recrudescenses. J Gen Virol 68: 2693-2698

Scala R (1991) Risk assessment. In: Amdur M, Doull J, Klaassen C (eds) Casarett and Doull's Toxicology, 3d edition. Pergamon Press, New York, pp 985-996

Selgrade MJK, Daniels MJ, Dean JH (1992) Correlation between chemical suppression of natural killer activity in mice and susceptibility to cytomegalovirus: Rationale for applying murine cytomegalovirus as a host resistance model and for interpretation of immunotoxicity in terms of risk of disease. J Toxicol Environm Health 37: 123-137

Slaper H, Den Elzen MJG, Van de Woerd W, De Greef J (1991) Ozone depletion and skin cancer incidence: an integrated modelling approach. Report 249104002, National Institute of Public Health and Environmental Protection, Bilthoven, pp 30

Taylor JR, Schmieder GF, Shimizu T, Streinlein JW (1993) UVB susceptibility is a risk factor for recurrent herpes labialis. Photochem Photobiol 57: 135-153

Vermeer M, Schmieder GJ, Yoshikawa T, Van Den Berg JW, Metzman MS, Taylor JR, Streilein JW (1991) Effects of ultraviolet B light on cutaneous immune responses of humans with deeply pigmented skin. J Invet Dermatol 97: 1-10

Sentinel Screening for Human Immunotoxicity

J. Descotes, B. Nicolas, E. Pham, T. Vial
Poison Information and Toxicovigilance Centre & Department of Pharmacology and Medical Toxicology, INSERM U80, Lyon, France.

The potential immunotoxicity of xenobiotics has been the focus of a growing concern during the past decade. Even though a large number of experimental studies, mainly in rodents, suggested that the immune system is a likely target organ of toxicity for many pharmaceutical products and occupational as well as environmental chemicals (Descotes 1988), little is actually known of their immunotoxic effects in man (Newcombe et al. 1992; Burrell et al. 1992). Among possible approaches to identify those chemical exposures more likely to induce immunotoxic effects, namely clinical consequences associated with immunosuppression and immunostimulation, hypersensitivity, and autoimmunity, biomarkers of immunotoxicity have been first considered for use. Sentinel screening is proposed as an alternative to biomarkers that are still poorly, if at all, standardized and validated.

Biomarkers of human immunotoxicity

As extensive efforts have been devoted to designing assays for the diagnosis of immune diseases in patients with primary or secondary immune deficiencies, hypersensitivity or autoimmune diseases, a number of potential biomarkers of immunotoxicity have been proposed to investigate whether drug treatments and chemical exposures are likely to be associated with clinically significant adverse effects on the human immune system (National Research Council 1992; Straight et al. 1994).

At the present time, it is both difficult and largely unrewarding to conduct immunotoxicity studies in exposed populations for the following reasons (Descotes et al. 1995) :

a) the vast majority of available immune tests have not been standardized and validated adequately;

b) the size of the functional capacity reserve of the human immune system is not known so that it has generally not been established to what extent a change in one given endpoint is relevant as regards immunotoxicity (e.g. what is the actual clinical relevance of a 20% decrease in antibody production ?).

c) the selection of appropriate control subjects is absolutely mandatory to avoid factors interfering with the immune competence such as age, gender, smoking, nutritional status or underlying illness; as so many confounding factors have to be taken into account, the size of the exposed and control populations in field immunotoxicity studies has to be much larger than in conventional studies.
d) the level of chemical exposure should be well documented and sufficiently high to generate interpretable results due to the usual lack of sensitivity of currently available immunological tests; as the level of exposure, either environmentally or occupationally, is fortunately decreasing in many areas of the world, this lack of sensitivity may well result in falsely negative results.
e) sample acquisition at sites geographically distant from the investigator's laboratory is another major difficulty to be overcome and this difficulty will essentially be overcome by the availability of tests requiring no immediate processing of cells.
f) finally and largely based on the various points raised previously, field studies are extremely expensive if all these criteria have to be met.

Therefore, field studies using immune endpoints are seldom conducted and when results are available, they often are controversial. It is not surprising that very few, if any, epidemiological studies have so far been published including a comparison of clinical adverse effects with changes in immune endpoints.

Use of sentinel events in toxicity assessment

Another approach has been recommended by several groups of experts to better identify chemical exposures the most likely to be associated with immunotoxic consequences (National Research Council 1992; Sundaran and Wing 1995).
The concept of sentinel events has already been used in other areas of medicine, e.g. infectious diseases, Parkinson's disease, cerebrovascular accidents and suicide, as well as in other areas of toxicology, e.g. occupational toxicology (Rutstein 1984; Baker 1989), veterinary toxicology (Hungerford et al. 1995), and mutagenicity (Czeizel 1989). A sentinel toxic event can be defined as an adverse event, the pathophysiology of which is understood well enough to show that a chemically-mediated injury to a given target organ/function has occurred, and to indicate that a relationship between this toxic injury and a documented chemical exposure may exist.

One major specificity of sentinel programmes is that they are based on spontaneous case reporting and therefore rely on less stringent and expensive procedures than epidemiological studies, but this has positive and negative consequences to be taken into account (Bartelds 1993). Although a wider fraction of the general population may be attained using a sentinel approach, it is not possible to calculate incidence rates. However, despite many biases, changes in rates or reported events may indicate a new trend possibly resulting from changes in chemical exposure. It is obviously easier for physicians with many unrelated medical activities to take part in a sentinel programme. However, drops in adherence to the programme may be due either to increased workload or diminished motivation, or to an actual drop in the occurrence of sentinel events under scrutiny.

Importantly, a sentinel programme whatever the efforts put in it, will never be able to achieve the completeness of disease registries or epidemiological studies. However, such a programme will prove useful to generate warning signals about adverse health events resulting in further targeted toxicological investigations (Pham et al. 1995)

French Regional network of immune-mediated sentinel events

Sentinel screening in immunotoxicology is considered to be a valuable tool. First of all, there is a current lack of knowledge on the human immunotoxicity of drugs and chemicals. The available information is essentially derived from rodent and *in vitro* human studies (Descotes 1988). Because the use of biomarkers for immunotoxicity is so much cost-ineffective and/or unrewarding, epidemiological studies can hardly include such markers. Secondly, the pathophysiology of several diseases is clearly immune-mediated and they have been shown to be associated in patients with exposure to drugs and/or chemicals. Several panels of experts recommended that sentinel events/diseases be used to better identify the immunotoxic effects of drugs and chemicals in humans.

A network of physicians has been created in two regions of France, namely Rhône-Alpes and Auvergne. Both regions from the East-Centre part of the country account for 15% of the French population, with approximately 7 million inhabitants. The network includes Poison Control and Toxicovigilance Centres and Post-Marketing Drug Surveillance Centres, together with hospital departments in Occupational Medicine, Internal Medicine, Rheumatology, Dermatology and Oncology, from the major cities of these regions, namely Lyon, Saint-Etienne, Grenoble and Clermont-Ferrand.

Ten immune-mediated diseases have been selected taking into account public health concern, well-established diagnosis criteria, and previous reports in the literature of diseases associated with documented drug or chemical exposure. These diseases include B-lymphomas, the organ-specific autoimmune diseases, myasthenia, pemphigus and thyroiditis, and the systemic autoimmune diseases lupus, Gougerot-Sjögren syndrome, rheumatoid arthritis, Sharp's syndrome, dermatopolymyositis and sclerodermia. Opportunistic infections, e.g. listeriosis or toxoplasmosis, have not been included because of expected difficulties in assessing the degree of immunosuppression and in relating it to the drug or chemical exposure. Hypersentivity reactions have not been included as the current knowledge on these reactions is far greater than for other immune-mediated diseases of drug and chemical origin.

All collaborating physicians were provided with specific forms to obtain information on the patient's identity and medical history, past drug and chemical exposure(s), clinical and biological diagnosis criteria of the disease. When a physician detects a sentinel disease in one of his (or her) patients, a survey of previous drug treatments and chemical (e.g. occupational) exposures is undertaken to identify whether a temporal, even though not necessarily causal, association exists between the onset of the disease and identified chemical exposure(s). When a sufficient amount of information is judged to be available on the disease and the associated drug and/or chemical exposure(s), it is decided whether the new disease can be considered as an immune-mediated sentinel event. The new file is then stored in a dataset specifically tailored to this programme using the dataset generator Microsoft's FoxPro° for the PC/Macintosh. Importantly, no procedure has been included to determine causal relationships (or imputation). Therefore, no firm conclusion can be expected to be drawn from stored data regarding the immunotoxic effects of drugs and chemical exposure. Each new sentinel event which is stored in the computer is considered as a new pair disease/exposure. When the number of pairs combining one given sentinel event with a documented drug or chemical exposure, exceeds a limit given by the random distribution of events in the dataset, then a warning signal is generated suggesting the pair is likely to be linked to an immunotoxic effect (Pham et al. 1995). Epidemiological studies and/or experimental studies will then be warranted to confirm the validity of the warning signal.

Conclusion

Immune-mediated sentinel events are suggested to be helpful tools to expand our current knowledge on the immunotoxic effects of drugs and chemicals in humans. Limitations can be found in the retrospective assessment of drug and chemical exposure(s) relying on estimates or memories of actual exposure, and the incomplete reporting of relevant cases because spontaneous reporting genuinely relies on motivated cooperation and not on preestablished criteria for patients' inclusion. On the opposite, valuable advantages of sentinel programmes are related to the largest fraction of the population being surveyed even though less systematically and extensively, to cost-effectiveness as sentinel programmes are far less expensive than disease registries or large epidemiological studies, and to the current lack of standardized and validated biomarkers of immunotoxicity. Nevertheless, even though this programme is expected to be helpful to expand our knowledge on the immunotoxicity of drugs and chemicals in man, other approaches still need be developped, e.g. the use of refined biomarkers and field studies in carefully selected populations.

References

Baker EL (1989) Sentinel event notification system for occupational risks (SENSOR) : the concept. Am. J. Public Health, 79-suppl: 18-20.
Bartelds AIM (1993) Validation of sentinel data. Gesundh Wes, 55: 3-7.
Burrell R, Flaherty DK, Sauers L eds. (1992) Toxicology of the immune system. A human approach. Van Rostrand Reinhold, New-York.
Czeizel A (1989) Population surveillance of sentinel anomalies. Mutat Res, 212: 3-9.
Descotes J ed (1988) Immunotoxicology of drugs and chemicals. 2nd edition. Elsevier Science, Amsterdam.
Descotes J, Nicolas B, Pham E, Vial T (1995) Assessment of immunotoxic effects in humans. Clin Chem, in press.
Hungerford LL, Trammel HL, Clark JM (1995) The potential utility of animal poisoning data to identify human exposure to environmental toxins. Vet Hum Toxicol, 37: 158-162.
National Research Council (1992) Biologic markers in immunotoxicology. National Academy Press, Washington DC.
Newcombe DS, Rose NR, Bloom JC eds (1992) Clinical immunotoxicology. Raven Press, New-York.
Pham E, Nicolas B, Vial T, Descotes J (1995) Modeling of warning procedures to detect the toxic effects of drugs and chemicals in man : application to a sentinel disease program in immunotoxicology. Toxicol. Model., in press.
Rutstein DD, Mullan RJ, Halperin WE, Melisu JM, Sestito JP (1984) Sentinel health events (occupational): A basis for physician recognition and public health surveillance. Arch Environ Health, 39: 159-168.
Straight JM, Kipen HM, Vogt RF, Ameler RW (1994) Immune function test batteries for use in environmental health field studies. Agency for Toxic Substances and Disease Registry, Atlanta.
Sundaran S, Wing M (1995) Conclusions of an international workshop on environmental toxicology and human health. Hum Exp Toxicol, 14: 85-6.

Impact of Environmental Pollution on Respiratory Sensitization and Asthma

I Kimber
Zeneca Central Toxicology Laboratory, Alderley Park, Macclesfield, Cheshire, UK.

Introduction

In western Europe and the USA the prevalence of asthma in both adults and children is rising; an increase that can not be reconciled simply by changes in diagnostic fashion (Cullinan, 1988; Gergen *et al.*, 1988; Hahhtela *et al.*, 1990; Burney *et al.*, 1990; Weitzman *et al.*, 1992). Such increased prevalence has been too rapid to be explained by alterations in the gene pool and for this reason there has been a growing interest in the possible association between the environment and asthma (Newman Taylor, 1995). There are many ways in which the external environment or environments encountered in the home or workplace could impact on the incidence of asthma and other respiratory diseases. These include alterations in exposure to environmental or occupational aeroallergens, respiratory virus infection, indoor air quality and tobacco smoking (Newman Taylor, 1995). In addition, it has been proposed that outdoor air pollution plays a role in the pathophysiology of asthma and it is this latter association upon which the present chapter will focus.

Asthma and environmental pollutants

Asthma is characterized by a chronic inflammatory process in the airways associated with the presence of activated T lymphocytes, mast cells and eosinophils (Barnes and Lee, 1992; Corrigan and Kay, 1992). Two broad divisions are recognized. Extrinsic or allergic asthma is provoked in sensitized individuals by inhalation exposure to the inducing protein or chemical allergen. Examples include allergy to house dust mite, mould spores, pollens and certain industrial chemicals such as toluene diisocyanate and trimellitic anhydride. The mechanisms whereby non-allergic or intrinsic asthma is induced remain unclear.

It has been proposed that there are four main ways in which environmental air pollutants may affect the prevalence and/or severity of asthma (Wardlaw, 1993; Barnes, 1994).

(a) Pollutants may act as inciters or triggers of asthmatic reactions in airways that are already hyperresponsive, without themselves causing inflammation. The result would be comparable with other inciters that cause a transient narrowing of the airways such as methacholine and bradykinin.

(b) The pollutant could itself serve to induce or augment airway inflammation and, as a result, cause airway hyperresponsiveness that persists after exposure has ceased.

(c) Pollutants may have a direct toxic influence on the respiratory epithelium leading to inflammation, airway hyperresponsiveness and the appearance of asthma-like symptoms in previously normal individuals.

(d) Certain pollutants may have the ability to augment or modify immune responses to inhaled antigens or to enhance the severity of reactions elicited in the respiratory tract following inhalation exposure of the sensitized individual to the inducing allergen.

There is evidence from laboratory studies that certain atmospheric pollutants have the potential to stimulate bronchoconstriction or airway inflammation. Exposure to sulphur dioxide is associated with chest tightness and bronchoconstriction, with the concentration required to induce a response being dependent upon the degree of hyperresponsiveness; non-asthmatics requiring a higher level to develop symptoms (Sheppard *et al.*, 1980; Lin *et al.*, 1987). On the basis of such investigations it is probable, given the concentrations of SO_2 likely to be encountered in the atmosphere, that, in practice, only asthmatics will be affected. It may be, however, that the effects of SO_2 will be augmented in the presence of other pollutants. It has been reported, for instance, that the sensitivity of mild asthmatics to SO_2 is increased during exercise by prior exposure to ozone (Koenig *et al.*, 1990). Ozone itself has been investigated extensively and has been found to cause increased bronchial responsiveness (Lippman, 1989). However, in contrast to SO_2, there is no evidence that asthmatics are more susceptible to O_3 than are normal subjects (Koenig *et al.*, 1987). It has been found also that exposure of asthmatic subjects to 0.12ppm ozone for 1 hour caused a small increase in airway responsiveness to inhaled allergen (Molfino *et al.*, 1991). The health relevance of oxides of nitrogen, and in particular nitrogen dioxide, has attracted some interest as the level of these pollutants, the products of car exhausts and fuel combustion, has increased in urban environments. Although some data are conflicting, the consensus view is that inhalation exposure to NO_2 at concentrations of up to 4ppm fails to induce any effects in either those with asthma or normal subjects (Wardlaw, 1993; Barnes, 1994). There is little evidence to suggest that under normal conditions and levels of exposure NO_2 will have any material influence on airway function in either normal or asthmatic individuals. Non-gaseous air pollutants are also of interest and while there have been no relevant clinical laboratory investigations there has been reported an association between allergic rhinitis to pollens and pollution with diesel exhaust particles (Kaneko *et al.*, 1980; Ishizaki *et al.*, 1987). The interesting question raised by this apparent association is whether certain air pollutants have the potential to influence the incidence or severity of allergic rhinitis and asthma secondary to modulation of immune function and sensitization (see d above).

Environmental pollutants and respiratory sensitization

The issue is whether prior or concurrent exposure to certain environmental pollutants will induce or enhance sensitization to aeroallergens. Although it is possible to identify several separate mechanisms whereby pollutants may contribute to the effectiveness of allergic sensitization, it is important to emphasize that these are not mutually exclusive and may operate in concert. Essentially these are processes that would be predicted to influence the quantity or quality of an immune response.

(a) Increased immunogenicity of the inhaled aeroallergen resulting from its association with particulate pollutants.

(b) Increased immunogenicity secondary to more effective antigen handling, antigen processing and/or antigen presentation in the respiratory tract.

(c) Increased immunogenicity resulting from the inhibition or elimination by pollutants of suppressive or regulatory cells that normally serve to constrain the vigour of immune responses in the respiratory tract.

(d) Preferential stimulation of the quality of immune response necessary for the production of IgE antibody, the class of antibody that effects sensitization of the respiratory tract.

Immunological mechanisms : antigen and antigen presentation

The effective stimulation of immune responses demands that the inducing antigen is presented in sufficient quantity and in an immunogenic form to responsive T lymphocytes. There are many cells that have the potential for antigen presentation. However, it is believed that for the activation of virgin (unprimed) T lymphocytes and the initiation of primary immune responses dendritic cells (DC), the most potent of antigen presenting cells, are required (Inaba and Steinman, 1984). Presumably the network of DC found within the airway epithelium will provide this function within the respiratory tract (Holt *et al*, 1989; 1990a; b; Pollard and Lipscomb, 1990; Schon-Hegrad *et al*, 1991). The equivalent cells within the skin are epidermal Langerhans cells (LC). Following encounter with antigen at skin surfaces LC are induced to migrate, via afferent lymphatics, to lymph nodes draining the site of exposure (Kimber and Cumberbatch, 1992). A proportion of these cells bear high levels of antigen and reach the lymph node in an immunologically active form where they interact with and activate T lymphocytes. There is evidence that in a similar fashion pulmonary DC traffic to regional lymph nodes in response to antigen challenge in the respiratory tract (Xia *et al*, 1995). The migration and functional maturation of LC during the induction phase of cutaneous immunity is initiated and regulated by skin cytokines. The development of antigen presenting cell activity is effected by granulocyte/macrophage colony-stimulating factor (GM-CSF), a product of keratincoytes, possibly acting in concert with other epidermal cytokines (Kimber and Cumberbatch, 1992). The signal for LC to leave the skin and migrate to draining lymph nodes is provided by another keratinocyte cytokine, tumour necrosis factor α (TNF-α) (Kimber and Cumberbatch, 1995). It is apparent that the effective stimulation of cutaneous immune responses will be dependent upon the local availability of relevant cytokines. The same will undoubtedly be true for the migration and function of pulmonary DC and for the stimulation of immune activation in the respiratory tract. Certainly it is known that the respiratory epithelium is a source of cytokines, including those that have been shown to influence LC function in the skin (Cromwell *et al*, 1992; Churchill *et al*, 1992). Studies conducted *in vitro* with human bronchial epithelial cell cultures have revealed that exposure to comparatively low concentrations of nitrogen dioxide results in significant cell damage, decreased ciliary beat frequency and the release of several cytokines including GM-CSF, TNF-α and interleukin 8 (IL-8) (Devalia *et al*, 1993a; b). One may speculate therefore that enhanced immune responsiveness to aeroallergen may result from concurrent inhalation exposure to certain environmental pollutants that serve to stimulate the local production by airway epithelium of cytokines that promote the activity and migration of pulmonary DC. It has been proposed that an augmentation of skin immune responses by cutaneous irritants may operate via similar mechanisms (Cumberbatch *et al*, 1993). The impairment by NO_2 of bronchial epithelium ciliary beat frequency (Devalia *et al*, 1993b) may further augment pulmonary immune responses as the result reduced antigen clearance.

It is possible also that pollutant-induced enhancement of respiratory sensitization secondary to more effective local antigen presentation may operate by less direct mechanisms. The available evidence suggests that pulmonary alveolar macrophages function poorly as antigen presenting cells and indeed may suppress T lymphocyte responses (Holt, 1978; 1985). Consistent with this is the fact that the elimination *in vivo* of alveolar macrophages is associated with increased pulmonary immune function (Thepen *et al*, 1989; 1991). Of particular relevance to the induction of respiratory sensitization to airborne allergens is the demonstration that removal of alveolar macrophages increases selectively IgE antibody to inhaled antigens (Thepen *et al*, 1992). One important way in which pulmonary alveolar macrophages may down-regulate immune responses in the respiratory tract is by direct inhibition of the antigen presenting cell function of local DC (Holt *et al*, 1985; 1988; Holt, 1993). It is believed therefore that the effectiveness of antigen presentation to T lymphocytes in the respiratory tract is a function of a dynamic equilibrium between the immunosuppressive properties of mature alveolar macrophages and the immunostimulatory function of airway DC. Perturbation of this balance can have profound effects on pulmonary immune responses. Destruction of alveolar macrophages or inhibition of their function by inhaled pollutants would result in enhanced immune responsiveness favouring IgE antibody production.

Taken together it is apparent that, in theory at least, inhaled pollutants may facilitate antigen presentation in the respiratory tract in two ways. Firstly, the induced production or upregulation of proinflammatory cytokine expression by airway epithelium would provide the signals necessary for the migration and functional activation of pulmonary DC. Secondly, inhibition or elimination of alveolar macrophages by pollutants would free local DC from the immunosuppressive effects of these cells and facilitate more vigorous immune responses, including possibly enhanced IgE antibody production.

It has been shown recently that exposure of mice to atmospheres of formaldehyde results in enhanced IgE antibody responses to ovalbumin following intranasal sensitization to the protein (Tarkowski and Gorski, 1995). This augmentation could result from the influence of formaldehyde on airway epithelium and/or alveolar macrophages in the ways described above.

The form in which the inducing allergen reaches the respiratory tract may also have an important influence on subsequent antigen presentation and the development of immune responses. A suggestion is that the direct association between allergen and respirable particulate environmental pollutants may serve to enhance immunogenicity. Certainly studies using scanning and transmission electron microscopy have shown that pollen may bind to airborne particulate matter (Behrendt *et al*, 1992). The impact of such association on the induction of pulmonary immune responses is as yet unclear, however.

Immunological mechanisms : regulation of IgE antibody responses

Immediate-type hypersensitivity reactions are affected by IgE antibody and the influence of environmental pollutants on IgE responses has therefore attracted some attention. There is good evidence that some pollutants and tobacco smoke may act as adjuvants for IgE production. Several independent reports indicate that tobacco smoking is associated with an increased concentration of total plasma IgE (Gerrard *et al*, 1980; Zetterstrom *et al*, 1981; Burrows *et al*, 1981; Omenaas *et al*, 1994). It has been found also that the frequency of IgE sensitization and of occupational asthma to chemical and protein allergens is higher in those who smoke (Anon, 1985; Venables *et al*, 1985). The incidence of allergic disease is increased in the offspring of mothers who smoke cigarettes and maternal smoking is also associated with higher levels of IgE in cord blood (Magnusson, 1986).

Experimental studies in mice have demonstrated that diesel exhaust particles (DEP) enhance the production of specific IgE antibody. It was found by Muranka *et al* (1986) that the intraperitoneal injection of mice with a mixture of ovalbumin and DEP resulted in more vigorous IgE antibody responses than those exhibited by control animals injected with the protein allergen alone. Similar results were obtained when Japanese cedar pollen allergen was used in place of ovalbumin (Muranka *et al*, 1986). In a subsequent investigation by the same group it was demonstrated that DEP when administered to mice by the intranasal route also enhanced IgE anti-ovalbumin antibody responses (Takafuji *et al*, 1987). In exploring the mechanistic basis for an adjuvant effect of some environmental pollutants on IgE antibody production it is relevant to consider the immunological processes that initiate and regulate such responses. The production of IgE is controlled by the promoting and antagonistic influences of cytokines. The stimulation and maintenance of IgE responses in mice are dependent upon the availability of interleukin 4 (IL-4) (Finkelman *et al*, 1988b). Mice that lack a functional gene for IL-4 have no detectable IgE and fail to mount IgE responses (Kuhn *et al*, 1991). Conversely, possession of a transgene for this cytokine with constitutive production of IL-4 results in elevated concentrations of serum IgE (Tepper *et al*, 1990; Burstein *et al*, 1991). An opposing influence is interferon γ (IFN-γ), a cytokine that antagonizes IgE antibody responses (Finkelman *et al*, 1988a). In man also the effectiveness of IgE antibody production is regulated by the reciprocal effects of IL-4 and IFN-γ (Del Prete *et al*, 1988; Pene *et al*, 1988). These cytokines are products of phenotypically-distinct populations of differentiated $CD4^{+}$ T helper (Th) cells. Two main types of Th cells are recognized and have been designated Th1 and Th2. They differ with respect to cytokine production profiles. Both populations produce GM-CSF and interleukin 3 (IL-3). However, only Th1 cells express IFN-γ, interleukin 2 (IL-2) and tumour necrosis factor β (TNF-β) and only Th2 cells secrete interleukins 4, 5, 6 and 10 (IL-4, IL-5, IL-6 and IL-10) (Mosmann *et al*, 1986; 1991; Mosmann and Coffman, 1989). A similar heterogeneity among Th cells is found in humans (Romagnani, 1991; Romagnani *et al*, 1992).

A simple dichotomy of Th cell subtypes is an over-simplification. Th1 and Th2 cells should be regarded as the most differentiated forms of T helper lymphocytes that develop with time from a common precursor or precursors during the evolution of immune responses. The pathway of Th cell differentiation that predominates is dependent upon a number of factors including the nature and quantity of the inducing antigen, the characteristics of antigen presenting cells and the costimulatory molecules they express and, importantly, the availability of cytokines that influence selective growth and development of Th cells (Chang *et al*, 1990; Gajewsky *et al*, 1991; Coffman *et al*, 1991; Pfeiffer *et al*, 1991; Secrist *et al*, 1995; Kuchroo *et al*, 1995). Changes to one or more of these parameters induced by environmental pollutants could in theory alter the balance between Th1 and Th2 cell development and, as a consequence, the vigour of IgE antibody responses. Certainly $CD4^{+}$ T lymphocytes of Th2-type with respect to cytokine expression predominate in the sites of respiratory hypersensitivity, atopic dermatitis and other forms of IgE-mediated hypersensitivity reactions (van der Heijden *et al*, 1991; Kay *et al*, 1991; Robinson *et al*, 1992; Walker *et al*, 1992; Del Prete *et al*, 1993; Ricci *et al*, 1993).

There is evidence that exposure to DEP may augment IL-4 production. It was found that mice sensitized by intratracheal instillation of ovalbumin mixed with DEP exhibited increased levels of serum IgE anti-ovalbumin antibody compared with control animals exposed to the protein alone. Increased IgE antibody production was associated with elevated secretion of IL-4 by mediastinal lymph node cells following culture *in vitro* with the inducing allergen (Fujimaki *et al*, 1994). It is possible, however, that DEP or soluble extracts of DEP, may interact directly with activated B lymphocytes to enhance IgE production. Thus, it was found that nasal administration to normal individuals of DEP caused a significant increase in IgE, but not other immunoglobulin isotypes, in nasal washes. Moreover, exposure to DEP resulted in a 20-fold increase in the frequency of IgE secreting cells in nasal lavage fluid (Diaz-Sanchez *et al*, 1994). In several series of investigations DEP, or polyaromatic hydrocarbons extracted from DEP, were shown to act directly on B lymphocytes to enhance ongoing IgE transcription and to modify the patterns of epsilon mRNAs coding for distinct isoforms of the IgE heavy chain (Diaz-Sanchez *et al*, 1994; Takenaka *et al*, 1995).

Although there is no doubt that the balance between Th1 and Th2 cells and their cytokine products will exert an important influence of the stimulation of IgE antibody responses and the development of respiratory sensitization, other regulatory cells and molecules are relevant also. In this context $CD8^+$ T lymphocytes may be of greater significance than was recognized previously. It has been appreciated recently that $CD8^+$ T cells also display a functional heterogeneity with respect to cytokine production and that these populations may also serve to regulate IgE antibody production (Kemeny and Diaz-Sanchez, 1993; Kemeny *et al*, 1994; Le Gros and Erard, 1994; Seder and Le Gros, 1995). Several lines of evidence indicate that under standard conditions of inhalation exposure to antigen specific $CD8^+$ T lymphocytes that are of type 2 phenotype and produce IFN-γ are a normal feature of the immune response and constrain the expression of IgE antibody responses (Kemeny *et al* 1994; McMenamin and Holt, 1993; Renz *et al*, 1994). Such $CD8^+$ cells have been shown also to regulate negatively the immediate-type skin test reactivity and airway responsiveness that develops following sensitization of mice to ovalbumin by inhalation exposure (Renz *et al*, 1994). It has been proposed that as or more important than the balance between $CD4^+$ Th1 and Th2 cells in shaping the response to antigen is the interplay between Th1 cells and IFN-γ-producing $CD8^+$ cells. One theory is that in the normal individual the immune response to inhaled antigen is dominated by regulatory $CD8^+$ cells or Th1 cells that prevent or limit the production of IgE antibody via secretion of IFN-γ, while in atopics Th2-type immune responses predominate. If such balance is perturbed then the quality of induced immune responses will shift. It remains to be determined whether exposure to environmental pollutants at the time of contact with antigen can modify the function of cytokine producing T lymphocytes in such a way as to alter the characteristics of induced immune responses.

Concluding comments

Environmental pollution has been implicated as one possible cause of the increased prevalence of asthma in western Europe and the USA. Although some pollutants, such as ozone, have been found in laboratory studies to increase airway responsiveness, others, such as nitrogen dioxide, are considered not to influence airway function even at concentrations that match or exceed likely peak exposure levels. Neither is there any compelling evidence that normal exposure to environmental pollutants will induce or enhance allergic sensitization. Some experimental investigations indicate, however, that certain pollutants have the potential to influence sensitization and that such influences may operate through one or more of several mechanisms that will serve to augment or modify immune responses to inducing allergens. The relevance for human health of immune stimulation or immune deviation mediated by environmental pollutants remains to be clarified.

References

Anon (1985) Smoking, occupation and allergic lung disease. The Lancet i : 965

Barnes PJ (1994) Air pollution and asthma. Postgrad Med J 70 : 319-325

Barnes PJ, Lee TH (1992) Recent advances in asthma. Postgrad Med J 68 : 942-953

Behrendt H, Becker WM, Friedrichs KH, Darsow U, Tomingas R (1992) Interaction between aeroallergens and airborne particulate matter. Int Arch Allergy Immunol 99 : 425-428

Burney PGJ, Chinn S, Roma RJ (1990) Has the prevalence of asthma changed? Evidence from the national study of health and growth 1973-86. Br Med J 300 : 1306-1310

Burrows B, Halonen M, Barbee RA, Lebowitz MD (1981) The relationship of serum immunoglobulin E to cigarette smoking. Am Rev Respir Dis 124 : 523-525

Burstein HJ, Tepper RI, Leder P, Abbas AK (1991) Humoral immune functions in IL-4 transgenic mice. J Immunol 147 : 2950-2956

Chang T-L, Shea CM, Urioste S, Thompson RC, Boom WH, Abbas AK (1990) Heterogeneity of helper/inducer T lymphocytes. III. Responses of IL-2- and IL-4-producing (Th1 and Th2) clones to antigens presented by different accessory cells. J Immunol 145 : 2803-2808

Churchill L, Friedman B, Schleimer RP, Proud D (1992) Production of granulocyte-macrophage colony-stimulating factor by cultured human tracheal epithelial cells. Immunology 75 : 189-195

Coffman RL, Varkila K, Scott P, Chatelain R (1991) Role of cytokines in the differentiation of $CD4^+$ T-cell subsets in vivo. Immunol Rev 123 : 189-207

Corrigan CJ, Kay AB (1992) T cells and eosinophils in the pathogenesis of asthma. Immunol Today 13 : 501-507

Cromwell O, Hamid Q, Corrigan CJ, Barkans J, Meng Q, Collins PD, Kay AB (1992) Expression and generation of interleukin 8, IL-6 and granulocyte-macrophage colony-stimulating factor by bronchial epithelial cells and enhancement by IL-1β and tumour necrosis factor-α. Immunology 77 : 330-337.

Cullinan P (1988) Respiratory disease in England and Wales. Thorax 43 : 949-954

Cumberbatch M, Scott RC, Basketter DA, Scholes EW, Hilton J, Dearman RJ, Kimber I (1993) Influence of sodium lauryl sulphate on 2,4-dinitrochlorobenzene-induced lymph node activation. Toxicology 77 : 181-191

Del Prete GF, De Carli M, D'Elios MM, Maestrelli P, Ricci M, Fabbri L, Romanagni S (1993) Allergen exposure induces the activation of allergen-specific Th2 cells in the airway mucosa of patients with allergic respiratory disorders. Eur J Immunol 23 : 1445-1449

Del Prete G, Maggi E, Parronchi P, Chretien I, Tiri D, Macchia M, Ricci J, Banchereau J, De Vries J, Romagnani S (1988) IL-4 is an essential factor for the IgE synthesis induced in vitro by human T cell clones and their supernatants. J Immunol 140 : 4193-4198

Devalia JL, Campbell AM, Sapsford RJ, Rusznak C, Quint D, Godard P, Bousquet J, Davies RJ (1993a) Effect of nitrogen dioxide on synthesis of inflammatory cytokines expressed by human bronchial epithelial cells in vitro. Am J Respir Cell Mol Biol 9 : 271-278

Devalia JL, Sapsford RJ, Cundell DR, Rusznak C, Campbell AM, Davies RJ (1993b) Human bronchial epithelial cell dysfunction following exposure to nitrogen dioxide. Eur Respir J 6 : 1308-1316

Diaz-Sanchez D, Dotson AR, Takenaka H, Saxon A (1994) Diesel exhaust particles induce local IgE production in vivo and alter the pattern of IgE messenger RNA isoforms. J Clin Invest 94 : 1417-1425

Finkelman FD, Katona IM, Mosmann TR, Coffman RL (1988a) IFN-γ regulates the isotypes of Ig secreted during in vivo humoral immune responses. J Immunol 140 : 1022-1027

Finkelman FD, Katona IM, Urban JF Jr, Holmes J, Ohara J, Tung AS, Sample JG, Paul WE (1988b) IL-4 is required to generate and sustain in vivo IgE responses. J Immunol 141 : 2335-2341

Fujimaki H, Nohara O, Ichinose T, Watanabe N, Saito S (1994) IL-4 production in mediastinal lymph node cells in mice intratracheally instilled with diesel exhaust particulates and antigen. Toxicology 92 : 261-268

Gajewski TF, Pinnas M, Wong T, Fitch FW (1991) Murine Th1 and Th2 clones proliferate optimally in response to distinct antigen-presenting cell populations. J Immunol 146 : 1750-1758

Gerrard JW, Heiner DC, Ko CG, Mink J, Meyers A, Dosman JA (1980) Immunoglobulin levels in smokers and non-smokers. Ann Allergy 44 : 261-262.

Gergen PJ, Mullally DI, Evans R (1988) National survey of prevalence of asthma among children in the United States 1976 to 1980. Pediatrics 81 : 1-7

Hahhtela T, Linholm H, Bjorksten F, Koskenuvo K, Laitinen LA (1990) Prevalence of asthma in Finnish young men. Br Med J 301 : 266-268

Holt PG (1978) Inhibitory activity of unstimulated alveolar macrophages on T lymphocyte blastogenic response. Am Rev Respir Dis 118 : 791-793

Holt PG (1985) Downregulation of immune responses in the lower respiratory tract : role of alveolar macrophages. Clin Exp Immunol 63 : 261-270

Holt PG (1993) Macrophage : dendritic cell interaction in the regulation of the IgE response in asthma. Clin Exp Allergy 23 : 4-6

Holt PG, Degebrodt A, O'Leary C, Krska K, Plozza T (1985) T-cell activation by antigen presenting cells from lung tissue digests : suppression by endogenous macrophages. Clin Exp Immunol 62 : 586-594

Holt PG, Schon-Hegrad MA, McMenamin PG (1990a) Dendritic cells in the respiratory tract. Int Rev Immunol 6 : 139-149

Holt PG, Schon-Hegrad MA, Oliver J (1988) MHC class II antigen-bearing dendritic cells in pulmonary tissues of the rat : regulation of antigen presentation activity by endogenous macrophage populations. J Exp Med 167 : 262-274

Holt PG, Schon-Hegrad MA, Oliver J, Holt BJ, McMenamin PG (1990b) A continguous network of dendritic antigen-presenting cells within the respiratory epithelium. Int Arch Allergy Appl Immun 91 : 155-159

Holt PG, Schon-Hegrad MA, Phillips MJ, McMenamin PG (1989) Ia-positive dendritic cells form a tightly meshed network within the human airway epithelium. Clin Exp Allergy 19 : 597-601

Inaba K, Steinman RM (1984) Resting and sensitized T lymphocytes exhibit distinct (antigen presenting cell) requirements for growth and lymphokine release. J Exp Med 160 : 1717-1735.

Ishizaki T, Koizumi K, Ikemori R, Ishiyura Y, Kushibiki E (1987) Studies of prevalence of Japanese cedar pollinosis among the residents of a densely cultured area. Ann Allergy 58 : 265-270

Kaneko S, Shimada K, Noriuchi H, Endo A, Kodama M, Shinoda N, Imai T, Takiguchi K (1980) Nasal allergy and air pollution. Oto-Rhino-Laryngol, Tokyo 23 : 270-279

Kay AB, Ying S, Varney V, Gaga M, Durham SR, Moqbel R, Wardlaw AJ, Hamid Q (1991) Messenger RNA expression of the cytokine gene cluster, interleukin 3 (IL-3), IL-4, IL-5 and granulocyte macrophage colony-stimulating factor, in allergen-induced late-phase cutaneous reactions in atopic subjects. J Exp Med 173 : 775-778

Kemeny DM, Diaz-Sanchez D (1993) The role of $CD8^+$ T cells in the regulation of IgE. Clin Exp Allergy 23 : 466-470

Kemeny DM, Noble A, Holmes BJ, Diaz-Sanchez D (1994) Immune regulation : a new role for the $CD8^+$ T cell. Immunol Today 15 : 107-110

Kimber I, Cumberbatch M (1992) Dendritic cells and cutaneous immune responses to chemical allergens. Toxicol Appl Pharmacol 117 : 137-146

Kimber I, Cumberbatch M (1995) Langerhans cell migration : initiation and regulation. In : Moll H (ed) The Immune Functions of Epidermal Langerhans Cells. R G Landes, Austin pp103-117

Koenig JQ, Covert DS, Hanley GB, Pierson WE (1990) Prior exposure to ozone potentiates subsequent response to sulfur dioxide in adolescent asthamatic subjects. Am Rev Respir Dis 141 : 377-380

Koenig JQ, Covert DS, Marshall SG, Van-Belle G, Pierson WE (1987) The effects of ozone and nitrogen dioxide on pulmonary function in healthy and asthmatic adolescents. Am Rev Respir Dis 136 : 1152-1158

Kuchroo VK, Das MP, Brown, JA, Ranger AM, Zamvil SS, Sobel RA, Weiner HL, Nabavi N, Glimcher LH (1995) B7-1 and B7-2 costimulatory molecules activate differentially the Th1/Th2 developmental pathways : application to autoimmune disease therapy. Cell 80 : 707-718

Kuhn R, Rajewsky K, Muller W (1991) Generation and analysis of interleukin-4 deficient mice. Science 254 : 707-710

Le Gros G, Erard F (1994) Non-cytotoxic IL-4, IL-5, IL-10 producing $CD8^+$ T cells : their activation and effector functions. Curr Opinion Immunol 6 : 453-457

Linn WS, Avoi EL, Peng R, Shamoo DA, Hackney JD (1987) Replicated dose-response study of sulfur dioxide effects in normal, atopic and asthmatic volunteers. Am Rev Respir Dis 136 : 1127-1134

Lippman M (1989) Health effects of ozone. A critical review. JAPCA 39 : 672-695

Magnusson CG (1986) Maternal smoking influences cord serum IgE levels and increases the risk of subsequent infant allergy. J Allergy Clin Immunol 78 : 898-904

McMenamin C, Holt PG (1993) The natural immune response to inhaled soluble protein antigens involves major histocompatability complex (MHC) class I-restricted $CD8^+$ T cell-mediated but MHC class II-restricted $CD4^+$ T cell-dependent immune deviation resulting in selective suppression of immunoglobulin E production. J Exp Med 178 : 889-899

Molfino NA, Wright SC, Katz I, Tarlo S, Silverman F, McClean PA, Szalai JP, Raizenne M, Slutsky AS, Zamel N (1991) Effect of low concentrations of ozone on inhaled allergen in asthmatic subjects. The Lancet 338 : 199-203

Mosmann TR, Cherwinski H, Bond MW, Giedlin MA, Cofmann RL (1986) Two types of murine helper T cell clone. I. Definition according to profiles of lymphokine activities and secreted proteins. J Immunol 136 : 2348-2357

Mosmann TR, Coffman RL (1989) Heterogeneity of cytokine secretion patterns and functions of helper T cells. Adv Immunol 46 : 111-147

Mosmann TR, Schumacher JH, Street NF, Budd R, O'Garra A, Fong TAT, Bond MW, Moore KWM, Sher A, Fiorentino DF (1991) Diversity of cytokine synthesis and function of mouse $CD4^+$ T cells. Immunol Rev 123 : 209-229

Muranka M, Suzuki S, Koizumi K, Takafuji T, Miyamoto T, Ikemori R, Tokiwa H (1986) Adjuvant activity of diesel-exhaust particulates for the production of IgE antibody in mice. J Allergy Clin Immunol 77 : 616-623

Newman Taylor A (1995) Environmental determinants of asthma. The Lancet 345 : 296-299

Omenaas E, Bakke P, Elsayed S, Hanoa R, Gulsvik A (1994) Total and specific serum IgE levels in adults : relationship to sex, age and environmental factors. Clin Exp Allergy 24 : 530-539

Pene J, Rousset F, Briere F, Chretien I, Paliard X, Banchereau J, Spits H, De Vries JE (1988) IgE production by normal human B cells induced by alloreactive T cell clones is mediated by IL-4 and suppressed by IFN-γ. J Immunol 141 : 1218-1224

Pfeiffer C, Murray J, Madri J, Bottomly K (1991) Selective activation of Th1- and Th2-like cells in vivo. Response to human collagen IV. Immunol Rev 123 : 65-84

Pollard AM, Lipscomb MF (1990) Characterization of murine lung dendritic cells : similarities to Langerhans cells and thymic dendritic cells. J Exp Med 172 : 159-167

Renz H, Lack G, Saloga J, Schwinzer R, Bradley K, Loader J, Kupfer A, Larsen GL, Gelfand EW (1994) Inhibition of IgE production and normalization of airways responsiveness by sensitized CD8 T cells in a mouse model of allergen-induced sensitization. J Immunol 152 : 351-360

Ricci M, Rossi O, Bertoni M, Matucci A (1993) The importance of Th2-like cells in the pathogenesis of airway allergic inflammation. Clin Exp Allergy 23 : 360-369

Robinson DS, Hamid Q, Ying S, Tsicopoulos A, Barkans J, Bentley AM, Corrigan CJ, Durham SR, Kay AB (1992) Predominant Th2-like bronchoalveolar T-lymphocyte population in atopic asthma. New Eng J Med 326 : 295-304

Romagnani S (1991) Human TH1 and TH2 subsets : doubt no more. Immunol Today 12 : 256-257

Romagnani S, Del Prete G, Maggi E, Parronchi P, De Carli M, Macchia D, Manetti R, Sampognaro S, Piccinni M-P, Guidizi MG, Biagotti R, Almerigogna F (1992) Human TH1 and TH2 subsets. Int Arch Allergy Immunol 99 : 242-245

Schon-Hegrad MA, Oliver J, McMenamin PG, Holt PG (1991) Studies on the density, distribution and surface phenotype of intraepithelial class II MHC antigen (Ia)-bearing dendritic cells (DC) in the conducting airways. J Exp Med 173 : 1345-1356

Secrist H, DeKruyff RH, Umetsu DT (1995) Interleukin 4 production by $CD4^+$ T cells from allergic individuals is modulated by antigen concentration and antigen-presenting cell type. J Exp Med 181 : 1081-1089

Seder RA, Le Gros GG (1995) The functional role of $CD8^+$ T helper type 2 cells. J Exp Med 181 : 5-7

Sheppard D, Wong WS, Uehara CF, Nadel JA, Boushey HA (1980) Lower threshold and greater bronchomotor responsiveness of asthmatic subjects for sulfur dioxide. Am Rev Respir Dis 122 : 873-878

Takafuji S, Suzuki S, Koizumi K, Tadokoro K, Miyamoto T, Ikemori R, Muranka M (1987) Diesel-exhaust particulates inoculated by the intranasal route have an adjuvant activity for IgE production in mice. J Allergy Clin Immunol 79 : 639-645

Takenaka H, Zhang K, Diaz-Sanchez D, Tsien A, Saxon A (1995) Enhanced human IgE production results from exposure to the aromatic hydrocarbons from diesel exhaust : direct effects on B-cell IgE production. J Allergy Clin Immunol 95 : 103-115

Tarkowski M, Gorski P (1995) Increased IgE antiovalbumin level in mice exposed to formaldehyde. Int Arch Allergy Immunol 106 : 422-424

Tepper RI, Levinson DA, Stanger BZ, Campos-Torres J, Abbas AK, Leder P (1990) IL-4 induces allergic-like inflammatory disease and alters T cell development in transgenic mice. Cell 62 : 457-467

Thepen T, McMenamin C, Girn B, Kraal G, Holt PG (1992) Regulation of IgE production in presensitized animals : in vivo elimination of alveolar macrophages selectively increases IgE responses to inhaled allergen. Clin Exp Allergy 22 : 1107-1114

Thepen T, McMenamin C, Oliver J, Kraal G, Holt PG (1991) Regulation of immune response to inhaled antigen by alveolar macrophages : differential effects of in vivo alveolar macrophage elimination on the induction of tolerance vs. immunity. Eur J Immunol 21 : 2845-2850

Thepen T, van Rooijen N, Kraal G (1989) Aveolar macrophage elimination in vivo is associated with an increase in pulmonary immune response in mice. J Exp Med 170 : 499-509

van der Heijden FL, Wierenga EA, Bos JD, Kapsenberg ML (1991) High frequency of IL-4-producing $CD4^+$ allergen-specific T lymphocytes in atopic dermatitis lesional skin. J Invest Dermatol 97 : 389-394

Venables KM, Topping MD, Howe W, Luczynska CM, Hawkins R, Newman Taylor AJ (1985) Interaction of smoking and atopy in producing specific IgE antibody against a hapten protein conjugate. Br Med J 290 : 201-204

Walker C, Bode E, Boer L, Hansel TT, Blaser K, Virchow J-C (1992) Allergic and non-allergic asthmatics have distinct patterns of T-cell activation and cytokine production in peripheral blood and bronchoalveolar lavage. Am Rev Respir Dis 146 : 109-115

Wardlaw AJ (1993) The role of air pollution in asthma. Clin Exp Allergy 23 : 81-96

Weitzman M, Gortmaker SI, Sobol AM, Perrin JM (1992) Recent trends in the prevalence and severity of childhood asthma. JAMA 268 : 2673-2677

Xia W, Pinto CE, Kradin RL (1995) The antigen-presenting activities of Ia^+ dendritic cells shift dyamically from lung to lymph node after an airway challenge with soluble antigen J Exp Med 181 : 1275-1283

Zetterstrom O, Osterman K, Machado L, Johansson SG (1981) Another smoking hazard : raised serum IgE concentration and increased risk of occupational allergy. Br Med J 283 : 1215-1217

Functional Teratogenicity

(Chairs: L. Dencker, Sweden, and
O. Benešová, Czech Republic)

Introduction to Functional Teratology

L. Dencker
Department of Pharmaceutical Biosciences, Division of Toxicology, Uppsala University, P.O. Box 594, S-751 24 Uppsala, Sweden

Classically, the final manifestations of abnormal development mentioned in the literature are embryonic/fetal/neonatal death, malformations, growth retardations, and functional disorders. It is, however, reasonable to assume that chemically induced as well as "spontaneous" disturbances of embryonic and fetal development most often result in a continuum of defects from major malformations to minor functional defects, rather than any of the above mentioned defects as clear cut entities. For obvious resons, the focus since the thalidomide catastrophy in the early sixties has been on the risk for morphological malformations.

What functional teratology is concerned, as will be illustrated in this symposium, there has been an emphasis on defects in the nervous system. For further reading on developmental neurotoxicity, the reader is referred to a recent article and references therein (Rodier et al. 1994).

There may be several resons to focus on the nervous system in this context:

1. the central nervous system is considered to have a much longer period of development and maturation than other functional systems in the body
2. despite the great plasticity of the nervous system, it is also considered to have a limited repair capacity
3. most other organs are composed of functional units that are more or less identical and repeated thousands of times, giving a greater chance of redundancy
4. many chemicals, and pharmaceutical products and drugs in particular, act on receptors in the nervous system, which may in fact be more or less unique to the nervous system
5. there is a believe, right or wrong, that discrete effects on the nervous system leading to functional disturbances may be induced at lower concentrations of exposure, and thus more realistically mirror the real life situation than more massive exposures created in the laboratory
6. the intellectual, emotional or motoric capacity of an individual chararacterize him or her much more than e.g. his or her kidney function would do
7. there is an assumption that while other organ dysfunctions can be detected early in life, and possibly corrected surgically or by drug treatment, damage to the nervous system may be more insidious in its expression and come through as a negative surprise even as late as in adulthood.

For those engaged in drug development, this distinction made under (7) above is important, because in the former case, at least theoretically, detection of the side effect can occur earlier to stop the use of a teratogenic drug, while in the latter case it may even be impossible to connect a functional deficit in adulthood to an exposure to drugs decades earlier.

Relations between Morphological and Functional Teratology

It should be obvious that functional deficiencies can result from morphological aberrations. As far as drug induced teratogenesis in humans is concerned, in a historical perspective it seems as if functional deficiencies became obvious first after a malformation, or a syndrome of malformations, had been ascribed to that particular drug. As far as this is true, the perspective of detecting minor functional defects in the human population - in the absence of morphological "markers" - is not overwhelmingly bright.

An interesting example of this is diethylstilboestrol. Having an oestrogenic effect, it was used during pregnancy to prevent abortions in millions of women over two decades, especially in USA. In the early seventies, it was found that female offspring had an increased tendency to develope vaginal and cervical clear cell adenocarcinomas. Eventually, also functional defects in the form of reproductive failure in female as well as in male offspring of thus treated mothers were revealed. This example is of course not necessarily representative of all types of functional teratogens, since the disturbance can not be expected to be observed until after puberty.

Other examples are the use of ethanol and antiepileptics (notably hydantions) during pregnany. It took a long time before the facial dysmorphogenesis specific for each of these exposures were identified and generally accepted, and even later were the effects on CNS development recognized.

Offspring of thalidomide treated mothers have lately been systematically investigated with respect to specific functional disturbances, such as of vision, ocular motility disturbances and other cranial nerve failure leading to e.g. abnormal lacrimation. This indicates another important target of thalidomide during embryonic development, namely the hindbrain (Strömland and Miller, 1993).

Vision disturbances in the offspring are common also after other drug treatments, such as alcohol, antiepileptics, retinoic acid, warfarin, cocain etc. (Strömland and Miller 1993).

Importance of the Time of Exposure

As other organ systems, the nervous system has peaks of specific and increased sensitivity during development. Such peaks of increased sensitivity will appear repeatedly in each specific region of the CNS. Developmental events responsible for such peaks are the rolling up of the neural plate to form a neural tube, regional specifications along the neural tube, such as segmentation into rhombomeres (neuromeres) in the hindbrain as a mean e.g. of giving the cranial nerves their specific identities, the relative rate of cell division and cell migration, development of neurotransmitter systems, the "brain growth spurt", including growth and differntiation of axons and dendrites, and synaptogenesis, formation of myelin sheets, etc. For xenobiotics, also the time of development and the function of the blood-brain barrier is of great importance. Far too little is still known about the relation between developmental events and increased sensitivity, which goes hand in hand with our poor understanding of mechanisms behind disturbances of the particular processes.

How can CNS-Function be Studied?

By tradition, functional studies of the nervous system has been performed with different behaviour tests in combination with neurochemical and neuropathological evaluation. The combination of the three appears essential, for the actual experiment as well as for the gain of precedence, so that any change in behavioural pattern can be mechanistically related to an alterd histological appearance or perhaps regional changes in transmitter activities. The latter are increasingly often studied by means of autoradiographic localisation of receptors and/or immunohistochemical/in situ hybridization to localize synthesis and or storage of transmitter substances and their receptors. This has a great advantage over biochemical determinations when it comes to morhological specificity of any change that may be detected. At the same time it must be remembered that behavioural deficiencies most likely may be induced in the absence of apparent morphological changes, or detectable transmitter system aberrations. Conversely, morhological and physiological aberrations (of e.g. blood flow as revealed by PET studies or other non-invasive techniques in humans) may be seen in the absence of apparent behavioural or functional deficits.

What concerns behavioural studies, there is a battery of tests specifically designed or at least applied in the postnatal period, such as surface righting test, negative geotaxis test, rotating rod test and simple swim maze tests (e.g. M-maze). Even more important is of course if the nervous system is affected permanently, and then already established tests from the fields of behaviour pharmacology, experimental psychology etc. has been applied (Rodier et al. 1993 and references therein).

Can Birth Registers be Used to Detect Drug Induced Teratogenesis?

For obvious resons it is difficult to relate conditions of a human fetus during development to functional disturbances of that individual later in life. Information on exposure to chemicals, including pharmaceutical products, is often not properly registered even today, and in those cases when data may be collected, the information is not good enough to make it possible to relate exposure of an individual to its stage of development (i.e. sensitivity) during his/her mother´s pregnancy.

Lately, data from birth registers has been related to the health state of the individuals at middle age. It has turned out that there is a relation between smaller than normal abdominal circumference and raised serum LDL-cholesterol and apolipoprotein B (Parker et al 1993). The conclusion of the authors was that impaired liver growth in the fetal period may permanently alter LDL-cholesterol metabolism. This may be compared to animal experimental data on similar effects of prenatal exposure to cholesterol-lowering drugs (Innis 1983).

Similarly, low birth weight of human babies is related to a higher systolic blood pressure in adulthood, which became larger with increasing age (Law et al 1993). Whether such studies can be done in the future to relate chemical exposure to functional defects in humans remains to be shown.

Own Animal Experiments

In our laboratory, we have investigated the effects of pre- and perinatal exposure to metallic mercury, by maternal inhalation, on the adult behaviour in rats. Previous investigations had show that the fetal brain has the ability to oxidize metallic mercury to divalent mercuric mercury (Khayat and Dencker 1982). Since this oxidation is mediated by intracellular catalase, an intracellular accumulation of the divalent occurs with the perspective of a long term retention in the cell.

Funtional studies on spontaneous motor activity and performance in a radial arm maze and circular swim maze showed that both prenatal and postnatal exposure changed the various test parameters studied in a dose dependent and age dependent manner. Thus, prenatal exposure (day 11-14 plus 17-20 p.c.) at a dose level up to 1.5 mg Hg/m^3 (resulting in 0.2 mg/kg/day for the mother) not apparently affecting maternal behaviour, nor the pups in the standard developmental tests mentioned above, resulted in hypoactivity at the age of 3 months, but hyperactivity at 14 months af age (Danielsson et al 1993). These rats showed retarded aquisition in the radial arm maze but normal behaviour in the swim maze.

The postnatally treated rats (day 11-17) were exposed to around 50 microgram/m^3 for one or four hours per day (Fredriksson et al 1992). Again, as was observed after prenatal exposure, there was a dose- and time-dependent variation in total activity, and for the radial arm maze there was a retarded aquisition. No effect was observed in the swim maze.

Additional experiments have shown that there may be an additive effect between metallic mercury and methyl mercury. In these experiments, methylmercury was given by gavage, days 6-9 of gestation, 2 mg/kg/day, and then the dams were exposed to metallic mercury vapour by inhalation, 1.8 mg/m^3 air for 1.5 hours/day, day 14-19 of gestation. Proper control groups (including metallic mercury and methylmercury alone) were treated simultaneously. Calculations revealed that the inhaled dose corresponded to around 0.1 mg Hg/kg /day. Conventional developmental markers tested on the pups again showed no apparent deviations from normal. At 4-5 months of age, the metallic mercury group showed a small deviation from normal in both spontaneous and learned behaviour, suggesting some deficit in adaptive functions. Coexposure to methylmercury, the latter in itself not inducing any alterations in theses functions at the doses given, served to significantly aggravate the changes (Fredriksson et al. 1995).

Considering the ubiquitous distribution of metallic mercury in the fetal brain, it may be expected that a global effect on several physiological system in the brain may be affected, were enough high doses applied. Nerve growth factor (NGF) is a neurotrophic factor of importance for the normal development of the peripheral as well as cental nervous system. In the CNS, especially the targets, such as hippocampus, for axons of cholinergic neurons in the basal forebrain sythetize NGF. In a collaborative study with the group Ted Ebendal in Uppsala (Söderström et al. 1995), we have shown that exposure to metallic mercury in the early stages of development in the rat results in an increased level of NGF (but not of its mRNA) in the hippocampus, while there was a reduction of NGF in the basal forebrain at postnatal day 21. In addition, mRNA of the low affinity receptor for NGF, p75, was reduced in the basal forebrain, while at the same time the mRNA for the high affinity receptor p140^{trk} was unaffected. There was asignificant effect also on the mRNA coding for choline acetyltransferase (ChAT). Although the role of the low affinity receptor for NGF , p75, is not known, there is evidence that it may be involved in the axonal transport of NGF by the target cell, which may explain the present resuts of an increased level at the production site but a low level in the area of the target cell bodies.

References

Rodier PM, Cohen IR, Buelke-Sam J (1994) Developmental Neurotoxicology. In: Kimmel CA, Buelke-Sam J (eds) Developmental Toxicology. Target Organ Toxicology Series, second edition. Raven Press, New York.

Strömland K, Miller M (1993) Ocular Teratology. In: Tasman W, Jaeger EA (eds) Duane's Foundation of Clinical Ophthalmology, revised edition. J.B. Lippincott Company, Philadelphia.

Parker DJP, Martyn CN, Osmond C, Hales CN, Fall CHD (1993) Growth in utero and serum cholesterol concentrations in adult life. Brit Med J 307:1524-1527

Innis SM (1983) Influence of maternal cholestyramine treatment on cholesterol and bile acid metabolism in adults. J Nutr 113:2464-2470

Law CM, de Swiet M, Osmond C, Fayers PM, Barker DJP, Cruddas AM, Fall CHD (1993) Initiation of hypertension in utero and its amplification throughout life. Brit Med J 306:24-27

Khayat A, Dencker L (1982) Fetal uptake and distribution of metallic mercury vapor in the mouse: influence of ethanol and aminotriazole. Biol Res Pregn 3:38-46

Danielsson BR, Fredriksson A, Dahlgren L, Gårdlund AT, Olsson L, Dencker L, Archer T (1993) Behavioural effects of prenatal metallic mercury inhalation exposure in rats. Neurotoxicol Teratol 15: 391-396

Fredriksson A, Dahlgren L, Danielsson B, Eriksson P, Dencker L, Archer T (1992) Behavioural effects of neonatal metallic mercury exposure in rats. Toxicology 74:151-160

Fredriksson A, Dencker L, Archer T, Danielsson BRG (1995) Prenatal coexposure to metallic mercury vapour and methylmercury produce additive behavioural changes in adult rats. (in press)

Söderström S, Fredriksson A, Dencker L, Ebendal T (1995) The effect of mercury vapour on cholinergic neurons in the fetal brain: studies on the expression of nerve growth factor and its low- and high-affinity receptors. Dev Brain Res 85:96-108

The Neurobehavioural Toxicology and Teratology of Lead

Gerhard Winneke[1], Hellmuth Lilienthal[1] and Ursula Krämer[2]
[1] Department of Psychophysiology; [2] Department of Biostatistics; Medizinisches Institut für Umwelthygiene an der Heinrich-Heine-Universität Düsseldorf, Auf'm Hennekamp 50, D-40225 Düsseldorf

Introduction

Lead has been used by man since antiquity, and its detrimental health effects are well known for centuries. Inorganic lead enters the body by way of inhalation or ingestion; uptake through the gastrointestinal tract is the more important route of exposure. Absorption is better in infants than in adults. Lead-concentration in whole blood (PbB), normally expressed as µg/l or µmol/l, is a representative marker of current lead exposure, whereas tooth lead concentrations have also been used as markers of past exposure (Needleman et al. 1979; Winneke et al. 1983). Both placental transfer and transfer across the blood-brain barrier occur easily in the developing brain, so that prenatal exposure as well as CNS-involvement are possible. Lead serves no known physiological function. Almost all organs or organ systems are potential targets for lead: depending on the duration and degree of exposure, diverse biological effects have been described, the more critical of which are those on heme biosynthesis, erythropoiesis, and the nervous system.

Both the peripheral (PNS) and the central nervous system (CNS) are targets of lead, although PNS effects (wrist drop, slowing of nerve conduction velocities) have, so far, preferentially been described in occupational settings in adults. Lead encephalopathy has been reported to occur in cases of acute symptomatic lead poisoning in adults and in children and, in the latter case, appears to be associated with PbB exceeding 80-100 µg/dl resulting from swallowing of lead-based paint. The clinical picture of lead encephalopathy is characterized by some or all of the following signs and symptoms (CDC 1991): coma, seizures, ataxia, apathy, incoordination, vomiting, clouded consciousness, and loss of previously acquired skills. Since neurological and neuropsychological sequelae have been described in individuals surviving acute lead encephalopathy the hypothesis has been proposed that even asymptomatic lead exposure might be associated with subclinical neurobehavioural toxicity which, due to ist subtlety, may often go undetected. The following text describes typical neuroepidemiological studies in occupational and environmental exposure settings dealing with this hypothesis.

Neurobehavioural Toxicity in Humans

Cognitive as well as sensorimotor functions have typically been studied. In most studies cognitive functions are usually covered by means of results from standard intelligence tests, such as the "Wechsler Adult Intelligence Scale (WAIS)" or parts of such clinical tests. Despite their established psychometric quality in terms of reliability, validity and normative data bases such tests are of limited usefulness for studying cognitive functions in detail, because performance involves several subfunctions and it is, therefore, difficult to identify the basis for poor performance from such a global test (Bellinger 1995). For this reason, in some studies, more specific tests for measuring specific cognitive functions have been used. Tests for the assessment of sensory motor functions include different reaction time paradigms, tasks of sensory motor speed and coordination, as well as tasks of perceptual speed and sustained attention.

Neurobehavioural Toxicity In Adults: Several occupational exposure studies have been conducted generally not showing neurobehavioural deficit to occur at blood lead-levels below 40 - 50 µg/dl (Hogstedt et al. 1983; Stollery et al. 1989; 1991; Williamson & Teo 1986). From this set of studies those from Stollery et al. (1989; 1991) will be described in more detail, (1) because they selected tasks on theoretical considerations based on models of cognitive information processing models rather than on purely clinical grounds, and (2) because they looked at three rather than only one measure of lead-exposure.

Stollery et al. (1989) studied 91 workers of mean age 41.5 years from battery and printing industries with low (<20 µg/dl), medium (21-40 µg/dl) and high (41-80 µg/dl) blood lead concentrations (PbB). They were given computer-based tests assessing sensory motor serial reaction time, visual memory, attention, verbal reasoning, and spatial processing. In addition to PbB zinc protoporphyrin (ZPP) as well as aminolevulinic acid in urine (ALA-U) were measured as markers of lead exposure.

For the statistical analysis a number of non-lead (extraneous) variables, as e.g. years of lead exposure, years of schooling, alcohol consumption, work demands and stress at work, were controlled for by analysis of covariance, if they exhibited associations with outcome. Significant exposure-related neurobehavioural deficit was found for decision and movement times, as well as for decision gap rates, of the serial reaction time task and, although less compelling, for classification speed and accuracy in the category search task. In both cases only Ss with PbBs in excess of 40 µg/dl exhibited functional deficit. It was concluded that the sensory motor, rather than the cognitive requirements of the neurobehavioural tasks provided the most sensitive indicator of the effects of chronic occupational lead exposure, and that the effect on gap rates could be interpreted in terms of lead-related attentional difficulties. In no case were associations with ZPP or ALA-U significant.

In a follow-up study of the same cohort (Stollery et al. 1991) the same battery of computer-based neurobehavioural tests was used. The results of the previous study were confirmed. Impairment tended to be restricted to Ss with PbBs above

40 μg/dl. Again, the main deficit was a slowing of sensory motor reaction time, which was most evident if the cognitive task-load was low. From these findings the conclusion was drawn that the spectrum of lead-induced neurobehavioural impairment is mainly characterized by sensory motor slowing.

Neurobehavioural Toxicity In Children: Children rather than adults are considered a particularly sensitive risk group for lead because, on a body weight basis, their dietary intake, absorption and total retention is markedly higher. In addition, the blood-brain-barrier is not yet fully developed in young children, and specific behavioural characteristics, such as hand-to-mouth and playground activities, put them at higher risk. It should, furthermore, be mentioned that the developing rather than the mature brain is known to be particularly vulnerable to the effects of chemical exposure. For these reasons both cross-sectional and prospective epidemiological studies have been conducted to test the assumption, that children may be particularly at risk for neurobehavioural toxicity even at environmental levels of exposure.

Cross Sectional Studies: Following the pioneering work of Needleman et al. (1979) several cross sectional studies have primarily relied on PbT as a marker of longterm retrospective cumulative lead exposure (Smith et al. 1983; Winneke et al. 1983; Fergusson et al. 1988, Hansen et al. 1989; Rabinowitz et al. 1991), whereas others have primarily, used blood lead concentrations as the measure to characterize internal exposure (Yule et al. 1981; Winneke et al. 1985; Schroeder et al. 1985; Hatzakis et al. 1986).

Both cognitive as well as sensorimotor outcome measures have been used to characterize lead-induced neurobehavioural deficit in these studies. Cognitive measures in most instances were parts of or fullsize clinical tests of intelligence, such as the Wechsler Intelligence Scale for Children (WISC) or the Wechsler Preschool and Primary Scale of Intelligence (WPPSI), the British Ability Scales (BASC), and the Stanford-Binet Intelligence Scale. Measures of sensorimotor performance included different types of simple and choice reaction time-paradigms, tapping speed, and memory for design-tasks, such as the Benton-test or the Bender Gestalt-test.

The European Multicenter Study as described by Winneke et al. (1990) will be used to illustrate a typical cross-sectional approach.This study combined eight individual studies from eight European countries (Bulgaria, Denmark, Greece, Germany, Hungary, Italy, Roumania, Yugoslavia), which were tied together by a common study protocol with inherent quality assurance elements. 1879 six to eleven year old children were studied altogether, although for some of the teststhe number of children was reduced to only 971. The range of PbB-values was about 5 to 60 μg/dl. Tests used in all or part of the individual studies included the full or reduced WISC (four subtests), a serial choice reaction-time paradigm, the Bender Gestalt-test, the trailmaking test, as well as behaviour rating-scales for teachers and parents. The overall statistical analysis was done using a uniform pre-determined regression model with age, gender, occupational status of the father,

Table 1. Outcome of the WHO/CEC collaborative lead study (modified after Winneke et al. 1990)

RESULTS FROM THE EUROPEAN MULTICENTER STUDY					
Neurobehavioural Endpoint		**N of children**	**Slope of regression**	**Significance t-value**	**% Variance explained by PbB**
WISC		1698	− 0.53	−1.82*	0.19%
Bender Gestalt-Test (errors) 1)	easy	1698	1.64	2.17**	0.31%
	difficult	1684	2.03	2.64***	0.46%
Reaction Performance (errors)	easy	971	0.15	1.68**	0.37%
	difficult	971	0.20	2.37***	0.56%
Delayed Reaction Time 2)		986	0.01	0.41	0.02%
Behaviour Ratings by teachers (Needleman-Scales) 3)		1012	−0.00	−0.03	0.00%

*p<0.1; **p<0.05; ***p<0.01 (one-tailed p)
1) x'=log(x+0.5); 2) log msec; 3) sum of 8 items

and maternal education as confounders or covariates. The most salient findings of that study are given in table 1. An inverse association between PbB and IQ was of only borderline significance ($p < 0.1$), and an IQ-decrement of 3 points was estimated for a PbB-increase from 5 to 20 µg/dl, and of 5.3 points from 5 to 50 µg/dl. The degree of inverse association between the error-scores of the Bender Gestalt-test and the serial choice reaction-paradigm were statistically significant and more consistent across individual studies, although the outcome-variance explained by lead never exceeded 0.8% of the total variance. No obvious threshold could be located on the dose-effect curves.

Summary of Cross Sectional Studies: Clinical measures of intelligence, have been the preferred outcome measure in all of the cross sectional studies although, as mentioned above, this global measure does not carry information about the specific cognitive functions primarily influenced by lead. Since, however, the degree of comparability of findings between studies is higher for the IQ-measure relative to the different sensorimotor paradigms, efforts have been made to estimate the overall effect-size across studies by means of meta-analysis, concentrating on more recent blood lead-studies and on those studies, which have given regression coefficients as estimates of effect-size (Pocock et al. 1994; WHO 1995). Although, for the majority of studies, the reported associations between blood lead-concentrations and the IQ-measure were not statistically significant, the overall evaluation revealed a highly significant average IQ-decrement of 1-3 points for a change of PbB from 10 to 20 µg/dl (Pocock et al. 1994; WHO 1995).

Prospective Studies: Recent prospective studies have attempted to clarify the role of environmental lead-exposure as a contributing factor in developmental neurotoxicity. Such studies have been performed in Boston (Bellinger et al. 1987; 1991), Cincinnati (Dietrich et al.,1987; 1991; 1993), Cleveland (Ernhard et al. 1988; 1989), Glasgow (Moore et al. 1989), Port Pirie (McMichael et al. 1988; Baghurst et al. 1992), Sydney (Cooney et al. 1989; 1991), and the Kosovo-area in Yougoslavia (Wasserman et al. 1994). The strength of this set of studies, apart from their prospective design, is given by the fact, that initial agreement was reached on the main features of a core protocol covering independent, dependent and main confounding variables. Essential neurobehavioural test in these studies included the Bayley Scales of Infant development until age 2 years, the McCarthy Scales or the Kaufman Assessment Battery beyond age 3 years, and the WPPSI or the WISC-R at or beyond 5-6 years of age. Blood lead-levels were monitored at regular intervals in order to document the exposure history from pre-/perinatal stages until or beyond schoolage. Important confounders within the core protocol included maternal IQ and the HOME-scale for an assessment of the quality of the home environment.

Despite these efforts towards harmonization of study protocols the final results were only partly consistent across studies. Only four of these studies reported inverse associations between indeces of pre-/perinatal lead exposure (prenatal or cordblood Pb) and reduced MDI and/or PDI-scores of the Bayley-test up to 6 or 12 months (Ernhart et al. 1987; Dietrich et al. 1987), two years (Bellinger et al. 1987), or to four years of age (Wasserman et al. 1994), but these associations were no longer detectable at school-age (Bellinger et al. 1991; Dietrich et al. 1993). The other studies did not report significant negative MDI/PDI-associations with prenatal lead exposure after confounder control (Cooney et al. 1989; Moore et al. 1989; Wigg et al. 1988).

As for preschool age assessments, both the Boston- and the Port Pirie-studies, using the McCarthy Scales at 57 or 48 months, respectively, did report significant inverse associations between GCI-scores and blood lead-levels measured at different postnatal ages. None of the other studies did so after confounder adjustment. At school age, using the WISC-R, significant inverse associations between postnatal blood leads and IQ were observed in the Boston cohort (Bellinger et al. 1992), the Cincinnati sample (Dietrich et al. 1993), and in the Port Pirie cohort, as well (Baghurst et al. 1992).

Summary of Prospective Studies: Despite some inconsistencies as described above, the overall outcome of this ambitious set of prospective studies confirms the main outcome of the cross sectional studies (see above), namely an inverse association between environmental lead-exposure during development and cognitive ability in children. Quantitatively, as confirmed by meta-analyses (Pocock et al. 1994; WHO 1995) and expressed in probabilistic terms, an increase of developmental lead-exposure from 10 to 20 µg/dl is likely to be associated with a cognitive deficit of between 1-3 IQ-points. It should also be mentioned, however, that the specific strength of the developmental approach, namely the potential of identifying sensitive periods of exposure and of excluding the possibility of reverse causality, could not be demonstrated, because peak effects occurred at different time-points in different studies, and because postnatal rather than prenatal exposure was most consistently related to neurodevelopmental outcome. It is remarkable, however, that quantitatively both the crossectional and the prospective studies are in remarkably good agreement.

Conclusions from Human Studies: In comparing lead-related neurobehavioural findings in adults and children striking sensitivity-differences are obvious. Whereas in lead-exposed workers neurobehavioural dysfunction has not been demonstrated at blood lead-concentrations below 40 µg/dl, cognitive and sensorimotor deficit has been shown in children to be associated with PbBs as low as 10 - 15 µg/dl. The reasons for this difference in vulnerability are not entirely clear at present, although the developing blood brain-barrier in young children as well as the particular vulnerability of the developing brain are being discussed as contributing to the susceptibility to lead of the immature vs the mature central nervous system.

The subtlety of lead-associated cognitive impairment in children should also be emphasized. For both cross sectional and prospective studies average IQ-decrement between 1 to 3 points has been shown to be associated with increasing PbB from 10 to 20 µg/dl. This corresponds to 1/5th or less of a standard deviation of the typical IQ-distribution. This must be qualified as a small population-effect (Bellinger 1995), which applies to the individual child only in a probabilistic sense. It should be pointed out, however, that even small average effects, due to the normal variability of individual susceptibility, will be associated with larger IQ-deficit in individual cases of particular vulnerability. It has also been argued (Needleman et al. 1981) that even small average IQ-differences between populations of children with low and elevated lead-exposure, respectively, may have substantial effects at the upper and lower ends of the IQ-distribution. This "population-shift hypothesis" still awaits more general validation, however.

Observations on the neurobehavioural toxicity of lead at low levels of environmental exposure have been used to establish or revise exposure limits. The basic limitations of the epidemiological approach in arriving at unequivocal conclusions need careful consideration, however. This is particularly true if, as in the case of inorganic lead, the critical neurobehavioural effects are small and embedded in a complex background of confounding. In such cases questions as to the true cause-effect contingencies have been and are still being raised (Pocock et al. 1994). In these critical issues neurobehavioural studies in animal models have helped substantially to corroborate epidemiological observations, to contribute to elaborating the underlying neural mechanisms, and to identify critical periods of vulnerability. This important approach will be recovered in the following section.

Neurobehavioural Toxicity In Animal Models

Only the rat will be considered here because systematic studies comparing neurobehavioural effects of early neonatal with those following later adult lead exposure are almost lacking in the primate literature. Behavioural teratology or, more broadly speaking, developmental neurotoxicology deals with functional alterations in the postnatal nervous system following chemical exposure to the developing brain which, in most cases, means in utero or transplacental exposure. The concept and the term "behavioural teratology" was first introduced by Werboff & Gottlieb (1963) and rediscovered almost 20 years later by several authors (Vorhees 1986). Although, because of the continuity of ontogenetic development, the prenatal/postnatal dichotomy is an oversimplification it is, nevertheless, a useful distinction from a practical point of view (Vorhees 1986).

Species differences with respect to neurodevelopment relative to the time of birth have to be considered in any cross-species extrapolation effort. Relative to man the rat is less developed at birth. The rat's brain is in the same developmental stage as the human brain at 5 - 6 months of gestation. Therefore, the developmental stage of the human brain at birth is reached by the rat not until postnatal day 10 (PND 10). Cerebral neurogenesis in the rat is almost complete at birth and gliogenesis starts shortly after birth. In the human cerebral cortex neurogenesis is complete by the seventh month of pregnancy, and gliogenesis, neuronal differentiation, synaptogenesis, and cerebellar development start prior to birth and extend into the first year of postnatal development (Berry 1974).

In paying tribute to the increased vulnerability of the developing nervous system to inorganic lead Pentschev & Garro (1966) developed a rodent model of lead encephalopathy in the offspring of lactating dams subjected to substantial dietary lead exposure. This model and modifications of it has become the preferred mode of developmental lead exposure in rodents in many studies, although, in order to better approach environmental exposure conditions, extensions to include in utero as well as post-weaning developmental periods have also been used in rats (e.g. Winneke et al. 1977).

Longterm Effects Following Preweaning Exposure: Although some neurobehavioural effects of postweaning lead exposure in rats have been demonstrated to occur in different models, as e.g. fixed interval or multiple FI-FR schedules of operant behaviour (Cory-Slechta and Thompson 1979; Angell and Weiss 1982), spatial behaviour (Bushnell and Levin 1983), or probability learning in a runway (Davis et al. 1993), the longlasting nature of such deficit remains to be demonstrated. Preweaning exposure, however, has more consistently been shown to produce longlasting neurobehavioural alterations and, furthermore, some mechanistic information is available to support the particular vulnerability to lead of the early developmental stages of brain development.

By means of a cross-fostering approach delays in exploratory and locomotor activity were reported for the prenatal and the pre- plus postnatal rats, but not for the group with only postnatal exposure (Crofton et al. 1980). In drug-discrimination learning no difference was found between rats with exposure during lactation only and rats with subsequent lifetime exposure (Zenick and Goldsmith 1981). In other studies using the drug discrimination model differential effects of preweaning and postweaning exposure involving different dopamine receptors were found (Cory-Slechta and Widzowski 1991; Cory-Slechta et al. 1992).

Other observations suggest that, although effects of preweaning exposure in the rat are certainly longlasting, such that postweaning exposure does not aggravate the deficit induced by preweaning exposure alone, they may nevertheless not be totally irreversible: This was shown using spatial learning in rats with dietary lead exposure using the radial arm maze (Munoz et al. 1988; 1989). Lead was given as lead acetate (1380 ppm or 750 ppm Pb^{2+}) 50 d prior to mating to the dams, and subsequently through gestation and lactation. For half of the exposed animals lead exposure was discontinued at weaning (group 750 Co), whereas the other half was kept on the same diet until testing at adult age, namely PND 200 (Munoz et al. 1988) or PND 500 (Munoz et al. 1989). Steady state blood lead levels were between 340 and 320 µg/l at ages 110 and 500 days, respectively, but had been as low as 173 µg/l at weaning. Both acquisition and retention of spatial learning was tested in the radial arm maze (RAM). The most interesting observation with respect to lead-induced performance deficit occurred for the acquisition of spatial learning in the RAM (Fig. 1). Both in terms of "days - to- criterion" and "days-without-reward" animals with lifetime lead-exposure were significantly inferior to controls at postnatal ages 200 and 500 days: This was strikingly different for those animals having experienced lead exposure only indirectly via their dams until weaning. For both criteria of acquisition they, too, performed significantly inferior to controls at 200 days of age, but approached control level at 500 days of age. This observation is suggestive of longlasting, but not necessarily irreversible neurobehavioural impairment following early developmental lead-exposure.

In an effort to compare in a more systematic manner the differential vulnerability to lead-induced neurobehavioural toxicity of different developmental periods, two groups of female Wistar-rats were fed diets containing 0 or 750 mg

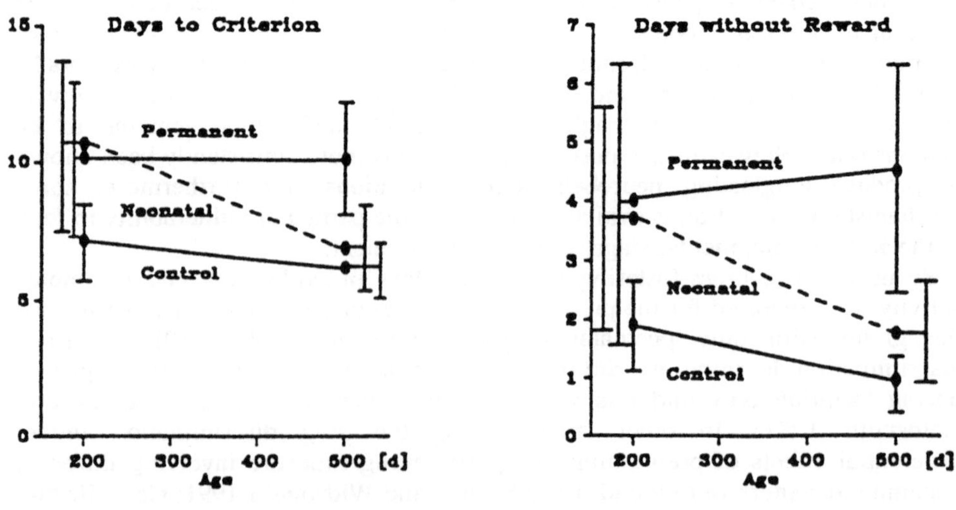

Fig.1. Acquisition in Radial Arm Maze (RAM) at age 200 and 500 days (modified from Munoz et al. 1988; 1989)

lead acetate/kg diet (408 mg Pb^{2+}/kg) for 50 d prior to mating and through gestation and lactation (Altmann et al. 1993). The male offspring were randomly assigned to the following four groups: Maternally exposed animals with only gestational and preweaning exposure (A); animals with only postweaning exposure (B); animals with lifetime pre- and postnatal lead-exposure (C); unexposed controls (D). Steady-state blood lead-levels at adult age were 26 µg/l (A), 162 µg/l (B), 143 µg/l (C), and 21 µg/l (D). Lead-concentrations in brain tissue were <0.01 (A), 0.09 (B), 0.16 (C), and <0.01 µg/g in control animals. Comparisons of groups B and C suggest a partially effective blood-brain barrier.

Animals were tested for two-way active avoidance learning in the shuttle-box as described previously (Lilienthal and Winneke 1991), as well as for synaptic plasticity by measuring hippocampal long-term potentiation (LTP) as described elsewhere (Altmann et al. 1991). Hippocampal LTP is a persistent synaptic enhancement induced by high-frequency stimulation (HFS) of afferents (Bliss and Lomo 1973), and has been suggested to be a component of the cellular basis of learning and memory (Lynch and Baudry 1984). Since the glutamatergic NMDA receptor has been shown to be involved in both LTP-blocking and learning-deficit (Morris et al. 1986) and since, furthermore, lead has been found to selectively block NMDA-activated channel currents (Alkondon et al. 1990), the comparison of hippocampal LTP and learning in the same lead-treated animals should offer some insight into possible mechanisms of lead-induced cognitive dysfunction.

A moderate correlation was found between lead-induced deficit of active avoidance-learning (AAL) and lead-induced impairment of LTP ($r = 0.31$; $p < 0.05$). AAL-deficit relative to controls was significant for groups C ($p < 0.02$) and A ($p < 0.05$), but only borderline for the postweaning group ($p < 0.08$). Likewise, the degree of LTP-induction was significantly reduced in group C relative to controls ($p < 0.03$), insignificantly reduced in group A (maternal exposure), and unchanged in the postweaning animals (group B). From these observations the conclusion was drawn that developmental brain dysfunction caused by low-level lead exposure persists long into adulthood, that the developing rather than the mature hippocampus is functionally impaired by lead, and that the findings are compatible with the hypothesis that the same cellular processes underlying lead-induced LTP-disruption may also be involved in lead-induced learning deficit (Altmann et al. 1993).

Little consistent information is presently available to explain the increased vulnerability to lead of the developing as opposed to the mature brain. One such piece of information relates to the inhibition by lead of NMDA-activated currents in hippocampal neurons, which was found to be age-dependent (Ujihara and Albuquerque 1992). It has, furthermore, been shown that the NMDA receptor density in the cerebral cortex of lead exposed rats was reduced at postnatal day 14 (Guilarte and Miceli 1992), whereas increased receptor densities were found in adult rats following pre- and postnatal exposure (Guilarte et al. 1993). Another promising line of evidence relates to disturbed calcium signaling in synaptosomes from brains of rats with either developmental or adult lead exposure: Whereas chronic low-level lead-exposure starting at PND 5 reduced the IP_3-induced mobilization of Ca^{2+} and decreased both the influx and the efflux of Ca^{2+} in synaptosomes, this was not true for animals with exposure starting not until 10 weeks after birth (Singh 1995). Since the regulation of the initiation and the growth of neurons from embryonic growth cones is crucially dependent on developmental changes in Ca^{2+} mobilization and influx it was hypothesized that lead-induced disruption of Ca^{2+} homeostasis might also disrupt the normal development of the immature brain in rats (Singh 1995). These interesting observations certainly deserve close attention in explaining the developmental neurotoxicity of inorganic lead.

Summary and Conclusions

When comparing neurobehavioural observations from occupational lead-exposure of adults on the one hand, and environmental lead exposure of children on the other, it appears that the developing relative to the mature brain is more at risk. Neurobehavioural toxicity in occupational lead-exposure has typically not been observed at blood lead-concentrations (PbBs) below 400 µg/l, whereas in environmentally exposed children such deficit has been reported to occur down to PbB of 100 - 150 µg/l and, perhaps, even below this range. Both cross-sectional and prospective studies have arrived at similar conclusions in this respect. The preferred endpoint in most such studies has been the IQ-measure, which has good psychometric qualities, is sufficiently well standardized to be comparable across studies, and exhibits attractive simplicity for the regulator in a public health context. At the same time, however, this IQ-focus has also interfered with systematic efforts to identify more specific lead-induced functional deficits by means of more detailed neurobehavioural analyses (Bellinger 1995).

Metanalyses on both cross sectional and prospective studies in lead-exposed children have concluded that a typical doubling of PbB from 100 to 200 µg/l is associated with an average IQ-loss of 1 - 3 points (Pocock et al. 1994; WHO 1995), and no threshold has as yet been identified. Since, however, cause-effect contingencies necessarily remain doubtful in epidemiological studies if the observed effects are as subtle as these, experimental studies in animals have become helpful in supporting the causative role of lead to produce neurobehavioural deficit at steady-state PbB down to about 150 to 200 µg/l. Such deficit has been demonstrated by means of a variety of learning/memory models with positive and negative reinforcement contingencies in the rat - and in primates as well. It has also been shown in such studies that neurobehavioural deficit subsequent to early developmental exposure extends long into adulthood after cessation of exposure at weaning. It, therefore, appears that the neurobehavioural teratology of lead has more convincingly been demonstrated in animal models than in human exposure conditions, so far. A coherent theory to explain the particular vulnerability to lead of the developing brain is still lacking. Recent data do suggest, however, that Pb-induced disruption of calcium homeostasis in the immature brain might interfere with normal brain development.

References

Alkondon M, Costa A, Radhakrishnan V, Aronstam R, Albuquerque E (1990) Selectiveblockade of NMDA-activa-ted channel currents may be implicated in learning deficits caused by lead. FEBS Lett 261:124-130
Altmann L, Sveinsson K, Wiegand, H (1991) Long-term potentiation in rat hippocampal slices is impaired following acute lead perfusion. Neurosci Lett 128:109-120
Altmann L, Weinsberg F, Sveinsson K, Lilienthal H, Wiegand H, Winneke G (1993) Impairment of long-term poten-tiation and learning following chronic lead exposure. Toxicol Lett 66:105-112
Angell NF, Weiss B (1982) Operant behavior of rats exposed to lead before and after weaning. Toxicol Appl Pharmacol 63:62-71
Baghurst PA, McMichael AJ, Wigg NR, Vimpani GV, Robertson EF, Roberts EJ, Tong SL (1992) Environmental exposure to lead and childrens's intelligence at the age of seven years: The Port Pirie cohort study. N Engl J Med 327:1279-1284
Berry M (1974) Development of the cerebral neocortex of the rat. In: Michael RP (ed) Studies on the Development of Behavior and the Nervous System: Aspects of Neurogenesis. Vol.2. Academic Press, New York, pp 7-67
Bliss TVP, Lomo T (1973) Long-lasting potentiation of synaptic transmission in the dentate area of the anaesthesized rabbit following stimulation of the perforant path. J Physiol 232:331-366
Bellinger D, Leviton A, Waternaux C, Needleman H, Rabinowitz M (1987) Longitudinal analyses of prenatal and postnatal lead exposure and early cognitive development. N Engl J Med 316:1037-1043
Bellinger D, Sloman, A, Leviton A, Rabinowitz M, Needleman HL, Waternaux C (1991) Low-level lead exposure and children cognitive functioning in the preschool years. Pediatrics 87:219-227
Bellinger DC (1995) Interpreting the literature on lead and child development: The neglected role of the "Experi mental System". Neurotox Teratol 17:201-212
Bushnell PJ, Levin ED (1983) Effects of zinc deficiency on lead toxicity in rats. Neurobehav Toxicol Teratol 5:283-288
Centers for Disease Control = CDC (1991) Preventing lead poisoning in young children. A statement by the Centers for Disease Control. US Dept of Health and Human Services
Cooney GH, Bell A, McBride W, Carter C (1989) Neuro-behavioral consequences of prenatal low level exposures to lead. Neurotoxicol Teratol 11:95-104
Cooney G, Bell A, Stavron C (1991) Low level exposure to lead and neurobehavioural development: the Sydney study at seven years. Proceedings: Heavy Metals in the Environment, Edinburgh.
Cory - Slechta DA, Thompson T (1979) Behavioral toxicity of chronic postweaning lead exposure in the rat. Toxicol Appl Pharmacol 47:151-159
Cory-Slechta DA, Widzowski DV (1991) Low level lead exposure increases sensitivity to the stimulus properties of dopamine D1 and D2 agonists. Brain Res 553:65-74
Cory-Slechta DA, Pokora MJ, Widzowski DV (1992) Postnatal lead exposure induces supersensitivity to the stimulus properties of a D2-D3 agonist. Brain Res 598:162-172
Crofton KM, Taylor DH, Bull RJ, Sivulka DJ, Lutkenhoff SD (1980) Developmental delays in exploration and locomotor activity in male rats exposed to low level lead. Life Sci 26:823-831
Davis SF, Nation JR, Mayleben MA (1993) The effects of chronic lead exposure on reactivity to frustrative non-reward in rats. Toxicol Lett 66:237-246

Dietrich KN, Krafft KM, Bornschein RL, Hammond PB, Berger OG, Succop PA, Bier M (1987) Low-level fetal lead exposure effect on neurobehavioral development in early infancy. Pediatrics, 80:721-730

Dietrich KN, Succop PA, Berger OG, Hammond PB, Borschein R (1991) Lead exposure and the cognitive development of urban preschool children: The Cincinnati lead study cohort at age 4 years. Neurotoxicol Teratol 13:203-211

Dietrich KN, Berger OG, Succop PA, Hammond PB (1993) The developmental consequences of low to moderate lead exposure: Intellectual attainment in the Cincinnati lead study cohort following school entry. Neurotoxicol Teratol 15:37-44

Ernhart CB, Morrow-Tlucak M, Marler MR, Wolf AW (1987) Low level lead exposure in the prenatal and early preschool periods: Early preschool development. Neurotoxicol Teratol 9:259-270

Ernhart CB, Morrow-Tlucak M, Wolf AW (1988) Low level lead exposure and intelligence in the preschool years. Science of the Total Environment 71:453-459

Ernhart CB, Morrow-Tlucak M, Wolf AW, Super D, Drotar D (1989) Low-level lead exposure in the prenatal and early postnatal periods: Intelligence prior to school entry. Neurotoxicol Teratol 11:161-170

Fergusson DM, Fergusson JE, Horwood LJ, Kinzett NG (1988) A longitudinal study of dentine lead levels, intelligence school performance and behaviour. Part I Dentine lead levels and exposure to environmental risk factors. J Child Psychol Psychiat 29:781-792

Guilarte TR, Miceli RC (1992) Age-dependent effects of lead on 3H- MK- 801 binding to the NMDA receptor-gated ionophore: In vitro and in vivo studies. Neurosci Lett 148:27-32

Guilarte TR, Miceli RC, Altmann L, Weinsberg F, Winneke G, Wiegand H (1993) Chronic prenatal and postnatal Pb2+ exposure increases [3H]MK801 binding sites in adult rat forebrain. Europ J Pharmacol Environ Toxicol Pharmacol Sec 248:273-275

Hansen ON, Trillingsgaard A, Beese I, Lyngbye T, Grandjean P (1989) A neuropsychological study of children with elevated dentine lead levels: Assessment of the effect of lead in different socioeconomic groups. Neurotoxicol Teratol 11:205-213

Hatzakis A, Kokkevi A, Maravelias C, Katsouyanni K, Salaminios F, Kalandidi A, Koutselinis A, Stefanis C, Trichopoulos D (1989) Psychometric intelligence deficits in lead exposed children. In: Smith M, Grant LD, Sors AI (eds) Lead exposure and child development: An international assessment. Kluwer, Dodrecht, pp 260-270

Hogstedt C, Hane M, Agrell A, Bodin L (1983) Neuropsychological test results and symptoms among workers with well-defined long term exposure to lead. Brit J Industr Med 49:99-105

Lilienthal H, Winneke G (1991) Sensitive periods for behavioral toxicity of polychlorinated biphenyls: Determination by cross-fostering in rats. Fund Appl Toxicol 17:368-375

Lynch G, Baudry M (1984) The biochemistry of memory: A new and specific hypothesis. Science 224:1057-1063

McMichael AJ, Baghurst PA, Wigg NR, Vampini GV, Robertson EF, Roberts RJ (1988) Port Pirie cohort study: Environmental exposure to lead and children's abilities at the age of four years. New Engl J Med 319:468-475

Moore MR, Bushnell IWR, Goldberg A (1989) A prospective study of the results of changes in environmental lead exposure in children in Glasgow. In: Smith MA, Grant LD, Sors AI (eds) Lead Exposure and Child Development. An International Assessment. Kluwer Acad.Publ, Dodrecht, pp 371-378

Morris RGM, Anderson E, Lynch G, Baudry M (1986) Selective impairment of learning and blockade of long-term potentiation by an N-methyl-D-aspartate receptor antagonist, AP5. Nature 319:774-776
Munoz C, Garbe K, Lilienthal H, Winneke G (1988) Significance of hippocampal dysfunction in low level lead exposure in rats. Neurotoxicol Teratol 10:245-253
Munoz C, Garbe K, Lilienthal H, Winneke G (1989) Neuronal depletion of the amygdala resembles the learning deficits induced by low level lead exposure in rats. Neurotoxicol Teratol 11:257-264
Needleman HL, Gunnoe C, Leviton A, Reed R, Peresie H, Maher C, Barrett P (1979) D Deficits in psychologic and classroom performance of children with elevated dentine lead levels. N Engl J Med 300:689-695
Needleman H, Bellinger D, Leviton A (1981) Does lead at low dose affect intelligence in children? Pediatrics 68: 894-896
Pentschew A, Garro F (1966) Lead encephalo-myelopathy of the suckling rat and its implications on the porphyrinopathic nervous diseases. Acta Neuropathol 6:266-278
Pocock SJ, Smith M, Baghurst P (1994) Environmental lead exposure and children's intelligence: A systematic review of the epidemiological evidence. BMJ 309:1189-1197
Rabinowitz MB, Wang J-D, Soong WT (1991) Dentine lead and child intelligence in Taiwan. Arch Environ Health 46:351-360
Schroeder SR, Hawk B, Otto DA, Mushak P, Hicks RE (1985) Separating the effects of lead and social factors on IQ. Environ Research 38:144-154
Singh AK (1995) Calcium signaling and the age-dependence of chronic low level lead neurotoxicity in rat. In: Association of clinical scientists (ed) Vth COMTOX Symposium on Toxicology and Clinical Chemistry of Metals Vancouver, Farmington
Smith M, Delves T, Lansdown R, Clayton B, Graham P (1983) The effects of lead exposure on urban childrem: The Institute of Child Health /Southampton Study. Dev Med Child Neurol Suppl 47:1-54
Stollery BT, Banks H-A, Broadbent DE, Lee WR (1989) Cognitive functioning in lead workers. Br J Ind Med 46:698-707
Stollery BT, Broadbent DE, Banks HA, Lee WR (1991) Short term prospective study of cognitive functioning in lead workers. Br J Ind Med 48:739-749
Ujihara H, Albuquerque EX (1992) Developmental change of the inhibition by lead of NMDA-activated currents in cultured hippocampal neurons. J Pharmacol Exp Ther 263:868-875
Vorhees CV (1986) Origins of behavioral teratology. In: Riley EP, Vorhees CV (eds) Handbook of behavioral teratology. Plenum Press, New York, pp 3-22
Wasserman G, Graziano JH, Factor-Litvak P, Popovac N, Morina N, Musabegovic A, Vrenezi V, Capuni-Paracka S, Lekic V, Preteni-Redjepi E, Hadzialjevic S, Slakovich V, Kline J, Shrout P, Stein Z (1994) Con-sequences of lead exposure and iron supplementation on childhood development at age 4 years. Neurotoxicol Teratol 16:233-240
Werboff J, Gottlieb JS (1963) Drugs in pregnancy: Behavioral teratolgy. Obstetr Gynecol Surv 18:420-423
World Health Organization = WHO (1995) Environmental health criteria on inorganic lead. International programme on chemical safety. WHO, Geneva (in press)
Wigg NR, Vimpani GV, McMichael AJ, Baghurst PA, Robertson EF, Roberts RJ (1988) Port Pirie cohort study: Childhood blood lead and neuropsychological development at age two years. J Epidemiol Commun Health 42:213-219
Williamson A, Teo RKC (1989) Neurobehavioural effects of occupational exposure to lead. Brit J Ind Med 43: 374-380
Winneke G, Brockhaus A, Baltissen R (1977) Neurobehavioral and systemic effects of longterm blood lead-elevation in rats. Arch Toxicol 37:247-263

Winneke G, Krämer U, Brockhaus A, Ewers U, Kujanek G, Lechner H, Janke W (1983) Neuropsychological studies in children with elevated tooth-lead concentrations. II. Extended study. Int Arch Occup Environ Health 51:231-252

Winneke G, Beginn U, Ewert T, Havestadt C, Krämer U, Krause C, Thron HL, Wagner HM (1985) Comparing the effects of perinatal and later childhood lead exposure on neuropsychological outcome. Environ Res 38:155-167

Winneke G, Brockhaus A, Ewers U, Krämer U, Neuf M (1990) Results from the European multicenter study on lead neurotoxicity in children: Implications for risk assessment. Neurotoxicol Teratol 12:553-559

Yule W, Lansdown R, Miller I, Urbanowicz M (1981) The relationship between blood lead concentrations, intelligence and attainment in a school population: A pilot study. Dev Med Child Neurol 23:567-576

Zenick H, Goldsmith M (1981) Drug discrimination learning in lead-exposed rats. Science 212:569-571

Teratogenic Effects of Alcohol: Current Status of Animal Research and *in vitro* Models

Consuelo Guerri
Instituto Investigaciones Citológicas (FIB), Amadeo de Saboya 4. 46010- Valencia, Spain

Alcohol consumption during pregnancy induces in the offspring a wide spectrum of adverse effects, the extreme consequence of which is the fetal alcohol syndrome (FAS), characterized by growth deficiency, altered morphogenesis, and central nervous system dysfunctions (Jones and Smith 1973). The incidence of FAS is estimated to be 1.9 per 1000 live births, and in Western societies alcohol consumption during pregnancy is considered to be the most prevalent cause of preventable mental retardation (Abel and Sokol 1987). Although in 20 years of basic and clinical research a wealth of information has been accumulated, there is a need for further definition and characterization of the prenatal effects of alcohol on the developing organism.

Animal research has played a major role in advancing our knowledge of the detrimental consequences of prenatal alcohol exposure. Although there is no animal analog for the full syndrome, experiments using several species and *in vitro* systems have firmly established the teratogenic actions of alcohol by eliminating potentially confounding variables, such as malnutrition, poor environmental conditions, disease, smoking, and other drug use. An effort has been made to answer questions such as: When is the fetus at greatest risk? What level of alcohol drinking is safe? What is the mechanism(s) of alcohol teratogenicity?.

This review will focus on the results obtained with different animal and *in vitro* models on the effects of alcohol, in terms of teratogenesis, critical periods of ethanol toxicity, and the mechanisms of ethanol action. In addition, the effects of alcohol on CNS development will be discussed.

Critical Periods for Inducing Teratogenesis with Alcohol

Care should be exercised in extrapolating to humans the results of animal studies because of species differences in the metabolism of alcohol, the timing of development and the complexity of the central nervous system (CNS). However, animal research has demonstrated that the deleterious effects of *in utero* alcohol exposure exist on a continuum ranging from gross morphological defects to more subtle cognitive/behavioral dysfunctions at the less severe end (Streissguth et al. 1980). Some of the variables determining this wide spectrum of effects include dose and pattern of alcohol consumption, as well as duration and period of exposure.

Regarding critical periods when the fetus may be particularly sensitive to alcohol exposure, animal research has shown that the different profiles of alcohol-related birth defects are related to the amount and levels of alcohol and the specific time of exposure during pregnancy. For example, administration of one or two high doses of ethanol to mice during an early stage of the embryogenesis was sufficient to produce craniofacial malformations resembling those of children with FAS (microcephaly, narrow forehead, short and upturned nose, long upper lip with absent philtrum, etc) (Sulik et al. 1981) as well as other central nervous system (CNS) anomalies such as exencephaly, microcephaly, pituitary dysplasia and maxillary hypoplasia (Webster et al. 1980, Blakley and Scott 1984, Kotch and Sulik 1992). Furthermore, exposure to ethanol during gastrulation also results in the appearance of other malformations such as hydronephrosis and limb malformations in mice (Webster et al. 1980, Gage and Sulik 1991).

These results suggest that binge drinking as well as chronic alcohol abuse during the early stage of human embryogenesis (corresponding to the third week of human gestation) are potentially dangerous. In fact, human epidemiological studies have shown that the critical period for alcohol teratogenicity is early after conception, and that the anatomic abnormalities in the offspring were clearly defined among infants whose mothers drank more than the equivalent of three ounces (100 ml) of absolute alcohol per day (Ernhart et al. 1987). Graham et al. (1988) also reported that anomalies at birth were related to heavy drinking in the very early period of gestation rather than at midpregnancy.

Low birth weight and postnatal growth retardation are some of the characteristics of prenatal ethanol exposure (Streissguth et al. 1980, Coles 1992). Studies on mice and rats show that the later stages of gestation appear to be more sensitive for ethanol-induced growth reduction than early gestation (e.g. Middaugh and Boggan 1991). Indeed, in humans it has been shown that when the mother stops drinking by the second trimester, the children tend not to differ in growth from children of nondrinkers (Coles 1992).

The third trimester of gestation in humans is also a critical period for the developing nervous system (early postnatal period in the rat), since the brain undergoes rapid growth and development during this time ("brain growth spurt"). Animal studies have shown that exposure to alcohol during this period induces microencephaly (Pierce and West 1986), the degree of which depends on the peak blood alcohol levels (PBAL) rather than on the amount of alcohol consumed per se (Pierce and West 1986). Thus, episodes of binge drinking producing high PBAL during the third trimester in humans may be particularly harmful to the developing brain.

Regarding the dose required to harm the fetus, no definitive answer is available as to the lowest amount of alcohol capable of causing damage, although it is clear that the effects are dose-related (Webster et al. 1980). Genetic or interindividual factors may increase the susceptibility to alcohol-induced damage, mediated either through maternal factors (e.g. alcohol metabolism) or through the fetal genotype. In fact, using different strains of mice, Chernoff (1980) demonstrated an inverse relationship between high alcohol dehydrogenase levels (yielding low PBAL levels) and low incidence of teratogenesis. Therefore, since the genetic variability of human alcohol metabolizing enzymes is greater than in animals (Bosron et al. 1993), greater variability in alcohol-inducing effects should be expected in humans than in animals. Thus, a clear threshold cannot be defined at present.

Adverse Effects of Prenatal Exposure to Alcohol

For the last 20 years investigators have attempted to develop animal models of FAS to study the mechanisms of alcohol damage *in utero*. Numerous reviews on animal studies of FAS have been published (e.g. Randall 1987,Blakley 1988). The different procedures of ethanol administration employed (intubation, inclusion of ethanol in the drinking water, and liquid diets containing ethanol) have advantages and disadvantages that are discussed elsewhere (Blakley 1988). Generally, the peak blood alcohol levels (PBAL) achieved by all these procedures (-70-200 mg/dl) are similar to those observed in pregnant human drinkers (e.g. Halmesmäki 1988).

Intrauterine growth retardation and high incidence of resorption, stillbirth, and a prenatal and perinatal mortality have been associated with maternal chronic alcohol intake (Streissguth et al. 1980), and have also been observed in a wide variety of laboratory animals including mice, rats, swine, sheep, ferrets and monkeys (see review, Blakley 1988). Intra and interspecies differences in susceptibility to *in utero* alcohol toxicity have been found. For example, rats are less susceptible than mice to the teratogenic effects of alcohol and the effects of alcohol on mice differ according to strain. A recent study with different strains of mice showed that the maternal genotype plays an important role on the fetal weigth deficits and malformations (Gilliam and Irtenkauf 1990).

Low birth weights and postnatal retardation are well documented characteristics of ethanol-exposed children (Streissguth et al. 1980, Coles 1992). Similar reductions in the body and brain weights of rats prenatally exposed to ethanol under carefully controlled nutritional conditions suggest that ethanol is the likely causative agent (e.g. Guerri 1987). Indeed, in some cases, the weight reduction persisted into adulthood (Guerri 1987, Middaugh and Boggan 1991).

Among the various proposed mechanisms of alcohol-induced brain and body growth deficits are the reductions of one or more endocrine trophic factors (see review, Weinberg 1993) and depression in the rate of protein synthesis and RNA content (Rawat 1980, Renau-Piqueras et al. 1989a). In hepatocytes, alcohol also induces damage in the Golgi apparatus and alteration in the process of glycosylation and/or transport of secretory proteins (Renau-Piqueras et al. 1989a). Thus, an impairment of the protein synthesis machinery might result in growth restriction.

Central nervous system dysfunctions

One of the most severe consequences of maternal alcohol consumption is the damage to the developing central nervous system (CNS), which can occur in the absence of the gross morphological defects associated with FAS. Thus, mental retardation, motor abnormalities, perceptual and attention deficits, and certain neuropsychological and behavioral abnormalities are some of the problems commonly found in the children of women who were heavy or moderate drinkers during pregnancy (Streissguth et al. 1980, 1990, Coles 1992). Animal studies have also shown numerous behavioral abnormalities following gestational alcohol exposure, and there appears to be good correspondence between the human and animal data (Driscoll et al. 1990). For example, rats prenatally exposed to alcohol (PBAL -30-180 mg/dl) exhibited numerous behavioral alterations such as hyperactivity, impaired feeding, learning disabilities, and deficits in spatial memory processes (e.g. Riley 1990). Some of these behavioral dysfunctions appear to diminish with age, but others such as hyperactivity and impaired learning are permanent, persisting into adulthood (e.g. Abel and Dintcheff 1986). The long-term behavioral effects are indicative of the long-lasting detrimental consequences of prenatal exposure to alcohol.

These behavioral dysfunctions may be the consequence of alcohol-induced alterations in the normal developmental patterns, including changes in the ontogeny of neurotransmitters (e.g. Vallés et al. 1995, Druse 1992), altered neuroendocrine function, inhibition of neurogenesis, cell death and abnormal neuronal migration etc. (see review, Druse 1992). Indeed, the presence of clusters or sheets of ectopic neurons in the cerebellar cortex and neocortex as well as alterations in the glial cytoarchitecture have been observed in children with FAS (Clarren et al. 1978). Animal studies have also shown that ethanol induces: depression of neurogenesis, cell death, delayed and aberrant neuronal migration, altered formation of axonal and dendritic projections, etc. (see review, Miller 1992). In addition, experimental evidence indicates that the stage of brain development and the amounts and levels of alcohol exposure are important factors influencing the type and severity of the damage. The vulnerability to alcohol-induced neuronal damage varied among regions and cell populations and depends on the PBAL (Bolthius and West 1991). Those regions whose development appears to be particularly sensitive to ethanol include the neocortex, hippocampus and cerebellum (see review, Druse 1992).

Data from our laboratory also suggest that ethanol-induced alterations in astrogliogenesis could be an important mechanism underlying the CNS dysfunctions observed after prenatal exposure to alcohol (Guerri et al. 1993). Using astrocytes in primary culture, we have demonstrated that ethanol reduces DNA and RNA synthesis (Guerri et al. 1990), alters the content and distribution of several cytoskeletal proteins such as glial fibrillary acidic protein (Renau-Piqueras et al. 1989b, Sáez et al. 1991) and decreases the levels of membrane-bound enzymes and plasma membrane glycoproteins (Renau-Piqueras et al. 1992). A particularly interesting finding was that astrocytes isolated from fetuses from alcohol-fed mothers, but cultured in absence of this toxin, showed an impairment of their proliferation and differentiation (Renau-Piqueras et al. 1989b,

Guerri et al. 1990, Sáez et al. 1991). This leads to a decrease in the production of molecules involved in neuronal survival such as nerve growth factor (Vallés et al. 1994). The advantage of this approach is that it permits assessment of the effect of alcohol on astrocyte precursor cells which appears at the beginning of fetal development.

Ethanol-induced alterations in astroglia development observed *in vitro* were also reflected *in vivo*. A significant decrease in the levels of two astrocyte markers, GFAP and glutamine synthetase, was observed in the brain of pups exposed to alcohol (Sáez et al. 1991; Guerri et al. 1993). Recently we have found that ethanol also depresses the GFAP-mRNA expression in brains from fetuses and pups exposed prenatally to alcohol as well as in astrocytes in primary culture and in radial glia (unpublished results). In addition, abnormal radial glia fibers and also fewer glia within the primary somatosensory cortex have been observed in brains from alcohol-exposed rats (Miller 1992). Regarding the mechanism of these alterations, we have recently proposed that an increase in free radical generation may be involved in the pathogenesis of brain and astrocyte alteration observed after prenatal ethanol exposure (Guerri et al. 1994, Montoliu et al. 1995).

These findings have potentially important implications, because astroglial cells are responsible for guidance of migrating neuronal bodies, promoting neurite outgrowth, and shaping of neuronal assemblies during development. In addition, these cells synthesize and secrete several neurotrophic factors which support the survival and growth of young neurons. Therefore, an alteration in astroglia could directly or indirectly explain some of the CNS abnormalities observed after *in utero* exposure to alcohol.

Pathogenesis of Alcohol-Related Developmental Abnormalities

Research on the pathogenesis of ethanol -induced its teratogenic effects has tradicionally focused on acetaldehyde toxicity, placental dysfunction and impaired fetal nutrition, fetal hypoxia, and abnormal prostaglandin metabolism (Schenker et al. 1990). More recently, it has also been proposed that retinoids play a role in the teratogenic effects of alcohol (Duester 1991).

Acetaldehyde toxicity

On the basis that acetaldehyde (the first product of alcohol metabolism) reaches the fetus (Guerri and Sanchis 1985, Karl et al. 1988) and is cytotoxic (Dreosti et al. 1981), mutagenic (Obe 1981) and induces embryolethality (growth retardation and malformations) (e.g. Dreosti et al 1981), some authors have postulated its possible role in the teratogenicity of ethanol. However, these results have been not confirmed by other groups (e.g. Blakley and Scott 1984). Discrepant results have also been reported in studies in which acetaldehyde was applied to the medium of rat embryos grown in culture (Campbell and Fantel 1983, Priscott 1985). Therefore, from the available literature, it appears that acetaldehyde may contribute to the teratogenicity of ethanol, but is not the main causative agent.

Malnutrition and placental dysfunctions

Nutritional and mineral deficiencies are known to accompany chronic alcohol exposure and many attempts have been made experimentally to separate the effects of ethanol from those of primary malnutrition. The role of malnutrition in FAS remains unsettled. In some studies protein supplementation has been found to ameliorate the adverse effects of alcohol, but in others it failed to improve the outcome of pregnancy (see review, Schenker et al. 1990).

Another potential factor in the pathogenesis of FAS is the development of placental dysfunctions. Acute and especially chronic alcohol intake decreases transplacental transport of several amino acids in rodents. Similar results were obtained with placental slices obtained from chronic alcoholic monkeys (e.g. Fisher et al. 1983). In contrast, acute exposure to ethanol, at concentrations found in humans, has no significant depressive effects on amino acid and glucose uptake by primate and human placentas (Schenker et al. 1989).

In conclusion, it appears that fetal malnutrition may potentiate the effects of ethanol, but it seems unlikely that it is primarily responsible for alcohol-induced developmental abnormalities. Indeed, alcohol has been shown to have a detrimental effect on growth and development in a chick embryo model (Pennington et al. 1983) and in rat embryos incubated in culture (e.g. Priscott 1982), two systems in which placental and nutritional factors play no role in the development of anomalies.

Ethanol-induced fetal hypoxia

Fetal hypoxia has been implicated in the pathophysiology of FAS. Most of the evidence for this hypothesis derives from experimental studies demonstrating that alcohol restricts the fetoplacental blood flow, thus reducing the supply of oxygen and nutrients to the fetus. Studies in monkeys demonstrated that intravenous administration of ethanol to pregnant animals (PBAL 250mg/100 ml) produced a transient collapse of the umbilical vessels. Blood gases in the mother remained normal, but the fetuses exhibited hypoxia and acidosis within a few minutes after ethanol administration, coinciding with changes in uterine vessels. Studies with human umbilical cords have demonstrated similar effects following *in vitro* alcohol administration (e.g. Altura et al. 1983). However, no effect on uterine blood flow was observed in a small samples of pregnant women given a low dose of ethanol (see review, Schenker et al. 1990). Clearly, more studies are needed to confirm this interesting hypothesis.

Role of the Prostaglandins

An ethanol-induced increase in prostaglandins has been suggested as a possible mechanism of FAS (Randall et al. 1987). This hypothesis is supported by the observations that pharmacologic agents such as aspirin and indomethacin, which inhibit the synthesis of prostaglandins, reduce alcohol-induced birth defects and attenuate alcohol-induced growth retardation in animal models (see review, Schenker et al. 1990). Since prostaglandins are critical for normal fetal growth, and some prostanoids (prostacyclin and thromboxane) are involved in placenta blood flow, it has been postulated that alcohol-prostaglandin interactions may be involved in the etiology of some of the deleterious effects of alcohol on fetal growth and development. Data from human pregnancies suggest that maternal alcohol exposure increases fetal prostanoid release (Ylikorkala et al. 1988). However, this field requires further study, especially to measure the prostaglandins levels in pregnant human alcoholics and evaluate the possible benefits of prostaglandin antagonists for the prevention of alcohol-induced birth defects and growth anomalies.

Possible role of retinoids

Ethanol is a competitive inhibitor of retinol oxidation catalyzed by human alcohol dehydrogenase (class I). Retinoic acid plays an important role in the growth and development of the embryonic nervous system and is a regulator of gene expression. In view of the critical developmental role of retinoic acid, Duester (1991) postulated that a reduction in fetal retinoic acid synthesis, due to the ethanol inhibition of retinol oxidation, may lead to some of the malformations in the developing central nervous system and limb defects observed in FAS (Duester 1991). Interestingly, this group also demonstrate that ADH3 transgene can be expressed in embryonic tissues sensitive to ethanol teratogenesis (e.g. neural tube, brain, heart) (Zgombic-Knight et al. 1994). Therefore, although this hypothesis may not explain all the defects documented in cases of FAS, further research in this area is warranted, as it may lead to our understanding of normal and abnormal embryonic development at the molecular level.

In summary, although many mechanisms have been proposed to explain the teratogenic effects of alcohol, it is quite possible that more than one mechanism is involved in FAS, and that different features on FAS may be the manifestation of different pathophysiologies. The use of animal research and *in vitro* models will help to obtain deeper understanding of the basic mechanism(s) , which will be important for the development of pharmacological interventions, and for the identification of individual risk factors that make the fetus susceptible to the deleterious effects of *in utero* alcohol exposure.

Acknowledgements

Work from this laboratory has been supported by CICYT, Grant SAF/94-0065 and by Institució Valenciana d'Estudis i Investigació, Valencia (Spain)

References

Abel EL, Dintcheff BA (1986) Effects of prenatal alcohol exposure on behavior of aged rats. Drug Alcohol Depend 16:321-330

Abel EL, Sokol RJ (1987) Incidence of fetal alcohol syndrome and economic impact of FAS-related anomalies. Drug Alcohol Depen 19:51-56

Altura BM, Altura BT, Carella A, Chatterjee M, Halevy S, Tejani N (1983) Alcohol produces spasms of human umbilical blood vessels: Relationship to fetal alcohol syndrome (FAS). Eur J Pharmacol 86:311-312

Blakley PM (1988) Experimental teratology of ethanol. In: Issues and Reviews in Teratology. Kalter H (ed) Plenum Publ. Corp, pp 237-282

Blakley PM, Scott WJ, Jr (1984) Determination of the proximate teratogen of the mouse fetal alcohol syndrome.1. Teratogenicity of ethanol and acetaldehyde. Toxicol Appl Pharmacol 72:355-363

Bonthius DJ, West JR (1991) Permanent neuronal deficits in rats exposed to alcohol during the brain growth spurt. Teratology 44:147-163

Bosron WI, Ehrig T, LI TK (1993) Genetic factors in alcohol metabolism and alcoholism. Semin liver Dis 13:126-135

Campbell MA, Fantel AG (1983) Teratogenicity of acetaldehyde *in vitro*: Relevance to the fetal alcohol syndrome. Life Sci 32:2641-2647

Clarren SK, Alvord EC, Sumi SM, Streissguth AD, Smith DW (1978) Brain malformations related to prenatal exposure to ethanol. J Pediatr 92:64-70

Coles, CD (1992) Prenatal alcohol exposure and human development. In: Development of the Central Nervous System: Effects of Alcohol and Opiates. Miller M (ed) Wiley Liss, NY, pp 9-36

Chernoff GF (1980) The fetal alcohol syndrome in mice: Maternal variables. Teratology 22:71-75

Dreosti IE, Ballard J, Bellin GB, Record IR, Manuel SJ, Hetzel BS (1981) The effect of ethanol and acetaldehyde on DNA synthesis in growing cells and foetal development in the Rat. Alcohol: Clin Exp Res 5:357-362

Driscoll CD, Streissguth AP, Riley EP (1990) Prenatal alcohol exposure: Comparability of effects in humans and animal models.Neurotoxicol Teratol. 12:231-238

Druse MI (1992)Effect of Maternal Alcohol Consumption on the Developing Nervous System. In: Alcohol and Neurobiology: Brain Development and Hormone Regulation. Watson RR (ed),CRC Press Boca Raton, pp 1-29

Duester G (1991) A hypothetical mechanism for fetal alcohol syndrome involving ethanol inhibition of retinoic acid synthesis at the alcohol dehydrogenase step. Alcoholism:Clin Exp Res 15:568-572

Ernhart CB, Sokol RJ, Martier S, Moron P, Nadler D, Ager JW, Wolf A (1987) Alcohol teratogenicity in the human: A detailed assessment of specificity, critical period, and threshold. Am J Obstet Gynecol 156:33-39

Fisher SE, Atkinson M, Jacobson M, et al. (1983) Selective fetal malnutrition: The effect of *in vivo* ethanol exposure upon *in vitro* placental uptake of amino acids in the non-human primate. Pediatr Res 17:704-707

Gage JC, Sulik KK (1991) Pathogenesis of ethanol-induced hydronephrosis and hydroureter as demonstrated following *in vivo* exposure of mouse embryos. Teratology 44:299-312.

Gilliam DM, Irtenkauf KT (1990) Maternal genetic effects on ethanol teratogenesis and dominance of relative embryonic resistance to malformations. Alcoholism: Clin Exp Res 14:539-545

Graham JM, Hanson JW, Darby BL, Barr HM, Streissguth AP (1988) Independent dysmorphology evaluations at birth and 4 years /of age for children exposed to varying amounts of alcohol *in Utero*. Pediatrics 81:772-778

Guerri C (1987) Synaptic membrane alterations in rats exposed to Alcohol. Alcohol and Alcoholism, suppl. 1, pp. 467-472

Guerri C, Sáez R, Sancho-Tello M, Martin de Aguilera E, Renau-Piqueras J (1990) Ethanol alters astrocyte development: A study of critical periods using primary cultures. Neurochem. Res. 15:559-565

Guerri C, Sáez R, Portolés M, Renau-Piqueras J (1993) Derangement of astrogliogenesis as a possible mechanism involved in alcohol-induced alterations of central nervous system development. Alcohol and Alcoholism. Suppl.2, pp.203-208

Guerri C, Montoliu C, Renau-Piqueras J (1994) Involvement of free radical mechanism in the toxic effects of alcohol: Implications for fetal alcohol syndrome. In: Free Radicals in Diagnostic Medicine. Armstrong D (ed) Plenum Press, New York, 291- 305

Guerri C, Sanchis R (1985) Acetaldehyde and alcohol levels in pregnant rats and their fetuses. Alcohol 2:267-270

Halmesmäki E (1988) Alcohol counselling of 85 pregnant problem drinkers: Effect on drinking and fetal outcome. Bri J of Obstet and Gynecol 95:243-247

Jones KL, Smith DW (1973) Recognition of the fetal alcohol syndrome in early Infancy. Lancet 2:999-1001

Karl PI, Gordon MHJ, Lieber CS, Fisher SE (1988) Acetaldehyde production and transfer by the perfused human placental cotyledon. Science 242:273-275

Koth LM, Sulik KK. (1992) Experimental fetal alcohol syndrome: proposed pathogenetic basis for a variety of associated facial brain anomalies. Am J Med Genet 44:168- 176

Middaugh LD, Boggan WO (1991) Postnatal growth deficits in prenatal ethanol-exposed mice: Characteristics and critical periods. Alcoholism: Clin Exp Res 15:919-926

Miller MW (1992) The effects of prenatal exposure to ethanol on cell proliferation and neuronal migration. In: Development of the Central Nervous System: Effects of Alcohol and Opiates Miller M (ed) A.R. Liss, New York, pp 47-69

Montoliu C, Sancho-Tello M, Azorin I, Burgal M, Vallés S, Renau-Piqueras J, Guerri C (1995) Ethanol increases CYP2E1 and induces oxidative stress in astrocytes. J Neurochem (in press)

Obe G (1981) Acetaldehyde not ethanol is mutagenic. In: Progress in Environmental Mutagenesis and Carcinogenesis Kappas A (ed) Elsevier/North-Holland, Amsterdam, pp 19-23

Pennington SN, Boyd JW, Kalmus GW, Wilson RW (1983) The molecular mechanism of fetal alcohol syndrome (FAS): I. Ethanol-induced growth suppression. Neurobehav Toxicol and Teratol 5:259-262

Pierce DR and West JR (1986) Alcohol-induced microencephaly during the third trimester equivalent: Relationship to dose and blood alcohol concentrations. Alcohol 3:185-191

Priscott PK (1982) The effects of ethanol on rat embryos developing *in vitro*. Biochem Pharmacol 31:3641-3643

Priscott PK (1985) Effects of acetaldehyde and 2,3-butanediol on rat embryos developing *in vitro*. Biochem Pharmacol 34:529-532

Randall CL, Anton RF, Becker HC (1987a) Alcohol, pregnancy and prostaglandins. Alcoholism (NY) 11:32-36

Rawat AK (1980) Biochemical aspects of neuroteratogenic effects of ethanol. Neurobehav Toxicol 2:259-265

Renau-Piqueras J, Guerri C, Burgal M, De Paz P, Sáez R, Mayordomo F (1992) Prenatal exposure to ethanol alters plasma membrane glycoproteins of astrocytes during development in primary culture as revealed by concanavalin A binding and 5'-nucleotidase activity. Glia 5:65-74

Renau-Piqueras J, Sancho-Tello M, Báguena Cervellera R, Guerri C (1989a) Prenatal exposure to ethanol alters the synthesis and glycosilation of proteins in fetal hepatocytes. Alcoholism: Clin Exp Res 13:817-823

Renau-Piqueras J, Zaragoza R, DePaz P, Baguena R, Mejias L, Guerri C (1989b) Effects of prolonged ethanol exposure on the glial fibrillary acidic protein-containing intermediate filaments of astrocytes in prilary culture: A quantitative immunofluorescence and immunogold electron microscopic Study. J Histochem Cytochem 37:229-240

Riley EP (1990) The long-term behavioral effects of prenatal alcohol exposure in rats. Alcoholism:Clin Exp Res 14:770-673

Sáez R, Burgal M, Renau-Piqueras J, Marqués A, Guerri C (1991) Evolution of several cytoskeletal proteins of astrocytes in primary culture: effect of prenatal alcohol exposure. Neurochem Res 16:737-747

Schenker S, Dicke JM, Johnson RF, Hays SE, Henderson GI (1989) Effect of ethanol on human placental transport of model amino acids and Glucose. Alcoholism: Clin Exp Res 13: 112-119

Schenker S, Becker HC, Randall CL, Phillips DK, Baskin GS, Henderson GI (1990) Fetal Alcohol Syndrome: Current status and pathogenesis. Alcoholism: Clin Exp Res 14:635-647

Streissguth AP, Landesman-Dywers S, Martin JC, Smith DW (1980).Teratogenic effects of alcohol in human and laboratory animals. Science 209:353-361

Streissguth AP, Barr HM, Sampson PD (1990) Moderate prenatal alcohol exposure: Effects on child IQ and learning problems at age 7 1/2 Years. Alcoholism: Clin Exp Res 14:662-659

Sulik KK, Johnston MC, Webb MA (1981) Fetal alcohol syndrome: embryogenesis in a mouse Model. Science 214:936-938

Valles S, Lindo L, Montoliu C, Renau-Piqueras J, Guerri C (1994) Prenatal exposure to ethanol induces changes in the nerve growth factor ans its receptor in proliferating astrocytes in primary culture. Brain Res 656:281-286

Vallés S, Felipo V, Montoliu C, Guerri C (1995) Alcohol exposure during brain development reduce 3H-MK-801 binding and enhances metabotropic-glutamat receptor-stimulated phosphoinositide hydrolysis in rathippocampus. Life Sciences 56:1373-1383

Webster WS, Walsh DA, Lipson AH, McEwen SE (1980) Teratogenesis after acute alcohol exposure in inbred and outbred mice. Neurobehav Toxicol 2:227-234

Weinberg J (1993) Neuroendocrine effects of prenatal alcohol exposure. Ann N.Y. Acad Sci 697:86-96

Ylikorkala O, Halmesmäki E, Viinikka L (1988) Urinary prostacyclin and thromboxane metabolites in drinking pregnant women and their infants. Relation to fetal alcohol syndrome.Obstet Gynecol 71:61-66

Zgombic-Knight M, Satre MA, Duester G (1994) Differential activity of the promoter for the human alcohol dehydrogenase (retinol dehydrogenase) gene ADH3 in neural tube of transgenic mouse embryos. J Biol Chem 269:6790-6795

Developmental Neurotoxicology in the Neonate - Effects of Pesticides and Polychlorinated Organic Substances

P. Eriksson
Department of Environmental Toxicology, Uppsala University, Norbyvägen 18 A, S-752 36 Uppsala, Sweden

During the development of an organism, there are periods which can be critical for its normal maturation. Vulnerable periods during the ontogenesis of the central nervous system (CNS) can be roughly divided into two main courses of events in the development of the brain. The first includes its early development, a period during which the general shape of the adult brain is acquired and the spongioblasts and neuroblasts, precursors of glia cells and neurons, respectively, multiply. Interference by xenobiotics during this period can cause malformation of the brain.

The second period coincides with the rapid growth of the brain, the so-called "brain growth spurt" (Davison and Dobbing, 1968), which in many mammalian species occurs during early postnatal life, a period when animals acquire many new motor and sensory faculties (Bolles and Woods, 1964) and increase the spontaneous motor behaviour (Campbell et al., 1969). During this period, the brain undergoes several fundamental developmental phases, e.g. maturation of axonal and dendritic outgrowth, establishment of neural connections, synaptogenesis, multiplication of glia cells with accompanying myelinization, and cell, axonal and dendritic death (Kolb and Whishaw, 1989). The brain growth spurt is also associated with numerous biochemical changes that will transform the feto-neonatal brain into that of the mature adult. One of the major signal substances in the CNS is acetylcholine (ACh), which acts as the transmitter in the cholinergic pathways. In rodents, this transmitter system in the CNS undergoes rapid development during the first 3-4 weeks after birth (Coyle and Yamamura, 1976; Fiedler et al., 1987) when gradually increasing numbers of muscarinic and nicotinic receptors are found in the cerebral cortex and hippocampus (Falkeborn et al., 1983; Kuhar et al., 1980, Slotkin et al., 1987; Fiedler et al., 1987). The cholinergic transmitter system is involved in many behavioural phenomena (Karczmar, 1975) and correlates closely with cognitive functions (Drachman, 1977; Bartus et al., 1982).

Environmental pollutants comprise a broad spectrum of substances. Many of the chlorinated hydrocarbons used in various industrial processes and pesticides, such as PCB (polychlorinated hydrocarbons) and DDT, respectively, have become known as persistent contaminants of the environment. Due to their accumulation in the food chain and storage in living organisms, they had serious biological effects and consequently also caused ecological disturbances. Other possible environmental hazards may be the short-acting substances that, though do not become bioconcentrated in biological systems, may nevertheless induce permanent disorders in organisms. In mammals, the fetus can be indirectly exosed during the fetal period via the mother's intake of toxic agents. During the neonatal period the offspring may be affected by toxic agents by ingesting mother's milk, or be directly exposed to xenobiotics.

In different reports we have shown that low-dose exposure to both persistent and non-persistent environmental agents during the neonatal period can lead to disruption of adult brain function. Among the toxicants shown to induce such neurotoxic effects are DDT, pyrethroids, organophosphates, nicotine, paraquat, MPTP and polychlorinated biphenyls (PCBs).

Neurotoxic effects after neonatal exposure to pesticides

Among the insecticides, DDT is known as a persistent environmental contaminant, while the pyrethroids belong to the group of short acting substances. One of the basic mechanisms of DDT and pyrethroid action involves interference with the sodium channels of the nerve membrane, leading to prolonged depolarization and induction of repetitive activity (Vijverberg and van den Bercken, 1990; Woolley, 1982), which can increase the release of a transmitter. In our experiments with DDT and pyrethroids we have observed that a single low peroral dose of DDT (1.4 μmol (0.5 mg/kg b.wt.) and repeated doses of bioallethrin [type I pyrethroid, given as a single oral dose (0.7 mg/kg b.wt.) for 7 days in order to simulate the exposure time effective for DDT (0.5 mg)] when given to 10-day-old mice affects the cholinergic system in the neonatal brain (Eriksson and Nordberg, 1986; 1990). The substances also caused regional differences within the brain, as the cerebral cortex was affected, though not the hippocampus. At a neonatal age of 17 days the density of MAChR (assayed with [^{3}H]quinuclidinylbenzilate(QNB)) was increased in cerebral cortex. The subpopulations of MAChR (assayed in an antagonist [^{3}H]QNB/agonist carbachol competition binding assay) were also changed, as the proportion of low-affinity (LA) binding sites was significantly increased, while that of high-affinity (HA) sites was reduced. This neonatal exposure to DDT and bioallethrin was shown to lead to permanent changes in the cholinergic system in animals reaching an adult age of 4 months (Eriksson and Fredriksson, 1991; Eriksson et al., 1992). The MAChR were still affected in the same brain region as in the 17-day-old mouse, namely the cerebral cortex. In contrast to the increase in the neonatal mouse there was a significant decrease in the amount of MAChR in the adult, which was not accompanied by any significant change in the subpopulations of MAChR, as was the case in the neonatal mouse. Exposure to DDT and bioallethrin during neonatal life was further shown to lead to permanent functional disturbances in the adult animals (Eriksson and Fredriksson, 1991; Eriksson et al., 1992). Testing of spontaneous behaviour indicated disruption of a simple non-associative learning process, viz. habituation, in 4-month-old mice exposed to DDT and bioallethrin as neonates. Habituation is defined here as a decrease in locomotion, rearing and total activity variables in response to the waning novelty of the test chambers over the 60-min test period, divided into three 20-min spells. In the insecticide-treated animals there was an absence of or delayed habituation to a novel environment.

Nicotine, which can be found in certain pesticides, makes its impact on human health as a component in tobacco products. Nicotine is also known as one of the most commonly used dependence producing substances. Nicotine is an agonist for nicotinic receptors, but may also mediate release of neurotransmitters such as ACh (Beani et al., 1989). The effects of neonatal exposure to nicotine (66 μg(as nicotine base)/kg b.wt., on days 10-16) on spontaneous and nicotine-induced behaviour in 4-month-old mice and on the development of nicotinic receptors in brain indicated that nicotine can prevent the development of LA nicotinic binding sites (assayed with (-)-[N-methyl-^{3}H]nicotine) in the cerebral cortex and that this exposure induces a different behavioural response to nicotine in adult animals (Nordberg et al., 1991). Nicotine doses known to cause hyperactive behaviour caused hypoactive behaviour in mice treated neonatally with nicotine.

Neurotoxic effects after neonatal exposure to PCBs

Polychlorinated biphenyls (PCBs) constitute a large category of chlorinated hydrocarbons; theoretically there are 209 different configurations of the PCB molecule. PCBs have been used in a wide variety of commercial and industrial products, such as hydraulic and heat transfer fluids, dielectric fluids in capacitors and transformers, plasticizers, flame retardants, etc., and are known persistent contaminants of the environment. However, human epidemiological studies and experimental investigations in animals have shown that PCBs, though not intended to be neurotoxic compounds, as is the case with most insecticides, can nevertheless produce neurotoxic effects (Tilson and Harry, 1994).

Neonatal exposure to single ortho-substituted PCB congeners 2,4,4'-trichlorobiphenyl (IUPAC 28), 2,2',5,5'-tetrachlorobiphenyl (IUPAC 52), 2,3',4,4',5-pentachlorobiphenyl (IUPAC 118) and 2,3,3',4,4',5-hexachlorobiphenyl (IUPAC 156) (one single dose (0.7 μmol - 14 μmol/kg b. wt. to 10-day-old mice) indicated persistent neurotoxic effects in the adult animal following exposure to PCB 28 and 52 (Eriksson and Fredriksson, 1995). Both compounds induced permanent aberrations in spontaneous motor behaviour. Furthermore, neonatal exposure to PCB 52 also affected learning and memory functions in the adult animal. In the animals showing deficits in memory and learning function, the cholinergic nicotinic receptors in the cerebral cortex were affected.

A defined critical time in neonatal brain development

The induction of the behavioural and cholinergic disturbances in the adult animal appears to be limited to a short period of time during neonatal development. DDT induces the effects on MAChR and on spontaneous behaviour when administered on postnatal day 10, but not on postnatal day 3 or 19 (Eriksson et al., 1992) (Fig. 1) . The lack of effect on mice exposed on postnatal day 3 or 19 appears not to be due to differences in the concentration of DDT in brain, as DDT has been found to be evenly distributed throughout the neonatal brain and its concentration appears not to be higher in mice exposed on postnatal day 10 than on to day 3 or 19 (Eriksson and Darnerud, 1985). An experiment with different doses of bioallethrin (0.21, 0.42, and 0.70 mg/kg b.wt) showed that the changes in neonatal and adult MAChRs (Fig. 3) and altered spontaneous behaviour (Fig. 2) were dose-dependent (Ahlbom et al., 1994). In the 17-day-old mice, bioallethrin exposure produced a significant dose-dependent increase in [^{3}H]QNB binding and a significantly increased proportion of LA-binding sites, compared with controls. In the adult mice there was a significant dose-dependent decrease in specific [^{3}H]QNB binding. The spontaneous behaviour test revealed a significant dose-dependent increase in hyperactive, non-habituating behaviour in the adult animal (Fig. 2). A limited period for induction of altered response to nicotine at adult an age has also been observed. Adult mice, exposed on postnatal days 10-14, displayed altered behaviour when provoked with renewed exposure to nicotine, which was not seen when the neonatal exposure occurred on days 3-7 or 19-23.

Increased susceptibility in adults neonatally exposed to pesticides

A circumstance likely to occur in nature is the combination of early (neonatal) and late (adult) exposure to persistent and non-persistent agents.

In recent experiments we have noticed that neonatal exposure to a neurotoxic agent such as DDT can lead to an increased susceptibility in adults to an agent with similar neurotoxic action, such as bioallethrin. Among the effects observed are a change in the density of MAChR and additional behavioural disturbances and learning disabilities (Eriksson et al., 1993; Johansson et al., 1995). In these experiment it was also observed that the behavioural disturbances observed 24 h after adult exposure to bioallethrin (in animals given vehicle as neonates and bioallethrin as adults) did not persist, or almost ceased after 2 months.

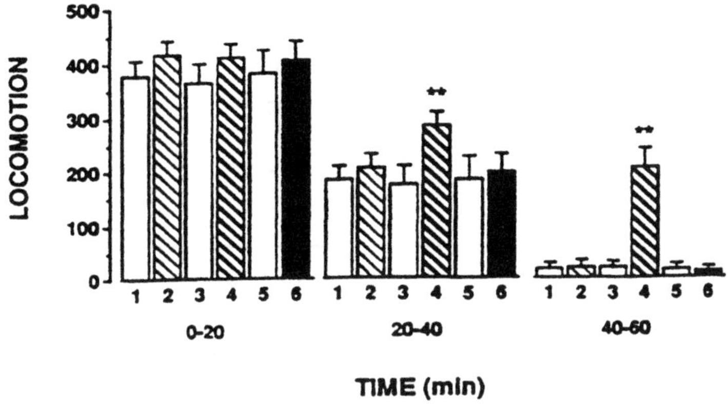

Fig. 1. Spontaneous behaviour (locomotion) in 4-month-old NMRI male mice exposed to a single oral dose of DDT (0.5 mg/kg b. wt.) or the vehicle at an age of either 3-, 10-, or 19-days. The different age-categories are indicated by: bars 1 and 2, 3-day-old; bars 3 and 4, 10-day-old; bars 5 and 6, 19-day-old, where plain bars denote controls, hatched bars denote DDT-treated mice (mean + SD). The statistical difference vs. control is indicated by asterisks, ** $p \geq 0.01$.
(Data taken from Eriksson et al., 1992)

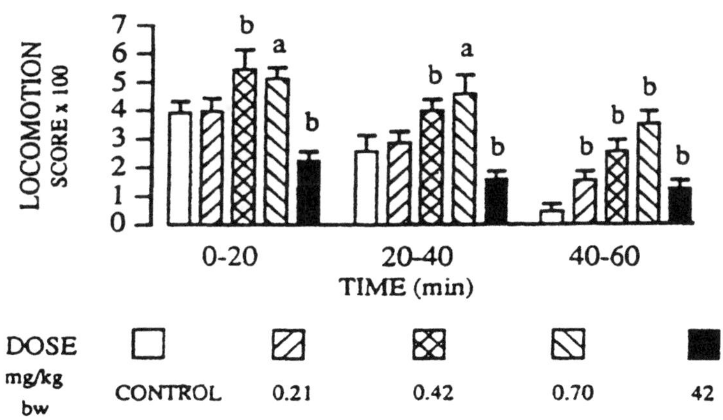

Fig. 2. Spontaneous behaviour (locomotion) in 4-month-old NMRI male mice exposed to repeated daily administration of one of the bioallethrin doses or the vehicle for 7 days, starting at the age of 10 days.
a) significant different vs. controls, $p \geq 0.01$.
b) significant different vs. closest lower dose, $p \geq 0.01$ and significantly different vs. controls, $p \geq 0.01$.
c) significant different vs. closest lower dose, $p \geq 0.05$ and significantly different vs. controls, $p \geq 0.01$.
(Data taken from Ahlbom et al., 1994)

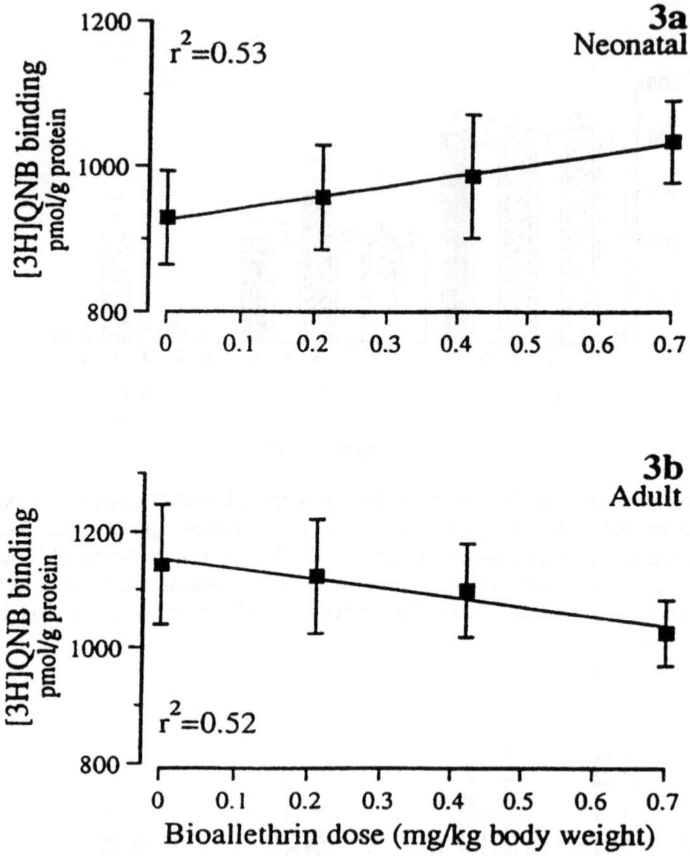

Fig. 3. Dose-response effect on muscarinic receptor density in the cerebral cortex of 17-day-old (3a) and 4-month-old (3b) NMRI male mice exposed to repeated daily administration of one of the bioallethrin doses or the vehicle for 7 days, starting at the age of 10 days. The slopes differ significantly from zero for both 17-day-olds (3a, p=0.0025) and 4-month-olds (3b, p=0.0324).
(Data taken from Ahlbom et al., 1994)

Concluding remarks

Considered together, these results indicate that low-dose exposure to environmental agents during the rapid development of the neonatal brain 'brain growth spurt' can lead to irreversible changes in adult brain function. Our studies have also indicated that there is a critical period in neonatal development in the mouse when the brain is vulnerable to the influence of pesticides at doses that apparently have no permanent effects when administered to the adult animal. The increased susceptibility to toxic agents at adult age in animals exposed during neonatal life indicates that neonatal exposure to toxic agents can potentiate and/or modify the reaction to adult exposure to xenobiotics.

References

Ahlbom, J, Fredriksson, A and Eriksson, P (1994) Neonatal exposure to a type-I pyrethroid (bioallethrin) induces dose-response changes in brain muscarinic receptors and behaviour in neonatal and adult mice. Brain Res. 645, 318-324

Bartus, RT, Dean III, RL, Beer, B and Lippa, AS (1982) The cholinergic hypothesis of geriatric memory dysfunction. Science 217, 408-417

Beani, L, Bianchi, C, Ferraro, L, Nilsson, L, Nordberg, A, Romanelli, L, Spalluto, P, Sundwall, A and Tanganelli, S (1989) Effect of nicotine on the release of acetylcholine and amino acids in the brain. In: Nordberg, A, Fuxe, K, Holmstedt, B and Sundwall, A (eds) Nicotinic receptors in the CNS. Their role in synaptic transmission. Elsevier, Prog. Brain Res. 79, 149-155

Bolles, RG and Woods, PJ (1964) The ontogeny of behaviour in the albino rat. Anim. Behav. 12, 427-441

Campbell, BA, Lytle, LD and Fibiger, HC (1969) Ontogeny of adrenergic arousal and cholinergic inhibitory mechanisms in the rat. Science 166, 635-637

Coyle, JT and Yamamura, HI (1976) Neurochemical aspects of the ontogenesis of cholinergic neurons in the rat brain. Brain Res. 118, 429-440

Davison, AN and Dobbing, J (1968) Applied Neurochemistry, Blackwell, Oxford pp. 178-221, 253-316

Drachman, DA (1977) Cognitive function in man. Does the cholinergic system have a special role? Neurology 27, 783-790

Eriksson, P, Ahlbom, J and Fredriksson, A (1992) Exposure to DDT during a defined period in neonatal life induces permanent changes in brain muscarinic receptors and behaviour in adult mice. Brain Res. 582, 277-281

Eriksson, P and Darnerud, PO (1985) Distribution and retention of some chlorinated hydrocarbons and a phthalate in the mouse brain during the pre-weaning period. Toxicology 37, 189-203

Eriksson, P and Fredriksson, A (1991) Neurotoxic effects of two different pyrethroids, bioallethrin and deltamethrin, on immature and adult mice: Changes in behavioral and muscarinic receptor variables. Toxicol. Appl. Pharmacol. 108, 78-85.

Eriksson, P and Fredriksson, A (1995) Developmental neurotoxicity of four ortho-substituted polychlorinated biphenyls in the neonatal mouse. (submitted for publ.)

Eriksson, P, Johansson, U, Ahlbom, J and Fredriksson, A (1993) Neonatal exposure to DDT induces increased susceptibility to pyrethroid (bioallethrin) exposure at adult age. - Changes in cholinergic muscarinic receptor and behavioural variables. Toxicology 77, 21-30

Eriksson, P and Nordberg, A (1986) The effects of DDT, DDOH-palmitic acid, and a chlorinated paraffin on muscarinic receptors and the sodium-dependent choline uptake in the central nervous system of immature mice. Toxicol. Appl. Pharmacol. 85, 121-127

Eriksson, P and Nordberg, A (1990) Effects of two pyrethroids, bioallethrin and d eltamethrin, on subpopulations of muscarinic and nicotinic receptors in the neonatal mouse brain. Toxicol. Appl. Pharmacol. 102, 456-463

Falkeborn, Y, Larsson, C, Nordberg, A and Slanina, P (1983) A comparison of the regional ontogenesis of nicotine- and muscarine-like binding sites in the mouse brain. Int. J. Dev. Neurochem. 1, 187-190

Fiedler, EP, Marks, MJ and Collins, AC (1987) Postnatal development of cholinergic enzymes and receptors in mouse brain. J. Neurochem. 49, 983-990

Johansson, U, Fredriksson, A and Eriksson, P (1995) Bioallethrin causes permanent changes in behavioural and cholinergic muscarinic receptor variables in adult mice exposed neonatally to DDT. Eur. J. Pharmacol. 293, 159-166

Karczmar, AG, (1975) Cholinergic influences on behaviour. In: Waser, PG (ed) Cholinergic mechanisms. Raven Press, New York, 501-529

Kolb, B and Whishaw, IQ (1989) Plasticity in the neocortex: mechanisms underlying recovery from early brain damage. Prog. Neurobiol. 32, 235-276

Kuhar, MJ, Birdsall, NJM, Burgen, ASV and Hulme, EC (1980) Ontogeny of muscarinic receptors in rat brain, Brain Res. 184, 375-383

Nordberg, A, Zang, X, Fredriksson, A and Eriksson, P (1991) Neonatal nicotine exposure induces permanent changes in brain nicotinic receptors and behaviour in adult mice. Dev. Brain Res. 63, 201-207

Slotkin, TA, Orband-Miller, L and Queen, KL (1987) Development of [3H]nicotine binding sites in brain, J. Pharmacol. Exp. Ther. 233, 361.

Tilson, HA and Harry, GJ (1994) Developmental neurotoxicology of polychlorinated biphenyls and related compounds. In: Isaacson, RL and Jensen, KF (eds), The vulnerable brain and environmental risks., Plenum Press, New York, vol. 3, 267-279.

Vijverberg, HPM and van den Bercken, J (1990) Neurotoxicological effects and the mode of action of pyrethroid insecticides. Crit. Rev. Toxicol. 21, 105-126

Woolley, DE (1982) Neurotoxicity of DDT and possible mechanism of action. In: Prasad, KN and Vernadakis, A (eds) Mechanism of Actions of Neurotoxic substances., Raven Press, New York, 95-141.

Drug Treatment in the Perinatal Period and the Risk of Functional Teratogenicity

O. Benešová
Psychiatric Center Prague, 181 03 Prague 8, Czech Republic

The recent success of perinatal medicine in protecting high-risk pregnacies and in saving the lives of very immature and injured fetuses/newborns is connected with a substantial increase in the use of drugs in obstetrics and neonatology. However, minimal attention has been paid so far to the potential teratogenic risk of these drugs for the developing fetal brain and immune system. After all, why care, when all drug information leaflets, and even textbooks of pharmacology, claim that the drug teratogenicity risk is restricted to the first trimester of gravidity. Is it really so?

The teratogenic effect of a drug - or of a chemical compound in general - depends not only on the quality and quantity of the noxious agent, but primarily on the developmental stage of the organism at the time of the interference (Jelínek and Rychter 1980). In the embryonic and early fetal period, in which organogenesis takes place and organs are differentiated, drug interference results in various gross or minor structural deformities, e.g. phocomely, known from the history of thalidomide, cleft palate, heart defects etc., which were studied by the *classical structural teratology*.

The perinatal period is characterized by intensive histogenesis and cytodifferentiation of already shaped organs, especially of the brain and the immune system which have a similar receptor-transmitter equipment. Even fine deviations induced by drugs in the programme of these developmental processes initiate disorders in neural networks, cytoarchitecture and receptor-transmitter communication systems. These alterations on the cellular/subcellular level are not evident at birth, but form the basis for various *functional defects* of the brain, and the immune and endocrine systems *(functional teratology)*. Disturbances of brain function appear gradually during further maturation in school age, adolescence or even in adulthood as various neuro-psycho-behavioural deviations such as hyperkinetic syndrome, dyslexia, dyspraxia, mental retardation, social maladaptation or psychic liability which may represent the predisposition to psychoses. The functional defects concern also the endocrine and immune systems which have mutual relations with hypothalamic regulatory centers.

It is thus evident that the risk of teratogenicity is not limited to the first three months of pregnancy. This still prevalent opinion of the medical public arises from the traditional concept of teratogenic defects as structural malformations. On the contrary, the risk of drug teratogenic sequelae exists during the entire gestation and lasts even throughout the postnatal development, only the manifestation of the teratogenic outcome is changing.

In contrast to the traditionally emphasized "caveat" for the first trimester of the pregnancy, there exists experimental evidence that some drugs may have a quite opposite trend of teratogenic hazard in the course of gestation (Jelínek et al. 1983).

The sensitivity of the organism to the teratogenic effect of *drugs with general cytotoxic effects* is highest during the first stages of development, in the period of rapid growth and primary differentiation with an extremely high rate of cell proliferation. On the contrary, the teratogenic potential of *drugs with receptor mediated effects* (psychotropic drugs, hormones) may be negligible on early stages of gestation, when relevant receptors, necessary for their action, are not yet present. However, it increases very quickly with advancing development along with the formation of relevant receptors. This refers especially to the perinatal period in which the majority of receptor systems are developing and formed. Consequently, drugs with receptor mediated effects may have a much higher teratogenic impact (i.e. functional teratogenic risk) during late pregnancy and even in the neonatal life.

Considering the social relevance of various congenital defects, functional deviations, especially when concerning brain and mental efficiency, may represent sometimes a heavier burden for the individuum, the family and the whole society than most congenital organ malformations which can be surgically repaired as e.g. cleft palate and heart defects.

The Recent System of Testing Drug Teratogenicity

Taking in account all mentioned aspects of drug teratogenicity, an urgent demand arises for reliable informations on the functional teratogenic potential of all perinatally administered drugs. And I ask: does the current system of official teratological tests provide such informations?

The validity of animal experiments for human therapeutic use depends on the adequacy and consistency of the animal model with the clinical condition. To date, the common approach to models in teratology is based on the assumption that drug administration to pregnant woman is simulated by drug administration to pregnant female animals. This approach, however, disregards the species specific differences in ontogenic development during the prenatal and postnatal period.

Figure 1 illustrates the differences in the timing of birth in relation to the vulnerable period of brain growth acceleration in man, rat and rabbit. In man, this phase is a real perinatal event: It begins in the third trimester of gravidity, culminates just before term and continues during the first two years of life. In the rat and rabbit, whose young are born very immature, the relevant ontogenic period of brain development is shifted to the first postnatal decade. This postponement of ontogenic development in the rat in comparison with man was described also on the level of various brain receptors by several authors (for review see Benešová et al. 1984).

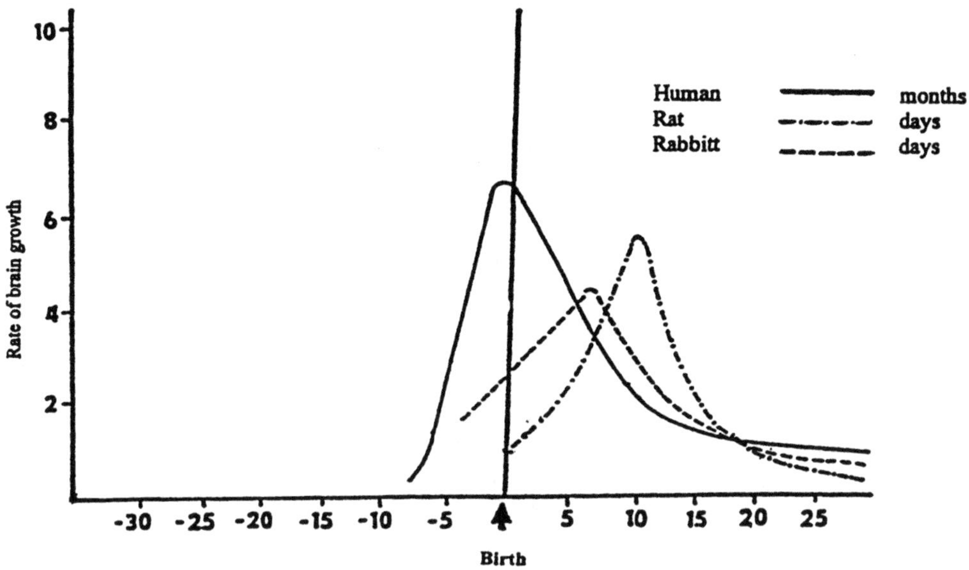

Fig. 1 Brain growth acceleration during ontogenesis in human, rat and rabbit (according to Dobbing1970)

Hence it follows, that teratological studies - as they are commonly performed by using drug administration in female rats or rabbits during pregnancy, i.e. prenatally - simulate the administration of the drug during the embryonal and early fetal ontogenic period in man. Consequently, they cannot provide any information concerning the effect of the drug given on a higher ontogenic stage, namely in the phase of human perinatal development which includes the highly vulnerable processes of the brain growth spurt and receptor-transmitter system formation. As it is evident, such information would be possible in model experiments with the rat or rabbit only by using postnatal drug administration.

A New Aproach to Testing Functional Teratogenicity of Drugs

It has to be confessed that the recent system of mandatory teratological tests does not ensure the safety of drugs prescribed during the perinatal period for saving risk pregnacies and immature neonates. It is evident that a new approach to functional teratogenicity testing needs to be adopted. In the rat - which is used in most teratogenic experiments - the human perinatal drug treatment should be simulated by a relevantly timed postnatal administration of the drug. This approach has also a technical advantage: the drug is injected directly into the target organism, eliminating all interfering variables of maternal body and placenta.

In our laboratory, the following experimental design is used. The tested drug is administered to rats in the first postnatal week as single or repeated subcutaneous injections. One half of the pups in a litter is injected with the drug, the other half with saline (or solvent) and serves as control. The animals are followed-up during development, sexual maturation, adulthood and senescence till the age of 2 years, using tests of behaviour, reproductive ability, immunoreactivity and brain biochemical analysis (Table 1).

It is very necessary to follow-up the delayed effects of perinatal drug treatment over the whole life of the experimental animal, since the manifestation of a functional teratogenic defect may vary during the lifespan. Our recent findings indicated that some functional deviations are not evident in young adult rats because they may be masked by compensatory mechanisms that are most effective during early adulthood, in the period of peak efficiency of the organism. The defect becomes apparent when the homeostatic balance breaks down along with the onset of aging processes. Such a critical period in the rat occurs usually at the age of 6 to 12 months when various functional deficits, both behavioural and neurobiochemical, were detectable. This phenomenon could be interpreted also as premature onset of brain degeneration or as accelerated aging (Benešová et al. 1992). Along with progressing senescence, however, the manifestation of functional defects may gradually decline due to parallel physiological degeneration of functions in age-matched control animals. It is thus very important to evaluate functional teratogenic potential of drugs in longterm animal studies covering the mentioned critical periods of life.

The Functional Teratogenic Risk of Three Perinatally Applied Drugs

Using the described comprehensive experimental procedure, the risk of functional teratogenicity was investigated in three drugs with receptor mediated effects which are used for the treatment of various risk situations in the perinatal period (Table 2). **Dexamethasone,** a synthetic glucocorticoid, is applied for inducing surfactant production in the immature fetal lung and is thus used in prevention or treatment of the neonatal respiratory distress syndrome. **Fenoterol**, a beta-adrenomimetic agent, is used as a tocolytic drug for the disruption of preterm uterine contractions. **Diazepam**, a benzodiazepine anxiolytic, is administered to pregnant women for its tranquilizing and myorelaxing activity in the situation of imminent premature parturition or in immature neonates when intubated. For every drug, the range of clinical therapeutic doses is indicated, as well as the doses used in experimental studies. The comparison of the therapeutic dosage and the effective experimental dose, inducing functional teratogenic sequelae (printed in bold letters), may serve as a quantitative indicator of the drug's teratogenic risk.

From this point of view, dexamethazone reveals the highest risk of functional teratogenicity: single administration of a dose 5 times lower than the clinical dosage induced serious functional teratogenic defects in behaviour, motor

Table 1. Experimental protocol for testing functional teratogenicity

Animals:
Rats, males and females, strain Wistar, breed Velaz.

Drug administration:
Tested drugs were administered subcutaneously in single or repeated injections in rat pups in the first postnatal week. One half of every 10-membered litter was applied by the drug, the other half by saline or solvent (control group).

Follow-up during development:
In the course of first two months of life, following developmental landmarks were registered: pinna and eye opening, incisor eruption, motor skill, sexual maturation in females (vagina opening) and males (preputium loosening).

Follow-up during adulthood and senescence:
Following tests were performed at the age 6, 12, 18 and 24 months.

A. Behavioural tests:

1) Exploration and defecation rate in the "open field"
Box 65x45cm,x 35cm; four 5-min-sessions with 24 h intervals.
The frequency and duration of 16 behavioural elements were registered.

2) Discrimination of familiar and new objects
Three 4-min-sessions with two objects A, fourth session with objects A and B.
Fine elements of approach and exploratory behaviour directed to the presented objects were checked.

3) Social recognition test ("social memory")
Repeated exposition of a rat pup in two 5-min-sessions with an interval of 30 min.
Elements of social interest (sniffing, touching, following) directed to the pup were checked.
All behavioural tests were recorded on videotape. The frequency and duration of various behavioural elements were registered using PC program "ACTIVITY" and evaluated by statistical programs BMDP.

4) Motor performance on rotating cylinder
Diameter 17 cm, rotation speed 270 cm/min; two 2-min-sessions with 24 h interval.
The ability of running or balancing on the cylinder is evaluated.

B. Reproductive function tests:

Females: estrous cycles, conception, natality, maternal behaviour.
Males: testis & epididymis weights, spermiogram, serum testosterone, ability of fertilize a female.

C. Immune function tests:

Cell mediated immune response:
Delayed type hypersensitivity to bovine serum albumin
Proliferative response of spleen lymphocytes to mitogens
Humoral immune response:
Serum titres of antibodies against sheep red blood cells

D. Brain biochemical analysis:

Carried out at the end of every series of experiments in brains dissected in 5 parts: cortex, hippocampus, hypothalamus, striatum, cerebellum.

1) Markers for central monoaminergic transmission:
Concentration of noradrenaline, dopamine, serotonine and their metabolites.

2) Markers for cholinergic neuronal activity:
High affinity choline uptake (HACU).

3) Markers of free radical induced damage:
Lipid peroxidation; Protein solubility

4) Markers of cell number and size:
Concentration of DNA, RNA and proteins

Table 2. Risk of functional teratogenicity in three perinatally applied drugs

Drug	Clinical therapeutic dosage	Experimental dosage in neonatal rat	Functional teratogenic effects
DEXAMETHAZONE	0.6 - 1 mg/kg	0.1 - **0.2** - 0.5 -1 mg/kg s.c. postnatal day 7	*Behavioural deviations:* **Hyperactivity with stereotypy, decreased adaptability, high emotional reactivity which induced lower scores in avoidance learning, deficit of motor coordination** *Defects of endocrine regulation:* **Deficits of male and female reproductive ability** *Neurobiological deviations:* **Permanent deficit of cerebellum weight** **Reduced size of cells in cerebellum and hippocampus** **Decreased concentration of noradrenaline in hypothalamus** *Deficits of immunocompetence:* **Depression of humoral and cell-mediated immune response** *REF: Benešová O.,Pavlík A.; Neuropharmacology 28: 89-97, 1989.*
PHENOTEROL	0.5 - 1 mg/kg	1 -**10** mg/kg s.c. postnatal days 5-7	**Slight deficit of passive avoidance acquisition joined with increased lipid peroxidation in the cortex of 1-year-old rats.** *REF: Tejkalová H.,Benešová O. et al.; Homeostasis 33: 89-93, 1991.*
DIAZEPAM	0.6 - 3 mg/kg	10 mg/kg s.c. postnatal day 7 **5** mg/kg s.c. postnatal days 5-6	*Behavioural deviations:* **Increased emotional reactivity, defects in novelty reaction** *Defects of endocrine regulation:* **Minor deviations in reproductive functions of females (irregular oestrous cycles)** *Neurobiological deviations:* **Decrease of serotonergic transmission in the hypothalamus of 1-year-old rats** **Slight transitional changes of dopaminergic transmission in the striatum** *Deficits of immunocompetence:* **Depression of humoral and cell-mediated immune response** *REF: Benešová O. et al.; Ann.N.Y.Acad.Sci. 717: 89-101, 1994* *Dostál M.,Benešová O. et al.; Reproductive Toxicol. 9: 115-121,1995.*

The threshold dosage for functional teratogenicity is printed in bold letters.

performance, reproduction and immunocompetence, jointly with neurobiological deviations in the brain. The teratogenic potential of fenoterol is significantly lower, since only a 10 times higher dose than the clinical dosage had some minor negative impact. However, its risk has to be taken in account when it is used in combination with dexamethasone as it occurs very often in perinatal therapeutic practice. The degree of functional teratogenic risk of diazepam seems to lie somewhere between the previous drugs, since it induced functional teratogenic sequelae following experimental doses relevant to maximal therapeutic dosage used in perinatal treatment. However, the finding of significantly depressed immunoreactivity persisting for the entire life has to be regarded as a serious risk in the use of this drug.

These results have shown the possibility to compare the functional teratogenic risk of various drugs, both qualitatively and quantitatively. Since some medication is necessary for the treatment of pregnancies at risk and newborns at risk, it seems very urgent to evaluate all perinatally used drugs for functional teratogenicity in adequate animal studies and to select such compounds for the clinical use which combine sufficient therapeutic potency with minimal functional teratogenic risk.

Conclusions

Summarizing the presented results, it is possible to draw several conclusions which are relevant not only from the scientific point of view, but contribute also to the clinical therapeutic practice in obstetrics and neonatology.

1. The risk of drug teratogenicity is not restricted to the first trimester of pregnancy, but exists throughout the entire gestation and even in the early postnatal period with the possibility of inducing delayed functional teratogenic defects of the brain and the immune system.

2. Drugs with receptor-mediated effects (psychotropics, hormones) seem to exert a high functional teratogenic potential, especially when given in the perinatal period in which the majority of receptor systems are developing and formed.

3. A new approach for testing functional teratogenic risk was suggested for the evaluation of perinatally applied drugs, using animal model experiments with drug administration in neonate rats and a lifelong follow-up of behaviour, learning, reproductive ability, immunoreactivity and brain neurobiochemical analysis.

4. Using this experimental protocol, the functional teratogenic risk was studied in three perinatally aplied drugs: *dexamethasone, fenoterol, and diazepam*. The highest risk of functional teratogenicity, concerning life-long disturbances in all tested markers, was found in dexamethasone, it was substantially lower in diazepam and very low in fenoterol.

Acknowledgement

The author wishes to express her thanks to all members of the research team: RNDr. H. Teikalová, Ing. Z. Krištofiková, RNDr. J. Klaschka (Psychiatric Center Prague), RNDr. V. Panajotovová CSc (Research Institute for Pharmacy and Biochemistry), Ing. P. Hušek CSc (Institute of Endocrinology), MUDr. M. Dostál CSc (Institute of Experimental Medicine CAV).

The research was supported by the grant IGA MZCR No 1811-3.

References

Benešová O, Pavlík A (1989) Perinatal treatment with glucocorticoids and the risk of maldevelopment of the brain. Neuropharmacology 28: 89-97.

Benešová O, Pavlík A, Pìtová J (1984) Brain Maldevelopment and Drugs. Avicenum, Prague.

Benešová O, Tejkalová H, Burešová M (1992) Accelerated brain aging in rats as a consequence of disturbed perinatal development induced by drugs. In: Trojan S, Lngmeier M (eds) Ontogenesis of the Brain. Vol.5. Universitas Carolina Pragensis, Prague, pp 135-139.

Benešová O, Tejkalová H, Krištofiková Z, Klaschka J, Dostál M (1994) Early postnatal diazepam treatment of rats and neuroimmunocompetence in adult-hood and senescence. Annals New York Academy of Sciences, 717: 89-101.

Dobbing J (1970) Undernutrition and the developing brain. In: Himwich WA (ed) Developmental Neurobiology. Thomas, Springfield, pp 241-261.

Dostál M, Benešová O., Tejkalová H, Soukupová D (1995) Immune response of adult rats is altered by administration of diazepam in the first postnatal week. Reproductive Toxicol. 9: 115-121.

Jelínek R, Pavlík A, Peterka M (1983) Glucocorticoid receptor mediated terato-genesis in the chick embryo. Teratogenesis, Carcinogenesis and Mutagenesis 3: 1-7.

Jelínek R, Rychter Z (1980) Morphogenetic systems and screening for embryo-toxicity. Arch. Toxicol. (Suppl.) 4: 267-273.

Tejkalová H, Benešová O, Krištofiková Z, Binková B, Dlohozková N, Burešová M (1991) The assessment of the behavioural teratogenic risk of the tocolytic drug fenoterol. Homeostasis 33: 89-93.

The Effects on Mammals of Pre- and Postnatal Environmental Exposure to PCBs[1]
The Dutch Collaborative PCB/Dioxin Study

A.E. Smits-van Prooije, D.H. Waalkens-Berendsen, D.C. Morse, C. Koopman-Esseboom, M. Huisman, P.J.J. Sauer, E.R. Boersma, J.H.C.M. Lammers, K.J. van den Berg, G.C. van der Paauw, B.M. Kulig and N.J. Snoeij
TNO Food and Nutrition Research Institute, Division of Toxicology, P.O. Box 360, 3700 AJ Zeist, the Netherlands

Introduction

Five institutes[2] are collaborating in the Netherlands in a national study concerning the long-term effects of foetal and neonatal exposure to polychlorobiphenyls and dioxins. This collaborative study started in 1990 and was funded by the Dutch Toxicology Research Promotion Programme and the Health Research Stimulation Programme. The study includes both human and animal studies.

Polychlorobiphenyls (PCBs) were already synthesized in the last century, but widespread production began in the 1950s. Because of their properties in conducting heat, resistance to high temperatures, solvability in fat and oils, chemical stability and electrical insulation, they were, and in some cases still

---

[1] This contribution was presented at the EUROTOX '94 Congress in Basel in the context of the OECD Symposium on Terrestial Ecotoxicology

[2] The participating institutes are: the University of Groningen (Academic Hospital, Dept of Obstetrics and Gynaecology: M Huisman, ER Boersma); Erasmus University of Rotterdam (University Hospital, Sophia's Children's Hospital, Dept of Pediatrics: C Koopman-Esseboom, PJJ Sauer), Agricultural University of Wageningen, (Dept of Toxicology: DC Morse, A Brouwer), RIKILT-DLO State Institute for Quality Control of Agricultural Products, Wageningen (LGMTh Tuinstra), TNO Nutrition and Food Research Institute, Zeist (Toxicology Division: DH Waalkens-Berendsen, AE Smits-van Prooije, JCHM Lammers, KJ van den Berg, BM Kulig, NJ Snoeij, and Analyses Division: GC van der Paauw)

are, used as plasticizers, fire retardants, hydraulic fluids, and dielectric fluids in capacitators and transformers. Dioxins are formed mainly as by-products during the manufacture or combustion of PCBs and other chlorine-containing chemicals, e.g. chlorination used for bleaching in paper mills.

In this way PCBs and dioxins could enter the environment. At the end of the sixties suspicion arose as to less favourable properties of these chemical agents, but only some disasters, among them the rice oil poisonings in Japan (Yusho in 1968: Urabe *et al.* 1979) and Taiwan (Yu-Cheng in 1979: Hsu *et al.* 1985), and the industrial accident in Italy (Seveso, 1976: Bertazzi *et al.* 1992), showed the full scope of their undesired side effects. Upon relatively short-term, high level exposure chloracne, hyperpigmentation and peripheral neuropathy were observed as the obvious phenomena in exposed adult humans. Children of exposed mothers also had chloracne, hyperpigmentation of skin and nails, natal teeth, and they were small for their age (Yu *et al.* 1991). In follow-up studies, that are still in progress, lower performances in motor function and intelligence tests were revealed (Rogan *et al.* 1988). Adverse oestrogenic effects were suggested a.o. by a shorter penis length (Guo *et al.* 1993). Elevated occurrence of carcinomas is in Italy suspected but has not yet been proved (Bertazzi *et al.* 1992).

As a result of these findings, production and (new) use of PCBs were banned in the western industrialized countries in the late seventies. However, they were manufactured in central Europe until the mid-eighties, and they are still in use in old-fashioned transformers, or have been dumped with those transformers. PCBs disappear only slowly from the environment. In western Europe, they are widely present in sludge, and from there enter the food-chain and accumulate, since they are stored in adipose tissues and are only slowly metabolized, or not metabolized at all. Dioxins are deposited in water or in the atmosphere and precipitate on the vegetation from where they enter and accumulate in the food chain which is, at present, probably the main direct source of exposure for both animals and humans. PCBs and dioxins can cross the placenta, thus exposing the offspring *in utero*, and will be mobilized from the adipose tissues of the mother and excreted in the breast-milk during lactation.

The effects of long-term exposure to low levels of PCBs and dioxins are, as yet, unclear. Epidemiological studies indicate an adverse effect on growth, neuromotor/psychomotor activity and cognitive functions (cf. a study on the effects of eating Lake Michigan fish, Jacobson *et al.* 1990a,b).

The industrialized western Europe belongs to the areas that have a history of a relatively high exposure level. The Dutch PCB/dioxin study aims to examine the effects of pre- and early postnatal exposure both in humans and in animals, in which the animals also serve as a model to study long-term effects.

Animal Studies

Animal studies, generally conducted with commercial mixtures of PCBs (e.g. Aroclor, Fenclor) and administered during organogenesis or after birth, revealed hepatoxicity, teratogenicity (e.g. cleft palate in mice: Abbott and Birnbaum 1990), neurotoxicity (e.g. spinning syndrome in mice: Chou *et al.* 1979), neurodevelopmental changes (Tilson *et al.* 1990), immunotoxicity (Davis and Safe 1989) and carcinogenic effects (Norback and Weltman 1985).

In the animal studies conducted in our Institute a planar (dioxin-like) PCB, 3,4,5,3',4',5'-hexachlorobiphenyl (PCB 169) was selected for administration because this particular PCB, although not abundantly found in humans, is rather toxic and contributes in man to some extent to the cumulative PCB toxicity as expressed in toxic equivalency factors (Safe 1990). PCB 169, moreover, is hardly metabolized which ensures that the effects found will be caused by the PCB itself and not by its metabolites. The test substance was administered once, to mated female rats on gestation day 1.

Several reproduction experiments with PCB 169 were conducted, in which the effects on the offspring were followed. One of the experiments was focused on neurobehavioural development: a wide range of tests was conducted concerning physical landmarks (e.g. tooth eruption, vaginal opening), motor and cognitive functions and neurochemistry. Effects were only found in the precocious development of a few physical landmarks, in male grip strength and in the dopamine levels in the striatum. All other parameters showed no effects, not even in the offspring whose dams had shown slight maternal toxicity. In a similar study that focused on thyroid hormone metabolism, dose-related effects were found in plasma (decrease T4) and in the brain (increase type II thyroxine 5'deiodinase) (Morse *et al.* 1993). Intrauterine and lactational exposure to PCB 169 also affected the reproduction capacity of rats, weaned from exposed females that showed a slight maternal toxicity. The most conspicuous phenomena were a reduction in pregnancy rate and an extension in duration of gestation of the rats, paired with partners from the same dose group (Smits-van Prooije *et al.* 1993). A consecutive cross-fostering study using the same experimental set-up showed that these effects were mainly caused during intrauterine development. Pairing of the treated offspring with untreated partners proved that the capability to reproduce was reduced both in male and female offspring.

Further investigations are presently conducted as to the underlying mechanisms using the same and some other organo halogen compounds among which another planar PCB, a mono- and a di-ortho PCB and a dibenzofuran, which will probably act through different mechanisms (Bouwman 1994).

Human Studies

In the human studies over 400 children from two different areas in the Netherlands were recruited, one being the heavily industrialized area of Rotterdam, the other the, for Dutch standards, rural area of Groningen. In each area 2 separate groups were selected: those that were breast-fed and those that were formula-fed (the formula-fed children all received formulated milk-powder from one and the same batch. In formula-milk all animal fats, supposedly containing PCBs and dioxins, are replaced by PCB/dioxin-free vegetable fats).

A blood sample was taken from the mothers around the 36th week of pregnancy to analyze the content of PCBs in maternal blood. Cord blood was collected for the analysis of PCBs in blood of the infants. Breast-feeding mothers collected representative 24 h milk samples 10 days, 6 weeks and 3 months *post partum*. PCB and dioxin levels in human milk are similar to levels in maternal adipose tissue. Accordingly, PCB/dioxin levels in milk reflect the maternal body burden and subsequently the pre- and postnatal exposure of the children.

The toxic equivalency factor according to Safe (1990), corrected by Ahlborg *et al.* (1994) was used to express the toxic potency of the blood and milk samples. The differences in body-burden between Rotterdam and Groningen were minimal. Increased levels of PCBs and dioxins in maternal milk samples correlated with relatively decreased TT3 and TT4 levels in maternal plasma, and with increased TSH levels in their children.

Neurological data collected in the second week after birth using the neonatal optimality score (Touwen *et al.* 1980) showed no relationship to the levels of PCBs/dioxins in maternal and cord plasma. Analysis of the breast-milk data showed a higher percentage of mild hypotonia with increasing levels of PCBs and dioxins in the milk. Positive as well as negative effects of perinatal exposure to breastfeeding have been found on early mental and psychomotor development. There was no relation with serious handicaps.

Conclusions

The animal studies conducted in the framework of the Dutch PCB/dioxin study show several adverse effects of the PCB tested; many effects were transient, such as the development of physical landmarks and the effects on thyroid hormone levels, whereas the effect on fertility was persistent. Effects on behaviour were hardly observed. The effect on fertility might raise concern, but the effect-levels of PCBs in the animal studies were high in comparison to the cumulative levels of PCBs/dioxins found in humans. In the human studies great similarity between the population of Groningen and Rotterdam was observed in the PCB/dioxin body burden. Preliminary results reveal a light correlation between pre- and postnatal exposure to PCBs/dioxins, and hypotonia and thyroid hormone levels which, however, appear to be still within the normal clinical ranges. The effects on fertility observed in animal studies (Mably *et al.* 1992a,b,c; Smits-van Prooije *et al.* 1993) may only be recognised in years to come. A careful follow-up of the present clinical cohort is therefore needed.

References

Abbott BD, Birnbaum LS (1990) TCDD-induced altered expression of growth factors may have a role in producing cleft palate and enhancing the incidence of clefts after coadministration of retinoic acid and TCDD. Toxicol Appl Pharmacol 106:418-432

Ahlborg UG, Becking GC, Birnbaum LS, Brouwer A, Derks HJGM, Feeley M, Golor G, Hanberg A, Larsen JC, Liem AKD, Safe SH, Schlatter C, Wærn F, Younes M, Yrjänheikki E (1994) Toxic equivalency factors for dioxin-like PCBs. Chemosphere 28:1049-1067

Bertazzi PA, Zocchetti C, Pesatori AC, Guercilina S, Consonni D, Tironi A, Laudi MT (1992) Mortality of a young population after accidental exposure to 2,3,7,8-tetrachlorodibenzodioxin. Ind J Epidemiol 21:118-123

Bouwman CA (1994) Modulation of vitamin K-dependent blood coagulation by chlorinated biphenyls and dioxins in rats: possible implications for breast fed newborn infants. Thesis, Utrecht, the Netherlands, ISBN 90-393-0581-1

Chou SM, Miike T, Payne WM, Davis GJ (1979) Neurology of "spinning syndrome" induced by prenatal in toxication with a PCB in mice. Ann NY Acad Sci: 373-395

Davis D, Safe S (1989) Dose-response immunotoxicities of commercial polychlorinated biphenyls (PCBs) and their interaction with 2,3,7,8-tetrachlorodibenzo-p-dioxin. Toxicol Lett 48:35-43

Guo YL, Lai TJ, Ju SH, Chen YB, Hsu CC (1993) Sexual development and biological findings in Yucheng Children. 13th International Symposium on Chlorinated Dioxins and related Compounds 14:235-238

Hsu ST, Ma CI, Hsu SK, Wu SS, Hsu NHM, Yeh CC, Wu SB (1985) Discovery and epidemiology of PCB poisoning in Taiwan: a four-year followup. Environ Health Perspec 59:5-10

Jacobson JL, Jacobson SW, Humphrey HEB (1990a) Effects of in utero exposure to polychlorinated biphenyls and related contaminants on cognitive function in young children. J Pediatr 116:38-46

Jacobson JL, Jacobson SW, Humphrey HEB (1990b) Effects of exposure to PCBs and related compounds on growth and activity in children. Neurotox Teratol 12:319-326

Mably TA, Moore RW, Peterson RE (1992a) *In utero* and lactational exposure of male rats to 2,3,7,8-tetrachlorodibenzo-p- dioxins. I. Effects on androgenic status. Toxicol Appl Pharmacol 114:97-107

Mably TA, Moore RW, Goy RW, Peterson RE (1992b) *In utero* and lactational exposure of male rats to 2,3,7,8- tetrachlorodibenzo-p-dioxins. II. Effects on sexual behavior and the regulation of luteinizing hormone secretion in adulthood. Toxicol Appl Pharmacol 114:108-117

Mably TA, Bjerke DL, Moore RW, Gendron-Fitzpatrick A, Peterson RE (1992c) *In utero* and lactational exposure of male rats to 2,3,7,8-tetrachlorodibenzo-p-dioxins. III. Effects on spermatogenesis and reproductive capability. Toxicol Appl Pharmacol 114:118-126

Morse DC, Groen D, Veerman M, Van Amerongen CJ, Koëter HBWM, Smits-van Prooije AE, Visser TJ, Koeman JH, Brouwer A (1993) Interference of polychlorinated biphenyls in hepatic and brain thyroid hormone metabolism in fetal and neonatal rats. Toxicol Appl Pharmacol 122: 27-33

Norback DH, Weltman RH (1985) Polychlorinated biphenyl induction of hepatocellular carcinoma in the Sprague-Dawley rat. Environ Health Perspec 60:97-105

Rogan WJ, Gladen BC, Hung KL, Koong SL, Shih LY, Taylor JS, Wu YC, Yang D, Ragan NB, Hsu CC (1988) Congenital poisoning by polychlorinated biphenyls and their contaminants. Science 241:334-336

Safe S (1990) Polychlorinated biphenyls (PCBs), dibenzo-p-dioxins (PCDDs), dibenzofurans (PCDFs) and related compounds: environmental and mechanistic considerations which support the development of toxic equivalency factors (TEFs). CRC Crit Rev Toxicol 21:51-88

Smits-van Prooije AE, Lammers JHCM, Waalkens-Berendsen DH, Kulig BM, Snoeij NJ (1993) Effects of the PCB 3,4,5,3',5'- hexaclorobiphenylon in the reproduction capacity of Wistar rats. Chemosphere 27:395-400

Tilson HA, Jacobson JL, Rogan WJ (1990) Polychlorinated biphenyls and the developing nervous system: cross species comparisons. Neurotox Teratol 12: 239-248

Touwen BCL et al (1980) Obstetrical condition and neonatal neurological morbidity. An analysis with the help of the optimality concept. Early Human Develop 4:207-228

Urabe H, Koda H, Asahi M (1979) Present state of Yusho patients. Ann NY Acad Sci 320: 273-276

Yu ML, Hsu CC, Gladen BC, Rogan WJ (1991) In utero PCB/PCDF exposure: relation of developmental delay to dysmorphology and dose. Neurotoxicol Teratol 13:195-202

Intercellular Communication

(Chairs: A.R. Boobis, UK, and
R. Jelinek, Czech Republic)

Connexin Genes and Cell Growth Control

H. Yamasaki, V. Krutovskikh, M. Mesnil and Y. Omori
Unit of Multistage Carcinogenesis, International Agency for Research on Cancer, Lyon, France

In multicellular organisms, individual cells not only maintain their own functions but also must behave as members of an orderly cellular society. Thus, cell growth is controlled both at the invididual cell level as well as by homeostasis within the cellular society. Known oncogenes and tumor-suppressor genes include not only genes involved in signal transduction and the cell cycle, but also those controlling growth factors and their receptors, which are important for individual cellular functions and cell-cell interaction, respectively.

Connexins are believed to be responsible for forming channels which are necessary for gap junctional intercellular communication (GJIC). Unlike growth factor- or hormone-mediated intercellular communication, GJIC mediates transfer of ions and molecules from the interior of one cell to that of the neighboring cells so that these factors are transferred without leaving the cytoplasmic membrane (Loewenstein, 1979). Since GJIC-diffusible molecules include those involved in signal transduction (Saez et al., 1989), e.g. calcium, cAMP and inositol triphosphate, this type of intercellular communication is considered to play a crucial role in the maintenance of homeostatis by keeping the level of growth-control signals at equilibrium among GJIC-connected cells.

Since cancer cells can be considered to be rebels among a cellular society, it is reasonable to assume that their ability to communicate is impaired. There is now increasing evidence which suggests that GJIC is indeed reduced in most tumors and that connexin genes form a family of tumor-suppressor genes (Trosko and Chang, 1988; Klaunig and Ruch, 1990; Yamasaki, 1990). Here we briefly review the role of connexin genes in cell growth control, including recent results obtained through a molecular genetic approach in our laboratory.

Abbreviations: GJIC, gap junctional intercellular communication; Cx26, connexin 26 (similar for Cx32 and others); SSCP, single strand conformation polymorphism; CMTX, X-linked Charcot-Marie-Tooth

Connexin expression and function in tumors

GJIC is usually measured by a dye-transfer assay and has been mainly applied to cultured cells. Based on *in vitro* studies, it has been established that transformed cells often lose their homologous (among themselves) and/or heterologous (with surrounding normal counterparts) GJIC (Loewenstein, 1979; Yamasaki, 1990). More recently, we have modified the dye-transfer assay and applied it to measure GJIC in freshly-removed liver samples, to better reflect *in vivo* situations (reviewed in Krutovskikh and Yamasaki, 1995). Applying this method to rat liver samples, we have shown that homologous GJIC is reduced in focal liver lesions (foci and nodules) as well as in hepatocellular carcinomas. About 30% of the GST-P-positive foci showed a significant decrease in the number of Cx32 protein spots per cell, in comparison with surrounding normal liver tissue. When we examined the heterologous GJIC between pre-neoplastic foci and surrounding cells, we found that more than 80% of GST-P-positive foci were not communicating with surrounding liver parenchyma. In a few cases, GST-P-positive foci formed functional GJIC with contiguous tissue. These results suggest that lack of heterologous GJIC is a general phenomenon, although some foci communicate with surrounding normal cells, and we hypothesize that cells which communicate with surrounding cells may not be able to progress towards more malignant stages (Krutovskikh et al., 1991).

The *ex vivo* dye-transfer assay was also applied to study GJIC in human liver tumour samples (Krutovskikh et al., 1994). As shown in Fig. 1, the dye-transfer assay revealed a strong reduction in GJIC in all liver tumors tested compared with surrounding non-tumorous tissue, regardless of the morphological structure. Samples of liver parenchyma surrounding tumors, used as internal controls, in all tested cases also showed much lower GJIC compared with liver samples showing no sign of disease. Primary human liver tumors were always separated from surrounding liver by a well-developed connective tissue capsula, and when Lucifer Yellow was injected into these capsula, dye always spread inside the connective tissue, but never penetrated the tumor nodule (Fig. 1C). Thus, while human tumors show a lack of heterologous GJIC with surrounding non-tumorous cells, as has been shown in rat liver tumors, this is apparently due to the absence of direct contact between them (Fig. 1).

One of the reasons why GJIC is almost always impaired in tumors may be that the regulatory mechanisms of connexin gene and protein expression are liable to disruption by a carcinogen and/or carcinogenic process at many points. Analysis of various types of tumors suggests that this is indeed the case. For example, the

Fig. 1 (see opposite page) GJIC of surgically-removed human liver tumor samples. Upper panel: extent of GJIC measured by the spread of Lucifer Yellow dye molecules microinjected into freshly removed liver slices. Middle panel: microphotographs of dye-spread; A, normal liver; B, liver tumor; C, encapsulated liver tumor (note physical as well as functional lack of IC between the carcinoma and surrounding cells). Lower panel: immunostaining by anti-Cx 32 antibody; D, normal liver; E-H, liver tumors (note aberrant localization of Cx32 proteins).

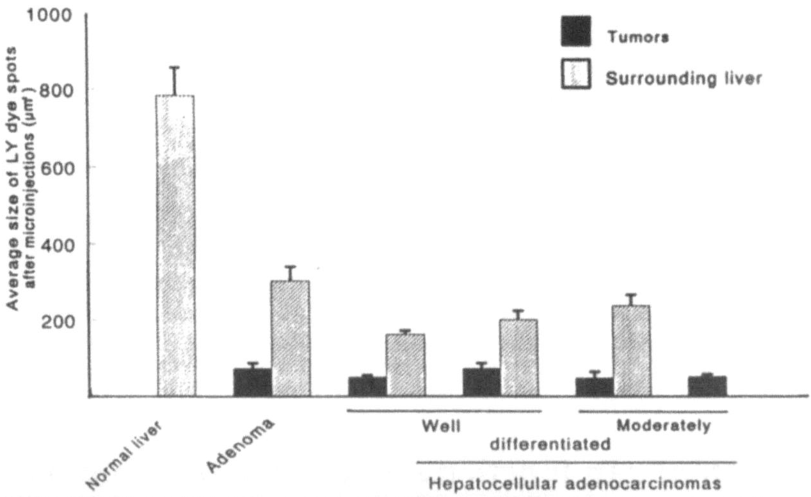
Average size of LY dye spots after microinjections (µm²)
1000
800
600
400
200
Tumors
Surrounding liver
Normal liver
Adenoma
Well
Moderately
differentiated
Hepatocellular adenocarcinomas

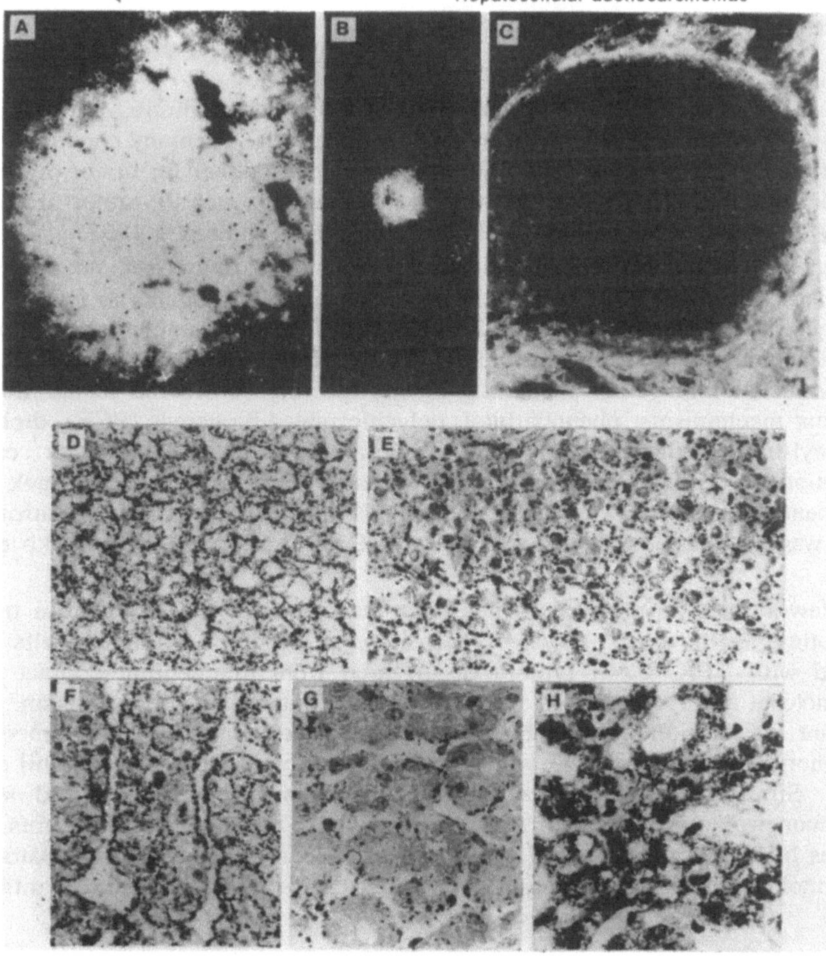
A
B
C
D
E
F
G
H

level of Cx32 gene expression was lower in rat liver tumors than in adjacent normal cells (Fitzgerald et al., 1989). On the other hand, the level of Cx32 mRNA was not changed in human hepatocellular carcinoma samples but that of Cx43 mRNA was increased (Oyamada et al., 1990). In addition to such changes of connexin mRNA levels, many human and rat liver tumors showed aberrant localization of Cx32 proteins in the cytoplasm (Fig.1). In human tumors, Cx43 proteins were also aberrantly localized (Krutovskikh et al., 1994).

Another mechanism involved in GJIC disorders in tumors is abnormal control of cell-cell recognition. In one mouse papilloma cell line that lacked GJIC, transfection of the E-cadherin gene restored it (Jongen et al., 1991). Aberrant expression and mutations of the E-cadherin gene have been reported in several types of human tumor (Shimoyama et al., 1989; Risinger et al., 1994).

Inhibition of gap junctional intercellular communication by tumor-promoting agents

Since the discovery of the inhibitory effect of the tumor-promoting phorbol esters on GJIC (Yotti et al., 1979; Murray and Fitzgerald, 1979), many chemicals have been tested to see whether GJIC inhibition is a good marker for tumor promotion (Trosko et al., 1984). The results suggest that many, but not all, tumor-promoting agents inhibit GJIC of cultured cells (Swierenga and Yamasaki, 1992).

Although many *in vitro* studies have demonstrated this effect, no systematic study has examined GJIC inhibition by tumor-promoting agents *in vivo*. Using our method to measure GJIC in freshly-removed tissue samples, we examined whether liver-specific tumor-promoting agents affect GJIC in the rat liver *in vivo*. We selected four tumor-promoting agents that are considered to act through different mechanisms: phenobarbital, polychlorinated biphenyls (PCB), dichlorodiphenyl-trichlorethane (DDT) and clofibrate. All four compounds caused inhibition of GJIC to different extents in rat liver as early as one week after treatment was started. These four compounds also induced cell proliferation, but there was no quantitative relationship with GJIC inhibition (Krutovskikh et al., 1995).

A few studies have revealed molecular mechanisms by which certain tumor-promoting agents block GJIC. When cultured rat liver epithelial cells were treated with TPA, Cx43 protein disappeared from intercellular contact areas (Asamoto et al., 1991). Cx43 was still present, but was in the cytoplasm. This aberrant localization induced by TPA was associated with the appearance of a phosphorylated form of Cx43 which is not present in treated cells (Mesnil et al., 1994). Similar aberrant localization of connexin molecules was induced *in vivo* by tumor-promoting agents. As described above, the treatment of rats with various liver tumor-promoting agents caused a decrease in GJIC in hepatocytes. Immunohistochemical staining of liver from treated rats with anti-Cx32 antibody revealed aberrant localization of the Cx32 molecules (Krutovskikh et al., 1995). These results suggest that aberrant localization of connexin proteins may be one of the major mechanisms leading to decreased GJIC capacity.

Tumor Growth Suppression by Direct Transfection of Connexin Genes

The most direct way to test the role of connexin genes in cell growth control is to transfect them into GJIC-deficient tumor cells and examine whether the growth of these cells is suppressed. As shown in Table 1, various connexins have been transfected into various tumorigenic cell lines and, in general, the results suggest that connexin genes control cell growth.

We have transfected Cx26, 40 and 43 genes into HeLa cells, which showed no detectable expression of these three connexin genes tested before transfection. All HeLa clones transfected with Cx26, 40 or 43 genes recovered GJIC to a similar extent, although Cx26-transfectants maintained their GJIC at similar level before and after confluence, while Cx40- and Cx43-transfectants lost GJIC at confluence. We also found that Cx26-transfectants did not grow in soft agar nor form tumors when injected into nude mice. In constrast, transfectants of Cx40 and 43 formed tumors, although the latent period was somewhat longer than with parental HeLa cells (Mesnil et al., 1995). In order to examine why the Cx26, but no other, connexin gene exerts strong negative growth regulation *in vitro* and *in vivo*, we have examined what connexin gene(s) is(are) expressed in the cervix. Since fresh human cervix biopsies were not yet available, we employed rat cervix for this study; human and rat tissue express the same types of connexin genes, though not always at the same level. We found that Cx26 was the major connexin gene expressed in the rat cervix, from which HeLa cells originated. These results support our hypothesis that different types of connexin genes exert differential growth control effects on cells depending on the cells' endogenous connexin species.

Connexin Gene Mutations and their Dominant Negative Effects on GJIC

While the inhibition of GJIC in toxicology has always been approached as a nongenotoxic process, it is possible that genotoxic mechanisms, such as mutations of connexin genes, lead to GJIC impairment. We have therefore started to search for Cx32 gene mutations in human and rodent tumors using SSCP. In 20 human liver tumors from France (Krutovskikh et al., 1994), and 22 human gastric tumors from Russia (Mironov et al., 1994), we found no sign of mutations in the coding regions by this analysis. In all of the same liver tumors, we found reduced GJIC and/or aberrant localization of Cx32 proteins.

When seven rat liver tumors induced by a nitrosamine were analysed, we found one sample with an aberrant SSCP pattern. Further analysis indicated that the mutation was a G to A transition (a type of mutation expected with this carcinogen) at the 220th codon of the Cx32 gene. This liver sample also showed aberrant localization of connexin molecules; Cx32 proteins were localized in the nuclear region rather than at cell-cell contact areas (Yamasaki et al., 1995). In order to see whether the mutation in the Cx32 gene was responsible for reduced GJIC, we cloned this mutant gene and transfected it into GJIC-deficient HeLa cells. The resulting transfectants communicated to a similar extent with cells transfected with the wild-type Cx32 gene, indicating that this mutation is functionally silent, at least in HeLa cells.

On the other hand, recent studies have begun to show that the mutations of connexin genes may be responsible for certain human diseases. For example, some types of viscero-atrial-heterotaxia syndrome are caused by the germ-line mutations in the Cx43 gene, resulting in loss of response to cAMP-dependent protein kinase (Britz-Cunningham et al., 1995). Moreover, germ-line mutations of the Cx32 gene have been reported to be responsible for X-linked Charcot-Marie-Tooth (CMTX) disease (Berghoffen et al., 1993; Fairweather et al., 1994; Ionasescu et al., 1994).

In order to see whether the Cx32 gene mutations found in CMTX patients impair GJIC, we introduced four different Cx32 mutations discovered in patients with CMTX disease into wild-type cDNA and transfected them into HeLa cells. When GJIC was estimated by direct microinjection of a fluorescent dye, those cells transfected with the wild-type Cx32 gene, but not those transfected with three different base substitution mutations (i.e. Cys 60 to Phe, Val 139 to Met and Arg 215 to Trp), restored GJIC of HeLa cells. Unexpectedly, however, the cells transfected with a nonsense mutant at codon 220 restored GJIC. These results suggest that most, if not all, Cx32 gene mutations associated with CMTX disease do indeed impair GJIC, which may be the cause of this disease.

Connexin proteins form homohexamers called connexons, which are transferred and anchored to cell membrane to form gap junctions (Beyer, 1993). If the mutants examined above, showing no GJIC, can form oligomers with wild-type connexin, they could prevent the connexon's function in a dominant negative manner (see Fig. 2). We have examined this possibility since mutant and wild-type connexin proteins would co-exist *in vivo* when one of two alleles of a connexin gene is mutated. For this purpose, we double-transfected the mutant constructs into HeLa cells which had already been transfected with wild-type Cx32 gene and thus were GJIC-proficient.Three base substitution mutants, Cys 60 to Phe, Val 139 to Met, and Arg 215 to Trp, did cause a reduction of the GJIC restored by the wild-type gene, while vector-transfected cells maintained the same level of GJIC, suggesting that these three mutants can erase the function of wild-type Cx32 in a dominant negative manner. The nonsense mutation at codon 220 did not show such a dominant negative effect. In the double transfectants, immunohistochemical analysis revealed less Cx32 proteins present at cell-cell contact areas. We thus suggest that base substitution mutations can down-regulate wild-type Cx32, presumably through producing non-functional chimeric connexons composed of mutant and wild-type connexins, which are not properly anchored to the cytoplasmic membrane. Our results suggest that a one-allele mutation of the connexin gene may be sufficient to eliminate GJIC through a dominant negative mechanism and cause diseases (Omori et al., unpublished results).

Although we found a Cx32 gene mutation in only one rat liver tumor and none in human liver and stomach tumors (Yamasaki et al., 1995), mutations of the Cx37 gene have been reported in cell lines derived from murine lung tumors (Mandelboim et al., 1994). Since a one-allele mutation of connexin genes may suffice to destroy GJIC and since connexins form a family of tumor-suppressor genes, as described above, it seems possible that future studies will reveal connexin gene mutations in certain types of tumors.

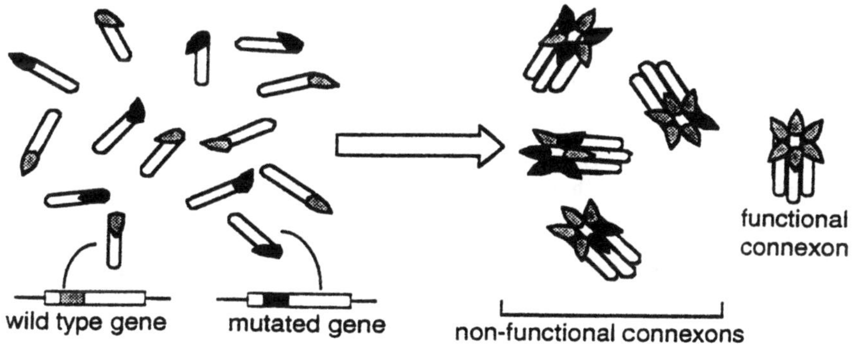

Fig.2 Schematic view of dominant-negative effect of mutant connexins on GJIC

It is also interesting to note that connexin gene mutations are implicated also in diseases other than cancers, such as CMTX and heart malformations. These results support the idea that GJIC is an essential mechanism in the maintenance of tissue homeostasis and that its impairment results in disturbed homeostasis, thus causing various forms of disease.

Acknowledgements

We are indebted to Mrs Chantal Déchaux for her secretarial help and to Dr John Cheney for editing the manuscript. Y.O. was the recipient of a Special Training Award from IARC.

References

Asamoto M, Oyamada M, El Aoumari A, Gros D, Yamasaki H (1991) Molecular mechanisms of TPA-mediated inhibition of GJIC; evidence for action on the assembly or function but not the expression of connexin 43 in rat liver epithelial cells. Mol Carcinogen, 4, 322-327

Berghoffen J, Scherer SS, Wang S, Oronzi Scott M, Bone J, Paul DL, Chen K, Lensch MW, Chance PF and Fischbeck KH (1993) Connexin mutations in X-linked Charcot-Marie-Tooth disease. Science, 262, 2039-2042

Beyer EC (1993) Gap junctions. Int Rev Cytol, 137C, 1-37

Bond SL, Bechberger JF, Khoo NK and Naus CC (1994) Transfection of C6 glioma cells with connexin 32: the effects of expression of a non-endogenous gap junction protein. Cell Growth Differ, 5, 179-86

Britz-Cunningham SH, Shah MM, Zuppan CW and Fletcher WH (1995) Mutations of the connexin 43 gap-junction gene in patients with heart malformations and defects of laterality. New Engl J Med, 332, 1323-1329

Chen SC, Pelletier DB, Ao P, Boynton AL (1995) Connexin 43 reverses the stenotype of transformed cells and alters their expression of cyclin/cyclin-dependent kinases. Cell Growth Differ, 6, 681-690

Eghbali B, Kessler JA, Raid LM, Roy C and Spray DC (1991) Involvement of gap junction in tumorigenesis: transfection of tumor cells with connexin 32 cDNA retards growth "in vivo". Proc Natl Acad Sci USA, 88, 10701-10705

Faiweather N, Bell C, Cochrane S, Chelly J, Wang S, Mostacciulo ML, Monaco AP and Haites NE (1994) Mutations in the connexin 32 gene in X-linked dominant Charcot-Marie-Tooth disease (CMTX1) Hum Mol Genet, 3, 29-34

Fitzgerald DJ, Mesnil M, Oyamada M, Tsuda H, Ito N and Yamasaki H (1989) Changes in gap junction protein (connexin 32) gene expression during rat liver carcinogenesis. J Cell Biochem, 41, 97-102

Goldberg GS, Martyn KD and Lau AF (1994) A connexin 43 antisense vector reduces the ability of normal cells to inhibit the foci formation of transformed cells. Mol Carcinogen, 11, 106-114

Ionasescu V, Searby C and Ionasescu R (1994) Point mutations of the connexin 32 (GJB1) gene in X-linked dominant Charcot-Marie-Tooth neuropathy. Hum Mol Genet, 3, 355-358

Jongen WMF, Fitzgerald DJ, Asamoto M, Piccoli, C, Slaga TH, Gros D, Takeichi M and Yamasaki H (1991) Regulation of connexin 43-mediated gap junctional interecellular communication by Ca++ in mouse epidermal cells is controlled by E-cadherin. J Cell Biol, 114, 545-555

Klaunig JE and Ruch RJ (1990) Role of inhibtion of intercellular communciation in carcinogenesis. Lab Invest, 62, 135-46

Krutovskikh VA, Oyamada M and Yamasaki H (1991) Sequential changes of gap-junctional intercellular communications during multistage rat liver carcinogenesis: direct measurement of communication in vivo. Carcinogenesis, 12, 1701-1706

Krutovskikh VA, Mazzoleni G, Mironov N, Omori Y, Aguelon A-M, Mesnil M, Berger F, Partensky C and Yamasaki H (1994) Altered homologous and heterologus gap junctional intercellular communication in primary human liver tumors associated with aberrant protein localization but not gene mutation of connexin 32. Int J Cancer, 56, 87-94

Krutovskikh VA, Mesnil M, Mazzoleni G and Yamasaki H (1995) Inhibition of rat liver gap junction intercellular communication by tumor-promoting agents *in vivo*: association with aberrant localization of connexin proteins. Lab Invest, 72, 571-577

Krutovskikh VA and Yamasaki H (1995) Ex-vivo dye transfer assay as an approach to study gap junctional intercellular communication disorders in hepatocarcinogenesis. Prog in Cell Res, 4, 137-140

Lin Z, Zhang Z, Han, Y, naus CC, Yu KR and Holtzer H (1995) Functional expression of gap junction gene Cx43 and the myogenic differentiation of rhabdomyosarcoma cells. Science China (Ser B), 38, 305-312

Loewenstein WR (1979) Junctional intercellular communication in the postimplantation mouse embryo. Cell, 18, 399-409

Mandelboim O, Berke G, Fridkin M, Feldman M, Eisenstein M and Eisenbach L (1994) CTL induction by a tumour-associated antigen octapeptide derived from a murine lung caricnoma. Nature, 369, 67-71

Mehta PP, Hotz-Wagenblatt A, Rose B, Shalloway D and Loewenstein WR (1991) Incorporation of the gene for a cell-cell channel protein into transformed cells leads to normalization of growth. J Memb Biol, 124, 207-225

Mesnil M and Yamasaki Y (1993) Cell-cell communication and growth control of normal and cancer cells: evidence and hypothesis. Mol Carcinogen, 7, 14-17

Mesnil M, Asamoto M, Piccoli C and Yamasaki H (1994) Possible molecular mechanisms of loss of homologous and heterologous gap junctionl intercellular communication in rat liver epithelial cell lines. Cell Adhes and Commun, 2, 377-384

Mesnil M, Krutovskikh VA, Piccoli C, Elfgang C, Traub O, Willecke K and Yamasaki H (1995) Negative growth control of HeLa cells by connexin genes: connexin species specificity. Cancer Res, 55, 629-639

Mironov N, Aguelon A-M, Potapova GI, Omori Y, Gorbunov OV, Klimenkov AA and Yamasaki H (1994) Alterations of (CA)n DNA repeats and tumor suppressor genes in human gastric cancer. Cancer Res, 54, 41-44

Murray AW and Fitzgerald DJ (1979) Tumor promoters inhibit metabolic cooperation in cocultures of epidermal and 3T3 cells. Biochem Biophys Res Commun, 91, 395-401

Naus C, Elisevich K, Zhu D, Belliveau D and Del Maestro R (1992) *In vivo* growth of C6 glioma cells transfected with connexin 43 cDNA. Cancer Res, 52, 4208-4213

Oyamada M, Krutovskikh V, Mesnil M, Partensky C, Berger F and Yamasaki H (1990) Aberrant expression of gap junction gene in primary human hepatocellular carcinomas: increased expression of cardiac-type gap junction gene connexin 43. Mol Carcinogen, 3, 273-278

Pitts JD and Finbow ME (1986) The gap junction. J Cell Sci, 4 (Suppl.), 239-266

Risinger JI, Berchuck A, Kohler MF and Boyd J (1994) Mutations of the E-cadherin gene in human gynecologic cancers. Nature Genet, 7, 98-102

Rose B, Mehta PP and Loewenstein WR (1993) Gap junction protein gene suppresses tumorigenicity. Carcinogenesis, 14, 1073-1075

Saez JC, Conner JA, Spray DC and Bennett MVL (1989) Hepatocyte gap junctions are permeable to a second messenger, inositol 1,4,5-triphosphate, and to calcium ions. Proc Natl Acad Sci USA, 86, 2708-2712

Shimoyama Y, Hirohashi S, Hirano Sh, Noguchi M. Shimosato Y, Takeichi M and Abe O (1989) Cadherin cell-adhesion molecules in human epithelial tissues and carcinomas. Cancer Res, 49, 2128-2133

Swierenga SHH and Yamasaski H (1992) Performance of tests for cell transformation and gap junction intercellular communication for detecting nongenotoxic carcinogenic activity. In: Vainio H, Magee PN, McGregor DB and McMichael AJ (eds) Mechanisms of Carcinogenesis in Risk Identification, IARC Scientific Publications No.116, Lyon, pp.165-193

Trosko JE, Jone C and Chang CC (1984) The use of *in vitro* assays to study and to detect tumor promoters. In: Börzsönyi M, Day NE, Lapis K and Yamasaki H (eds) Models, Mechanisms and Etiology of Tumour Promotion. IARC Scientific Publications No.56, Lyon, pp.239-252

Trosko JE and Chang CC (1988) Nongenotoxic mechanisms in carcinogenesis: role of inhibited intercellular communication. Banburry Rep, 31, 139-170

Yamasaki H (1990) Gap junctional intercellular communication and carcinogenesis. Carcinogenesis, 11, 1051-1058

Yamasaki H, Mesnil M, Omori Y, Mironov N and Krutovskikh V (1995) Aberrant control of connexin expression and functions in multistage rat and human hepatocarcinogenesis. Prog in Cell Res, 4, 79-82

Yotti LP, Chang CC and Trosko JE (1979) Elimination of metabolic cooperation in Chinese hamster cells by a tumor promoter. Science, 206, 1089-1091

Zhu D, Caveney S, Kidder GM and Naus CCG (1991) Transfection of C6 glioma cells with connexin 43 cDNA: analysis of expression, intercellular coupling and cell proliferation. Proc Natl Acad Sci USA, 88, 1883-1887

Disruption of Polarized Function in Hepatocytes

J.K. Chipman, R. Coleman, J.C. Wilton, M.J. Guppy, V. Stone and D.J. Lankester
School of Biochemistry, The University of Birmingham, Edgbaston, Birmingham. B15 2TT UK

Introduction: Polarized function in hepatocytes

Liver parenchymal epithelial cells (hepatocytes) are highly polarized in terms of both transport systems and intercellular communication (Handler, 1989; Figure 1).

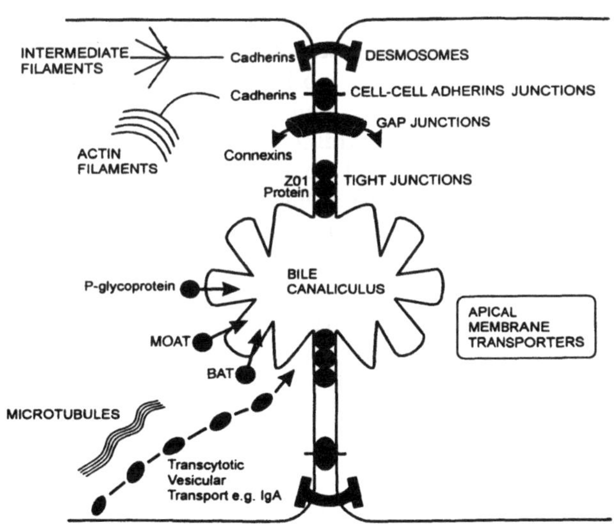

Figure 1: Summary of major polarity determinants in hepatocytes

Polarization is governed predominantly by transport systems that direct agents (both endogenous and exogenous) from the sinusoidal or basolateral membrane and from intracellular sites to the apical (canalicular) membrane, contributing to the formation of primary bile. Microtubules contribute to transcytotic vesicular transport of some components of bile such as IgA (Feldmann 1989). For many other substances there exist various transport proteins such as the P-glycoprotein transporter protein in the canalicular membrane. Microfilaments also are important in bile formation through their role in canalicular motility and contraction and the coordination of these functions between neighbouring cells (Oshio and Phillips 1981). In addition there are numerous cell-cell interactions that regulate communication between neighbouring polarized cells in relation to the maintenance of canalicular function (tight junctions), inter-cell anchorages (adherens junctions), intermediate filament attachment sites (desmosomes) and signal-transfer communication (gap junctions) (Feldmann 1989; Saez *et al.* 1989). There is an integrated control of polarity as the cytoskeleton and cadherins influence the maintenance of tight junctions (Schneeberger and Lynch 1992) and the gap junctions (Jongen *et al.* 1991).

An appreciation of this polarized nature is important in studies of hepatic toxicity, not only because xenobiotics that undergo biliary excretion rely on transport sytems and the integrity of the tight junctions, but also because the polarized function and associated structures provide numerous potential targets via which hepatic toxicity may be mediated. Conventional techniques for the isolation of hepatocytes from liver for use in suspensions or monolayer culture exploit the use of calcium-free perfusate to disrupt calcium-dependent intercellular and cell-matrix contacts followed by treatment with collagenase (Seglen, 1976). Singlet cells clearly retain limited polarized function, however when cultured to form a monolayer reorganization of the apical membrane can occur allowing cellular junctions and polarized canalicular membranes to reform. Particularly promising has been the work of Kawahara *et al.* (1989) who demonstrated canalicular membrane transport of fluorescein diacetate and vesicular transport of horseradish peroxidase in hepatocytes (from 14-day old rats) incubated in the presence of insulin and dexamethasone.

One problem associated with the use of hepatocyte monolayers, however, is the considerable loss of certain xenobiotic metabolising enzymes, specifically cytochrome P450 isoenzymes, with time of culture (see Paine 1990). The culture period required to regain polarity may therefore restrict the study of hepatotoxic agents that rely on metabolic activation (Gonzalez and Gelboin 1994). There are other approaches which enable polarity to be retained in fresh liver preparations such as the use of precision cut liver slices, which have the added advantage of including other cell types (Azri *et al.* 1990) or hepatocyte couplets (Graf *et al.* 1984; Gautam *et al.* 1987). The latter system provides an accessible functional biliary unit in which bile transport (Graf *et al.* 1984) and gap junctional intercellular communication (Spray *et al.* 1986; Guppy *et al.*, 1994) are both operative at approximately 4 h after preparation thus limiting the time available for the loss of P450 function (Lankester *et al.* 1994a).

In hepatocyte couplets, within only a few hours of culture, membrane domains which are stained for Mg^{2+}-ATPase relocate to the canalicular pole and the canalicular vacuole increases in size due to secretion of biliary components (retained by the tight junctions) (Gautam *et al.* 1987). Couplets also exhibit normal pericanalicular localization of microfilaments and microtubules (Watanabe *et al.* 1991; Stone *et al.* 1994) and apical localization of the Golgi apparatus, lysosomes and autophagic vacuoles as occurs *in vivo* (Graf and Boyer

1990). A transmission electron micrograph of a hepatocyte couplet showing the canalicular structure and intact tight junctions is shown in Figure 2. Couplets also maintain uninterrupted gap-junctional intercellular communication (GJIC) which has been observed by measurement of both electrical coupling (Spray *et al.* 1986) and by the transfer of the low molecular weight fluorochrome lucifer yellow (Guppy *et al.* 1994).

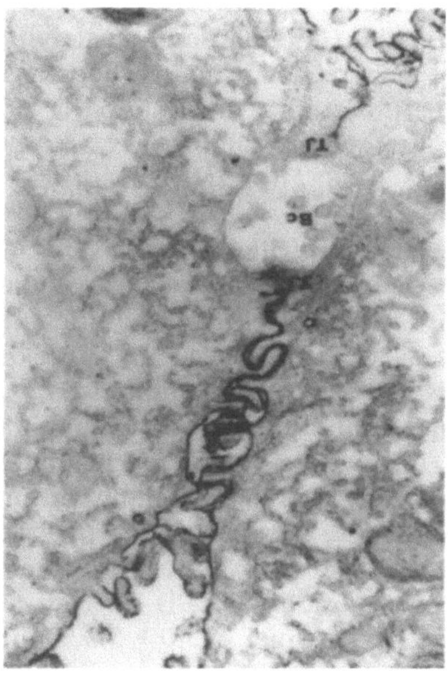

Figure 2: Transmission electron micrograph of a rat hepatocyte couplet following 4h in culture. Bc indicates the bile canaliculus, TJ indicates the tight junctions labelled with lanthanum.

Disruption of Gap Junctional Intercellular Communication (GJIC)

Disruption of GJIC has been associated with the loss of normal growth control and is also potentially an important deliberate process in the avoidance of the spread of cell toxicity (Lowenstein and Rose 1992). The connexins provide the protein subunit of the connexons which are major structural components of the gap junctions. There is substantial evidence that the connexins can play an important role in tumour suppression (see Chipman 1995) therefore highlighting the potential importance of the ability of various non-genotoxic carcinogens to inhibit their function (Trosko *et al.* 1990; Yamasaki 1991).

Hepatocyte GJIC is known to be inhibited in rodents by a number of liver-specific tumour promoting agents including phenobarbitone (Ruch and Klaunig 1988; Mesnil *et al.* 1993), and the peroxisome proliferator nafenopin (Leibold *et al.* 1994). In the analysis of the effect of PB, as an example, there are a number of potential mechanisms of subversion of GJIC. PB causes a transient centrilobular proliferation in rat liver followed by a reduction of the capacity of normal hepatocytes to proliferate in response to mitogenic stimuli (Jirtle *et al.* 1991). It is important to emphasise that inhibition of GJIC by PB in hepatocytes is one of several major events which have been associated with tumour promotion. These include modification of growth factor response (Jirtle *et al.*,1991), modification of growth inhibitory response to TGFbeta (Jirtle *et al.* 1994) and inhibition of apoptosis (Schulte-Hermann *et al.* 1990) in addition to inhibition of GJIC. Since these events are intimately related in the control of cell number, their co-incident deregulation is not surprising. The relative importance of inhibition of GJIC as a necessary component in tumour promotion is, however, not known. Also the significance of reduced connexin 32 (Cx32) immunoreactivity and gene expression as a possible marker of foci related to early hepatocarcinogenesis following certain tumour promoting protocols (including the use of PB in conjunction with the initiator N,N-diethylnitrosamine in rats) remains to be elucidated (Pitot *et al.* 1991; Beer *et al.* 1988; Fitzgerald *et al.* 1989; Neveu *et al.* 1990, 1994b).

There is a complex pattern of response to PB with respect to GJIC. The effect of PB on Cx32 protein and gene expression in the liver has been studied by Mesnil *et al.* (1988) and Neveu *et al.* (1994a). Cx32 mRNA was significantly reduced in rat liver at 3 days (but not at 60 days (Neveu *et al.* 1994) after PB exposure (0.1% in the drinking water). There was also no detectable quantitative change in Cx32 protein expression at 60 days although a diffuse pattern of immunoreactive Cx32 was associated with PB exposure (Neveu *et al.* 1994). The latter effect was postulated to be due to modification of the transport, assembly or turnover of the gap junction components. In cultured mouse hepatocytes *in vitro*, PB also has a rapid inhibitory effect on GJIC as assessed by dye-transfer (Ruch & Klaunig 1988). Exposure to non-cytotoxic concentrations of PB effected a decrease in GJIC which was significant at 60 min and maximal (25.0 ± 5.3 % inhibition) at 4 h in short-term cultures of rat hepatocyte couplets (Guppy *et al.* 1994). This effect was reversible and independent of protein synthesis. Both in mouse hepatocyte cultures (Ruch & Klaunig 1988) and in rat hepatocyte couplets (Guppy *et al.* 1994), the relatively non-specific P450 inhibitor SKF525A inhibited the PB-induced decrease in dye-coupling.

These findings do not necessarily however implicate a metabolite of PB in the effect on GJIC. Indeed the major metabolite *p*-hydroxy-PB was not effective in rat couplets (Guppy *et al.* 1994). Leibold and Schwarz (1993) have suggested that the short term effect of PB on GJIC may be mediated by reactive oxygen species possibly produced via the uncoupling of cytochrome P450. These workers showed that the antioxidants vitamin E and superoxide dismutase prevented the rapid loss of GJIC by PB. We have also found that glutathione depletion exacerbated the inhibitory effect of PB and found evidence for the production of reactive oxygen species by PB in glutathione depleted rat hepatocyte couplets as indicated by the oxidation of the fluorescent indicator dichlorodihydrofluorescein (unpublished data). Interestingly, following 14 day and 28 day- pretreatment of rats with PB (0.1 % in drinking water) subsequently prepared hepatocyte couplets were not only functional but also refractory to the GJIC-inhibitory effects of PB in culture (Guppy *et al.* 1994). Whilst the latter might reflect the influence of enzyme induction it was also noted, via preliminary immunolocalisation studies, that Cx32 was lost from the inter-cellular localization (compared to controls; see

Figure 3) to give reduced immunodetection and aberrant location in perivenous hepatocyte couplets at 28 days. It was also interesting that there was however an apparent increase in the immunodetection of Cx26 at inter-cellular domains. This preliminary finding is being investigated further, however, since no such adaptive change was reported following PB administration by gavage (Krutovskikh et al. 1995).

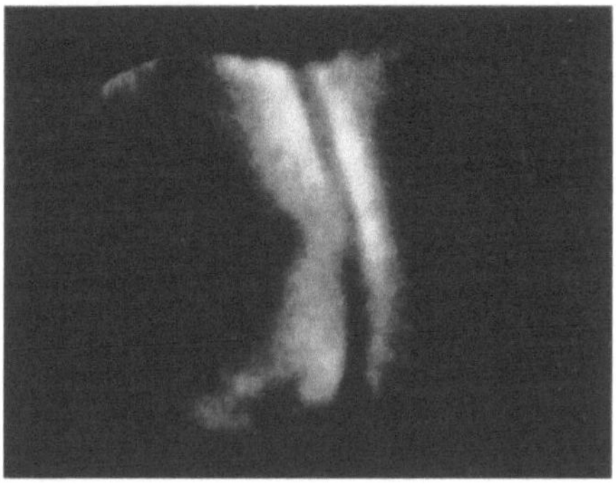

Figure 3: High resolution deconvolution micrograph of control rat hepatocyte couplet indicating fluorescent immunolocalisation of connexin 32 at inter-cell domains (connexin 32 antibody was kindly provided by D L Paul, Harvard Medical School, Boston).

In the same way that normal rodent hepatocytes exposed to PB appear to adapt to down-regulation of growth factor responses (see above) they may also adapt to overcome inhibition of GJIC. If it wese possible to see how "initiated" cells respond and adapt, this would contribute to the testing of the "resistant hepatocyte" model in tumour promotion (see Tsuda *et al.*, 1980).

Canalicular function and its disruption

The effect of toxic agents on hepatic canalicular primary bile formation is difficult to study *in vivo* due to the inaccessibility of the bile canaliculus. Canalicular function can be assessed in hepatocyte couplets by monitoring their ability to secrete fluorescent markers such as cholyl lysyl fluorescein (CLF) (Wilton *et al.* 1993; Stone *et al.* 1994). The study of the potential biliary excretion of xenobiotics in this paired-cell system is limited by the available detection systems for canalicular accumulated material. While fluorescence is readily monitored by microscopy (Figure 4) and by flow cytometry (Figure 5; Lankester *et al.* 1994b), radiolabelled drug analyses are less applicable.

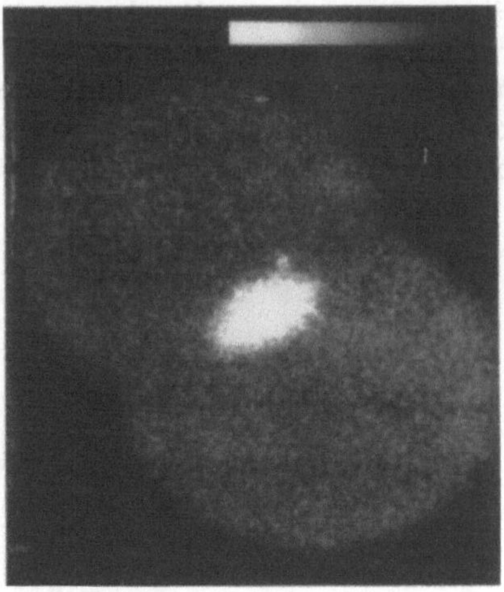

Figure 4: Fluorescence confocal microscopic image of rat hepatocyte couplet showing the canalicular accumulation of cholyl lysyl fluorescein (CLF). Following 4h culture, CLF (1mM) was added for 15 min.

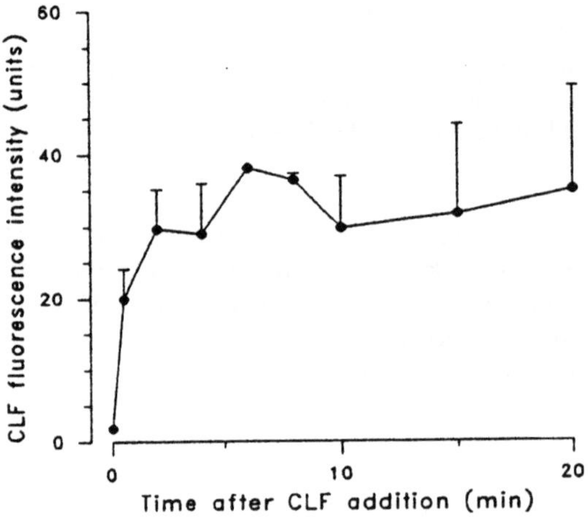

Figure 5: Accumulation of CLF (1μM) by 'viable' (propidium iodide-excluding) rat hepatocyte couplets measured by flow cytometry

These studies indicate the potential of hepatocyte couplets to model biliary excretion. In addition, however, interference by agents of specific transport systems can also be identified. Takeguchi *et al.* (1993) showed that the canalicular accumulation of the indicators Fura 2 and BCECF (which fluoresce in the presence of Ca^{2+} and H^{+} respectively) was inhibited by agents such as verapamil, vinblastine and daunomycin indicating the role of the relatively non-specific P-glycoprotein transporter (multidrug resistance pump) in their secretion.

If we are to understand the basis of cholestasis as produced by substances that interfere with specific transport systems, it is important that the transport processes for both normal constituents and for the offending chemicals are characterised. Lithocholate is a cholestatic bile salt whose mechanism of action is unclear. Lithocholate (tagged with a lysyl-fluorescein moiety) is transported across the hepatocyte to the apical membrane by an apparent transcytotic mechanism in contrast to the apparent cytoplasmic route taken by fluorescent analogues of cholate and chenodeoxycholate at physiological concentrations (Wilton *et al.* 1994). The accumulation of the fluorescent-labelled analogue of lithocholate into the canalicular vacuole of rat hepatocyte couplets was inhibited by brefeldin A (an inhibitor of the formation of transport vesicles) and by colchicine (which blocks tubulin assembly) suggesting the importance of microtubules and a transcytotic route. Interestingly, the glycine conjugate of lithocholate produced a marked and rapid redistribution of annexin II to the apical cell membrane via a perinuclear location. This was not seen with glycine conjugates of cholate or chenodeoxycholate and suggests an apical membrane fusion event and exocytosis which may be linked to vesicular movement towards the canalicular vacuole. Whether or not the cholestatic effect of lithocholate is related to the transcytotic route of this agent is not known, however the process of vesicular transport is clearly a possible target in hepatobiliary toxicity. Mallory bodies formed from altered intermediate filaments via disrupted microtubules have been associated with alcoholic liver disease, and reduction of polymerized tubulin by alcohol appears to contribute to disruption of bile formation (Lieber, 1984). This may be mediated by acetaldehyde (Kawarahara *et al.* 1989).

We have also studied other changes in hepatocyte couplets associated with cholestasis. Menadione (2-methyl-1,4-naphthoquinone) is a hepatotoxic quinone which also induces cholestasis (Akerboom *et al,.* 1988). In line with the latter, the ability of rat hepatocyte couplets to accumulate CLF into canalicular vacuoles was severely inhibited by menadione at concentrations (10-30 μM for 15 min) which were not cytotoxic or able to deplete cellular concentrations of ATP (Wilton *et al.*, 1993; Stone *et al.*, 1994). The effect was greater in perivenous compared to periportal hepatocyte couplets (Wilton *et al.* 1993). Although raised intracellular calcium mediated by A23187 (5-30 μM, 15 min.) was able to reduce canalicular vacuole accumulation of CLF and to reduce canalicular vacuole volume, calcium did not appear to be the only mediator of the effects of menadione. This was evidenced by the greater effect of menadione than thapsigargin (which depletes endoplasmic reticulum calcium stores) and by the fact that nickel chloride (which blocks receptor-mediated extracellular calcium influx) did not infuence the effect of menadione (Stone *et al.* 1994). There was however an association with glutathione depletion (Figure 6).

The rise in intracellular calcium produced by the calcium ionophore A23187 and a pathophysiological concentration of vasopressin appears to be important in increased paracellular permeability via changes in the tight junction and reduced bile flow (Lowe *et al.* 1988; Kan & Coleman 1988; Nathanson *et al.* 1990). These effects appear to be independent of effects on protein kinase C although activation of the latter can also reduce bile formation by a number of possible mechanisms (see Nathanson and Boyer 1991).

Menadione is both an arylating agent and is also able to undergo redox cycling and production of superoxide radical (see Thor *et al.* 1988). In comparison of the effects of a predominantly arylating quinone (*p*-benzoquinone) and a predominantly redox-cycling quinone (2,3-dimethoxy-1,4-naphthoquinone) (Thor *et al.* 1988; Gant *et al.* 1988) it was shown that the former (arylating) agent was the more effective at inhibition of canalicular function in rat hepatocyte couplets

(Stone *et al.* 1995). The oxidants H_2O_2 and *t*-butyl hydroperoxide also cause cholestasis in the perfused rat liver. This appeared to be due to an effect on the tight junctions and paracellular permeability since the bile-to-perfusate ratio of sucrose was increased (Ballatori and Truong 1989).

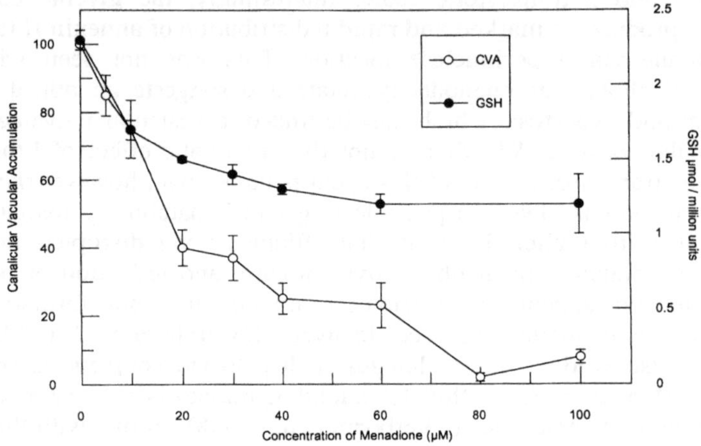

Figure 6: Inhibition of canalicular accumulation of CLF and depletion of glutathione by menadione

The peri-canalicular localisation of F-actin microfilaments has also been studied in relation to cholestasis. Menadione (30 µM and above) reduced the fluorescence of phalloidin-FITC-labelled F-actin in both the total and peri-canalicular cytoskeleton (Stone *et al.* 1994). Mirabelli *et al.* (1988) also showed that menadione formed large molecular weight aggregates from F-actin. Thibault *et al.* (1992) treated rat hepatocyte couplets with taurolithocholate or phalloidin for a relatively long period of 2h. In this study, an increase in peri-canalicular F-actin was seen in relation to total couplet fluorescence intensity. Under these conditions, depolymerisation of actin may be affected or an adaptive stress response (Yi *et al.* 1987) may be occurring. Other studies have shown that disruption of the canalicular cytoskeleton elements, eg by phalloidin (Naramoto *et al.* 1991), cytochalasin B and colchicine (Phillips *et al.* 1983; Kawahara *et al.* 1989) and cyclosporin A (Roman and Coleman 1994) has been associated with cholestasis. Also cytochalasin D, through causing microfilaments to condense, disrupts the structure of tight junctions in intestinal cells (Madara *et al.* 1986). Cholestasis induced by raised intracellular calcium (see above) is likely also to involve microfilaments since microinjection of calcium into hepatocyte triplets caused contraction of canalicular vacuoles (Watanabe *et al.* 1991) which is thought to involve F-actin (Watanabe *et al.* 1985). Cytoskeletal disruption therefore appears to be an important component and may contribute to some mechanisms of cholestasis.

This overview has highlighted the various facets of polarised function in hepatocytes which are sometimes overlooked in studies of hepatic toxicity.

neighbouring cells but may also be important in carcinogenesis. Interference with the maintenance or function of the biliary pole can also lead to sublethal hepatotoxicity, manifested as cholestasis.

Acknowledgements

Various aspects of the authors' work discussed herein were supported by the Wellcome Trust, Unilever and SmithKline Beecham.

References

Akerboom T, Bultmann T and Helmet S (1988) Inhibition of biliary taurocholate excretion during menadione metabolism in perfused rat liver. *Arch Biochem Biophys* **263**,10-18.

Azri S, Gandolfi AJ and Brendel K (1990). Precision-cut liver slices: an "*in* vitro" system for profiling poential hepatotoxicants . In *Vitro Toxicol* **3** 309 320.

Ballatori N and Truong AT (1989) Altered hepatic junctional permeability, bile acid excretion and glutathione efflux during oxidant challenge. *J Pharmacol Exp Ther* **251** 1069-1075.

Beer DG, Neveu MJ, Paul DL, Rapp UR and Pitot HC (1988) Expression of the c-raf proto oncogene, g-glutamyltranspeptidase, and gap junction protein in rat liver neoplasms. *Cancer Res* **48** 1610-1617.

Chipman JK (1995) Connexin genes as tumour suppressor genes? "Toxicology Elsewhere" *Human Exptl Toxicol* **14** 539-540.

Feldmann G (1989) The cytoskeleton of the hepatocyte. *J Hepatol* **8** 380-386.

Fitzgerald DJ, Mesnil M, Oyamada M, Tsuda H, Ho N and Yamasaki M (1989) Changes in gap junction protein (connexin 32) gene expression during rat liver carcinogenesis. *J Cell Biochem* **41** 97-102.

Gant TW, Ramakrishna Rao DN, Mason RP and Cohen GM (1988) Redox cycling and sulphydryl arylation; their relative importance in the mechanism of quinone cytotoxicity to isolated hepatocytes. *Chem Biol Interac* **65** 157-173.

Gautam A, Ng O-C and Boyer JL (1987) Isolated rat hepatocyte couplets in short-term culture: structural characteristics and plasma membrane reorganisation. *Hepatology* 7 216-223.

Gonzalez FJ and Gelboin HV (1994) Role of human cytochromes P450 in the metabolic activation of chemical carcinogens and toxins. *Drug Metab Revs* **26** 165-183.

Graf J, Gautam A and Boyer JL (1984) Isolated rat hepatocyte couplets: a primary secretory unit for electrophysiological studies of bile secretory function. *Proc Natl Acad Sci* USA **81** 6516-6520.

Graf J and Boyer JL (1990) The use of isolated rat hepatocyte couplets in hepatobiliary physiology. *J Hepatol* **10** 387-394.

Guppy MJ, Wilton JC, Sharma R, Coleman R and Chipman JK (1994) Modulation of phenobaritone-induced loss of gap junctional intercellular communication in hepatocyte couplets. *Carcinogenesis* **15** 1917-1921.

Handler JS (1989) Overview of epithelial polarity. *Annu Rev Physiol* **51** 729-740.

Jirtle RL, Handkins GR, Reisenbichler H and Boyer IJ (1994) Regulation of mannose 6-phosphate/insulin-like growth factor II receptors and transforming factors b during liver tumour promotion with phenobarbital. *Carcinogenesis* **15** 1473-1478.

Jirtle RL, Meyer SA and Brockenbrough JS (1991) Liver tumour promoter phenobarbital: a biphasic modulator of hepatocyte proliferation. In: Chemically induced Cell proliferation: Implications for Risk Assessment. Wiley-Liss pp209-216.

Jongen WMF, Fitzgerald DJ, Asamoto M, Piccoli C, Slaga TJ, Gros D, Takeichi M and Yamasaki H (1991) Regulation of connexin 43-mediated gap junctional intercellular communication by Ca++ in mouse epidermal cells is controlled by E-cadherin. *J Cell Biol* **114** 545-555.

Kan KS and Coleman R (1988) The calcium ionophore A23187 increases the tight-junctional permeability in rat liver. *Biochem J* **256** 1039-1041.

Kawahara H, Marceau N and French SW (1989) Effects of agents which rearrange the cytoskeleton *in vitro* on the structure and function of hepatocyte canaliculi. *Lab nvest* **60** 692-704.

Krutovskikh V.A., Mesnil M., Mazzoleni G and Yamasaki H. (1995) Inhibition of rat liver gap junction intercellular communication by tumor-promoting agents *in vivo.* Lab. Invest. 72 571-577

Lankester DJ, Chipman JK, Coleman R and Chenery RJ (1994a) Modelling of biliary excretion by metabolism and accumulation studies in hepatocyte couplets. *Br J Clin Pharmacol* **38** 170-171.

Lankester DJ, Paul F, Chenery R, Coleman R and Chipman JK (1994b) Flow cytometric analysis of biliary excretion of cholyl-lysyl-fluorescein in rat hepatocyte couplets. Abstract B15. 14th European Drug Metabolism Workshop.

Leibold E and Schwarz LR (1993) Inhibition of intercellular communication in rat hepatocytes by phenobarbital, 1,1,1-trichloro-2,2-bis (p-chlorophenyl) ethane (DDT) and hexachlorocyclohexane (lindane): modification by antioxidants and inhibitors of cyclooxygenase. *Carcinogenesis* **14** 2377-2382.

Leibold E, Greim H and Schwarz LR (1994) Inhibition of intercellular communication of rat hepatocytes by nafenopin: involvement of protein kinase C. *Carcinogenesis* **15** 1265-1269.

Lieber CS (1984) Alcohol and the liver: 1984 update *Hepatology* **4** 1243-1260.

Lowe PJ, Miyai K, Steinbach JH and Hardison WGM (1988) Hormonal regulation of hepatocyte tight junctional permeabilty. *Am J Physiol* **255** G454-G461.

Lowenstein WR and Rowe B (1992) The cell-cell channel in the control of growth. Seminars *Cell Biol* **3** 59-79.

Madara JL, Barenberg D and Carlson S (1986) Effects of cytochalasin D on occluding junctions of intestinal absorptive cells: Further evidence that the cytoskeleton may influence paracellular permeability and junctional charge selectivity. *J Cell Biol* **102** 2125-2136.

Mesnil M, Fitzgerald DJ and Yamasaki H (1988) Phenobarbital specifically reduces gap junction protein mRNA level in rat liver. *Mol Carcinog* **1** 79-81.

Mesnil M, Piccoli C and Yamasaki H (1993) An improved long-term culture of rat hepatocytes to detect liver tumour-promoting agents: results with phenobarbital. *Eur J Pharmacol* **24** 59-66.

Mirabelli F, Salis A, Marinoni V, Finardi G, Bellomo G, Thor H and Orrenius S (1988) Menadione-induced bleb formation in hepatocytes is associated with the oxidation of thiol groups in actin. *Arch Biochem Biophys* **264** 261-269.

Naramoto A, Ohno S, Kiyoshi F, Itoh N, Nakazawa K, Nakamo M and Shigematsu H (1991) Ultrastructural studies in hepatocyte cytoskeletons of phalloidin-treated rats by quick-freezing and deep-etching method. *Hepatology* **13** 222-229.

Nathanson MH, Gautam A and Boyer JL (1990) Effects of vasopressin and phorbol dibutyrate on bile secretion in isolated rat hepatocyte couplets (Abstract). *Gastroenterology* **98** A615.

Nathanson MH and Boyer JL (1991) Mechanisms and regulation of bile secretion. *Hepatology* **14** 551-569.

Neveu MJ, Babcock KL, Hertzberg EL, Paul DL, Nicholson BJ and Pitot HC (1994a) Colocalized alterations in connexin 32 and cytochrome P450 IIB1/2 by phenobarbital and related liver tumour promoters. *Cancer Res* **54**, 3145-3152

Neveu MJ, Hully JR, Babcock KL, Hertzberg EL, Nicholson BJ, Paul DL and Pitot HC (1994b) Multiple mechanisms are responsible for altered expression of gap junction genes during oncogenesis in rat liver. *J Cell Science* **107** 83-95.

Neveu MJ, Hully JR, Paul DL and Pitot HC (1990) Reversible alteration in the expression of the gap junctional protein connexin 32 during tumour promotion in rat liver and its role during cell proliferation. *Cancer Commun* **2** 21-31.

Oshio C and Phillips MJ (1981) Contractility of bile canaliculi: implications for liver function. *Science* **212** 1041-1042.

Paine AJ (1990) The maintenance of cytochrome P450 in rat hepatocyte culture; some applications of liver cell cultures to the study of drug metabolism, toxicity and the induction of the P450 system. *Chem-Biol Interact* **74** 1-31.

Phillips MJ, Oshio C, Miyiari M and Smith CR (1983) Intrahepatic cholestasis as a canalicular motility disorder. Evidence using cytochalasin. *Lab Invest* **48**, 205-211

Pitot HC, Dragan Y, Sargent and Xu Y-H (1991) Biochemical markers associated with the stages of promotion and progression during hepatocarcinogenesis in the rat. *Env Health Perspect* **93** 181-189.

Roman ID and Coleman R (1994) Disruption of canalicular function in isolated rat hepatocyte couplets caused by cyclosporin A. *Biochemical Pharmacology* **48**, 2181-2188

Ruch RJ and Klaunig JE (1988) Kinetics of phenobarbital inhibition of intercellular communication in mouse hepatocytes. *Cancer Res* **48** 2519-2523.

Saez JC, Conner JA, Spray DC and Bennett MVL (1989) Hepatocyte gap junctions are permeable to a second messenger, inositol 1,4,5-triphosphate and to calcium ions. *Proc Natl Acad Sci* USA **86** 2708-2712.

Schneeberger EE and Lynch RD (1992) Structure, function and regulation of cellular tight junctions. *Am Physiol Soc* L647-L661.

Schulte-Hermann R, Timmermann-Trosiener I, Barthel G and Bursch W (1990) DNA synthesis apoptosis and phenotypic expression as determinants of growth of altered foci in rat liver during phenobarbital promotion. *Cancer Res* **50** 5127-5135.

Seglen PO (1976) Preparation of isolated rat liver cells. *Methods Cell Biol* XIII 29-83.

Spray DC, Ginzberg RD, Morales EA, Gatmaitan Z and Arias IM (1986) Electrophysiological properties of gap junctions between dissociated pairs of rat hepatocytes. *J Cell Biol* **103** 135-144.

Stone V, Coleman R and Chipman JK (1995) Comparison of the effects of redox cycling and arylating quinones on hepatobiliary function and glutathione homeostasis in rat hepatocyte couplets. *Toxicol Apppl Pharmacol.* In Press.

Stone V, Johnson GD, Wilton JC, Coleman R and Chipman JK (1994) Effect of oxidative stress and disruption of Ca++ homeostasis on hepatocyte canalicular function *in vitro. Biochem Pharmacol* **47** 625-632.

Takeguchi N, Ichimura K, Koike M, Matsui W, Kashiwagura T and Kawahara K (1993) Inhibition of the mutlidrug efflux pump in isolated hepatocyte couplets by immunosuppressants FK506 and cyclosporine. *Transplantation* **55** 646-650.

Thibault N, Claude JR and Ballet F (1992) Actin filament alteration as a potential marker for cholestasis: a study in isolated rat hepatocyte couplets. *Toxicology* **73** 269-279.

Thor H, Mirabelli F, Salis A, Cohen GM, Bellomo G and Orrenius S (1988) Alterations in hepatocyte cytoskeleton caused by redox cycling and alkylating quinones. *Arch Biochem Biophys* **266** 397-407.

Trosko JE, Chang CC, Madhukar BV and Oh SY (1990). Modulators of gap junction function: the scientific basis of epigenetic toxicology. *In Vitro Toxicol* **3** 9-26.

Tsuda H, Lee G and Farber E (1980) Induction of resistant hepatocytes as a new principle for a possible short term test for carcinogens. *Cancer Res* **40** 1157-1164.

Watanabe N, Tsukada N, Smith CR, Edwards V and Phillips MJ (1991) Permeabolised hepatocyte couplets. Adenosine triphosphate-dependent bile canalicular contractions and a circumferential pericanalicular microfilament belt demonstrated. *Lab Invest* **65** 203-213.

Watanabe S, Smith CR and Phillips MJ (1985) Coordination of the contractile activity of bile canaliculi: evidence from calcium microinjection of triplet hepatocytes. *Lab Invest* **53** 275-279.

Wilton JC, Matthews GM, Burgoyne RD, Mills CO, Chipman JK and Coleman R (1994) Fluorescent choleretic and cholestatic bile salts take different paths across the hepatocyte: transcytosis of glycolithocholate leads to an extensive redistribution of annexin II. *J Cell Biol* **127** 401-410.

Wilton JC, Chipman JK, Lawson CJ, Strain AJ and Coleman R (1993) Periportal and perivenous-enriched hepatocyte couplets: differences in canalicular activity and in response to oviduthe stress. *Biochem J* **292** 773-779.

Yamasaki H (1991) Aberrant expression and function of gap junctions during carcinogenesis. *Environ Hlth Perspect* **93** 191-197.

Yi W, Zhao Y and Chou I-N (1987) Paraquat-induced cytoskeletal injury in cultured cells. *Toxicol Appl Pharmacol* **91** 96-106.

Cytokines in human lung fibrosis

Yves Martinet, Olivier Menard, Pierre Vaillant, Jean-Michel Vignaud, and Nadine Martinet
INSERM U 14, C.O. n° 10, F-54511 Nancy-Vandoeuvre, and Fédération Médico-Chirurgicale de Pneumologie, CHU de Nancy-Brabois, F-54511 Nancy-Vandoeuvre, Tel : (33) 83 15 34 00, Fax : (33) 83 15 35 41

Summary

Fibrosis is a pathological process characterized by the replacement of normal tissue by mesenchymal cells and the extracellular matrix produced by these cells.

The sequence of events leading to fibrosis of an organ involves the subsequent processes of injury with inflammation and disruption of the normal tissue architecture, followed by tissue repair with accumulation of mesenchymal cells in the area of derangement. The same sequence of events occurs in wound healing with normal granulation tissue and scar formation, but, while normal scar formation is very localized and transient, in contrast, in fibrosis, the repair process is exaggerated and usually widespread and can be chronic.

Inflammatory cells (mainly mononuclear phagocytes), platelets, endothelial cells, and type II pneumocytes play a direct and indirect role in tissue injury and repair. The evaluation of three human fibrotic lung diseases, two diffuse [idiopathic pulmonary fibrosis (IPF), and the adult respiratory distress syndrome (ARDS)], and one focal (tumor stroma in lung cancer), has shown that several cytokines participate to the local injury and inflammatory reaction [interleukin-1 (IL-1), interleukin-8 (IL-8), monocyte chemotactic protein-1 (MCP-1), tumor necrosis factor-α (TNF-α)], while other cytokines are involved in tissue repair and fibrosis [platelet-derived growth factor (PDGF), insulin-like growth factor-1 (IGF-1), transforming growth factor-ß (TGF-ß), and basic-fibroblast growth factor (b-FGF)].

A better understanding of the cytokines and cytokine networks involved in lung fibrosis leads to the possibility of new therapeutic approaches.

Abbreviations : AMDGF: alveolar macrophage-derived growth factor; ARDS: adult respiratory distress syndrome; BAL: bronchoalveolar lavage; b-FGF: basic-fibroblast growth factor; EGF: epidermal growth factor; ELAM-2: endothelium-leucocyte adherance molecule-2; G-CSF: granulocyte-colony stimulating factor; GM-CSF: granulocyte macrophage-colony stimulating factor; ICAM-1: intercellular adherance molecule-1; IGF-1: insulin-like growth factor-1; IL: interleukin; IL-1-ra: IL-1-receptor antagonist; IPF: idiopathic pulmonary fibrosis; MCP-1: monocyte chemotactic protein-1; NSCLC : non-small cell lung cancer; PDECGF: platelet-derived endothelial cell growth factor; PDGF: platelet-derived growth factor; PGE_2: prostaglandin E_2; PMN: polymorphonuclear neutrophil; SMC: smooth muscle cell; TGF: transforming growth factor; TNF-α: tumor necrosis factor-α.

Cytokines (*kytos*, cell ; *kinesis*, movement) are cell-produced and -released proteic mediators modifying cellular biological behaviour. Each cytokine exerts its activity by binding specific receptor(s) on target cells, and, according to the types of cells producing one cytokine and/or expressing its receptor, the mode of activity can be autocrine, paracrine, and/or endocrine. Most cytokines are free in the local milieu but some can be cell-associated (and, then, be part of cell membrane proteins ; in this case, these cytokines can only be active through cell to cell contacts) or bound to the extracellular matrix. Thus, cytokines carry information between cells, and play a central role in the development of living organisms, and in most physiological processes, including inflammation and tissue repair.

The everyday increasing number of cytokines and the multiple, complex, and intricated interactions among them, the producing cells, and the target cells, lead to the concept of cytokine networks. In this respect, their global *in vivo* functions are not yet always fully understood.

Fibrosis and wound healing

(1) **Fibrosis** is a pathological process characterized by the replacement of normal tissue by mesenchymal cells and the extracellular matrix produced by these cells. The fibrotic process can occur in most organs : lung, liver, artery, ..., and leads to the alteration of the organ structure and, eventually, to its functional impairment. While the clinical result depends upon the organ involved, the general process leading to fibrosis is the same and thus, the observations made in the lung can be relevant to the pathogeny of, for example, liver fibrosis or atherosclerosis. The sequence of events leading to fibrosis of an organ involves the subsequent process of (1) injury with inflammation and disruption of the normal tissue architecture, followed by (2) tissue repair with accumulation of mesenchymal cells in the area of disruption and their local release of extracellular matrix (Martinet and Crystal 1987; Gauldie *et al* 1993; Rochester and Elias 1993). Once these two processes have taken place, fibrosis is inevitable, i.e., the normal tissue has been replaced by mesenchymal cells which deposit connective tissue in the local milieu. For most organs, the class of cells that produce the tough, fibrosis matrix called "scar" includes fibroblasts, myofibroblasts, smooth muscle cells (SMC), pericytes and undifferentiated mesenchymal cells. Mesenchymal cells synthesize and secrete connective molecules, including (to a varying degree depending on the mesenchymal cell) collagen, elastic fibers, fibronectin, and proteoglycans.

(2) The biologic phenomenon of mesenchymal cell accumulation and production of extra cellular matrix is the same that occurs in **wound healing** with normal granulation tissue and scar formation after, for example, a skin wound. In this sense, "fibrosis" is essentially a pathological process that utilizes normal biological mechanisms in an abnormal fashion. The only real difference is its extent. While normal scar formation is very localized and transient, in contrast, fibrosis is usually widespread within an organ, and is generally chronic. In normal healing, the parallel migration and replication of epithelial cells can lead to a progressive reepithelialization. In the lung, type II pneumocyte proliferation and differentiation into type I pneumocytes can result in the

restauration of normal alveoli. When the initial injury is limited, this process of repair can be efficient, and a functional recovery can be observed. In contrast, when mesenchymal cell accumulation is too explosive, as in fibrosis, a functional reepithelialization is not possible and, furthermore, a poor reepithelialization leads to an exuberant mesenchymal cell accumulation (Terzaghi *et al* 1978).

Human lung fibrosis

(1) **Lung fibrosis** is frequent and can result from known (asbestos inhalation, bleomycin toxicity, ...) or unknown causes and can be, according to the type of injury, acute (adult respiratory distress syndrome, acute silicosis, ...) or chronic (farmers' lung, collagen vascular diseases, ...), and focal (tumor stroma in lung cancer, radiation-induced fibrosis, ...) or diffuse (idiopathic pulmonary fibrosis, paraquat toxicity, ...) (Crystal *et al* 1984). Several animal models have been proposed to evaluate lung fibrosis. However, since in respect to cytokines, the observations made in humans are very detailed, we will focus on them. Two classical "models" of human pulmonary fibrosis have been thoroughly evaluated: idiopathic pulmonary fibrosis and the adult respiratory distress syndrome. The evaluation of the pathogeny of the fibrotic disorders of the lung has been markedly helped by the use of bronchoalveolar lavage (BAL). Another model, tumor stroma formation in lung cancer, allows the description of the importance of cytokine receptors in fibrosis. Other models have been evaluated in respect to the role of cytokines: coal workers' pneumoconiosis, scleroderma, Hermansky-Pudlak syndrome, ... (Vanhee *et al* 1994; Harrison *et al* 1994; Harmon *et al* 1994) with observations consistent with the concepts described in this paper.

(2) **Idiopathic pulmonary fibrosis** (IPF) is a chronic fibrotic lung disorder of unknown origin, characterized by a progressive interstitial fibrosis leading to respiratory failure with a progressive dyspnoea with dry cough, bilateral crackles, a restrictive functional impairment with low DLCO and reduced lung volumes, and bilateral diffuse reticulonodular opacities on chest X-ray. The evolution is progressive with worsening dyspnoea, and a 5 year mean survival after the onset of symptoms (Hance and Crystal 1983). The lungs are characterized by the coexistence of normal looking areas and fibrotic patches; as the disease progresses, fibrotic lesions become more important. Each lesion consists of three main pathological aspects: cuboidal metaplasia of alveolar epithelial cells with progressive elimination of type I pneumocytes, diffuse interstitial and intra-alveolar inflammation mainly characterized by the presence of mononuclear cells (macrophages, lymphocytes), and progressive thickening of the alveolar walls by accumulation of fibroblasts, SMC and their extracellular matrix.

(3) **The adult respiratory distress syndrome** (ARDS) is characterized by an acute lung failure following, within 24 - 72 h, a lung aggression of known or unknown cause, with severe hypoxaemia (despite mechanical ventilation with oxygen enriched air) due to a right/left shunt, often leading to death (Hanley and Repine 1994). In ARDS, the first cells to be altered are endothelial cells, due to the adhesion of platelets and polymorphonuclear neutrophiles (PMN) and to their release of active metabolites. This initial step is rapidly followed by an epithelial

lung cell injury with a parallel disruption of the alveolar basement membrane, leading to the rapid development of intra-alveolar oedema and fibrosis by migration of mesenchymal cells from the interstitium into the alveolar space. While death is frequent, in some cases, a recovery with partial regression of the fibrotic lesions can be observed.

(4) **Tumor stroma**, a perequisite for tumor growth, develops simultaneously with the proliferation of cancer cells in all cancers and specifically in non-small cell lung cancer (NSCLC). Tumor stroma is composed of different cell types including endothelial, mesenchymal, and inflammatory cells (lymphocytes, mononuclear phagocytes). In this respect, close similarities have been observed between tumor stroma generation and normal healing (Dvorak 1986).

Pathogeny of lung fibrosis

The sequence of events leading to tissue fibrosis involves an injury with an inflammatory reaction followed by an excessive repair process. Both reactions will be described separately while, in fact, they are highly intricate.

(1) Tissue injury and inflammation

This first stage can be widespread and acute as in ARDS, or localized to some scattered areas of the lung as in IPF. In this last case, no specific clinical symptom in relation to tissue injury and inflammation can be recorded.

The first cells to be injured are vascular endothelial cells and type I pneumocytes. **This injury** can lead to their death with the release of several biological mediators. Vascular and alveolar basal membranes are also injured and, in conjonction with cell injury, it results in an abnormal vascular permeability with cells (platelets, inflammatory cells) and active metabolites (growth factors, coagulation factors, fibrin, complement fractions, ...) leaking into the alveoli and the interstitium. Platelet aggregation leads to the local release of several potent cytokines [platelet-derived growth factor (PDGF), transforming growth factor-β (TFG-β), epidermal growth factor (EGF), TGF-α, platelet-derived endothelial cell growth factor (PDECGF), ...], these cytokines contributing to the inflammatory reaction. A progressive regeneration of endothelial cells is possible, starting from intact areas. However, a chronic cycle of injury and endothelial cell regeneration (repair) initiates progressive vessel dysfunction. In parallel, type I pneumocytes are very fragile cells that can be injured by the local presence of oxidants. Type II pneumocytes are more resistant and contribute to the repair process by dividing, migrating, and differentiating into type I pneumocytes. However, a chronic state of injury ends in a cuboidal metaplasia with the abnormal presence in the alveoli of type II pneumocytes and/or epithelial cells migrating from the terminal bronchioles.

Inevitably, the sequence of tissue injury is accompanied by a local **inflammation**. The inflammatory reaction following an aggression is comprised, first, within hours/days of PMN, and, second, within days/weeks of mononuclear phagocytes (blood monocytes locally transforming into tissue macrophages). This observation has been made in most tissues and, specifically, in the lung. When the local aggression persists, mononuclear phagocytes can accumulate in large numbers, as it is the case, for example, in pulmonary fibrosis. Inflammatory cells participate in processes of defenses such as phagocytosis and opsonisation of foreign bodies and cellular debris, and this accumulation of inflammatory cells soon becomes a dominant feature that accompanies all forms of injury. The importance of the inflammation in the development of fibrosis is two-fold. First, inflammatory cells amplify the disruption of the normal architecture by contributing to the local injury. Second, inflammatory cells direct, in part, the accumulation of the mesenchymal cells in the local milieu, and thus contribute to the development of the repair process and of fibrosis itself.

The inflammatory cells amplify the disruption of the tissue by several mechanisms. Inflammatory cells take up space, thus crowding the cells that comprise the tissue. More importantly, they release mediators, such as oxidants (H_2O_2, $O_2^{\cdot-}$, OH·) and proteases, that injure the local cells and extracellular components and cytokines, such as TNF-α (and IL-1), involved in tissue injury. While PMN contribute to tissue injury by releasing oxidant and proteases, mononuclear phagocytes contribute mainly by the release of cytokines.

The recruitment of PMN and macrophages depends upon the nature of the initial pathological process (infection, pneumoconioses, ...), the expression on endothelial cells of adhesion molecules [intercellular adherence molecule-1 (ICAM-1), endothelium-leucocyte adherence molecule-2 (ELAM-2)], the local presence of chemotactic factors [complement fractions (C_5a), ...] and the local release of cytokines chemotactic for PMN and blood monocytes (IL-8, MCP-1). Furthermore, PMN are able to cleave fibronectin with the release of a 120 kDa fragment chemotactic for blood monocytes.

Other inflammatory cells are involved: (1) eosinophils with a role closely similar to PMN'; (2) following an injury; lymphocytes accumulate after blood monocytes; they play an indirect role in the fibrotic process by releasing activating factors for macrophages and γ-interferon; (3) the role of mast cells is poorly understood; they can release cytokines such as TNF-α.

(α) **IPF** is characterized by an intense local inflammation with presence in the alveoli of PMN and alveolar macrophages (AM) and, in the interstitium, of mononuclear cells, mainly macrophages and lymphocytes.

In IPF, AM release a chemotactic activity for PMN. In this respect, **interleukin-8** (IL-8) is a 8.4 kDa cytokine released by different cells including mononuclear phagocytes, epithelial cells, endothelial cells, fibroblasts, ... IL-8 receptor is a 58 kDa surface receptor present in high numbers on PMN and to a much lower extent on lymphocytes and blood monocytes. IL-8 is a chemotactic and activating factor leading to PMN migration and the release of their granule content. In IPF, AM expression of IL-8 gene is increased and AM release of IL-8 is also increased in comparison to normal AM (Carré *et al* 1991), resulting in a high concentration of this cytokine in BAL fluid and a correlation between these levels and the number of PMN present in BAL fluid has been reported (Car *et al* 1994).

The local accumulation of macrophages can result from two mecanisms:

- recruitment from circulating blood monocytes: this biological event is likely to be important in view of the "monocyte-like" phenotype of these cells (Kiemle-Kallee *et al* 1991), and of the presence of chemotactic factor(s) for blood monocytes in the lavage fluid of these patients. In this respect, different cytokines can be involved. **Monocyte chemoattractant protein-1** (MCP-1) is a monomere of 8.7 kDa with a specific chemotactic activity for blood monocytes. In IPF, the gene coding for MCP-1 is specifically expressed by pulmonary epithelial cells, and also by mononuclear phagocytes, vascular endothelial cells and smooth muscle cells (Antoniades *et al* 1992). Furthermore, high levels of MCP-1 have been observed in the BAL fluid of IPF patients (Car *et al* 1994). Interestingly, PDGF and TGF-β, two cytokines involved in tissue repair and fibrosis (see below), are also chemotactic factors for blood monocytes, and could thus could play a role in monocyte recruitment.

- local proliferation: however, some other data suggest that AM and interstitial macrophages could also accumulate, although to a lower extent, due to their local replication (Bitterman *et al* 1984). Since **granulocyte macrophage-colony stimulating factor** (GM-CSF) is able to induce AM replication and since GM-CSF can be released by several cell types present in the lung (activated T-cells, epithelial cells, fibroblasts, ...), it is possible, but not yet demonstrated, that this cytokine could locally participate in the replication of these cells (Gauldie *et al* 1993, Vaillant *et al* 1993).

(β) **In ARDS**, the very initial phase is characterized by an acute intraalveolar oedema with the presence of an elevated number of platelets, the rapid accumulation and activation of PMN and, within a few hours, of mononuclear phagocytes. High levels of **IL-8** are present early in the lavage fluid of patients with ARDS while their blood levels are normal, leading to a concentration gradient of this chemotactic factor inducing PMN recruitment in the alveoli (Miller *et al* 1992 ; Rochester and Elias 1993). Furthermore, TNF-α and IL-1 have been shown to induce IL-8 production and release by endothelial cells and, thus, contribute to the recruitment of PMN (Strieter and Kunkel 1994).

By releasing oxidants, activated macrophages may add to the local cell injury, but this direct macrophage cytotoxic activity is limited. In contrast, macrophages (and blood monocytes) are involved, at least in septic ARDS, in lung cell injury by the release of tumor necrosis factor-α (TNF-α). **TNF-α/cachectin** is a mediator produced by different types of cells, mainly, in the lung, by mononuclear phagocytes and mast cells. TNF-α consist of three identical 17 kDa chains and it binds to two different receptors (p55, p75), and soluble TNF receptors have been described. TNF-α is a mediator of cachexia and a cytotoxic factor for certain tumor cell lines. It promotes PMN activation and adhesion to endothelial cells with a local cytotoxicity leading to an increased capillary permeability with fluid leakage and shock. High BAL and seric levels of TNF-α have been observed by different groups in patients with ARDS and they correlate with the prognosis (Hanley and Repine 1994). In animal models, the prevention of the release of TNF-α or of its binding to its receptors has led to the prevention, at least in part, of infectious ARDS. **Interleukin-1 α (IL-α) and IL-1β** are two 17.5 kDa cytokines, sharing little sequence homology, produced by multiple types of cells (mononuclear phagocytes, endothelial cells, fibroblasts, ...) and with two types of receptor, IL-1α binding with high affinity to type I receptor and IL-1β to type II. IL-1α is mainly cell associated, while IL-1β is released freely in the local milieu. The role of IL-1 is multiple with effect on inflammation, the immune system and different cell functions. Soluble receptors for IL-1 have been isolated as well as an IL-1 receptor antagonist (IL-1-ra) competing with IL-1 for its receptors resulting in the inhibition of IL-1 biological activity. High levels of BAL IL-1 have been observed in patients with ARDS (Jacobs *et al* 1989, Suter *et al* 1992) and are likely to be produced by AM (Jacobs *et al* 1989). IL-I activity is very similar to that of TNF-α.

(2) Tissue repair and fibrosis

Fibrosis is the result of three different mechanisms : (1) mesenchymal cell recruitment, (2) their exaggerated local replication, and (3) their production of extra cellular matrix. This process leads to the formation of a granulation tissue with mesenchymal cells and extra cellular matrix allowing the growth of new vessels. In normal repair, this granulation tissue undergoes a remodelling with possible reepithelialization and functional recovery. An excessive development of the granulation tissue leads to fibrosis.

Cytokines direct the accumulation of mesenchymal cells by recruiting mesenchymal cells to the site of inflammation and by signaling mesenchymal cells to replicate and produce extra cellular matrix. Mononuclear phagocytes, particularly tissue macrophages, but also alveolar epithelial cells and, at least in acute lung injury, vascular endothelial cells, and platelets can release several of these cytokines.

The recruitment by **migration of mesenchymal cells** is strongly suggested in lung fibrotic diseases characterized by the accumulation of mesenchymal cells in areas where these cells are normally absent as this is the case in IPF and ARDS. In ARDS, SMC migrating through the gaps of the basal membranes into the alveoli can be observed. The recruitment of mesenchymal cells is induced by the release of cytokines with chemotactic activity inducing them to migrate along a concentration gradient to the point of maximum concentration of the chemoattractant. Transforming growth factor-β (TGF-β), and platelet-derived growth factor (PDGF) are two potent chemotactic factors for these cells.

The accumulation by **replication of mesenchymal cells** in the local milieu is directly related to the presence of cytokines signaling them to proliferate. The most direct evidence for this process is the observation, in fibrotic disorders, of mesenchymal cells undergoing mitosis at a rate significantly greater than that observed in normal tissues. Normal tissue mesenchymal cell divisions are regulated in order to allow for tissue structure maintenance. In this aspect, most mesenchymal cells, in normal tissues, are resting cells. The initiation and regulation of fibroblast division result from the combined action of several mediators called growth factors (most of them being cytokines), interacting with specific receptors on the target cell. In order to accomplish a cellular division, quiescent fibroblasts need to receive two signals. The first signal allows the cell to leave its quiescent stage and to enter the mitotic cycle ; this signal can be delivered by several cytokines called "competence" growth factors [e.g. basic-fibroblast growth factor (b-FGF), PDGF]. If the cell is then exposed to a second signal given by a "progression" growth factor [e.g., insulin-like growth factor-1 (IGF-1)] the fibroblast will proceed through its complete cycle and divide. In the absence of "competence" growth factor, optimal doses of "progression" growth factor will not induce fibroblast division. **Platelet-derived growth factor** (PDGF) is a 31 kDa glycoprotein composed of two chains (A and B) sharing 60 % amino-acid sequence homology, and linked by disulfide bonds. The gene coding for PDGF B-chain is the *c-sis* proto-oncogene.

PDGF is a chemotactic and a "competence" growth factor for fibroblasts and SMC. PDGF interacts on target cells with a receptor composed of two subunits α and β. The α subunit binds PDGF A-and B-chains, while the β subunit only binds PDGF B-chain with high affinity. **Insulin-like growth factor-1** (IGF-1) is a 25 kDa peptide with "progression" activity leading to the division of a fibroblast already primed by a "competence" growth factor such as PDGF.

The **local production of extracellular matrix** by these accumulating mesenchymal cells leads to the progressive deposition of collagens, elastin, fibronectin, and proteoglycans. The production of extracellular matrix by each mesenchymal cell is increased in fibrotic lung diseases, with a frequent shift in collagen type production. Some cytokines, such as **transforming growth factor-β** (TGF-β), are directly involved in the control of the production of these molecules. TGF-β (several subtypes of TGF-β are known, TGF-β1, being the most prominent) is an homodimer (one disulfide bond) of 25 kDa able to inhibit several functions of inflammatory cells (macrophages, lymphocytes) and promote tissue repair. TGF-β is released as an inactive molecule and must be locally activated by acid treatment or enzymatic action. TGF-β induces mesenchymal cell migration, and increases their production of extracellular matrix (collagen I, III and V, fibronectin, proteoglycans, ...), while inhibiting the production of connective tissue proteases and stimulating the production of protease inhibitors. According to the experimental conditions, TGF-β inhibits mesenchymal cell replication or stimulates the production of PDGF by these cells resulting in an autocrine simulation of these cell replication.

The local release of cytokines is not the only mechanism leading to mesenchymal cell accumulation. Indeed, lung mesenchymal cells from these patients have a higher replication rate than mesenchymal cells from normal lungs in the presence of the same amounts of cytokines (Jordana *et al* 1988, Raghu *et al* 1988, Chen *et al* 1992). The reason for this difference is not clearly understood; one possible explanation is the progressive selection, due to the presence of high amounts of cytokines, of mesenchymal cells with high replication rate.

Reepithelialization is poorly known but likely to be important. In fibrosis this process is not successful because of the exuberant development of the granulation tissue leading to fibrosis. But, on the other hand, the absence of reepithelialization allows the development of a disproportionate granulation tissue (Terzaghi *et al* 1978). Indeed, normal bronchial cells can limit granulation tissue formation by their release of prostaglandin E_2 (PGE_2) a factor inhibiting fibroblast replication. Several cytokines, including TGF-α, epidermal growth factor (EGF), and IGF-1, could play a role in the reepithelialization process by inducing epithelial cell migration and replication.

(α) **In IPF**, the production of cytokines by different cell types has been well demonstrated:

- alveolar macrophages: IPF AM recovered by BAL spontaneously produce four times more PDGF than normal AM (Martinet *et al* 1987), and their spontaneous expression of PDGF B-chain gene (and also, but to a much lower level, of PDGF A-chain gene) is markedly increased in comparison to normal AM (Nagaoka *et al* 1990). In parallel, AM spontaneous production of IGF-1 [previously described as alveolar macrophage-derived growth factor (AMDGF) (Bitterman *et al* 1983)] is elevated in IPF, while normal AM do not spontaneously release IGF-1 (Rom *et al* 1988); furthermore, in patients with systemic sclerosis and lung involvement (a disease close to IPF), IGF-1 is present in high amounts in BAL fluid and lavage fluid is able to induce fibroblast replication (Harrison *et al* 1994);

- Lung interstitial macrophages: evaluation, by immunohistochemistry and *in situ* hybridization, demonstrates an increased production, in comparison to normal lung, of PDGF by these cells (Vignaud *et al* 1991). Furthermore, the number of interstitial macrophages producing PDGF is increased in normal looking areas of the lung suggesting that the local production of PDGF precedes fibrosis ;

- Hyperplastic type II pneumocytes: PDGF B-chain gene expression and PDGF production by alveolar epithelial cells (hyperplastic type II pneumocytes) is present in IPF (while normal pneumocytes do not express PDGF genes) (Vignaud *et al* 1991). These cells also express the gene coding for TGF-β (Khalil *et al* 1991);

- Lymphocytes: the amount of γ-interferon present in BAL fluid is increased in only < 1/5 patients, suggesting a limited role for this cytokine (Robinson and Rose 1990); γ-interferon is a 20 kDa cytokine produced mainly by lymphocytes and macrophages and playing a central role in the cellular immune response : γ-interferon can activate macrophages and suppress fibroblast replication.

- Finally, the presence of extracellular TGF-β, detected by immunohistochemistry, and the expression of collagen genes, detected by *in situ* hybridization, have been observed in the same fibrotic areas of the lung suggesting the direct role of this cytokine in the production of ECM by mesenchymal cells (Broekelmann *et al* 1991). However, TGF-β production by AM is decreased and TGF-β BAL levels are normal (Vanhee *et al* 1994);

(β) **In ARDS**, several cells are involved in the local production of cytokines related to tissue repair and fibrosis. Most data come from the observation of PDGF and b-FGF:

- PDGF is present, within a few hours/days after clinical onset, in BAL fluid in ARDS, and is likely to result from its local release by platelets. Furthermore, endothelial cells can produce PDGF, and they could participate in the early release of PDGF in ARDS. Interestingly, a 14 kDa biologically active PDGF has been described in BAL fluid of these patients (Snyder *et al* 1991). Obviously, other cytokines present in platelets α-granules are also likely to be locally released: TGF-β, but also TGF-α, and platelet-derived endothelial cell growth factor (PDECGF);

- After several days, AM of these patients release several times more PDGF than AM from control patients, and these cells are likely to participate in the production of the 14 kDa PDGF (Vaillant *et al* 1994).

- Within a few days after lung injury, BAL fluid from patients with ARDS is characterized by the presence of a chemotactic activity for endothelial cells due, in part, to b-FGF while AM express the gene coding for b-FGF (Henke *et al* 1991, Henke *et al* 1993). b-FGF is a 18 kDa cytokine able to induce fibroblast ("competence" factor) and endothelial cell replication. In this respect, this cytokine can play a direct role in the accumulation of two main cell types forming the granulation tissue.

(γ) **In tumor stroma** formation in non small cell lung cancer, cancer cells of 1/3 tumors express the genes coding for PDGF A- and/or B-chains, and cancer cells of 1/2 tumors present PDGF receptor α and/or β subunit on their surface. Thus, only 1/6 tumors are characterized by the simultaneous coexpression of PDGF and its receptor. More interestingly, in respect to this paper, in almost all tumors, tumor associated macrophages express PDGF A- and/or B-chain genes and mesenchymal cells, as well as endothelial cells, express PDGF receptor α and β subunits (but not PDGF A-and B-chain genes), and this process is markedly increased in the periphery of the tumor where new tumor stroma is formed by accumulation of mesenchymal cells. A sequence of events leading to tumor stroma formation has been suggested with first recruitment of tumor associated macrophages and then recruitment and local proliferation of mesenchymal and endothelial cells (Vignaud *et al* 1994).

Conclusion

Several other cytokines could be involved in lung fibrosis. For example, members of the epidermal growth factor (EGF) family (EGF, TGF-α) could be important, since EGF presence is increased in the airways of children dying of severe bronchopulmonary dysplasia with lung fibrosis, and epithelial cell expression of the gene coding for TGF-α is increased in bleomycin-induced lung fibrosis in animals. Furthermore, human TGF-α transgenic mice are characterized by the progressive development of lung fibrosis (Korfhagen *et al* 1994). GM-CSF and granulocyte-CSF (G-CSF) could also be involved, in respect to their production by different cells present in the lung and to their biological activities on phagocytic cells and mesenchymal cells (Vaillant *et al* 1993).

If these diseases are characterized by the increased local production of several cytokines, they could also result, in part, from the lack of production of cytokines controlling (down-regulating) the inflammatory reaction and/or the repair process. No direct proof of this hypothesis has been published.

The local production of a cytokine does not necessary implies its role; for example:

- TGF-β needs to be activated,
- PDGF can be inactivated by binding with α_2-macroglobulin,
- the specific receptor of a cytokine may not be expressed,
- specific inhibitors can be present: IL-1-ra, soluble TNF-α receptor, ...
- some cytokines are bound to the extracellular matrix
- the effects of one cytokine can be altered, and even reversed, by the presence of another cytokine.

New therapeutic approaches may be undertaken to suppress the role of these cytokines in lung fibrosis (Hunninghake and Kalica 1995). This includes the use of: cytokine inhibitors (IL-1-ra, α_2-macroglobulin, ...), products inhibiting cytokine and/or cytokine receptor synthesis (antisens mRNA and oligonucleotides), cytokine antagonists (PGE_2, soluble receptors, ...), antibodies directed against cytokines or their receptors, antagonist cytokines (IL-4, ...), drugs inhibiting the interaction of a cytokine with its receptor (suramin, ...), ... However, two concepts are important: (1) wound healing is a fundamental physiological process and its suppression could be detrimental to the body; thus only a focal, instead of a systemic, treatment should be undertaken; (2) as suggested by this paper, the suppression of the biological activity of only one cytokine is likely to be poorly efficient in view of the high number of involved cytokines.

Acknowledgments: we want to thank Nathalie Thomas for her patient and talented typing. This work was supported by a grant from the Comités Lorrains de la Ligue Française Contre le Cancer.

References

Antoniades HN, Neville-Golden J, Galanopoulos T, Kradin RL, Valente AJ and Graves DT (1992) Expression of monocyte chemoattractant protein-1 mRNA in human idiopathic pulmonary fibrosis. Proc Natl Acad Sci USA 89, 5371-5375

Bitterman PB, Adelberg S and Crystal RG (1983) Mechanisms of pulmonary fibrosis. Spontaneous release of alveolar macrophage-derived growth factor in interstitial lung disorders. J Clin Invest 72, 1801-1813

Bitterman PB, Saltzman LE, Adelberg S, Ferrans VJ and Crystal RG (1984) Alveolar macrophage replication: one mechanism for the expansion of the mononuclear phagocyte population in the chronically inflamed lung. J Clin Invest 74, 460-469

Brœkelmann TJ, Limper AH, Colby TV and Mac Donald JA (1991) Transforming growth factor ß1 is present at sites of extracellular matrix gene expression in human pulmonary fibrosis. Proc Natl Acad Sci USA 88, 6642-6646

Car BD, Meloni F, Luisetti M, Semenzato G, Gialdroni-Grassi G and Walz A (1994) Elevated IL-8 and MCP-1 in the bronchoalveolar lavage fluid of patients with idiopathic pulmonary fibrosis and pulmonary sarcoidosis. Am J Respir Crit Care Med 149, 655-659

Carré PC, Mortenson RL, King TE, Noble PW, Sable CL and Riches DWH (1991) Increased expression of the interleukin-8 gene by alveolar macrophages in idiopathic pulmonary fibrosis. J Clin Invest 88, 1802-1810

Chen B, Polunovsky V, White J, Blazar B, Nakleh R, Jessurun J, Peterson M and Bitterman P (1992) Mesenchymal cells isolated after acute lung injury manifest an enhanced proliferative phenotype. J Clin Invest 90, 1778-1785

Crystal RG, Bitterman PB, Rennard SI, Hance AJ and Keogh BA (1984) Interstitial lung diseases of unknown cause. N Engl J Med 310, 154-166 and 235-244

Dvorak HF (1986) Tumors: wounds that do not heal. Similarities between tumor stroma generation and wound healing. N Engl J Med 315, 1650-1659

Gauldie J, Jordana M and Cox G (1993) Cytokines and pulmonary fibrosis. Thorax 48, 931-935

Hance AJ and Crystal RG (1983) Idiopathic pulmonary fibrosis. In: Recent advances in respiratory medicine. Flenley DC, Petty TL. (eds) Edinburgh, Churchill Livingstone (Vol 3) 249-287

Hanley ME and Repine JE (1994) Pathogenic aspects of the adult repiratory distress syndrome. Sem Respir Crit Care Med 15, 260-270

Harmon KR, Witkop CJ, White JG, King RA, Peterson M, Moore D, Tashjian J, Marinelli WA and Bitterman PB (1994) Pathogenesis of pulmonary fibrosis: platelet-derived growth factor precedes structural alterations in the Hermansky-Pudlak syndrome. J Lab Clin Med 123, 617-627

Harrison NK, Cambrey AD, Myers AR, Southcott AM, Black CM, Dubois RM, Laurent GJ and Mac Anulty RJ (1994) Insulin-like growth factor-I is partially responsible for fibroblast proliferation induced by bronchoalveolar lavage fluid from patients with systemic sclerosis. Clin Sci 86, 141-148

Henke C, Fiegel V, Peterson M, Wick M, Knighton D, MacCarthy J and Bitterman P (1991) Identification and partial characterization of angiogenesis bioactivity in the lower respiratory tract after acute lung injury. J Clin Invest 88, 1386-1395

Henke C, Marineili W, Jessurun J, Fox J, Harms D, Peterson M, Chiang L and Doran P (1993) Macrophage production of basic fibroblast growth factor in the fibroproliferative disorder of alveolar fibrosis after lung injury. Am J Pathol 143, 1189-1199

Hunninghake GW and Kalica AR (1995) Approaches to the treatment of pulmonary fibrosis. Am J Respir Crit Care Med 151, 915-918

Jacobs RF, Tabor DR, Burks AW and Campbell GD (1989) Elevated interleukin-1 release by human alveolar macrophages during the adult respiratory distress syndrom. Am Rev Respir Dis 140, 1686-1692

Jordana M, Schulman J, Mac Sharry C, Irving LB, Newhouse MT, Jordana G and Gauldie J (1988) Heterogenous proliferative characteristics of human adult lung fibroblast lines and clonally derived fibroblasts from control and fibrotic tissues. Am Rev Respir Dis 137, 579-584

Khalil N, O'Connor RN, Unruh HW, Warren PW, Flanders KC, Kemp A, Bereznay OH and Greenberg AH (1991) Increased production and immunohistochemical localization of transforming growth factor-ß in idiopathic pulmonary fibrosis. Am J Respir Cell Mol Biol 5, 155-162

Kiemle-Kallee J, Kreipe H, Radzun HJ, Parwaresch MR, Auerswald U, Magnussen H and Barth J (1991) Alveolar macrophages in idiopathic pulmonary fibrosis display a more monocyte-like immuno-phenotype and an increased release of free oxygen medicals. Eur Respir J 4, 400-406

Korfhagen TR, Swantz RJ, Wert SE, Mac Carty JM, Kerlakian CB, Glasser SW and Whitsett JA (1994) Respiratory epithelial cell expression of human transforming growth factor-α induces lung fibrosis in transgenic mice. J Clin Invest 93, 1691-1699

Martinet Y and Crystal RG (1987) Pathogenesis of tissue fibrosis: inflammatory cell modulation of mesenchymal cell accumulation. In: Tissue fibrosis: immune cells and mediators. Revillard, JP, Wierzbicki N. (eds) Suresnes, Fondation franco-allemande 3, 13-31

Martinet Y, Rom WN, Grotendorst GR, Martin GR and Crystal RG (1987) Exaggerated spontaneous release of platelet-derived growth factor by alveolar macrophages from patients with idiopathic pulmonary fibrosis. N Engl J Med 31, 202-209

Miller EJ, Cohen AB, Griffith D, Maunder J, Martin TR, Weiner-Kronisch JP, Stickerling M, Christophers E, Matthay MA and Nagao S (1992) Elevated levels of NAP-1/Interleukin-8 are present in the airspaces of patients with the adult respiratory distress syndrome and are associated with increased mortality. Am Rev Respir Dis 146, 427-432

Nagaoka I, Trapnell BC and Crystal RG (1990) Upregulation of platelet-derived growth factor-A and -B gene expression in individuals with idiopathic pulmonary fibrosis. J Clin Invest 85, 2025-2027

Raghu G, Chen Y, Rusch V and Rabnovitch PS (1988) Differential proliferation of fibroblasts cultured from normal and fibrotic human lungs. Am Rev Respir Dis 138, 703-704

Robinson BWS and Rose AH 1990 Pulmonary γ interferon production in patients with fibrosing alveolitis. Thorax 45, 105-108

Rochester CL and Elias JA (1993) Cytokines and cytokine networking in the pathogenesis of interstitial and fibrotic lung disorders.Semin Respir Med 14, 389-416

Rom WN, Basset P, Fells GA, Nukiwa T, Trapnell BC and Crystal RG (1988) Alveolar macrophages release an insulin-like growth factor 1-type molecule. J Clin Invest 82, 1685-1693

Snyder LS, Hertz MI, Peterson MS, Harmon KR, Marinelli WA, Henke CA, Greenheck JR, Chen B and Bitterman PB (1991) Acute lung injury. Pathogenesis of intraalveolar fibrosis. J Clin Invest 88, 663-673

Strieter RM and Kunkel SL (1994) Acute lung injury: the role of cytokines in the elicitation of neutrophils. J Invest Med 42, 640-651

Suter PM, Suter S, Girardin E, Roux-Lombard P, Graw GE and Dayer JM (1992) High bronchoalveolar levels of tumor necrosis factor and its inhibitors, interleukin-1, interferon, and elastase, in patients with adult respiratory distress syndrome after trauma, shock, or sepsis. Am Rev Respir Dis 145, 1016-1022

Terzaghi M, Nettesheim P and Williams ML (1978) Repopulation of denuded tracheal grafts with normal, preneoplasic, and neoplasic epithelial cell populations. Cancer Res 38,4546-4553

Vaillant P, Muller V, Martinet Y and Martinet N (1993) Human granulocyte- and granulocyte-macrophage-colony stimulating factors are chemotactic and «competence» growth factors for human mesenchymal cells. Biochem Biophys Res Commun 192, 879-885

Vaillant P, Peterson M, Skubitz K and Bitterman PB (1994) Alveolar macrophages produce a 14 kDa platelet-derived growth factor related peptide: potential role in intraalveolar fibrosis following acute lung injury. Am J Respir Crit Care Med 149, A 706

Vanhee D, Grosset P, Wallaert B, Voisin C and Tonnel AB (1994) Mechanisms of fibrosis in coal workers'pneumoconiosis. Increased production of platelet derived growth factor, insulin-like growth factor type I, and transforming growth factor ß and relationship to disease severity. Am J Respir Crit Care Med 150, 1049-1055

Vignaud JM, Allam M, Martinet N, Pech M, Plenat F and Martinet Y (1991) Presence of platelet-derived growth factor in normal and fibrotic lung is specifically associated with interstitial macrophages, while both interstitial macrophages and alveolar epithelial cells express the c-*sis* proto-oncogene. Am J Respir Cell Mol Biol 5, 531-538

Vignaud JM, Marie B, Klein N, Plénat F, Pech M, Borrelly J, Martinet N, Duprez A and Martinet Y (1994) The role of platelet-derived growth factor production by tumor-associated macrophages in tumor stroma formation in lung cancer. Cancer Res 54, 5455-5463

The Glio-Vascular Toxicity of *m*-Dinitro-Benzene and Related Agents: Modulation of Toxicity by Neuronal Activation

D. E. Ray, J. L. Holton, T. Lister and C. C. Nolan.
Medical Research Council Toxicology Unit, Hodgkin Building, Lancaster Road, Leicester, LE1 9HN, United Kingdom.

Since function and dysfunction in the central nervous system are only expressed as a result of the activity of neurones, it is easy to overlook the contribution of other cell types. The specialised brain vasculature, ventricular system, and glial network make a vital contribution to brain function, and also represent additional targets for neurotoxic agents. Given the complex physiological relationships between these different cell types one should expect to find correspondingly complex toxicological relationships also. We set out to examine the extent to which functional changes in one cell type (neurones) could influence the susceptibility of another cell type (astrocytes) to toxic agents.

Astrocytes have many functions, but a predominant theme in the mature nervous system is the maintenance of the local extracellular micro-environment. Thus astrocytic ion pumps and channels help to maintain potassium and sodium equilibrium. Astrocytic glutamine synthetase allows removal of neuronal ammonia. Most P-450 mediated drug metabolism and most glutathione conjugation is in non-neuronal cells (Lowndes *et al.* 1995). Neurotransmitter uptake systems are also expressed, and it is possible that glutamate released from glia may modulate neuronal excitability (Memmerick & Zorumski, 1994; Parpura *et al.* 1994). Production of glial-cell line derived neurotrophic factor also appears to contribute to survival of a number of neuronal cell types (Lindsay, 1995). In the case of the human neurotoxicant MPTP, glial uptake and metabolic activation via the MAO-A system may contribute to the regionally selective neuronal toxicity seen in the substantia nigra and related areas (Kitahama *et al.* 1994).

A number of chemical agents, most notably the industrial intermediate *m*-dinitrobenzene, the antibiotic and radiosensitiser metronidazole, and the sugar analogue α-chlorohydrin produce astrocytic and oligodendrocytic death with a characteristic topography within the nuclei of the brainstem (Cavanagh, 1993). In addition the first two agents produce perivascular oedema and micro-haemorrhage. We have previously shown that dinitrobenzene lesions within the central auditory pathway can be both increased and decreased in severity by alterations in the level of sound stimulation within the physiological range (Ray *et al.* 1992). Although a decrease in auditory stimulation would have acted primarily upon the neurones of the auditory pathway, the decrease in lesion severity was seen in the glial cells of the auditory nuclei. This led us to propose a toxicological link between neurones and glia via either local neuronal function or local metabolic demand.

We have tested the generality of this observation by conducting similar experiments with metronidazole-induced lesions and with 6-amino-nicotinamide. We also extended our observations beyond the auditory system by producing either a general decrease in activity by anaesthesia, or an increase in motor activity by inducing tremor. These conditions were tested against dinitrobenzene intoxication.

In addition we examined the reciprocal communication back from glia to neurones, since the glio-vascular lesions produced by dinitrobenzene can produce a degree of secondary neuronal death and also ataxia. We attempted to quantify the relationship between glial lesion severity and neuronal function by monitoring auditory responses in the inferior colliculus.

Materials and Methods

Animals: Male F344 rats supplied by Harlan-Olac and weighing between 240 and 260g were used for all experiments.

Toxic agents: Metronidazole (Sigma chemical company) was given to 10 rats intraperitoneally in dimethylsulphoxide (400 mg/ml) twice daily at a dose of 400 mg/kg (800 mg/kg/day) for 10 to 12 days. *m*-Dinitrobenzene (Sigma) was given to rats intraperitoneally in dimethylsulphoxide (10 mg/ml) as three doses at 0, 4 & 24 hours. Total dinitrobenzene dose ranged from 24 to 30 mg/kg. 6-Aminonicotinamide (Sigma) was given intraperitoneally in dimethylsulphoxide to 4 rats as a single dose of 8 mg/kg and to 4 others at 9 mg/kg.

For control experiments separate groups of rats were given 1 ml/kg dimethylsulphoxide solvent alone plus the various experimental manipulations used to produce the three models of altered brain functional state.

Models of altered brain functional state: We established three test conditions: unilateral reduction in auditory input, motor hyperactivity, and anaesthesia. As the period of pathogenesis of the lesions is 24 to 48 hours after the first dose of dinitrobenzene, these conditions should to be sustained for most of this time.

A permanent reduction in auditory input was achieved by rupture of the left tympanic membrane (under recovery anaesthesia). A seven day recovery period was allowed before the onset of dinitrobenzene, metronidazole or 6-aminonicotinamide dosing, and animals were examined regularly for signs of local infection (which was never seen).

Motor hyperactivity was induced by oral administration of the type I pyrethroid bifenthrin (a gift from Zeneca plc) in arachis oil. Two oral doses of bifenthrin were given, the first (20 mg/kg) 6 hours after the first dose of dinitrobenzene, and the second (10 mg/kg) 18 hours later. Dinitrobenzene was given as three doses of either 8, 9, or 10 mg/kg.

Prolonged anaesthesia was produced either by giving a single intraperitoneal injection of urethane at 1·4 g/kg, or by continued inhalation of 1·5% isoflurane in 2:1 nitrous oxide:oxygen. Urethane was given 24·5 hours after the first dose of dinitrobenzene, or isoflurane from 24·5 to 36 hours.

Rats from all experiments were perfuse-fixed via the ascending aorta with formalin acetic acid fixative under terminal anaesthesia 24 hours after the last dose of dinitrobenzene or metronidazole. All procedures were conducted under the ethical guidelines of the U.K. Home Office regulations for experimentation on laboratory animals.

Monitoring of neuronal and vascular function after dinitrobenzene: Eleven rats were implanted with pairs of platinum recording electrodes in the central nuclei of the left and right inferior colliculi 7-10 days before dosing. Auditory evoked potentials were measured under light isoflurane anaesthesia, using 100 dB SPL 40 kHz tone bursts of 1 msec duration delivered via an ear phone. Inferior collicular blood flow was also measured, in the conscious state, by hydrogen polarography using the same electrodes (Ray, 1982).

Measurements were made both two days before, and daily up to 8 days after administration of DNB. Rats were killed for histological evaluation 8 days after the first dose of dinitrobenzene.

Assessment of lesion severity: Step serial sections of brain and cervical cord were stained with haematoxylin and eosin. For all experimental groups and in all brain areas lesions were scored using a uniform scale of 0 to 3, with 0 representing no damage, and 3 representing a severe lesion involving the entire nucleus. Histological assessment was carried out independently of the other measurements by one of us (J.L.H.) who was unaware of nature of the individual animal treatments.

Statistical analysis: For assessing the effect of tremor and dose on lesion severity across brain areas, multivariate analysis of variance was used. For assessing the effect of anaesthesia on qualitative lesion severity Wilcoxon's unpaired test was used, and for assessing the asymmetry of lesions in animals with tympanic rupture the paired test was used. Student's paired "t" test was used for assessing the effect on blood flow and auditory evoked response slope.

Results

Neurotoxic effects of metronidazole and 6-aminonicotinamide: Metronidazole produced gliovascular lesions in the brainstem closely resembling those previously described by Rogulja *et al.* (1973). These also closely resembled those produced by dinitrobenzene (Philbert *et al.* 1987) both in nature and topography. Animals showed a rapid onset of motor incoordination over the last 1 to 3 days of dosing, which was accompanied by a weight loss of 10 to 14%. Daily single intraperitoneal doses of 400 mg/kg for 21 days (n=4), or 20 daily oral doses of 800 mg/kg in Tween-80 detergent (n=8) however failed to produce any lesions or motor effects.

The lesions produced by 6-aminonicotinamide resembled those previously described by Cavanagh (1993).

Effect of reduction in auditory input on glial lesions: Previous work in this laboratory has established that tympanic rupture produces a 40 to 50 dB increase in auditory threshold (Ray *et al.* 1992). Both metronidazole (n=10) and dinitrobenzene (n=11) lesions developed asymmetrically in the cochlear nuclei and inferior colliculi, although not elsewhere. Mean lesion severity scores (out of 3) and the significance of left-right asymmetries are summarised in Table 1.

Table 1: Effect of left tympanic membrane perforation (perf.) on metronidazole and dinitrobenzene mean lesion severities:

Metronidazole:		Dinitrobenzene:	
Cochlear nn. perf. : intact	Inferior Colliculi perf. : intact	Cochlear nn. perf. : intact	Inferior Colliculi perf. : intact
0.5 : 1.1	1.2 : 2.1	0.0 : 0.8	0.2 : 1.2
($p<0.02$)	($p<0.01$)	($p<0.002$)	($p<0.0005$)

In marked contrast the lesions produced by 6-aminonicotinamide were symmetrical in all 8 rats, having a mean severity of 2·1 in both cochlear nuclei.

Effect of reduction in auditory input on neuronal function and blood flow: The amplitudes of the auditory responses on the side of the intact pathway (left colliculus) were significantly depressed over 24 to 120 hours but returned to normal at 8 days. The individual peak decreases are shown in Fig. 1. The auditory threshold increased significantly by 12 to 15 dB at 24 hours on both sides, and on the unprotected left side the slope of the stimulus-response curve (linear over 80 to 120 dB stimulation) remained unchanged at 99±6.9% of the predose value, resulting in a decreased amplitude. However in the protected right colliculus the effect of the increased threshold was offset by a compensatory increase in slope to 132±11% of the predose value ($p<0·01$ relative to the left side).

The corresponding inferior collicular blood flows showed a large increase of from 1·0±0·04 ml/g/min to 12 hour values of 3·4±0·5 on the unprotected side, but a significantly smaller increase to 1·5±0·1 on the protected side. The time course was similar to that seen in our previous study of intact rats (Romero *et al.*, 1991), with flow returning to predose values after 72 hours.

Effect of anaesthesia: Anaesthesia sufficient to suppress the eye blink reflex was sustained from 0.5 until 12 - 18 hours post injection of urethane. A similar depth of anaesthesia was seen throughout the 12 hour period of administration of isoflurane, with rapid recovery on discontinuing the anaesthetic. Lesions were significantly reduced in severity in four brain areas in the urethane group and in two areas in the isofluorane group, as summarised in Table 2.

Table 2: Effect of anaesthesia with urethane or isoflurane on dinitrobenzene lesion severity:

	Cerebellar nuclei	Vestibular nuclei	Superior olives	Inferior colliculi	Red nuclei
Conscious (n=6):	2·5	2·5	2·5	1·7	0·8
Urethane (n=7):	1·3 (p=0·04)	0·6 (p=0·01)	0·1 (p=0·007)	0·1 (p=0·007)	0·0 (N.S.)
Isoflurane (n=6)	2·8 (N.S.)	2·7 (N.S.)	0·8 (p=0·02)	0·3 (p=0·01)	0·3 (N.S.)

Neither agent produced a significant reduction in lesion severity in the red nuclei, trigeminal motor nuclei or trapezoid bodies (mean lesion severity 0·8 for all three areas in the conscious group).

Effect of tremor: Bifenthrin produced a fine generalised tremor which began 12 hours after the first dose of dinitrobenzene, lasted until 42 hours, and was unaffected by dinitrobenzene. Bifenthrin alone produced no lesions or deaths, but 3/8 of the high dose DNB + bifenthrin rats died with severe ataxia within 48 hours. Mean lesion severities in the survivors are summarised in Table 3.

Table 3: Effect of bifenthrin-induced tremor on dinitrobenzene lesion severity:

DNB dose:	Cerebellar nn. control : tremor	Vestibular nn. control : tremor	Sup.Oliv. nn. control : tremor
3 x 8 mg/kg	0.0 : 0.4	0.0 : 0.4	0.0 : 0.0
3 x 9 mg/kg	1.4 : 2.8	1.2 : 2.8	1.6 : 2.0
3 x 10 mg/kg	2.5 : [3.0][a]	2.5 : [3.0][a]	2.5 : [2.0][a]
(n = 8 per group)		[a] = 5 survivors only	

Analysis of variance of lesion severity showed a significant dinitrobenzene dose effect ($p<0.001$), and a significant sensitising effect of tremor on lesion severity in the cerebellar nuclei ($p=0.03$) and vestibular nuclei ($p=0.02$), but not in the superior olives.

Correlations with blood flow and auditory response: Although in this experiment all rats were given the same dose of dinitrobenzene, lesion severity was variable, ranging from 0/3 to 2.5/3 in the inferior colliculi and 0/9 to 7/9 in all three auditory nuclei. The corresponding blood flow increases (intact pathway) ranged from 41 to 457 % and the fall in auditory responses ranged from 14 to 67 %. This variability allowed correlations to be calculated over the entire range of survivable lesion severities. The magnitude of the flow increase correlated with the severity of the subsequent lesion, linear regression giving a correlation coefficient of 0·676 ($p=0·022$), as shown in the first part of Fig. 1.

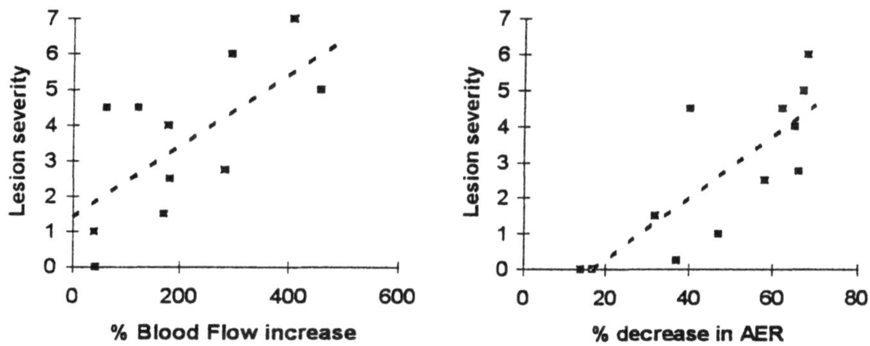

Fig.1 Correlations between individual inferior collicular flow increases at 12 hours and lesion severities, and between peak decreases in collicular auditory evoked response (AER) to 100 dB stimuli and lesion severities. Lesion severity for each animal was calculated as the sum of the lesion scores in the cochlear, superior olivary and inferior collicular nuclei on the unprotected sides.

The peak auditory deficit was also well correlated with the summed auditory pathway lesion, linear regression giving a correlation coefficient of 0·756 ($p=0·007$), as shown in the second part of Fig. 1. The auditory deficit was less well correlated with just the local lesion in the inferior colliculus ($r=0·627$, $p=0·039$ data not shown).

Discussion

The severity of both metronidazole and dinitrobenzene lesions can be decreased unilaterally by reduction in auditory input. The dinitrobenzene result is essentially the same as we found previously in animals with 24 hour survival (Ray *et al.* 1992), although the magnitude of the protection is slightly greater, possibly because the somewhat lesser severity of the lesions. This protection must be a specific function of decreased neuronal activity on glial cells, since the unilateral nature of the effect allows generalised changes to be discounted. This unilateral protection was reflected in all three measures: blood flow, auditory function, and lesion severity. In contrast the lesions produced by 6-aminonicotinamide were not altered by auditory deprivation, possibly because of its more widespread metabolic effects, or the lack of a major vascular component in its toxicity (Cavanagh, 1993).

In order to reproduce the metronidazole lesions described by Rogulja *et al.* (1973) we had substitute twice daily intraperitoneal for daily oral dosing. This may have been due to incomplete intestinal absorption by our animals. The nature of the lesions was however identical to that described by these authors.

Reduction in functional activity by anaesthesia produced a similar degree of protection against dinitrobenzene lesions as auditory deprivation. In this case the effect was bilateral and it is possible that systemic or pharmacokinetic changes might have contributed to the protection. However protection was produced by two very different anaesthetic agents, and was not uniform across brain areas, which would suggest a local action. The somewhat greater effect of urethane may relate to its longer duration of anaesthesia.

Conversely the induction of intense motor activity by pyrethroid intoxication increased the severity of dinitrobenzene lesions in the cerebellar motor and vestibular nuclei. Again this effect was bilateral and could have been due to a systemic action of the pyrethroid, such as an increase in blood pressure or catecholamine secretion (Cremer *et al.* 1982), but this is unlikely as the increase was significant only in motor nuclei. The tremor produced was typical of that seen after dosing with type I pyrethroids. This tremor results from activity generated in the brainstem (Forshaw and Ray, 1986), and is associated with increased glucose phosphorylation (Cremer *et al.* 1981).

None of the test conditions used were so severe as to produce neuropathological changes in themselves, and they changed only the severity and not the nature of the dinitrobenzene or metronidazole damage. Hence we conclude that the susceptibility of glia to such damage can be modulated by the level of local functional activity current in the brain at the time of poisoning.

As regards the reciprocal influence of the glio-vascular lesions on neuronal function, there were good correlations between acute neuronal dysfunction (as estimated by auditory response) and the glial lesion severity. Astrocytic and oligodendrocytic death would be expected to disrupt ionic homeostasis and neurotransmitter re-uptake, leading to loss of neuronal function. The poor recovery of the auditory response in some animals may have reflected neuronal death, but this was difficult to quantify in the presence of massive glial cell loss. A detailed analysis of these auditory response changes will be the subject of a subsequent paper.

The blood flow increases may have reflected a local reactive response to energy deprivation (Romero *et al.* 1995), and provide an *in vivo* measure that parallels both the time course and the severity of the glial lesion (Romero *et al.* 1991). The eight day lesions showed a marked resolution of œdema compared to previously studied one day lesions, although loss of glial cells was still a prominent feature. Although such indirect measures provided a good index of glial dysfunction during intoxication, we are currently evaluating antioxidant status and extracellular potassium homeostasis as more direct indices.

In conclusion we have demonstrated that both increases and decreases in activity of neuronal cells lead to corresponding changes in the susceptibility of glial cells to damage *in vivo*. This applies to both dinitrobenzene and metronidazole, but not to 6-aminonicotinamide damage. We have also shown that this glio-vascular damage in turn causes physiological impairment of neuronal function.

References

Cavanagh JB (1993) Selective vulnerability in acute energy deprivation syndromes. *Neuropathol Appl Neurobiol* **19:**461-470

Cremer JE, Ray DE, Sarna GS, Cunningham VJ (1981) A study of the kinetic behaviour of glucose based on simultaneous estimates of influx and phosphorylation in brain regions of rats in different physiological states *Brain Res* **221:**331-342

Cremer JE, Seville MP (1982) Comparative effects of two pyrethroids, deltamethrin and cismethrin, on plasma catecholamines and on blood glucose and lactate. *Toxicol Appl Pharmacol* **66:**124-133

Forshaw PJ, Ray DE (1986) The effects of two pyrethroids, cismethrin and deltamethrin, on skeletal muscle and the trigeminal reflex system in the rat. *Pestic Biochem Physiol* **25:**143-151

Kitahama K, Maeda T, Denney RM, Jouvet M (1994) Monoamine oxidase: distribution in the cat brain studied by enzyme- and immunohistochemistry: recent progress. *Prog in Neurobiol.* **42:**53-78

Lindsay RM (1995) Neurone saving schemes. *Nature* **373:**289-290

Lowndes HE, Philbert MA, Beiswanger M, Kauffman FC, Reuhl KR (1995) Xenobiotic metabolism in the brain as mechanistic bases for neurotoxicity. In Chang LW, Dyer RS (eds) Handbook of neurotoxicology. Marcel Dekker, New York, pp 1-28

Mennerick S, Zorumski CF (1994) Glial contributions to excitatory neurotransmission in cultured hippocampal cells. *Nature* **368:**59-62

Parpura V, Basarsky TA, Liu F, Jeftinja K, Jeftinja S, Haydon PG (1994) Glutamate-mediated astrocyte-neuron signalling *Nature* **369:**744-747

Philbert MA, Nolan CC, Cremer JE, Tucker D, Brown AW (1987) 1,3-Dinitrobenzene-induced encephalopathy in rats. *Neuropathol Appl Neurobiol* **13:**371-389

Ray DE (1982) Changes in brain blood flow associated with deltamethrin-induced choreoathetosis in the rat. *Exptl Brain Res* **45:**269-276

Ray DE, Brown AW, Cavanagh JB, Nolan CC, Richards HK, Wylie SP (1992) Functional/metabolic modulation of the brain stem lesions caused by 1,3-dinitrobenzene in the rat. *Neurotoxicology* 13:379-388
Rogulja PV, Kovac W, Schmid H (1973) Metronidazol-encephalopathies der ratte. *Acta Neuropathologica* 25:36-45
Romero I, Brown AW, Cavanagh JB, Nolan CC, Ray DE, Seville MP (1991) Vascular factors in the neurotoxic damage caused by 1,3-dinitrobenzene in the rat. *Neuropathol Appl Neurobiol* 17:495-508
Romero IA, Lister T, Richards HK, Seville MP, Wylie SP, Ray, D.E. (1995) Early metabolic changes during *m*-dinitrobenzene neurotoxicity and the possible role of oxidative stress *Free Radical Biol Med* 18:311-319

Mechanistical studies of the inhibition of intercellular communication by organochlorine compounds

L. Wärngård, Y. Bager, Y. Kato, K. Kenne and U.G. Ahlborg.
Institute of Environmental Medicine, Karolinska Institutet, Box 210, S-171 77 Stockholm, Sweden

Abstract

Many hydrocarbons are environmental pollutants that, due to their lipophilicity and chemical stability , accumulate in biological systems including milk and body fat. A number of investigations have demonstrated that many organochlorine compounds can act as tumour promoters *in vivo* and inhibit gap junctional intercellular communication between cells in culture. In the present study we have investigated the dioxin 2,3,7,8-tetrachlorodibenzo-p-dioxin (TCDD), different polychlorinated biphenyls, chlorinated paraffins and the pesticide endosulfan.

Using techniques of scrape loading dye/transfer and Western blot analysis the function, expression and phosphorylation of different connexins *in vitro* and *in vivo* were studied. The results show a good correlation between the ability to act as a tumour promoter and to interfere with gap junctional intercellular communication. All tested compounds inhibited the intercellular communication in a liver derived cell line (IAR 20). However, the results show that the time to inhibition varies between the different agents. Endosulfan and chlorinated paraffins inhibit the communication within one hour, whereas dioxin like substances need to expose the cells for 48 hours before the communication is affected.

Introduction

Gap junctions are located in the plasma membrane and consist of six subunits, named connexins, which form channels with adjacent cells. They permit intercellular exchange of ions and low molecular weight molecules. The expression of the different connexin genes is a specifically regulated process depending on cell type, tissue and species. Until now, at least 12 different connexins have been isolated and sequenced (reviewed by Willecke *et al.* 1991). The most studied connexins are cx 26, cx 32 and cx 43. It has been shown that several connexins can exist in the same tissue and it is also possible that different connexins can be expressed in the same cell (Nicholson et al.1987). Different connexins are also expressed differently during embryonal and postnatal maturation suggesting an important role during embryogenesis (Willecke and Traub, 1990; Dermietzel *et al.*1989) as well as during transformation of a normal cell to tumour cells. Inhibition of gap junctional communication is suggested to play an imported role in the carcinogenic process since many tumour promoters have been shown to inhibit gap junctional communication *in vitro*.

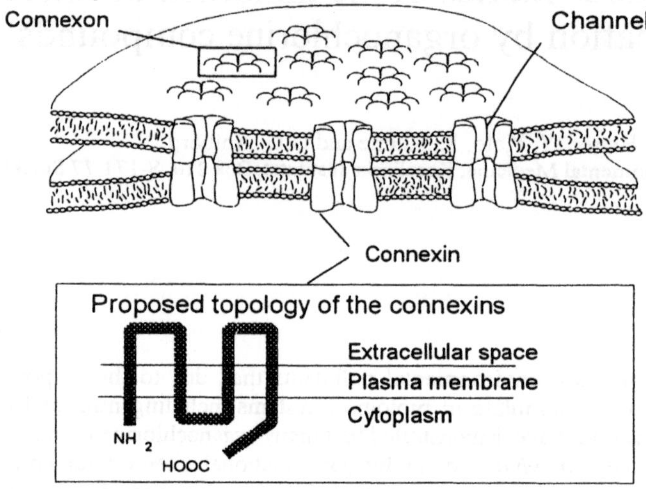

Figure 1. A schematic illustration of the gap junctions in the plasma membrane and the proposed topology of the protein subunit.

The hypothesis, on which *in vitro* short term assays measuring gap junctional communication is founded, is that an altered, initiated cell is isolated from the normal growth regulating properties exerted by surrounding, normal cells. If the communication is inhibited it is possible for initiated cells to proliferate and to express their altered phenotype (Figure 2). Since it is known that the loss of normal growth control is a characteristic tumour phenotype and that tumours and tumour cells deviate from normal cells regarding cellproliferation, differentiation and sensitivity to cytotoxic agents, it is considered that altered gap junctional communication may play an essential role in carcinogenesis. The altered phenotypes may be a mechanism whereby this cell and the subsequent clone of cells, constituting foci and nodules, can grow in an hostile environment. This growth advantage also results in an increased number of target cells, susceptible for further alterations towards malignancy (Yamasaki *et al* 1990).
The skin tumour promoter, 12-*O*-tetradecanoylphorbol-13-acetate (TPA), is a potent inhibitor of cell-cell communication in various cell lines. A proposed mechanism for this promoter is based on activation of protein kinase C. The elevated activity cause phosphorylation of connexin 43 (cx43) leading to loss of the gap junction function. The insecticide endosulfan is a broad-spectrum insecticide in agriculture. In a conducted oral two-stage initiation/promotion bioassay we have demonstrated that endosulfan cause a significant enhancement of altered hepatic foci in rats. Exposure of cells with endosulfan result in a rapid inhibition of cell communication (Fransson-Steen *et al.* 1992; Ruch *et al.* 1990).

The polychlorinated biphenyls (PCBs) and dioxins are groups of substances with a variety of chlorine substituents. Due to their persistence and stability they are found in the environment. Commercial mixtures of PCBs consist of a number of different congeners. The most toxic congeners have been suggested to mediate their biological effect via the

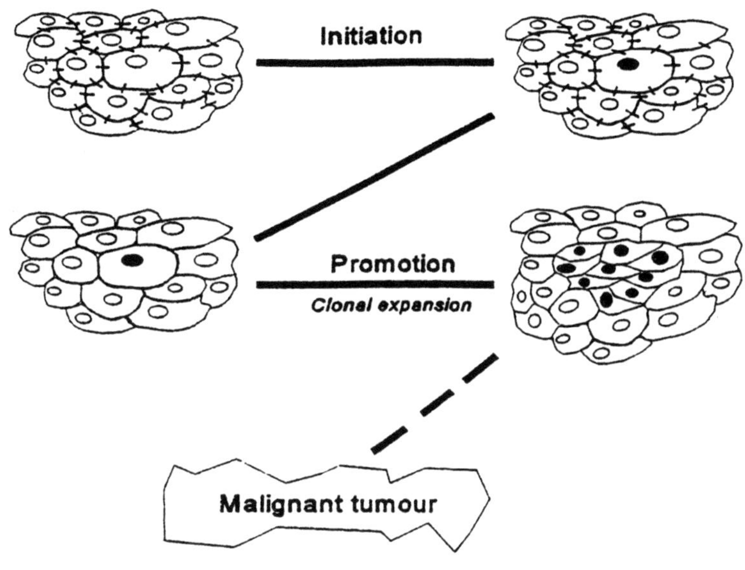

Figure 2. The current model for the role of gap junctional communication in tumour promotion.

same receptor as 2,3,7,8-tetrachlorodibenzo-*p*-dioxin (TCDD) (Poland *et al* 1991)

Chlorinated paraffins are a group of widely used industrial chemicals, which are extensively used as plasticisers, pressure additives in metal working fluids and as flame retardants. Due to the large amount used worldwide chlorinated paraffins have been detected in environmental samples like sediment and biota. Chlorinated paraffins is a complex mixtures with a large number of isomers. In the present investigation we have focused on one short and one intermediate carbon chain length chlorinated paraffins.

Using techniques of scrape loading dye/transfer and Western blot analysis the function, expression and phosphorylation of different connexins *in vitro* and *in vivo* were studied. The results shows that down regulation of gap junctional communication after treatment with organochlorine compounds is mediated via different mechanisms, a result which emphasizes the importance to understand the mechanisms for different group of chemicals.

Materials and methods

Chemicals: The PCBs used were from Firma S. Ehrenstrofer, Augsburg, Germany. TCDD (lot no MLB-682-49) was from CIL (Woburn, Mass. USA). Analytical grade of endosulfan (1,4,5,6,7,7-hexachloro-5-norbornene-2,3-dimethanol cyclic sulfite; α/β isomer ratio 70:30) was purchased from Dr Ehrenstrorfer GmbH (Augsburg, Germany). Commercial chlorinated paraffins were kindly provided by Professor Bo Jansson, Institute of Applied Environmental Research, Stockholm University, Sweden.

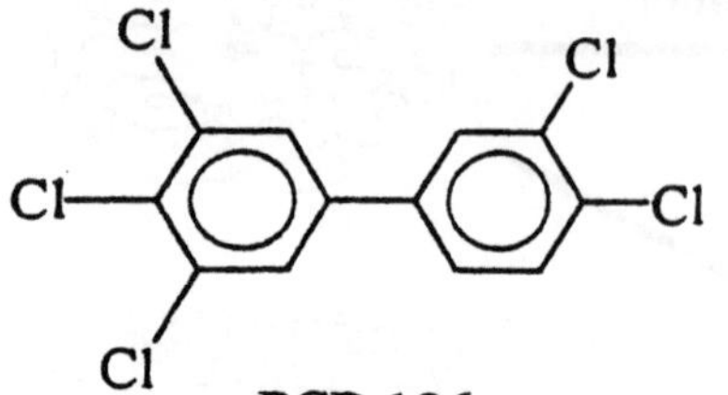

PCB 126

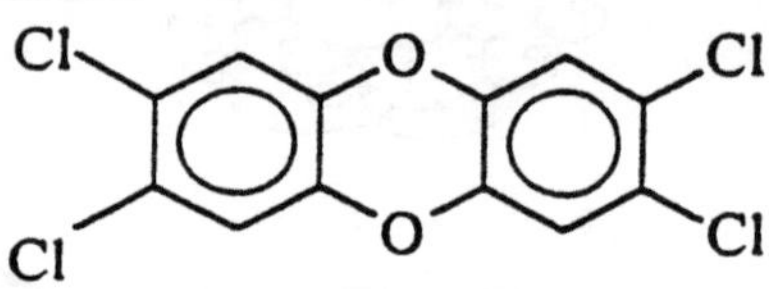

TCDD

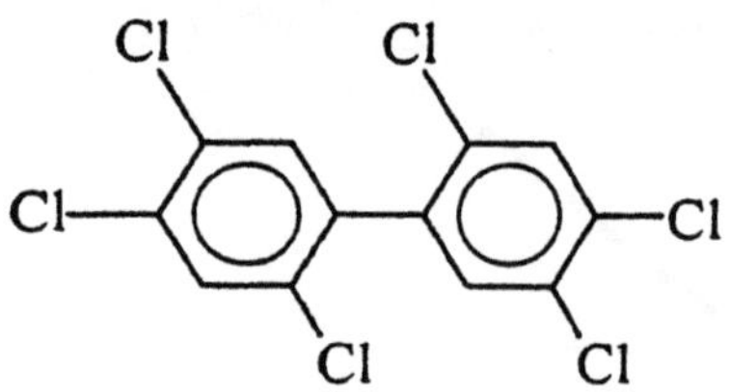

PCB 153

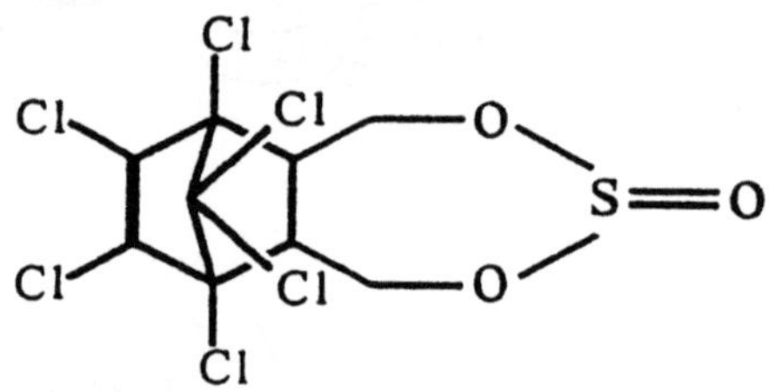

Endosulfan

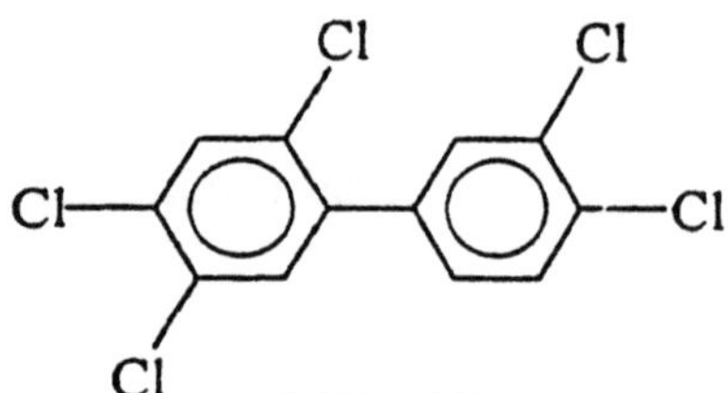

PCB 118

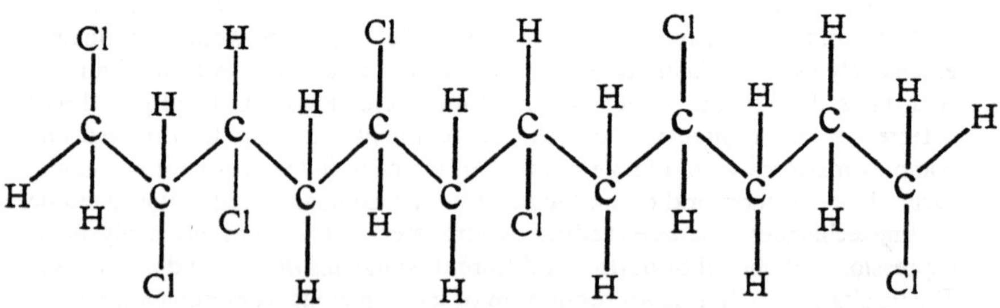

Chlorinated paraffin (C_{12}, 60% Cl)

Figure 3. Molecular structures of the tested organochlorine compounds

The substances were dissolved in dimethyl sulphoxide (DMSO) and the maximum concentration of DMSO in medium was 0.35% (v/v). Polyclonal rabbit antiserum raised against a synthetic oligopeptide corresponding to the sequence of amino acids 363-381 of the C-terminal end of cx 43 was kindly provided by Dr Edgar Rivedal (Montebello, Oslo, Norway).

Cell culture and incubation procedure: IAR 20 rat liver epithelial cells were kindly provided by Dr Hiroshi Yamasaki (IARC, Lyon, France). The cells were cultured in Williams' medium E supplemented with 10% FCS, penicillin and streptomycin (all from GIBCO BRL, Life Technologies, Inchinnan, UK). All studies were conducted on confluent cell layers. The viability of the cells was tested by the Trypan Blue exclusion test and the viability of treated cells was comparable with control cells. The concentrations used were not cytotoxic and did not influence the cloning efficiency during 48 hours

Intercellular communication assay: The scrape loading/dye transfer experiments were performed mainly as reported by El-Fouly *et al.* (1987) with some modifications (Hemming *et al.* 1991). Cells were incubated with the test substances in 3 dishes, and control dishes with DMSO were identically treated. The experimental protocol was repeated at least once .

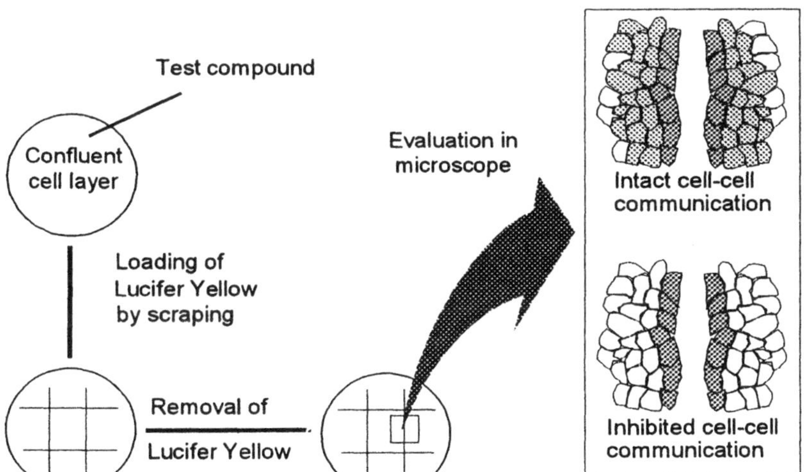

Figure 4. A schematic illustration of the scrape loading dye transfer assay.

Immunoblotting: After treatment with the test substances , the cells were immunoblotted for cx 43 as previously reported by Kenne *et al.* (1994). Cells were washed with PBS and lysed in 100 µl of SDS-sample buffer (2 % SDS , 60 mM Tris-HCl, 0.1 M dithiothreitol (DTT), pH 6.8). Subsequently, after sonication the cell debris was precipitated by centrifugation. Protein concentration was determined using BioRad D_c protein assay kit (Bio-Rad Laboratories, Hercules, CA) after removal of DTT by iodoacetate. Equal amounts of protein (8 µg) were resolved on 10 % SDS-polyacrylamide gels and electrophoretically transferred to nitrocellulose membranes. The membranes were first incubated with anti-cx 43-antibodies diluted 1/1000, then with peroxidase-conjugated anti-rabbit IgG (1/1000, Daco A/S, Glostrup, Denmark). Finally, the membranes were stained with peroxidase substrate, 3,3'-diaminobenzidine tetrahydrochloride (Sigma chemical Co, St Louis, MO) according to the procedures recommended by the manufacturer.

For molecular weight determination, protein mol. wt markers (Amersham, Buckinghamshire, UK) were used.

Results and Discussion

The multistep nature of the cancer process has been noted, both in experimental animals and by studying human carcinogenesis. Conceptually, the initiation, promotion and progression stages have been developed to explain the multistep nature of carcinogenesis. These operational steps does not point out any specific mechanism and today little is known what is the mechanisms are each of these steps. Many studies have shown a good correlation between chemically-induced inhibition of gap junctions and tumour promotion *in vivo*.

Table 1. A summary of the effects on cx 43 induced by the chlorinated paraffins in IAR 20 cells.

Treatment (dose µg/ml)	GJIC[a]	Phosphorylation[b]		
		NP	P_1	P_2
1 h				
Control	+++	+++	+++	+++
Hüls 60 (15)	-	+++	+++	+
Crereclor S52 (35)	-	+++	+++	+
24 h				
Control	+++	+++	+++	+++
Hüls 60 (15)	-	++++	+++	-
Cereclor S52 (35)	-	NT	NT	NT

[a]GJIC = gap junction intercellular communication: +++ control level, ++ low, + medium and - complete inhibition.

[b]Phosphorylation as determined by immunoblotting. Relative band intensity of the different cx 43 bands: ++++ increased relative to the control, +++ control level, ++ slightly , + medium and - highly reduced level.

Table 2. A summary of the effects on cx 43 induced by endosulfan in IAR 20 cells extracted from Kenne *et al* (1994).

Treatment (dose μM)	GJIC[a]	Phosphorylation[b] NP	P_1	P_2
1 h				
Control	+++	+++	+++	+++
Endosulfan (50)	-	+++	++	+
24 h				
Control	+++	+++	+++	+++
Endosulfan (50)	-	++	+	+

[a]GJIC = gap junction intercellular communication: +++ control level, ++ low, + medium and - complete inhibition.
[b]Phosphorylation as determined by immunoblotting. Relative band intensity of the different cx 43 bands: ++++ increased relative to the control, +++ control level, ++ slightly , + medium and - highly reduced level.

Polychlorinated biphenyls and dioxins are highly lipofilic compounds which have been demonstrated to act as tumour promoters. The PCB congeners are divided into three groups, regarding their enzyme induction pattern, i.e., dioxin like PCBs (induces CYP1A1), not-dioxin like PCBs (induces CYP2B1/B2) and mix inducing PCBs. In addition, a group of congeners that do not induce CYP1A1 or CYP2B1/2 exists. The PCBs are believed to act primary as tumour promoters but with different mechanistical pathways. It has been shown that many congeners affect the intercellular communication *in vitro* as shown by Hemming *et al.* (1991). In addition TCDD and PCB 126 decrease the amount of connexin 32 in rat liver in two stage tumour promotor assays in rats (Bager *et al* 1994; Wärngård *et al* 1994).

The results from the scrape loading dye/transfer experiments are shown in Table 3. After one hour treatment, PCB 153 inhibited the gap junctional intercellular communication completely and PCB 118 also decreased the communication. No effect after one hour treatment with PCB 126 or TCDD was observed. Prolonged treatment-duration with PCB 153 caused a reversible inhibition of the communication.. The extended treatment with PCB 126, PCB 118 or TCDD caused a low inhibition after 48 hours incubation with the substances.

Chlorinated paraffins is a group of widely used industrial chemical. In chronic toxicity animal experiments exposure to short carbon length chain chlorinated paraffins increase the incidence of hepatic tumours in rats and mice (NTP, 1986; Bucher *et al* 1987). Investigations concerning the mutagenic potential has not given evidence for genotoxic activity (Birtley *et al* 1980; Serrone *et al* 1987). These to properties lead us to investigate whether short and medium carbon chained chlorinated paraffins inhibit gap junctional intercellular communication.Exposure to chlorinated paraffins down-regulated the

Table 3. A summary of the effects on cx 43 induced by PCBs and TCDD in IAR 20 cells.

Treatment (dose μM)	GJIC[a]	Phosphorylation[b] NP	P_1	P_2
1 h				
Control	+++	+++	+++	+++
PCB 126 (1)	+++	+++	+++	+++
PCB 153 (50)	-	++	+	-
PCB 118 (50)	+	+++	+++	+
TCDD (0.1)	+++	+++	+++	+++
48 h				
Control	+++	+++	+++	+++
PCB 126 (1)	++	++	++	++
PCB 153 (50)	+++	+++	+++	++
PCB 118 (50)	++	++	++	+
TCDD (0.1)	++	++	++	++

[a]GJIC = gap junction intercellular communication: +++ control level, ++ low, + medium and - complete inhibition.
[b]Phosphorylation as determined by immunoblotting. Relative band intensity of the different cx 43 bands: ++++ increased relative to the control, +++ control level, ++ slightly , + medium and - highly reduced level.

intercellular communication between IAR 20 cells. Creveclor 50LV (C50LV) and Huls (H60) abolished the communication at noncytotoxic concentrations within one hour. The dye transfer was inhibited at 20 ug/ml of C50LV and 15 ug/ml of H60. Positive control plates treated with TPA showed complete inhibition of dye transfer, and solvent control plates had no effect on cell-cell communication (data not shown). The amount of cx 43 in extracts of IAR cells was detected on immunoblots by using peroxidase-conjugated antibodies as detection system. The results show a reduction of the phosphorylated forms of cx 43 as indicated in table 1. Three major bands in the mol. wt range of 38-44 kDa in extracts of control cells were observed and in accordance with Musil *et al.* (1990) they were termed cx 43-NP and cx 43-P_1 to P_2. The bands, observed in the IAR 20 cells, are likely to correspond to different phosphorylated forms of cx 43 since they were sensitive to alkaline phosphate-treatment (Kenne *et al* 1994). The phosphorylation pattern in the extracts from H60 or CS52 treated cells suggest that these compounds cause a dephosphorylation of the connexin.

In a recent study (Kenne *et al* 1994) it was demonstrated that the insecticide endosulfan completely inhibited gap junctional communication between IAR 20 cells after treatment with 50 or 100 uM endosulfan. These doses gave full inhibition and were noncytotoxic , as determined by measuring cell survival after staining with trypan blue at different time points. In this study the amounts of cx43 in extracts of IAR 20 cells were detected in immunoblots. The results showed that exposure to endosulfan gave rise to gradually decreased phoshorylation. Dephosphorylation of cx 43 was evident starting at 60 minutes exposure (table 2).

Interestingly the same pattern is seen in this study after treatment with chlorinated paraffins suggesting that dephosphorylation of the connexin 43 could be a mechanistically important pathway beside the generally accepted hyperphoshorylation induced by agents like the classical tumour promoter TPA. We have in earlier studies showed that TPA induced downregulation in many aspects differ from the inhibition of intercellular communication induced by chlorinated environmental pollutants (Wärngård and Flodström 1989; Wärngård *et al* 1989; Kenne *et al* 1994).

In the recovery study after incubation with PCB 153 for one hour the communication was comparable with the controls one 1 hour after the removal of the test substance. Similarly in the 48 hours TCDD treated cells the communication also returned to normal after one1 hour of removal.

After one hour treatment with PCB 153 the P_2 band was nearly completely erased, indicating a hypophosphorylation of cx 43. Treatment with PCB 118 caused a reduction of the P_2 band, with a reduced level after one hour treatment which was reversible after prolonged treatment. However, after 48 hours treatment the band was decreased again. PCB 126 and TCDD treatment caused a parallel response after 48 hours treatment with a slight reduction of the P_2 band after this time of incubation. The immunoblots after removal of PCB 153 showed a transient reduction of the relative level of the P_2 band and after 1 hour of removal the level was back to normal.

The results clearly demonstrate that all three PCB congener and TCDD inhibited the communication, but further support the assumption that TCDD and the dioxinlike congeners like PCB 126 act by a mechanism different from the other PCBs which may involve an activation of the Ah-receptor leading to an indirect downregulation of the intercellular communication.

In summary all three groups of chemicals tested inhibited intercellular communication between IAR 20 cells. The results also suggest that there are different mechanisms involved in the downregulation of cx 43 such as phosphorylation, dephosphorylation or receptor binding. This stress the importance to elucidate the underlying mechanisms for tumour promoter induced inhibition of intercellular communication. If the molecular mechanisms involved were characterized it could be possible to assess the ability to induce one or more critical changes important for the regulation of gap junction function and cell growth.

Acknowledgements

We acknowledge Sirpa Honkasalo, Gunilla Scheu and Margareta Ståhlberg for technical assistance. This research was supported by grants from the Swedish Work Environmental Fund (91-0236), the National Swedish Environmental Protection Board (5323091-9), the Swedish Natural Science Research Council (TM-ER 6305-300), the CEC programme BIOT-CT91-0261, the National Chemicals Inspectorate, and by funds at the Karolinska Institute.

References

Bager Y, Kenne K, Krutovskikh V, Mesnil M, Traub O, Wärngård L (1994) Alteration in expression of gap junction proteins in rat liver after treatment with the tumour promoter 3,4,5,3',4'-Pentachlorobiphenyl. Carcinogenesis 15:2439-2443

Birtley RDN, Conning DM, Daniel JW, Ferguson DM, Longstaff E, Swan AAB (1980) The toxicological effects of chlorinated paraffins in mammals. Toxicol Appl Pharmacol 54:514-525

Bucher JR, Alison RH, Montgomery CA, Huff J, Haseman JK, Farnell D, Thompson R, Prejean JD (1987) Comparative toxicity and carcinogenicity of two chlorinated paraffins in F344/N rats and B6C3F1 mice. Fund Appl Toxicol 9:454-468

Dermietzel R, Traub O, Hwang TK, Beyer E, Bennett MVL, Spray DC, Willecke K (1989) Differential expression of three gap junction proteins in developing and mature brain tissues. Proc Natl Acad Sci, USA 86:10148-10152

El-Fouly MH, Trosko JE, Chang CC (1987) Scrape-loading and dye transfer. A rapid and simple technique to study gap junctional intercellular communication. Exp Cell Res 168:422-430

Fransson-Steen R, Flodström S, Wärngård L (1992) The insecticide endosulfan and its two stereoisomers promote the growth of altered hepatic foci in rats. Carcinogenesis 13: 2299-2303

Hemming H, Wärngård L, Ahlborg UG (1991) Inhibition of dye transfer in rat liver WB cell culture by polychlorinated biphenyls. Pharm Tox 69:416-420

Kenne K, Fransson-Steen R, Honkasalo S, Wärngård L (1994) Two inhibitors of gap junctional intercellular communication, TPA and endosulfan: Different effects on phosphorylation of connexin 43 in the rat liver epithelial cell line, IAR 20. Carcinogenesis 15:1161-1165

Musil LS, Goodenough DA (1993) Multisubunit assembly of an integral plasma membrane channel protein, gap junction connexin 43, occurs after exit from the ER. Cell 74:1065-1077

Nicholson B, Dermietzel R, Teplow D, Traub O, Willecke K, Revel JP (1987) Two homologous protein component of hepatic gap junctions. Nature 329:732-734

NTP (National Toxicology Program) (1986a) Toxicology and carcinogenesis studies of chlorinated paraffins (C12, 60% chlorine) (CAS NO. 63449-39-8) in F344/N rats and B6C3F1 mice (gavage studies). NTP-TR-308. (Technical report series). Research Triangle Park, NC

NTP (National Toxicology Program) (1986b) Toxicology and carcinogenesis studies of chlorinated paraffins (C23, 43% chlorine) (CAS NO. 63449-39-8) in F344/N rats and B6C3F1 mice (gavage studies). NTP-TR-305. Technical Report Series. Research Triangle Park, NC

Poland A (1991) Receptor-mediated toxicity: Reflections on a quantitative model for risk assessment. In: Biological basis for risk assessment of dioxins and related compounds. Banbury Report 35, Cold Spring Harbor Laboratory Press, pp 417-426

Ruch RJ, Fransson R, Flodström S, Wärngård L, Klaunig JE (1990) Inhibition of hepatocyte gap junctional intercellular communication by endosulfan, chlordane and heptachlor. Carcinogenesis 11:1097-1101

Serrone DM, Birtley RDN, Weigand W, Millischer R (1987) Toxicology of chlorinated paraffins. Food Chem Toxicol 25:553-562

Willecke K, Hennemann H, Dahl E, Jungbluth S, Heynkes R (1991) The diversity of connexin genes encoding gap junctional proteins. Eur J Cell Biol 56:1-7

Wärngård L, Hemming H, Flodström S, Duddy S and Kass G (1989) Mechanistical studies on DDT-induced inhibition of intercellular communication. Carcinogenesis 10:471-476

Wärngård L, Flodström S (1989) Mechanistical studies on TPA-, pyrethroid- and DDT-induced effects in the V79 metabolic cooperation assay. Cell Biol Toxicol 5:67-75

Wärngård L, Bager Y, Hemming H, Honkasalo S, Kenne K (1994) The effect of 3,4,5,3',4-Pentachlorophenyl and 2,3,7,8-Tetrachlorodibenzo-p-dioxin on gap junctional intercellular communication in vitro and in vivo. Progress in Cell Research 4:119-122

Yamasaki H, (1990) Gap junctional intercellular communication and carcinogenesis. Carcinogenesis 11:1051-1058

Renal Toxicology

(Chairs: F. Nagelkerke, Netherlands,
and J. Květina, Czech Republic)

Biotransformation and Renal Processing of Nephrotoxic Agents

W. Dekant
Department of Toxicology, University of Würzburg, Versbacher Str. 9, 97078 Würzburg, FRG

Abstract

Nephrotoxicity is often observed as an endpoint in animal toxicity studies. In recent years, the mechanisms of biotransformation, which often provide the basis for renal toxicity, have been elucidated for a variety of compounds. These studies showed that nephrotoxicity of chemicals is either due to accumulation of certain metabolites in the kidney and further bioactivation or due to intrarenal bioactivation of the parent xenobiotic. Both types of mechanisms will be discussed using two relevant samples. The polychlorinated olefin hexachlorobutadiene and other haloolefins cause necrosis of the S-3 segment of the proximal tubules; their nephrotoxicity is dependent on bioactivation reactions. In the liver, hexachlorobutadiene is transformed by conjugation with glutathione to (S-pentachlorobutadien-yl)glutathione. This S-conjugate is processed by the enzymes of mercapturic acid formation to give N-acetyl-(S-pentachlorbutadienyl)-L-cysteine, which is accumulated in the proximal tubule cells and deacetylated there to give (S-pentachlorobutadienyl)-L-cysteine. Further bioactivation is catalyzed by renal cysteine conjugate ß-lyase. Both the renal accumulation by the organic anion transporter and the topographical distribution of cysteine conjugate ß-lyase along the nephron are major determinants of organ and cell selectivity. Vinylidene chloride (VDC) is nephrotoxic in mice after inhalation, but not after oral or intraperitoneal administration. The nephrotoxicity of VDC is due to the selective expression of an androgene-dependent cytochrome P450 in the proximal tubules of male mice. This enzyme oxidizes VDC to an electrophile and is not present in female mice, but can be induced be androgene treatment. The observation of nephrotoxicty of VDC after inhalation only is due to the high blood flow to the kideny and thus high concentrations of VDC delivered to the kidney after inhalation. After oral or intraperitoneal application, hepatic first-pass metabolism efficiently reduces the amount of VDC delivered to the kidney. The results demonstrated here demonstrate that prior to in vitro nephrotoxicity screening, toxicokinetics and biotransformation pathways for a chemical have to be elucidated and metabolites have to be included into the testing regimen.

Introduction

Renal transport and xenobiotic metabolism play an important role in the detoxication and excretion of potentially toxic xenobiotics. However, recent experimental evidence has demonstrated that renal xenobiotic metabolism and renal transport processes also play an important role in the nephrotoxicity of xenobiotics and xenobiotic metabolites. The high blood flow to the kidney combined with its ability to concentrate solutes may expose the kidney to high concentrations of xenobiotics and xenobiotics metabolites present in the systemic circulation. Recently, it has been demonstrated that xenobiotic metabolites formed in the liver and other organs may be targeted to the kidney by selective transport systems; many xenobiotics require enzymatic transformation to proximate reactive metabolites to elicit their toxic and carcinogenic effects. The enzymatic formation of reactive metabolites is termed bioactivation. The bioactivation mechanisms for many nephrotoxicants have, at least in part, been elucidated in the past 15 years. Many ultimate toxicants formed in the kidney are electrophiles whose interaction with cellular macromolecules may cause a perturbation of normal cell function resulting in necrosis and/or cancer (Anders 1988). Electrophilic metabolites may bind to nucleophilic sites in cellular macromolecules; the importance of covalent modification of protein and DNA in cell killing and in the induction of tumors is established (Miller and Miller 1981; Nelson and Pearson 1990; Hinson and Roberts 1992). The objective of this review is to summarize new information about renal transport, renal bioactivation and their relation to nephrotoxicity using two relevant example for the basic mechanisms outlined above.

The cysteine conjugate ß-lyase pathway: The selective proximal tubule toxicity observed with several xenobiotic-derived glutathione S-conjugates is an excellent example for contribution of the physiological transport functions of the kidney to nephrotoxicity.

Several halogenated alkenes and the alkyne dichloroacetylene, are selectively nephrotoxic in rats and induce proximal tubular damage (Reichert et al. 1975; Potter et al. 1981; Ishmael et al. 1982). Moreover, the widely used solvents trichloroethene (NCI 1986a) and tetrachloroethene, and also dichloroacetylene (NCI 1986b) and hexachlorobutadiene (Kociba et al. 1977), induced chronic nephrotoxicity and/or carcinomas of the proximal tubules in rats after giving high doses. Glutathione-dependent pathways have been implicated in the renal toxicity of these compounds (Fig. 1). Nephrotoxic haloalkenes are metabolized to glutathione S-conjugates by microsomal and cytosolic glutathione S-transferases to fluoroalkyl glutathione S-conjugates (fluoroalkenes) and chloroalkenyl glutathione S-conjugates (chloroalkenes); the microsomal fraction of rat liver generally exhibits a 2- to 10-fold higher activity toward haloalkenes than does the cytosolic fraction (Wolf et al. 1984; Dekant et al. 1987, 1988a; Vamvakas et al. 1989; Koob and Dekant 1990). The rates of glutathione conjugation for these compounds observed in renal subcellular fractions are an order of magnitude less than those observed in liver fractions. These observations and the protective effects of biliary canulation on hexachlorobutadiene nephrotoxicity suggest that.

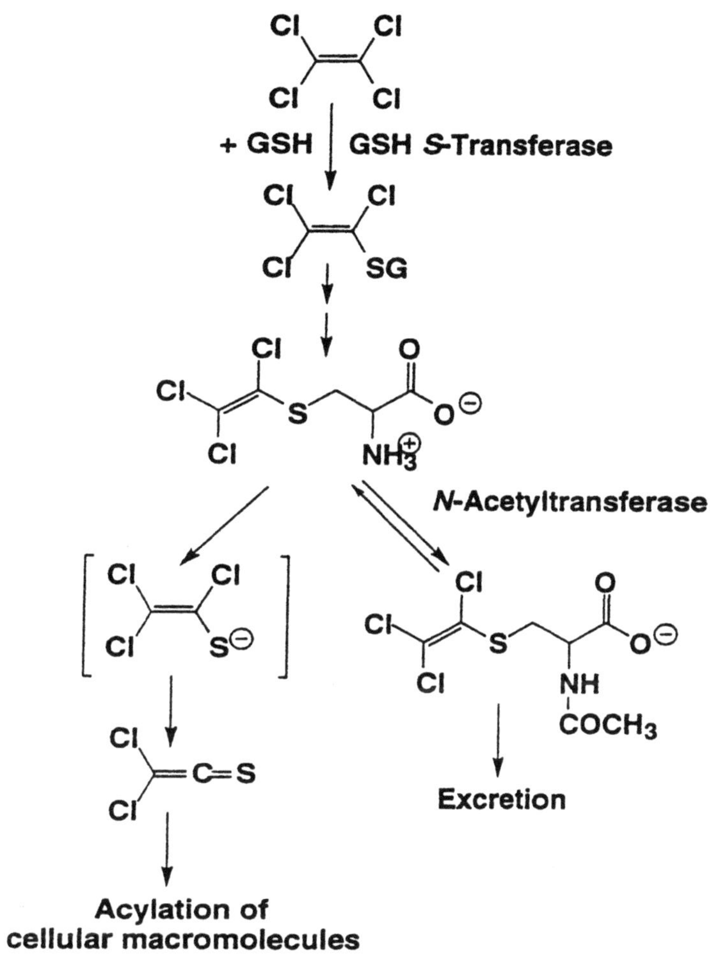

Fig. 1: Bioactivation of nephrotoxic polychlorinated alkenes by glutathione S-conjugate formation and metabolism of formed cysteine S-conjugates by renal cysteine-conjugate ß-lyase, as exemplified by tetrachloroethene. SG, Glutathionyl.

the metabolites responsible for the nephrotoxic effects are biosynthetized in the liver

Biotransformation of glutathione S-conjugates to excretable metabolites or conversion to reactive intermediates occurs elsewhere in the body. Therefore, transport systems exist to deliver glutathione S-conjugates and their metabolites to these extrahepatic sites (Okajima et al. 1983; Inoue et al. 1984). Efflux of glutathione S-conjugates occurs mainly across the canalicular membrane by active transport mechanisms into bile (Vore 1993, 1994); sinusoidal elimination of S-conjugates has been observed only at high intracellular glutathione S-conjugate concentrations. The molecular weight of most glutathione S-conjugates biosynthesized in the liver is above the threshold for biliary transport in rats and the presence of polar groups in the molecules prevents passive diffusion through the membrane of the hepatocytes (Inoue et al. 1984a, 1984b; Awasthi et al. 1989, 1990). The importance of biliary excretion is further supported by the observation that biliary cannulation protects male rats from the nephrotoxicity of hexachlorobutadiene (Nash et al. 1984), indicating that glutathione S-conjugate formation and biliary excretion are the first steps in the disposition of nephrotoxic S-conjugates.

The role of the organs involved in the disposition of S-conjugates is difficult to evaluate due to the complexicity of the pathways and the lack of adequate experimental systems. However, the following pathways are supported by experimental evidence.

After secretion into bile, glutathione S-conjugates derived from halogenated alkenes are either transported intact to the small intestine or degraded to the corresponding cysteine S-conjugates by the sequential action of g-glutamyltransferase and dipeptidases, which are present in the luminal membrane of the bile duct epithelium and in the bile canalicular membrane of hepatocytes (Anderson et al. 1980; Inoue et al. 1983; Meister 1988). S-Conjugates translocated to the small intestine may either be reabsorbed from the gut and after passage through the liver, be translocated to the kidney, may undergo enterohepatic circulation, or may be excreted with the feces (Dekant et al. 1988b). Glutathione S-conjugates excreted in the bile may be also reabsorbed by the liver from bile after breakdown to the cysteine S-conjugates and further transformed to the corresponding mercapturic acid. Intrahepatic mercapturic-acid formation may contribute to the disposition of S-conjugates in species other than the rat that have higher levels of hepatic g-glutamyltransferase. Unfortunately, the contribution of this pathway to the disposition of nephrotoxic S-conjugates is not known, since only few experiments on S-conjugate formation and toxicity have been performed in species other than the rat.

After passage through the intestinal wall, S-conjugates enter the systemic circulation and thus the kidney directly or after absorbtion and resecretion from the hepatocytes. Glutathione S-conjugates absorbed by the hepatocytes at the luminal surface will be excreted with bile and thus undergo enterohepatic circulation. Due to their molecular weight, cysteine S-conjugates taken up by the liver may be excreted again via the sinusoidal membrane into the blood circulation. Cysteine S-conjugates may also be N-acetylated by hepatic N-acetyltransferases and may reach the systemic circulation as mercapturic acids (Green and Elce 1975; Duffel and Jakoby 1982).

Glutathione S-conjugates leaving the liver via the sinusoidal membrane and glutathione S-conjugates biosynthesized in organs other than the liver may be delivered intact to the kidney. The high blood flow to the kidneys and their capacity to process and activate S-conjugates are major determinants of S-conjugate toxicity (Rush et al. 1984; Monks and Lau 1987, 1989; Monks et al. 1990). Once in the kidney, S-conjugates may reach their target in the straight portion of the proximal tubular cells by glomerular filtration or by transport across the basolateral membrane, or both. The kidney is rich in g-glutamyltransferase activity, dipeptidase activity, and aminoacylase activity; thus glutathione and cysteinylglycine S-conjugates and mercapturic acids are efficiently converted to cysteine S-conjugates in the proximal tubular cells (Hughey et al. 1978; Guder and Ross 1984; Guder and Wirthensohn 1985; Anderson et al. 1990).

Glutathione S-conjugates present in blood may be filtered or may enter the peritubular circulation. Studies with S-(pentachlorobutadienyl)glutathione in the isolated perfused rat kidney demonstrate that this glutathione S-conjugate is preferentially removed by non-filtering mechanisms (Schrenk et al. 1988).

The filtered glutathione S-conjugates may be metabolized to the corresponding cysteine S-conjugates by g-glutamyltransferase and cysteinylglycine dipeptidase or aminopeptidase M in the renal brush-border membrane (Hughey et al. 1978; Moldeus et al. 1978; Jones et al. 1979a,b; Tsao and Curthoys 1980). The glutathione S-conjugates in the peritubular capillaries may either be metabolized to cysteine S-conjugates by g-glutamyltransferase present in the renal vasculature or basolateral membrane. Transport of the intact tripeptide into renal epithelial cells by an electrogenic, sodium-coupled and probenecid-sensitive transport with a broad specificity for g-glutamyl compounds has also been observed (Rankin and Curthoys 1982; Lash and Jones 1983, 1984).

The probenecid-sensitive organic-anion transporter present on the basolateral side of the proximal tubular cells seems to play the most important role in the accumulation of S-conjugates in proximal tubular cells and in the organ-selective toxicity. Probenecid is a selective inhibitor of the organic-anion transporter without effects on energy metabolism. Haloalkene-derived mercapturates have the highest affinity for the organic-anion transporter, but glutathione and cysteine S-conjugates with lipophilic substituents on sulfur are also substrates (Ullrich and Rumrich 1988; Ullrich et al. 1988, 1989a,b). The inhibitory effects of probenecid on the toxicity of several haloalkene cysteine S-conjugates and mercapturic acids support a central role for the renal organic-anion transporter in the renal accumulation of S-conjugates.

Inside the renal cell, cysteine S-conjugates may have three fates: they may be secreted unchanged into the plasma and then returned to the liver for further metabolism (renal-hepatic circulation); they may be N-acetylated to form the corresponding mercapturic acids and excreted finally with urine (Heuner et al. 1991); or they may undergo ß-lyase-catalyzed metabolism to yield reactive intermediates.

Cysteine S-conjugates biosynthesized from haloalkenes are cleaved to unstable thiols by cysteine conjugate ß-lyase. The S-conjugates S-(1,2-dichlorovinyl)-L-cysteine, S-(1,1,2-trichlorovinyl)-L-cysteine, and S-(pentachlorobutadienyl)-L-cysteine are transformed by cysteine conjugate ß-lyase to yield halovinylthiols, and these vinylthiols give rise to electrophilic species (identified as thioketenes) whose interaction with cellular macromolecules is responsible for the cytotoxicity and mutagenicity of S-(1,2-dichlorovinyl)-L-cysteine, S-(1,1,2-trichlorovinyl)-L-cysteine, and S-(pentachlorobutadienyl)-L-cysteine.

Intrarenal bioactivation of xenobiotic to reactive metabolites: Cytochrome P450 dependent bioactivation reactions. Renal bioactivation enzymes are usually localized to discrete sites in the kidney. Several nephrotoxicants such as chloroform or 1,1-dichloroethene are converted to reactive intermediates in the kidney itself, the localisation of bioactivating enzymes is generally well correlated with the sites of damage.

Cytochromes P450 are a group of hemoproteins which catalyse the oxidation of C-H bounds and the epoxidation of carbon-carbon double bounds in xenobiotics. In the kidneys, cytochromes P450 are present in the cortical microsomal fractions and several isozymes of cytochromes P450 from renal cortex have been purified and characterized (Anders 1980). However, not all P450 genes are expressed in kidney. Moreover, the mechanisms of P450 gene expression in kidney are different from those established in the liver and sex hormones seems to play a major role in the regulation of renal P450s. For example, rat renal P450 4A2 is expressed at much higher levels in the kidneys of male rats as compared to female rats; in males, this P450 enzymes seems to account for approximately 90 % of total renal cytochrom P450 (Henderson and Wolf 1991; Wolf 1991; Bell et al. 1992; Imaoka et al. 1992; Schuetz et al. 1992). The localization of the cytochromes P450 isozymes along the nephron is heterogenous as indicated by immunohistochemical techniques; however, the proximal tubules seem to contain the highest concentrations of P450 enzymes. With respect to the bioactivation of nephrotoxicants, the renal expression of an enzyme closely related to hepatic cytochrome P450 2E1 in male mouse kidney seems most relevant. P450 2E1 is involved in the bioactivation of many low molecular weight toxicants and carcinogens (Guengerich et al. 1991).

The nephrotoxicity of 1,1-dichloroethene (VDC) and several other low molecular weight xenobiotics in male mice, which is dependent on their bioactivation in the kidney are relevant examples (Smith et al. 1984, 1985; Hu et al. 1993; Speerschneider and Dekant 1995). In male mice, renal lesions induced by chloroform, acetaminophen and VDC are primarily localized to the proximal tubules. Nephrotoxicity is not observed in female mice and in both sexes of rats under these exposure conditions. Moreover, the observed sex differences in VDC and chloroform induced nephrotoxicity in mice are closely correlated to the renal activities of cytochromes P450. Male mouse kidney contains much higher activities of a renal cytochrome P450 and is much more susceptible to VDC induced nephrotoxicity than female mice. This renal cytochrome P450 has substrate specificities similar to rat liver cytochrome P450 2E1 and crossreacts with an antibody to this P450 enzyme. Moreover, its localisation in the kidney as demonstrated by immunohistochemistry coincidences with the renal lesions in the proximal tubular cells seen after VDC, chloroform and acetaminophen. The sensitivity of male mice to chloroform, VDC and acetaminophen nephrotoxicity is decreased by castration, the effect of castration is reversed by testosterone administration as are cytochrome P450 concentrations.

Furthermore, kidney microsomes from male mice metabolized chloroform, VDC and acetaminophen much more intensivly to reactive intermediates than kidney microsomes obtained from female mice, the extent of metabolism in kidney microsomes is modulated by the hormonal status of the animals. The expression of this P450 enzyme is different in male mice of several strains, the relative activities seen in males of these strains are reflected by their relative susceptibilities to VDC and chloroform nephrotoxicity and their relative rates of microsomal VDC and chloroform bioactivation in the kidney. The data show that a P450 enzyme with similar substrate specificity as rat hepatic P450 2E1 is present in the kidney of male mice. The absence of this enzyme in rat kidney explains the relatively lower nephrotoxicity of these compounds in rats.

Acknowledgement.

The work was supported by the Deutsche Forschungsgemeinschaft, the Bundesministerium für Forschung und Technology and BASF AG.

References

Anders MW (1980) Metabolism of drugs by the kidney. Kidney Int 18: 636-647

Anders MW (1988) Bioactivation mechanisms and hepatocellular damage. In: Arias IM, Jakoby WB, Popper H, Schachter D, Shafritz DA (eds) The Liver: Biology and Pathology, Second Edition. Raven Press, Ltd., New York, pp 389-400

Anderson ME, Bridges RJ, Meister A (1980) Direct evidence for inter-organ transport of glutathione and that the non-filtration renal mechanism for glutathione utilization involves g-glutamyl transpeptidase. Biochem Biophys Res Commun 96: 848-853

Anderson ME, Naganuma A, Meister A (1990) Protection against cisplatin toxicity by administration of glutathione ester. FASEB J 4: 3251-3255

Awasthi YC (1990) The interrelationship between p-glycoprotein and glutathione S-conjugate transporter(s). TiPS 15: 376-377

Awasthi YC, Singh SV, Ahmad H, Wronski LW, Srivastava SK, Labelle EF (1989) ATP dependent primary active transport of xenobiotic-glutathione conjugates by human erxthrocyte membrane. Cell Biochem 91: 131-136

Bell DR, Bars RG, Elcombe CR (1992) Differential tissue-specific expression and induction of cytochrome P450IVA1 and acyl-CoA oxidase. Eur J Biochem 206: 979-986

Dekant W, Martens G, Vamvakas S, Metzler M, Henschler D (1987) Bioactivation of tetrachloroethylene - role of glutathione S-transferase-catalyzed conjugation versus cytochrome P-450-dependent phospholipid alkylation. Drug Metab Dispos 15: 702-709

Dekant W, Vamvakas S, Henschler D, Anders MW (1988a) Enzymatic conjugation of hexachloro-1,3-butadiene with glutathione: formation of 1-(glutathion-S-yl)-1,2,3,4,4-pentachlorobuta-1,3-diene and 1,4-bis(glutathion-S-yl)-1,2,3,4-tetrachlorobuta-1,3-diene. Drug Metab Dispos 16: 701-706

Dekant W, Schrenk D, Vamvakas S, Henschler D (1988b) Metabolism of hexachloro-1,3-butadiene in mice: in vivo and in vitro evidence for activation by glutathione conjugation. Xenobiotica 18: 803-816

Duffel MW, Jakoby WB (1982) Cysteine S-conjugate N-acetyltransferase from rat kidney microsomes. Mol Pharmacol 21: 444-448

Green RM, Elce JS (1975) Acetylation of S-substituted cysteines by rat liver and kidney microsomal N-acetyltransferase. Biochem J 147: 283-289

Guder WG, Ross BD (1984) Enzyme distribution along the nephron. Kidney Int 26: 101-111

Guder WG, Wirthensohn G (1985) Enzyme distribution and unique biochemical pathways in specific cells along the nephron. In: Bach PH, Lock EA (eds) Renal heterogeneity and target cell toxicity. Wiley and Sons, Chinchester, pp 195-198
Guengerich FP, Kim D-H, Iwasai M (1991) Role of human cytochrome P-450 IIE1 in the oxidation of many low molecular weight cancer suspects. Chem Res Toxicol 4: 168-179
Henderson CJ, Wolf CR (1991) Evidence that the androgen receptor mediates sexual differentiation of mouse renal cytochrome P450 expression. Biochem J 278: 499-503
Heuner A, Dekant W, Schwegler JS, Silbernagl S (1991) Localization and capacity of the last step of mercapturic acid biosynthesis and the reabsorption and acetylation of cysteine S-conjugates in the rat kidney. Eur J Physiol 417: 523-527
Hinson JA, Roberts DW (1992) Role of covalent and noncovalent interactions in cell toxicity: effects on proteins. Annu Rev Pharmacol Toxicol 32: 471-510
Hu JJ, Lee M-J, Vapiwala M, Reuhl K, Thomas PE, Yang CS (1993) Sex-related differences in mouse renal metabolism and toxicity of acetaminophen. Toxicol Appl Pharmacol 122: 16-26
Hughey RP, Rankin BB, Elce JS, Curthoys NP (1978) Specificity of a particulate rat renal peptidase and its localization along with other enzymesof mercapturic acid synthesis. Arch Biochem Biophys 186: 211-213
Imaoka S, Yamazoe Y, Kato R, Funae Y (1992) Hormonal regulation of rat renal cytochrome P450s by androgen and the pituitary. Arch Biochem Biophys 299: 179-184
Inoue M, Akerboom TP, Sies H, Kinne R, Thao T, Arias IM (1984a) Biliary transport of glutathione S-conjugate by rat liver canalicular membrane vesicles. J Biol Chem 259: 4998-5002
Inoue M, Kinne R, Tran T, Arias IM (1984b) Glutathione transport across hepatocyte plasma membranes. Analysis using isolated rat-liver sinusoidal-membrane vesicles. Eur J Biochem 138: 491-495
Inoue M, Kinne R, Tran T, Biempica L, Arias IM (1983) Rat liver canalicular membrane vesicles - isolation and topological characterization. J Biol Chem 258: 5183-5188
Inoue M, Okajima K, Morino Y (1984) Hepato-renal cooperation in biotransformation, membrane transport, and elimination of cysteine S-conjugates of xenobiotics. J Biochem Tokyo 95: 247-254
Ishmael J, Pratt I, Lock EA (1982) Necrosis of the pars recta (S3 segment) of the rat kidney produced by hexachloro-1:3-butadiene. J Pathol 138: 99-113
Jones DP, Moldeus P, Stead AH, Ormstad K, Joernvall H, Orrenius S (1979a) Metabolism of glutathione and a glutathione conjugate by isolated kidney cells. J Biol Chem 254: 2787-2792
Jones DP, Sundby G-B, Ormstad K, Orrenius S (1979b) Use of isolated kidney cells for study of drug metabolism. Biochem Pharmacol 28: 929-935
Kociba RJ, Keyes DG, Jersey GC, Ballard JJ, Dittenber DA, Quast JF, Wade LE, Humiston CG, Schwetz BA (1977) Results of a two year chronic toxicity study with hexachlorobutadiene in rats. Am Ind Hyg Assoc J 38: 589-602
Koob M, Dekant W (1990) Metabolism of hexafluoropropene - evidence for bioactivation by glutathione conjugate formation in the kidney. Drug Metab Dispos 18: 911-916
Lash LH, Jones DP (1983) Transport of glutathione by renal basal-lateral membrane vesicles. Biochem Biophys Res Com 112: 55-60
Lash LH, Jones DP (1984) Renal glutathione transport. Characteristics of the sodium-dependent system in the basal-lateral membrane. J Biol Chem 259: 14508-14514
Meister A (1988) Glutathione metabolism and its selective modification. J Biol Chem 263: 17205-17208
Miller EC, Miller JA (1981) Searches for ultimate chemical carinogenes and their reactions with cellular macromolecules. Cancer 47: 2327-2345

Moldeus P, Jones DP, Ormstad K, Orrenius S (1978) Formation and metabolism of a glutathione-S-conjugate in isolated rat liver and kidney cells. Biochem Biophys Res Commun 83: 195-200

Monks TJ, Anders MW, Dekant W, Stevens JL, Lau SS, van Bladeren PJ (1990) Glutathione conjugate mediated toxicities. Toxicol Appl Pharmacol 106: 1-19

Monks TJ, Lau SS (1987) Commentary: renal transport processes and glutathione conjugate-mediated nephrotoxicity. Drug Metab Dispos 15: 437-441

Monks TJ, Lau SS (1989) Sulphur conjugate-mediated toxicity. Rev Biochem Toxicol 10: 41-90

Nash JA, King LJ, Lock EA, Green T (1984) The metabolism and disposition of hexachloro-1:3-butadiene in the rat and its relevance to nephrotoxicity. Toxicol Appl Pharmacol 73: 124-137

NCI, National Cancer Institute (1986a) Carcinogenesis bioassay of trichloroethylene. National Toxicology Program Technical Report 311

NCI, National Cancer Institute (1986b) Carcinogenesis bioassay of tetrachloroethylene. National Toxicology Program Technical Report 232

Nelson SD, Pearson PG (1990) Covalent and noncovalent interactions in acute lethal cell injury caused by chemicals. Annu Rev Pharmacol Toxicol 30: 169-195

Okajima K, Inoue M, Itoh K, Horiuchi S, Morino Y (1983) Interorgan cooperation in enzymic processing. In: Sakamoto Y, Higashi T, Tateishi N (eds) Glutathione: Storage, Transport and Turnover in Mammals. Japan Scientific Societies Press, Tokyo, pp 129-144

Potter CL, Gandolfi AJ, Nagle R, Clayton JW (1981) Effects of inhaled chlorotrifluoroethylene and hexafluoropropene on the rat kidney. Toxicol Appl Pharmacol 59: 431-440

Rankin BB, Curthoys NP (1982) Evidence for the renal paratubular transport of glutathione. FEBS Lett 147: 193-196

Reichert D, Ewald D, Henschler D (1975) Generation and inhalation toxicity of dichloroacetylene. Fd Cosmet Toxicol 13: 511-515

Rush GF, Smith JH, Newton JF, Hook JB (1984) Chemically induced nephrotoxicity: role of metabolic activation. Crit Rev Toxicol 13: 99-160

Schrenk D, Dekant W, Henschler D (1988) Metabolism and excretion of S-conjugates derived from hexachlorobutadiene in the isolated perfused rat kidney. Mol Pharmacol 34: 407-412

Schuetz EG, Schuetz JD, Grogan WM, Naray-Fejes-Toth A, Fejes-Toth G, Raucy J, Guzelian P, Gionela K, Watlington CO (1992) Expression of cytochrome P450 3A in amphibian, rat, and human kidney. Arch Biochem Biophys 294: 206-214

Smith JH, Hewitt WR, Hook JB (1985) Role of intrarenal biotransformation in chloroform-induced nephrotoxicity in rats. Toxicol Appl Pharmacol 79: 166-174

Smith JH, Maita K, Sleight SD, Hook JB (1984) Effect of sex hormone status on chloroform nephrotoxicity and renal mixed function oxidases in mice. Toxicology 30: 305-316

Speerschneider P, Dekant W (1995) Renal tumorigenicity of 1,1-dichloroethene in mice: the role of male specific expression of cytochrome P450 2E1 in the renal bioactivation of 1,1-dichloroethene. Toxicol Appl Pharmacol 130: 48-56

Tsao B, Curthoys NP (1980) The absolute asymetry of orientation of g-glutamyltranspeptidase and aminopeptidase on the external surface of the rat renal brush border membrane. J Biol Chem 255: 7708-7711

Ullrich KJ, Rumrich G (1988) Contraluminal transport systems in the proximal renal tubule involved in secretion of organic anions. Am J Physiol 254 453: 462-

Ullrich KJ, Rumrich G, Kloess S (1988) Contraluminal para-aminohippurate (PAH) transport in the proximal tubule of the rat kidney. Pfluegers Arch 413: 134-146

Ullrich KJ, Rumrich G, Klöss S (1989a) Contraluminal organic anion and cation transport in the proximal renal tubule: V. Interaction with sulfamoyl- and phenoxy diuretics, and with ß-lactam antibiotics. Kidney Int 36: 78-88
Ullrich KJ, Rumrich G, Wieland T, Dekant W (1989b) Contraluminal para-aminohippurate (PAH) transport in the proximal tubule of the rat kidney. Europ J Physiol 415: 342-350
Vamvakas S, Herkenhoff M, Dekant W, Henschler D (1989) Mutagenicity of tetrachloroethylene in the Ames-test - metabolic activation by conjugation with glutathione. J Biochem Toxicol 4: 21-27
Vore M (1993) Canalicular transport: discovery of ATP-dependent mechanisms. Toxicol Appl Pharmacol 118: 2-7
Vore M (1994) Phase III elimination: another two-edged sword. Environ Health Perspect 102: 422-423
Wolf CR (1991) Individuality in cytochrome P450 expression and its association with the nephrotoxic and carcinogenic effects of chemicals. IARC Scientific Publication 115: 281-287
Wolf CR, Berry PN, Nash JA, Green T, Lock EA (1984) Role of microsomal and cytosolic glutathione S-transferases in the conjugation of hexachloro-1:3-butadiene and its possible relevance to toxicity. J Pharmacol Exp Ther 228: 202-208

Methods for Studying Renal Toxicity

Marieke Kruidering, Bob van de Water and J.Fred Nagelkerke
Division of Toxicology, Leiden Amsterdam Center for Drug Research, Sylvius Laboratories, University of Leiden, P.O.Box 9503, 2300 RA, Leiden, The Netherlands

Most nephrotoxic compounds, such as drugs and industrial compounds, exert their noxious effect in a specific region of the kidney. Within the kidney, the proximal tubular cells (PTC) are exposed to the highest concentrations of xenobiotics due to their transporting, concentrating and drug metabolizing abilities. Therefore mechanisms of nephrotoxicity are best studied in these particular cells. For this purpose, well defined in vitro systems are needed. Many in vitro systems to study renal toxicity have been described, ranging from isolated perfused kidneys, kidney slices, isolated tubules, primary cultures of renal cells, kidney cell iines to isolated organelles, mainly derived from rabbit, rat and mouse; reviews of the various available systems have been published in the past (Gstraunthaler 1988,Boogaard et al. 1990). In the first part of the present paper new developments in the application of kidney slices, isolated proximal tubular cells and cell lines will be reviewed. In the second part new types of renal toxicity studies in individual living or fixed cells using video microscopy, confocal laser scan microscopy and flow cytometry will be described and discussed.

Renal Slices

One of the in vitro techniques that has been used extensively and has proven its usefulness is the renal slices technique; a substantial part of our present knowledge on renal toxicology and physiology has come from the use of renal slices. However, the system suffered from two major disadvantages. Firstly, slices contain a heterogeneous cell population, which makes it cumbersome to assess functional changes in a distinct cell type. Secondly, many of the tubules on the exposed surface are damaged during the preparation and lack cellular integrity to various degrees. Fortunately, the use of new commercially available automatic slicers to prepare precision-cut slices has certainly improved the quality of the slices. Moreover, these slices are of a uniform thickness which makes comparison easier (Phelps et al 1987). Also improvements in incubation conditions now allow exposure of 24 hours or more (Ruegg et al 1987)and it appears that this system has the potential to become the easy-to-use system for routine-screening of newly developed compounds. However, the slice is still a complex conglomerate of different cells that interact with each other and also not every cell is exposed to nutrients and oxygen to the same extent. Therefore for mechanistic studies the single cell suspension or culture remains superior. Single cells in suspension form the simplest system in which cellular integrity is maintained. The cells are uniformly exposed to the solutes in the incubation media. Much effort has been directed towards the isolation of pure proximal tubular cells. This has resulted in a number of studies which deal with the isolation and subsequent purification procedures of proximal tubular cells from various animals and man.

Renal Cells

General Isolation and Purification of Renal Cells: In principle, all procedures for the isolation of tubules and single cells from the intact kidney belong to one of two methods: either mechanical or enzymatic disruption, followed by one or more purification steps. Many procedures combine the two methods and use mechanical force in order to disperse enzymatically digested tissue, or digest tissue that has been minced previously by mechanic means. The simplest mechanical treatment to obtain renal tissue fragments is chopping the kidney into slices and pushing these through a strainer. This method is quite successful for the isolation of glomeruli, because they are protected by Bowman's capsule (Norgaard 1976). However, for other renal cell types the yield of viable cells is rather low.

The most widely applied method to obtain tubular fragments and isolated PTC is proteolytic digestion. Collagenase is considered to be the enzyme of choice for proteolytic digestion. The commonly used collagenase preparations are mixtures of proteases, prepared from Clostridium histolyticum. They contain in addition to collagenase activity a number of hydrolytic enzymes, such as hyaluronidase, clostripain, neutral protease and even some tryptic activity. These contaminations seem needed for optimal digestion, since highly purified collagenase (clostridiopeptidase A, E.C. 3.4.24.3) exhibits very poor tissue-digestion characteristics. We use the collagenase, recommended by several companies for the isolation of pancreatic islets.

For the preparation of single cell suspensions it is necessary to loosen the tight junctions between the cells in the tubule. In contrast to the isolation procedures for hepatocytes (Seglen 1976) simple perfusion with a Ca^{2+}-free buffer is not sufficient for renal tissue. Ca^{2+}-free perfusion before collagenase treatment results in a suspension of intact tubules from which single cells can be obtained only by subsequent incubation with other proteases. Consequently, a number of Ca^{2+}-chelating agents have been tested. Sodium citrate (2--3 mM) has been used (Heidrich and Dew 1977), as well as EDTA (Vandewalle et al 1986).

Perfusion via the renal arteries, in situ or with excised kidneys, is the best way to remove blood from the tissue and to deliver chelators and proteases. A breakthrough in the preparation of single cell suspensions of PTC was the method developed by Jones 1979 and Ormstadt 1980.

Rat Kidney Cells: Rat kidneys are perfused in a recirculating system, first with EGTA and subsequently with collagenase. The perfusion with EGTA is started in situ, immediately after ligation of the aorta. The kidneys are excised while being perfused, and collagenase perfusion follows EGTA perfusion without interruption. Thus, the tissue is optimally supplied with nutrients and oxygen while the collagenase has optimal access to the tissue. The method appeared to be superior to other methods based on proteolytic digestion, both in cell quality and yield.

The different cell types of the kidney are highly intermingled as a consequence of the complex renal architecture. This makes it impossible to isolate a single cell type without a purification procedure, irrespective of the isolation method. Several techniques have been developed for the separation of cell types, from simple filtration and (density) centrifugation to more complex methods such as free flow electrophoresis and immunological selection.

Almost all methods include separation of the medulla from the cortex and removal of the glomeruli from the dispersed tissue by mechanical means. Separation of medulla from cortex is easily done when sliced or minced tissue is used, because the distinction can easily be made visually. When the intact kidney is perfused, the distinct difference between cortex and medulla as seen in fresh kidneys fades. Nevertheless, the cortex can usually be separated without difficulties because during collagenase digestion it becomes much softer than the medulla. Therefore, it can gently be scraped off and separation does not need any special procedures (Boogaard et al. 1989).

The most common way to remove glomeruli from a crude tissue suspension is by sieving. The diameter of the rat glomerulus is approximately 100 mm (depending on strain and age), while the diameter of the proximal tubule measures only 62--66 mm. The other cortical tubular parts (distal tubule, collecting duct) are even smaller in diameter. Thus by simple sieving through a 70--100 mm gauze all glomeruli and large tissue fragments are separated from single cells and small tubular fragments. As a consequence, sieving through such gauze or a tuft of glass-wool is included in almost every isolation method.

Thus, glomeruli, larger tissue fragments, blood vessels and connective tissue are removed for the greater part by sieving. However, although enrichment in proximal tubule cells is obtained, a significant amount of more distal cell types and small glomerular fragments remains (Toutain et al 1989). These cells can be removed in several ways. A simple but very effective technique to separate the PTC from the various other cell types in a crude cell suspension is isopycnic centrifugation. Both continuous and discontinuous gradients from various density media have been tested. The impure tissue suspensions, first only from rat kidney, but more recently from other species as well, were separated in two or more fraction. Both the silica-coated polymer of vinylpyrrolidone (Percoll) and especially the poly-iodinated low molecular weight density media (Metrizamide, Nycodenz) have been successfully used to separate cells (Boogaard et al 1989, Kruidering et al 1993).

Primary Cultures: In order to study one cell type specifically, primary cultures of cells derived from defined parts of the nephron have been used. Primary cultures of proximal tubular cells derived from various species have been developed, i.e.

mouse (Taub and Sato 1980), rat (Chen et al. 1989, Boogaard et al 1990), rabbit (Aleo et al 1989) and pig (Kruidering et al 1994). Although cells in primary culture also dedifferentiate, the characteristics of those cells are usually closer to the in vivo situation compared to cell lines. In the past years several attempts have been made to approach the in vivo situation as much as possible. Improved culture conditions, such as shaking for better oxygenation (Nowak and Schnellmann 1995), omission of glucose and serum (Jung et al 1992 , Blais et al 1992), addition of specific metabolic substrates (Tang and Tannen 1994) and the use of specific matrigel (Taub et al 1990) or permeable supports all contributed to better retention of in vivo characteristics.

Human Kidney Cells: Since the aim of most in vitro models is to extrapolate the obtained data to man, an in vitro model derived from human kidney would be the most realistic. Indeed, primary cultures of human kidney cells have been described. Kidneys were either from adults (nephrotectomy specimens, kidneys judged unsuitable for transplantation (Wilson et al 1985) and cadaver kidneys (Trifilis et al 1985)), or fetal kidneys obtained by therapeutic abortion (Briere 1987). The use of fetal kidney has the advantage of rapid growth, but fetal kidney never is fully differentiated at the time of isolation, resulting in different response to hormones and dependency on growth factors in culture. This dependency differs with the age of the fetus (Briere et al 1992). Therefore, fetal kidneys are mostly used to study differentiation (Briere and Chailler 1993). Optimalization of isolation- and culture - methods of human material paralleled culture of animal kidney cells. The first suspensions of kidney tubular cells were obtained by microdissection (Wilson et al 1985) or by incubation of minced cortex with digestive enzymes, after which the tubular cells were isolated by sieving the suspension (Yang et al. 1987). Optimalisation of the isolation procedure, yielding very pure suspensions of human proximal tubular cells was achieved by incubation the cell suspension with PTC specific antibodies and sorting the cells by flow cytometry (Vanderbiest et al 1994). Very pure cultures of human proximal tubular cells can be obtained by seeding proximal tubules in very selective growth media (Courjault-Cartier et al 1995). As with primary cultures derived from animal kidneys, the human PTC retained more characteristics of the PTC in situ by the use of serum-free media, specific hormones, basement membrane substrates (Yang et al 1987) and omission of glucose in the medium (Courjault-Gautier et al 1995).

Porcine Kidney Cells as an Alternative for Human Kidney Cells: One disadvantage of the use of human material is the limited availability. This can be overcome by the use of porcine kidney. Humans are physiologically more closely related to the pig than to the rat, rabbit or mouse, as indicated by cardiovascular, renal, and immunological studies (Mount and Ingham 1974). Therefore, an in vitro model using porcine kidney can provide a good model for the study of human nephrotoxicity, without the drawback of limited material. In addition, when pigs from the slaughterhouse are used, the need for laboratory animals will be reduced.

Recently, we developed an in vitro model to study nephrotoxicity using porcine kidney (Kruidering et al 1993). The isolation procedure, consisting of incubation of minced cortex with collagenase and purification of proximal tubular cells by means of a discontinuous density gradient yielded a suspension consisting of at least 83 % porcine proximal tubular cells (PPTC). PPTC in primary culture expressed keratin, characteristic for epithelia and brush border specific glycoproteins (FX1A) (Fig 1A and 1B). In addition, all cells were negative for the endothelial marker pal-E, less than 0.1 % expressed the Tamm-Horsfall protein, characteristic for the distal tubule, or desmin, characteristic for connective tissue and mesangial cells (not shown). Ultrastructural analysis revealed microvilli, tight junctions and abundant mitochondria and lysosomes, all characteristic for proximal tubular cells.

Freshly isolated PPTC were validated as in vitro model to detect nephrotoxicity, by studying the effect of mercuric chloride, cis-platin, para-aminophenol and the halogenated alkenes 1,2 dichlorovinyl-l-cysteine (DCVC), S-(1,1-difluoro-2,2-dichloroethyl)-L-cysteine (DCDFE-cys) and the glutathione conjugate of DCDFE (DCDFE-GSH) on viability. The sensitivity of the porcine cells to detect toxic effects corresponded favourably with in vitro systems derived from other animals (Kruidering et al 1994).

Established cell lines

Established cell lines have proven to be a very powerful tool in the study of nephrotoxicity in vitro. Besides many advantages, such as the possibility of long term exposure under controlled conditions, the unlimited life span, the lack of time-consuming isolation procedures, cell lines suffer from some drawbacks. Cell lines dedifferentiate in culture with as a result that cells, claimed to be of proximal tubular origin, exhibit a combination of functional properties, characteristic of different parts of the nephron. Two cell lines that are often used as model for the proximal tubule are the porcine cell line LLC-PK1 and the OK cell line from opossum kidney. The LLC-PK1 cells lack the enzyme fructose-1,6-biphosphatase, rendering the cells incapable of gluconeogenenis, which is the main energy source in proximal tubular cells situ. In addition, LLC-PK1 cells do not respond to parathyroid hormone and lack a probenecid sensitive organic anion transporter (Courjault-Gautier et al 1995). OK cells display very little gamma-glutamyl-transpeptidase activity and lack alkaline phosphatase, both specific enzymes for the proximal tubule (Gstraunthaler 1988). In order to improve the models, gluconeogenic strains of both cell lines have been developed by selective adaptation to glucose free media (Gstraunthaler 1993).

Immortalization: Combination of the advantages of a continuous cell lines with the better differentiation of primary cultures can be obtained by immortalization of the latter. Immortalization of proximal tubular cell cultures has been achieved in several ways. Cartier at al isolated proximal tubular cells from transgenic mice

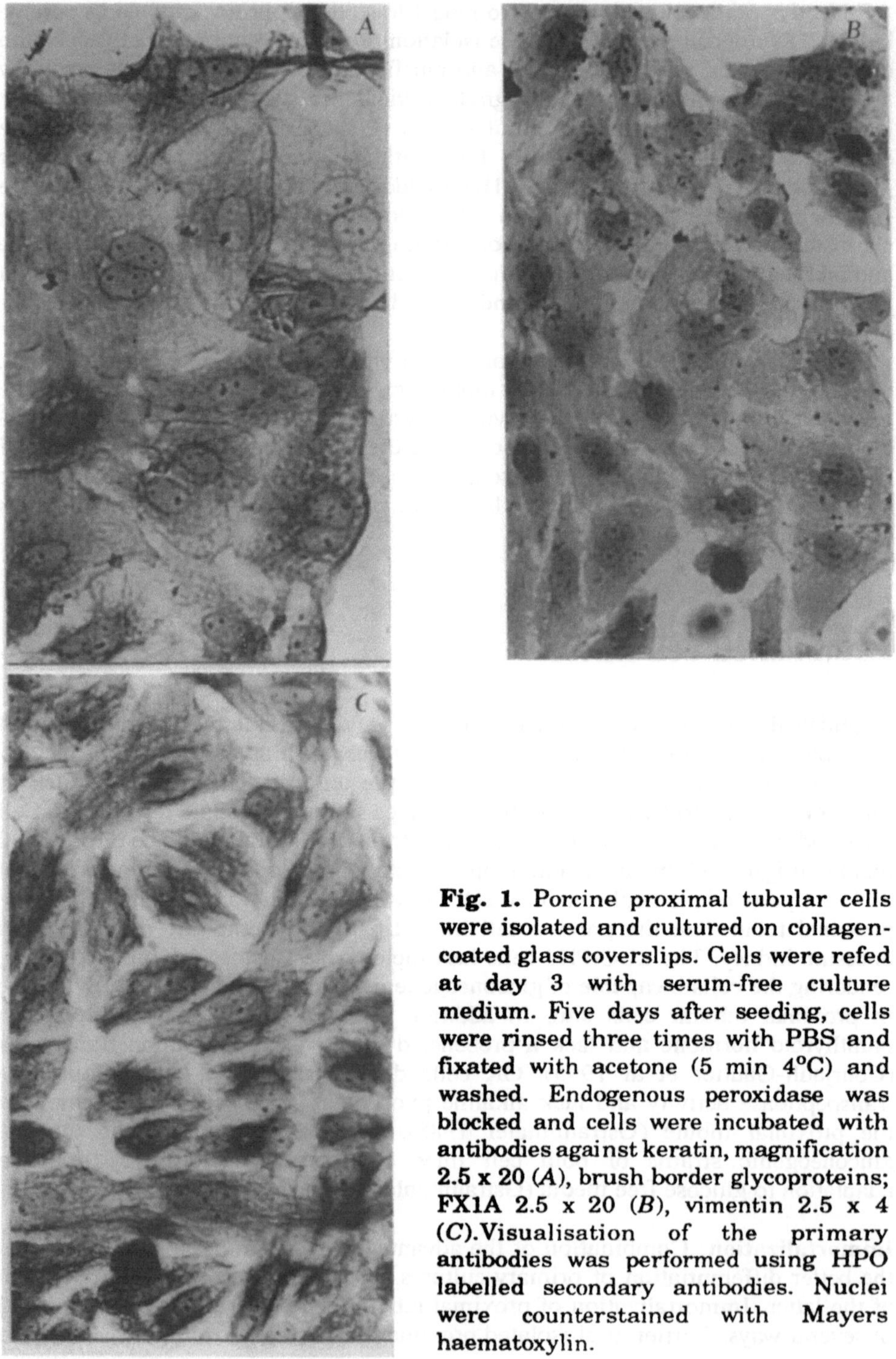

Fig. 1. Porcine proximal tubular cells were isolated and cultured on collagen-coated glass coverslips. Cells were refed at day 3 with serum-free culture medium. Five days after seeding, cells were rinsed three times with PBS and fixated with acetone (5 min 4°C) and washed. Endogenous peroxidase was blocked and cells were incubated with antibodies against keratin, magnification 2.5 x 20 (*A*), brush border glycoproteins; FX1A 2.5 x 20 (*B*), vimentin 2.5 x 4 (*C*). Visualisation of the primary antibodies was performed using HPO labelled secondary antibodies. Nuclei were counterstained with Mayers haematoxylin.

(Cartier et al. 1993), whereas others infected primary cultures of rabbit PTC either directly by addition of the virus or by coculturing the PTC with fibroblasts, excreting a recombinant retrovirus (Vandewalle et al 1989, Romero et al 1992). However, further research is needed to confirm that the cell lines keep their characteristics through multiple passages.

Toxicity Studies In Individual Proximal Tubular Cells

Recent technological innovations have made it possible to determine simultaneously multiple biochemical parameters in individual living cells. These new techniques, commonly designated as Video Microscopy (VM) and Confocal Laser Scan Microscopy (CLSM), make use of fluorescent probes, specific for a certain cellular parameter. Cells are loaded with the probes and placed under a fluorescence microscope. After excitation at a specific wavelength the probes emit light of a longer wavelength. The intensity of the emitted light is proportional to the magnitude of the specific parameter. Emission is recorded with a video-camera or a photo multiplier tube and transferred to an image processor for extraction of quantitative data from the image, or to improve image quality.

In these techniques the fluorescence of probes or antibodies is used to measure biochemical parameters or detect the cellular localisation of the probe. A large variety of probes for biochemical parameters is available nowadays and development has become a specific branch of science, see table 1 for some examples. In principle a fluorescent tag can be attached to every antibody and therefore cellular localisation of every type of antibody is possible. The strong point of VM and CLSM is that multiple parameters can be studied within one cell and in this way it is possible to make correlations between parameters i.e. separate causes from consequences. With conventional fluorescence microscopy the image is always blurred due to out-of-focus fluorescence. With confocal laser scan microscopy this fluorescence is not recorded and therefore crisp images are obtained. In addition because a laser is used as a light source, images from below the surface of the specimen can be obtained. In general, in biological tissue it is possible to visualise planes as deep as 120 mm.

Flow-cytometry is similarly based on fluorescence of specific dyes. In this technique cells are guided through a detector after excitation of the probe with a laser. From every cell three fluorescent and two scattering signals are recorded. Registration of these signals from a few thousands of cells takes no more than 10 seconds. Management of the enormous amount of data that is generated, is through user-friendly software.

We used these techniques to determine the mechanisms of toxicity of halogenated alkenes in isolated rat PTC. We determined in single cells the intracellular [Ca^{2+}] (with Fura-2), the mitochondrial membrane potential (MMP, with rhodamine 123), the formation of reactive oxygen species (ROS, with dichlorofluorescin) and the death of the cell (with propidium iodide) (van de Water et al 1993, 1994). We could conclude from these studies that the

Table 1 Examples of fluorescent probes available for studies with living cells

Indicators for	Probes
Ca^{2+}, Na^{+}, K^{+}, Mg^{2+}, Cl^{-}, Zn^{2+}.	Quin-2, Fura-2, Indo-2, Fluo-3.
Thiol groups	Bimanes, 5-chloromethyl-fluoresceinediacetate
membrane potentials	Rhodamine 123, tetramethylrhodamine
Lipid peroxidation	Dihydrofluoresceines dihydrorhodamines
Viability	Propidium Iodide, calceine
Localisation of organelles	
acidic organelles	Neutral Red
E.R.	Carbocyanines
Golgi	Ceramides
mitochondria	Rhodamines
nucleus	Acridines, propidium iodide etc.
membranes	Fluorescent phospholipids

Table 1 only contains examples of the available probes. For a complete overview consult: Molecular Probes Inc. P.O.Box 22010 Eugene, OR 97402-0414 USA for a copy of "Handbook of fluorescent probes and research chemicals".

halogenated alkene 1,2-dichlorovinylcysteine increases intracellular and mitochondrial [Ca^{2+}]. This was accompanied by a decrease of the MMP and the onset of the formation of ROS. Prevention of the increase in [Ca^{2+}] inhibited the decrease in the MMP but could not prevent it, suggesting a dual effect of DCVC.

Also the effect of the widely used antitumour drug cisplatin on freshly isolated porcine proximal tubular kidney cells (PPTC) in suspension was investigated in more detail using these techniques. Apart from the decrease in MMP and cell death, formation of ROS was observed within 20 minutes. Prevention of ROS formation with the antioxidant diphenyl-p-phenylene-diamine (DPPD) or desferrioxamine had no effect on the cisplatin-induced effects on MMP and cell death, implying that cisplatin-induced ROS formation was not a cause of cisplatin-induced cell death. We investigated whether the ROS formation was related to mitochondrial damage. Therefore we determined the effect of cisplatin on enzymatic activities of NADH:ubiquinone reductase (complex I) and succinate:ubiquinone reductase (complex II) of the respiratory chain. Exposure of the PPTC to cisplatin resulted in a time- and dose-dependent inhibition of both complex I and II activity. The inhibition of these activities as well as the depletion of ATP could not be prevented by the antioxidants, indicating that these were not a consequence of ROS formation. The results suggested that direct damage to the mitochondria may be a key event in cisplatin-induced cell death (Kruidering et al 1995). Further in depth investigation to elucidate the mechanism of cisplatin-induced toxicity at (sub) cellular level is now ongoing in our laboratory.

References

Aleo MD, Taub ML, Nickerson PA, Kostyniak PJ (1989) Primary cultures of rabbit renal proximal tubule cells: I growth and biochemical characteristics. In Vitro Cell Dev Biol 25: 776-783

Blais A, Jalal F, Crine P, Paiement J, Berteloot A (1992) Increased functional differentation of rabbit proximal tubule cells cultured in glucose-free media. Am J Physiol 263: F152-F162

Boogaard PJ, Mulder GJ and Nagelkerke JF (1989) Isolated proximal tubular cells from rat kidney as an in vitro model for studies on nephrotoxicity. I An improved method for preparation of proximal tubular cells and their functional characterization by a-methylglucose uptake. Toxicol Appl Pharmacol 101: 135-143

Boogaard PJ, Zoetwey JP, vanBerkel TJC, van 'tNoordende JM, Mulder GJ and Nagelkerke JF (1990) Primary cultures of proximal tubular cells from normal rat kidney as an in vitro model to study mechanisms of nephrotoxicity. Biochem Pharmacol 39:1335-1345

Boogaard PJ, Mulder GJ and Nagelkerke JF (1990) Renal proximal tubular cells in suspension or in primary culture as in vitro models to study nephrotoxicity. Chem-Biol Interactions 76: 251-292

Briere N (1987) Human fetal kidney explants in serum-free organ culture. Anat Embryol 176: 105-114

Briere N, Chailler P, Morin D (1992) Age-dependent regulation of insulin, transferrin and epidermal growth factor influence on proliferation of human etal kidney in culture. Growth Regulation 2: 108-114

Briere N, Chailler P (1993) Synergistic bombesin and insulin stimulation of DNA synthesis in human fetal kidney in serum-free culture Biofactors. 4: 133-137

Cartier N, Lacave R, VAllet V, Hagege J, Hellio R, Robine S, Pringault E, Cluzeaud F, Briand P, Kahn A, Vandewalle A (1993) Establishment of renal proximal tubule cell lines by targeted oncogenesis in transgenic mice using the l-pyruvate kinase-SV-40 (T) antigen hybrid gene. J Cell Science 104: 695-704

Chen TC, Curthoys NP, Lagenaur CF, Puschett JB (1989) Characterization of primary cell cultures derived from rat renal proximal tubules. In Vitro Cell Dev Biol 25: 714-722

Courjault-Gautier F, Chevalier J, Abbou CC, Chopin DK, Toutain HJ (1995) Consecutive use of hormonally defined serum-free media to establish highly diffrentiated human renal proximal tubule cells in primary culture J Am Soc Nephrol 5: 1949-1963

Gstraunthaler G, Thurner B, Weirich-Schwaiger H, Pfaller W (1993) A novel gluconeogenic strain of OK cells with metabolic properties different from gluconeogenic LLC-PK1 cells. Cellular Physiology and Biochemistry 3: 78-88

Gstraunthaler G (1988) Epithelial cells in tissue culture Renal Physiol and Biochem 11: 1-42

Heidrich HG and Dew ME (1974) Homogeneous cell populations from rabbit kidney cortex proximal, distal tubule, and renin-active cell isolated by free-flow electrophoresis. Cell Biol 74: 780--788

Jones DP, Sundby GB, Ormstad K and Orrenius S (1979) Use of isolated kidney cells for study of drug metabolism Biochem Pharmacol 28: 929-935

Jung JC, Lee SM, Kadakia N, Taub M (1992) Growth and function of primary rabbit kidney proximal tubule cells in glucose-free serum-free medium. J Cell Physiol 150: 243-250

Kruidering M, Prins FA, de Heer E, Mulder GJ, Nagelkerke JF (1993) Isolation,and culture of proximal tubular cells from porcine kidney. ATLA 21: 457-465

Mount LE and Ingam DL (1974) The pig as a laboratory animal. Academic Press London, pp1-21

Kruidering M, Maasdam DH, Prins FA, de Heer E, Mulder GJ, Nagelkerke JF (1994) Evaluation of nephrotoxicity in vitro using a suspension of highly purified porcine proximal tubular cells and characterization of the cells in primary culture Experimental Nephrology 2: 334-344

Kruidering M, Van de Water B, de Heer E, Mulder GJ, Nagelkerke JF (1995)The use of porcine proximal tubular kidney cells to study nephrotox-icity in vitro; The role of oxidative stress in cisplatin-induced cell death. Sumitted

Norgaard JOR (1976) A new method for the isolation of ultrastructurally preserved glomeruli. Kidney Int 9: 278-285

Nowak G, Schnellmann RG (1995) Improved culture conditions stimulate gluconeogenesis in primary cultures of renal proximal tubule cells. Am J Physiol: 268 C1053-C1061

Ormstad K, Orrenius Sand Jones DP (1980) Preparation and characteristics of isolated kidney cells. Methods Enzymol 77: 137--146

Phelps JS, Gandolfi AF, Brendel K and Dorr RT (1987) Cisplatin nephrotoxicity: in vitro studies with precision-cut rabbit renal cortical slices. Tox and Appl Pharmacol 90: 501-512

Prie D, Friedlander G, Coureau C, Vandewalle A, Cassingena R, Ronco PM (1995) Role of adenosine on glucagon-induced cAMP in a human cortical collecting duct cell line. Kideny Int 47:1310 -1318

Romero MF, Douglas JG, Eckert RL, Hopfer U, Jacobberger JW (1992) Development and characterization of rabbit proximal tubular epithelial cell lines. Kidney Int 42: 1130-1144

Ruegg CE, Gandolfi AF, Nagle RB and Brendel K (1987) Differential patterns of injury to the proximal tubule of renal cortical slices following in vitro exposure to mercuric chloride, potassium chromate or hypoxic conditions. Tox and Appl Pharmacol 90: 261-273

Seglen PO(1976) Preparation of isolated rat liver cells. Methods Cell Biol 13: 29-83

Tang MJ, Tannen RL (1994) Metabolic substrates alter attachment and differentiated functions of proximal tubule cell culture. J Am Soc Nephrol 4: 1908-1911

Taub M, Wang Y, Szczesny TM, Kleinman HK (1990) Epidermal growth factor or transforming growth factor a is required for kidney tubulogenesis in matrigel cultures in serum-free medium. Proc Natl Acad Sci USA 87: 4002-4006Taub M, Sato G (1980) Growth of Functional Primary Cultures of Kidney Epithelial Cells in Defined Medium Journal of Cellular Physiology. 105: 369-378

Toutain H, Vauclin JN, Fillastre JP and Morin JP (1989) Isolation of a pure suspension of proximal tubular cells from the rat kidney: morphological, histochemical and metabolic assessment. Cell Biol Int Rep 18: 701--710

Trifillis AL, Regec AL, Trump BF (1985) Isolation, culture and characterization of human renal tubular cells. J Urol 133: 324-329

Van der Biest I, Nouwen EJ, van Dromme SA, de Broe ME (1994) Charac-terization of pure proximal and heterogenous distal human tubular cells in culture Kidney Int. 45: 85-94

VandeWalle A, Lelongt B, Geniteau-Legendre M, Baudouin B, Antoine M, Estrade S, Chatelet F, Verroust P, Cassingena R, Ronco P (1989) Maintenance of proximal and distal cell functions in SV-40 transformed tubular cell liens derived from rabbit kidney cortex. J Cell Physiol 141: 203-221

Vandewalle A, Taub M, Cluzeaud F, Ronco R, Chatelet F, Verroust P and Poujeol P (1986) Indirect immunoselection of late distal cell populations from rabbit kidney cortex. Am J Physiol 250: F387-F395

Williams PD (1989) The application of renal cells in culture in studying drug-induced nephrotoxicity. In vitro Cell Dev Biol 25: 800-805

Wilson PD, Dillingham MA, Breckon R, Anderson RJ (1985) Defined human renal tubular epithelia in culture: growth characterization, and hormonal response. Am J Physiol 248: F436-F443

Yang A, Gould-Kosta J, Oberley TD (1987) In vitro growth and differentiation of human kidney tubular cells on a basement membrane substrate. In vitro Cell Dev Biol 23: 34-46

Site Specific Drug and Xenobiotic Induced Renal Toxicity

G.M.Hawksworth, R. McCarthy, T. McGoldrick, V. Stewart, K. Tisocki and E.A. Lock[1]

Departments of Medicine and Therapeutics and Biomedical Sciences, Polwarth Building, Foresterhill, Aberdeen AB9 2ZD, UK

[1]Research Toxicology Section, Zeneca, Central Toxicology Laboratory, Macclesfield, Cheshire SK10 4TJ, UK

The marked morphological, functional and biochemical heterogeneity of the kidney accounts for the site-specific toxicity of several drugs and xenobiotics. Puromycin aminonucleoside, daunomycin and adriamycin exert a direct effect on the glomerular epithelial cells, resulting in foot process fusion, accompanied by increased proteinuria, whereas the proximal tubule, in particular the S_2 and S_3 segments, is the target site for aminoglycoside, cephalosporin, cisplatin, halogenated alkene and heavy metal toxicity (Commandeur and Vermuelen 1990).This is assumed to be due to the presence of specifc transport processes and/or enzymes e.g. cysteine conjugate C-S lyase, located in the proximal tubule. Ischaemic damage to the proximal tubule is usually confined to the S_3 segment or *pars recta* in the outer medulla, whereas chemically induced proximal tubular injury is dependent on the physicochemical properties of the toxicant and the localisation of transport systems that result in cell-specific accumulation. For example Cd-metallothionein is specifically reabsorbed by the S_1 and S_2 segments of the proximal tubule, whereas cephaloridine is toxic to the S_2 segment, due to the localisation of the organic anion transporter. The cytotoxicity of cysteine S-conjugates is determined by the activities of several enzymes, including cysteine conjugate C-S lyase, localised in the S_1 and S_3 segments.

The long-term administration of analgesics and therapeutic doses of non-steroidal anti-inflammatory drugs has been implicated in renal papillary necrosis, the target cell being the medullary interstitial cell (MIC). Several of these drugs undergo metabolic activation via the prostaglandin synthetase and/or lipoxygenase systems in the kidney and this has been proposed as a mechanism responsible for the papillary necrosis (Bach and Bridges, 1984).

This paper will focus on the mechanisms responsible for drug-induced glomerular epithelial cell damage and on the proximal tubular damage caused by halogenated alkenes and drug-induced papillary necrosis.

Drug-induced glomerular epithelial cell toxicity

Glomerular podocyte dysfunction in disease states leads to marked changes in capillary permselectivity and subsequent proteinuria. Normal glomerular visceral epithelial structure i.e. tertiary foot process and slit diaphragm configuration is considerably altered in human minimal lesion disease, primary focal segmental glomerulosclerosis and membranous glomerulonephritis. In rat models of glomerular epithelial cell toxicity, i.e. in puromycin aminonucleoside (PAN) and adriamycin-treated animals, the same structural and functional abnormalities are observed. Although PAN-induced nephrotic syndrome has been used as an experimental model of the human minimal change nephrotic syndrome for many years, the pathogenesis remains unclear. Macrophage involvement, increase in TNFα mRNA and TGFβ mRNA have been implicated in the development of the PAN-induced glomerular sclerosis (Nakamura et al, 1993, Jones et al, 1992), but the early events may reflect an alteration in *de novo* protein synthesis of extracellular matrix protein, changes in cytoskeletal aggregation, cell-to-cell attachment or attachment to the glomerular basement membrane (Whiteside et al, 1993).

Recently it has been shown that some N-hydroxyurea derivatives being developed as 5-lipoxygenase inhibitors for the treatment of certain allergic and inflammatory conditions caused a dose-dependent severe glomerular nephropathy in the rat (Read et al, 1995). The earliest changes detected in the kidney by transmission electron microscopy were noted in the glomeruli, in which the visceral cells appeared enlarged and showed varying degrees of foot process loss. Soon after the onset of foot process loss, decreases in the total plasma protein and albumin and increases in the plasma cholesterol, triglycerides, urea and creatinine were noted (Table 1).

Table 1. 7-day toxicitv study of N-OH urea 5-LO inhibitor

Parameters measured	Control	100 mg/kg
Total plasma protein (g/L)	60±3	51±9
Plasma cholesterol (mM/L)	1.9±0.5	3.8±2.1
Urinary total protein (mg/5 h)	0.5 ± 0.1	25.1±30.9
% kidney wt/body wt	0.76± 0.06	0.88± 0.13

Read et al. Arch. Toxicol. 1995, 69, 480.

These changes are all characteristic of „minimal change nephrotic syndrome". No effect was seen when the related hydroxamic acid 5-lipoxygenase inhibitors were administered to Wistar rats. We are currently investigating the early events responsible for glomerular epithelial cell toxicity after administration of the N-hydroxyurea compounds using isolated rat and human glomeruli.

Haloalkene Cysteine S-Conjugate Proximal Tubular Toxicity

Halogenated alkenes such as hexafluoropropylene, tetrafluoroethylene, trichloroethylene, hexachlorobutadiene and chlorotrifluoroethylene have been implicated in glutathione-mediated nephrotoxicity in rats. The S-cysteine conjugates cause necrosis in the straight portion of the proximal tubule, a region which has been shown to contain C-S lyase by immunohistochemical localisation (Jones et al, 1988, Macfarlane et al, 1989). The mechanism of toxicity involves C-S lyase, which cleaves the S-cysteine conjugate thio-ether linkage by β-elimination to yield ammonia, pyruvate and a toxic thiol species (Lock, 1988). Human exposure to haloalkenes may be through chronic industrial or long-term low-dose environmental exposure (Green, 1990, Mutti et al, 1992). It is therefore important not only to use animal models in which toxicity is documented, but to quantify C-S lyase activity in human kidney tissue.

Cytosolic C-S lyase activity was approximately ten fold less in human renal cortical samples compared with rat samples and less than five fold interindividual variation was observed. Cytosolic activity was greatest with S-(1,2-dichlorovinyl)-L-cysteine (DCVC) and S-(1,1,2,2-tetrafluoroethyl)-L-cysteine (TFEC). In contrast to the rat, activity towards S-(2-chloro-1,1-difluorethyl)-L-cysteine (CDFEC) and S-(1,2,2-trichlorovinyl)-L-cysteine was very low (Table 2).

Table 2. C-S lyase activity in rat and human cytosolic fractions

Conjugate	Rat	Human
	nmol pyruvate/mg protein/120 min	
TFEC	707±91	74±38
CDFEC	304±16	9±4
DCVC	192±29	72±27
TCVC	536±41	3±2

No cleavage of aromatic conjugates was detectable. Structure activity relationships showed that increasing the fluorinated carbon chain length of the halogenated hydrocarbon species decreased conjugate cleavage by C-S lyase (McCarthy et al 1994). Mitochondrial activity was lower than cytosolic activity,

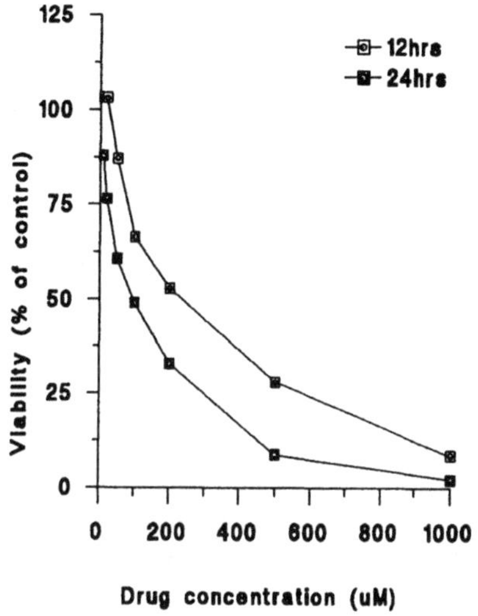

Figure 1. Toxicity of DCVC. Human proximal tubular cells were grown to confluence (120 hours) on 96 well microtitre plates. Toxicity was assessed by the MTT assay

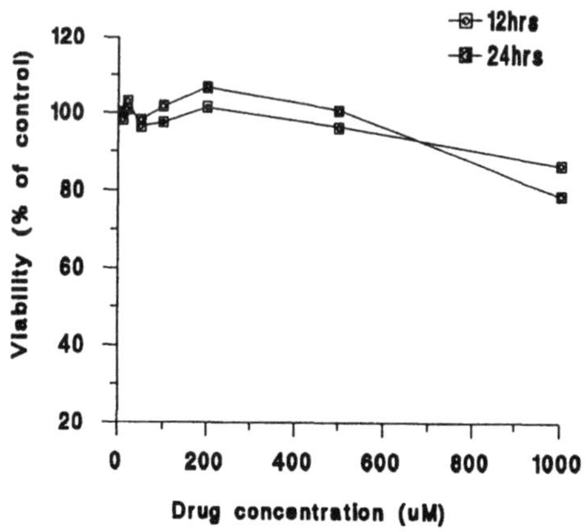

Figure 2. Toxicity of TFEC. Human proximal tubular cells were grown to confluence (120 hours) on 96 well microtitre plates. Toxicity was assessed by the MTT assay.

but there was no constant ratio in paired samples from human kidney cortices.

When the toxicity of the four conjugates to human proximal tubular cell cultures was compared, DCVC was the most toxic, with TFEC showing slight toxicity at 24h, whereas CDFEC and TCVC exhibited negligible toxicity (Figs.1 and 2). Toxicity measurements were carried out after 120 hr in culture, by which time C-S lyase activity had fallen to approximately 40% of fresh cell values.

In short term toxicity tests DCVC was again the most toxic to human proximal tubular cell suspensions, lactate dehydrogenase leakage being 37% and 43% after 1 hr incubation with 50 and 500 μM DCVC respectively. It should be noted that DCVC also undergoes oxidative metabolism, which may contribute to the toxicity when C-S lyase levels are low. The activities of N-acetyltransferase, acylase, cysteine S-conjugate and L-α-hydroxy-acid oxidase will also contribute to the toxicity in the intact cell. Chen et al (1990), however, showed that the toxicity of DCVC to human proximal tubular cell cultures was completely inhibited by the addition of aminooxyacetic acid, suggesting that even 24 hr after the cells reached confluence C-S lyase is mainly responsible for the cytotoxicity observed.

Toxicity is also dependent on the transport mechanisms. *In vitro* cysteine conjugates may be taken up into the cell by both sodium-dependent and -independent systems (Lash and Anders, 1989, Zhang and Stevens, 1989). However *in vivo* it is possible that uptake of the mercapturates (the main circulating metabolites of halogenated alkenes) by the proximal tubular cells may play an important role in the toxicity of these compounds.

Experiments using isolated human proximal tubular cell suspension and human proximal tubular cells cultured on semipermeable membranes, together with expression of the human C-S lyase in systems exhibiting or co-expressing the carrier systems found *in vivo* will clarify the molecular mechanisms involved.

Drug-induced renal papillary necrosis (RPN)

Several hypotheses have been postulated to account for the development of drug-induced RPN. Counter-current concentration and medullary ischaemia have been implicated, but an attractive hypothesis is metabolic activation of the drugs by prostaglandin synthetase (PGS) (Eling and Curtis 1992) or lipoxygenases (LO) (Naidu and Kulkarni 1994) present in the renal papilla. Bioactivation of xenobiotics can occur via a peroxidase mediated pathway, a peroxyl radical mediated pathway or by means of co-substrate-derived oxidant. The oxidised species is often a highly reactive free radical. Arachidonic acid (AA) dependent cooxidation of xenobiotics in the kidney exhibits a corticopapillary gradient (Fig 3). In addition distinct interspecies variation is seen in terms of renal PGS and LO activities. PGS is primarily responsible for AA-dependent metabolism of xenobiotics in the rabbit, whilst a combination of PGS and LO catalysed drug oxidation in rat kidney (Stewart et al, 1993). In contrast in man, AA-dependent co-oxidation of drugs in renal tissue occurred via lipoxygenase. Use of selective inhibitors confirmed that the enzyme was a 5-lipoxygenase and not a 12- or 15-lipoxygenase (Table 3).

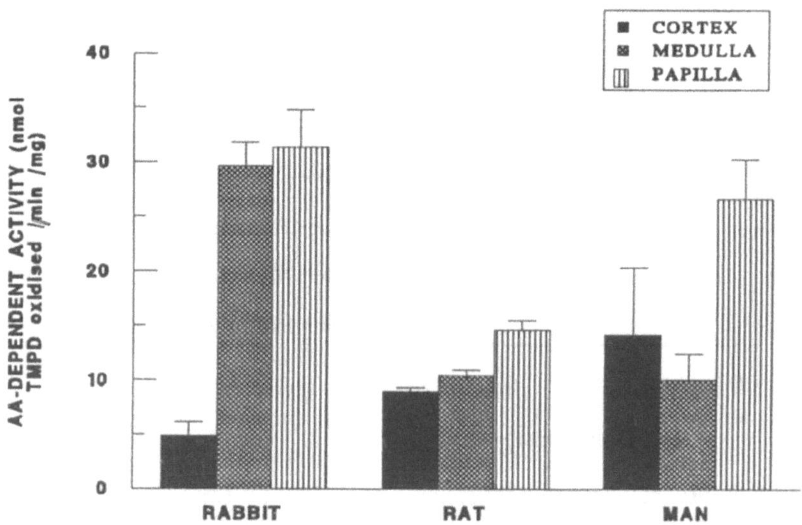

Figure 3. Arachidonic acid dependent cooxidation of TMPD by renal microsomes

Table 3. Inhibition of partially purified human papillary lipoxygenase

Inhibitor	% inhibition
10 μM BWA4C	90
10 μM NDGA	89
50 μM BW755C	67
100 μM Baicalein	5
50 μM ETYA	0
100 μM Indomethacin	0

Activity of the purified human papillary enzyme was dependent on ferrous ions, but in contrast to other lipoxygenases was microsomally located and was calcium and ATP independent.

We have isolated and cultured human renal medullary interstitial cells (hRMIC) (Tisocki and Hawksworth, 1996), necrosis of these cells being one of the first degenerative changes seen in vivo, and demonstrated PGE_2, hydroxy-eicosatetraenoic acid and hydroperoxytetraenoic acid production by these cells. PGE_2 formation was enhanced in the presence of arachidonic acid and the calcium ionophore A12387. When the cells were exposed to a range of non-steroidal and anti-inflammatory drugs, paracetamol did not cause a significant decrease in viability after 24hr exposure, but showed dose-dependent cytotoxicity on prolonged exposure (7 days). Aspirin, mefenamic acid and phenylbutazone were cytotoxic to hRMIC and this effect was enhanced after prolonged exposure (Figs 4 and 5).

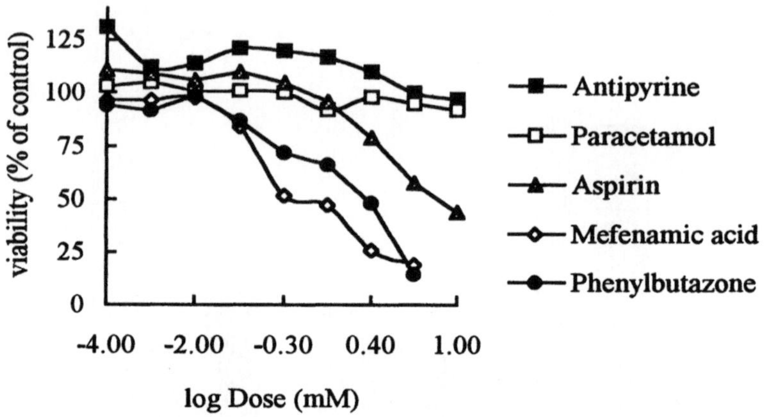

Figure 4. Cytotoxicity of various analgesics after 24 hour treatment. Human RMIC were grown to near confluence on 96 well plates and viability was assessed by the MTT assay.

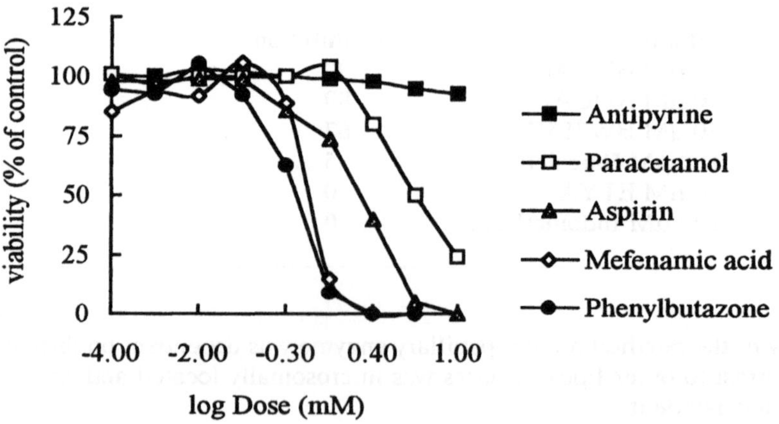

Figure 5. Cytotoxicity of various analgesics after 7 days treatment. Human RMIC were grown to near confluence on 96 well plates and viability was assessed by the MTT assay.

5-LO may be implicated in the development of RPN in man, either by metabolic activation of drugs or production of proinflammatory products, including the leukotrienes LTB_4, LTC_4, LTD_4 and LTE_4, causing alteration of the renal eicosanoid balance, leading to altered renal haemodynamics. It is possible therefore that inhibitors of 5-LO or 5-lipoxygenase activating protein (FLAP) may provide the means to protect against, or reverse, drug-induced papillary necrosis.

Acknowledgements

This work was supported by the Wellcome Trust, Glaxo Wellcome UK, the National Kidney Research Fund and an MRC Collaborative studentship with Zeneca, Central Toxicology Laboratory, UK.

References

Bach PH and Bridges JW (1984) The role of metabolic activation of analgesics and non-steroidal antiinflammatory drugs in the development of renal papillary necrosis and upper urothelial carcinoma. Prostaglandins Leukotrienes Med 15: 251-274

Chen IC, Stevens JL, Trifillis AL and Jones TW (1990) Renal cysteine conjugate β-lyase-mediated toxicity studied with pnmary cultures of human proximal tubular cells. Toxicol Appl Pharm 103:463-473

Commandeur JNM and Vermuelen NPE (1990) Molecular and biochemical mechanisms of chemically induced nephrotoxicity: a review. Chem Res Toxicol 3:171-194

Eling TE and Curtis FJ (1992) Xenobiotic metabolism by prostaglandin synthase. Pharmac Ther 53:261-73

Jones TW, Qin C, Schaeffer VH and Stevens JL (1988) Immunohistochemical localization of glutamine transaminase K, a rat kidney cysteine conjugate betalyase, and the relationship to the segment specificity of cysteine conjugate nephrotoxicity. Mol Pharmacol 34:621-627

Jones CL. Buch S, Post M, McCulloch L, Liu E and Eddy AA (1992) Renal extracellular matrix accumulation in acute puromycin aminonucleoside nephrosis in rats. Am J Path 141:1381-1396

Green T (1990) Chloroethylenes: a mechanistic approach to human risk evaluation Ann Rev Pharmacol Toxicol 30:73-89

Lash LH and Anders MW (1989) Uptake of nephrotoxic S-conjugates by isolated rat renal proximal tubular cells. J Pharm Exp Ther 248:531-537

Lock EA (1988) Studies on the mechanism of nephrotoxicity and nephrocarcinogenicity of halogenated alkenes. Crit Rev Toxicol 19:23-42

Macfarlane M, Foster JR, Gibson GG, King LJ and Lock EA (1989) Cysteine conjugate beta-lyase of rat kidney cytosol: characterization, immunochemical localization, and correlation with hexachlorobutadiene nephrotoxicity. Toxicol Appl Pharmacol 98: 185-197

McCarthy RI, Lock EA and Hawksworth GM (1994) Cytosolic C-S lyase activity in human kidney samples - relevance for the nephrotoxicity of halogenated alkenes in man. Toxicol Ind Health 10:103-112

Mutti A, Alinovi R, Bergamaschi E et al (1992) Nephropathies and exposure to perchloroethylene in dry-cleaners. Lancet 340:189-193

Naidu KA and Kulkami AP (1994) Lipoxygenase: a non-specific oxidative pathway for xenobiotic metabolism. Prostagl Leukotr and Ess Fatty A 50:155-159

Nakamura T, Ebihara I, Fukui M, Tabahashi T. Tomino Y and Koide H (1993) Altered glomerular steady-state levels of tumour necrosis factor-α mRNA during nephrotic and sclerotic phases of puromycin aminonucleoside nephrosis in rats Clin Sci 84:349-356

Read NA, Astbury PJ, Evans GO, Goodwin DA and Rowlands A (1995) Nephrotic syndrome associated with N-hydroxyureas, inhibitors of 5-lipoxygenase Arch Toxicol 69:480-490

Stewart VC, Tisocki K, Bell JA, Whiting PH and Hawksworth GM (1993) Species differences in prostaglandin synthetase/lipoxygenase mediated cooxidation in the kidney. Br J Clin Pharmac 36 165-165P

Tisocki K and Hawksworth GM (1996) Culture of human renal medullary interstitial cells, in: Methods in Molecular Biology, Human Cell Culture Protocols Ed. Jones GE, Humana Press Inc (in press)

Whiteside CI, Cameron R, Munk S and Levy J (1993) Podocytic cytoskeletal disaggregation and basement-membrane detachment in puromycin aminonucleoside nephrosis Am J Pathol 142:1641-1653

Zhang G and Stevens JL (1989) Transport and activation of S-(1,2dichlorovinyl)-L-cysteine and N-acetyl-S-(1,2-dichlorovinyl)-L-cysteine in rat kidney proximal tubules. Toxicol Appl Pharmacol 100:51-61

Metal Antidotes-
New Developments

(Chairs: M. Johnson, UK, and
K. Bláha, Czech Republic)

Mobilizers of Stored Cadmium

K. Bláha[1] and M. Cikrt[2]
[1]Ministry of the Environment of the Czech Republic, Vršovická 65, 100 10 Prague 10
[2]National Institute of Public Health, Srobárova 48, 100 42 Prague 10

Introduction

Cadmium occupies a special position among all toxic metals found in the environment and used in industry (Nordberg et al. 1985). It has a long biological half-life and is excreted very slowly from its soft tissue deposits. The ability to induce metallothionein (MT) and to form stable intracellular MT complexes makes the mobilization of Cadmium by chelation, particularly with regard to chronic exposure, extremely difficult.

Numerous reviews dealing with the chelation of cadmium have been published in the past 15 years (May and Bullman 1983, Andersen 1984). In the most recent one (Jones and Cherian 1990) the authors not only trace the history of the attempts to develop effective antagonists for chronic cadmium intoxication, but also conclude that compound(s) suitable for use in humans may become available in the near future.

This contribution is focused on dithiocarbamates (DTCs), which have been found quite promising chelators of „aged“ cadmium within the past decade. Because of the limited space available for this contribution, it cannot be comprehensive; it should rather aim at encouraging a critical overview which may facilitate further experimental work in this field.

History of cadmium chelation

Almost 50 years ago, 2,3-dimercapto-1-propanol (BAL) has been found efficient for the treatment of acute poisoning (Gilman et al. 1947). Later on, BAL was proven to be also effective in chronic intoxication; however, severe adverse effects were observed (Dalhamn and Friberg 1955). Extensive investigations have been conducted worldwide on vicinal dithiols, namely on water-soluble analogs of BAL (Cherian et al. 1982, Jones et al. 1982) and on polyaminopolycarboxylates like EDTA or DTPA (Eybl et al. 1966). Despite of considerable efforts, very little success was achieved. Many chelators could increase the biliary or urinary excretion of cadmium only when administered no later than 30 min after the cadmium application (i.e. before MT synthesis). If this interval was increased to 24 hours, only BAL was effective (Jones and Cherian 1990). It became obvious that the presence of vicinal thiol groups and a relatively high lipophilicity of the compound, as well as the ability to cross cellular membranes to reach intracellular cadmium binding sites were the major requirements for activity.

Dithiocarbamates (DTCs) - chelators of choice

Several derivatives of the general formula shown in Fig. 1 have been synthesized in the early 80ies and tested for their use in acute cadmium intoxication (Gale et al. 1981, Cantilena and Klaassen 1981, Jones et al. 1982). In these early studies a distinct relationship between the polarity of R1 and R2 (Fig. 1) and the mobilizing efficacy was observed (Jones and Jones 1984).

$$R_1 - N - C = S$$

$$\quad\;\; | \quad\;\; |$$

$$\quad R_2 \quad S^{(-)} \qquad Na^{(+)}$$

$$R_1(R_2)NH + CS_2 \xrightarrow[\text{NaOH aq.}]{} DTC$$

Fig.1 General formula and synthesis of DTCs

Some DTCs (listed in Table 1) were also tested in rats exposed chronically (Cikrt et al. 1984), and a QSAR study confirmed the relationship metioned above (Bláha et al. 1984).

This QSAR study revealed that the substitution by polar, non-ionic groups increases the efficacy (DE vs DD). On the other hand, charged groups (as e.g. in IDA, SAR) have a negative effect. It was found that an optimal balance between hydrophilic and hydrophobic properties of the substituents should be aimed at.

The investigation of N-methyl-D-glucamine-DTC (Shinobu et al. 1984) showed that the additon of a hexose moiety is a good choice. Several derivatives (Table 2) were tested both for acute and for chronic cadmium intoxication in mice and rats (Kojima et al. 1987, Jones et al. 1991).

Table 1. Changes in polarity of R_1 and R_2

R_1	R_2	Abbr.
CH_3	CH_3	DM
CH_2CH_3	CH_2CH_3	DD
CH_2CH_2OH	CH_2CH_2OH	DE[a]
CH_3	CH_2CH_2OH	MH
CH_3	CH_2COONH_4	IDA
CH_2COONH_4	CH_2COONH_4	SAR
N	O	M
N	$CONH_2$	INA[b]
H	CH_2CH_3	(H)E[c]
CH_3	n-C_8H_{15}	MnO[c]

[a] The most efficient in enhancement of biliary excretion.
[b] The most efficient in enhancement of urinary excretion.
[c] Highly toxic.

Table 2. Hexose containing DTCs

R_1	R_2	Abbr.
CH_3	glucose	MG
C_6H_5-CH_2	glucose	BG
CH_3-C_6H_4-CH_2	glucose	MeBG
CH_3O-C_6H_4-CH_2	glucose	MeOBG
C_6H_5-CH_2	galactose	BGA
C_6H_5-CH_2	lactose	BLA
C_6H_5-CH_2	mannose	BMN

All of these DTCs were more effective than, and were effective even at lower doses as, the DTCs listed in Table 1. Moreover, neither signs of toxicity nor of adverse effects (i.e. redistribution of cadmium to the brain) were observed. The above findings led to the synthesis of DTCs containing two hexose molecules (Jones et al. 1990). These derivatives are shown in Fig. 2.

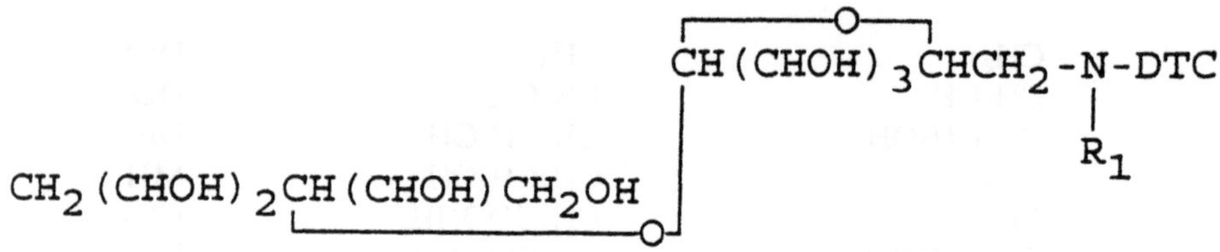

R_1	Abbr.
CH_3-C_6H_4-CH_2	MeBL
CH_3O-C_6H_4-CH_2	MeOBL
C_6H_5-CH_2	BL

Fig. 2. DTCs containing two hexoses

Recently we have found (Bláha et al. 1995) that MeBL-DTC is the most efficient chelator for increasing the biliary excretion of cadmium in chronically exposed rats. Parallel to the experimental investigations, QSAR studies were carried out in an attempt to optimize the DTC structure (Jones et al. 1988, 1990). It was found that the mobilizing efficacy (Eq. 1) is a function of, at least, five variables (Eq. 2)

(1) $E = [\text{cadmium}]T \,/\, [\text{cadmium}]C$

[cadmium]T= cadmium concentration in tissue of DTC treated animal
[cadmium]C= cadmium concentration in tissue of control animal

(2) $E = E\ \{ K_{eff}, {}^{a}\times, M, Z, HB \}$

Keff = effective stability constant of cadmium-DTC
ª × = sum of Hansch's constants of N-substituents
M = molecular weight of DTC
Z = sum of ionic charges of N-substituents
HB = factor relating to hydrogen bonding ability of N-substituents

We believe that the application of this analysis should lead to compounds suitable for clinical use.

Further perspectives

Changes in the structure of DTCs may be limited for example by synthetic problems. Therefore, other approaches to improve the mobilizing ability of these chelators were examined.

The concept of *mixed ligand chelation*, i.e. the simultaneous application of two chelators with a similar mode of action (Cherian, 1980) might be advantageous for achieving the goal. Simultaneous administration of DE (increasing biliary excretion) and INA (enhancing urinary excretion, see Table 1) led to a *synergistic effect* on the biliary excretion of cadmium in chronically exposed rats (Bláha et al. 1988). This effect was even more pronounced after the administration of a mixture of MEBL and INA. In this case, however, the effect on urinary excretion was antagonistic, i.e. resulting in a lower urinary excretion in comparison with the sum of effects for both DTCs, administered individually (Bláha et al. 1995). Alternatively, enhancement of the *binding ability* may lead to more efficient chelators. The examination of bis-DTC, a bidentate ligand, chelating cadmium as shown in Fig. 3, demonstrated that such similar structures may be promising (Singh et al. 1994).

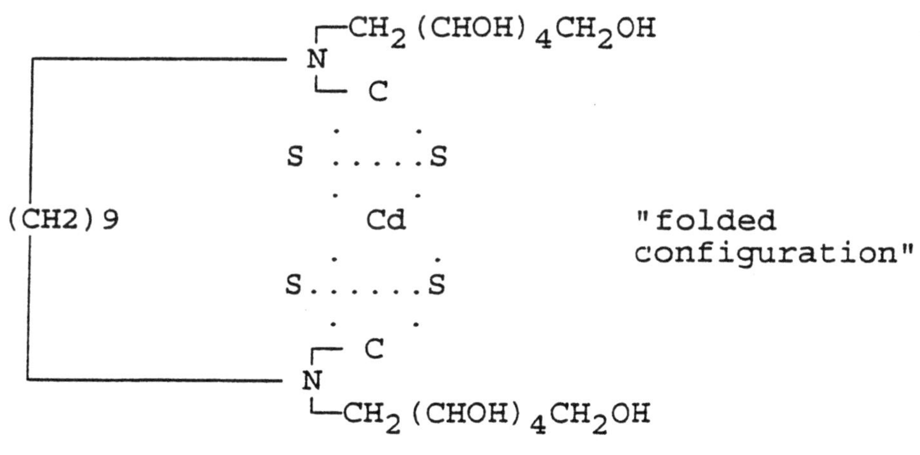

Fig. 3. Disodium N,N'-bis-(D-glucosyl)-1,9-nonanediamine-N,N'-bis-dithiocarbamate (C9G2 DTC)

In conclusion, this overview of the progress achieved in the past decade shows that the belief of having in the foreseeable future an efficient antagonist, suitable for use in chronic cadmium intoxication in humans, looks real.

Acknowledgement

We owe many thanks to Prof. Mark M. Jones, Vanderbilt University, Nashville, TN, to Prof. V. Eybl, Medical Faculty of the Charles University, Pilsen, and especially to all coworkers from the National Institute of Public Health, Prague for the friendly and fruitful collaboration during the past 13 years.

References

Andersen O (1984) Chelation of cadmium. Environ. Health Perspect., 54 249.

Bláha Jr K, Cikrt M, Kašparová L and Jones MM (1988) Synergistic enhancement of the biliary excretion of cadmium by the simultaneous administration of two dithiocarbamates. Toxicology Lett. 42: 207

Bláha K, Nerudová J, Jehlicková H, Cikrt M, Jones MM and Singh PK (1995) *In vivo* and *in vitro* efficacy of a new carbodithioate for the mobilization of cadmium. J. Tox.Environ. Health, 44: 87

Cantilena Jr LR, Irwin G, Preskorn S and Klaassen cadmium (1982) The effect of diethyldithiocarbamate on brain uptake of cadmium. Toxicol. Appl. Pharmacol. 63: 338.

Cherian MG (1980) Chelation of cadmium with BAL and DTPA. Nature 287: 871

Cherian MG, Onosaka S, Carson GK and Dean PAW (1982) Biliary excretion of cadmium in rat V. Effects of structurally related mercaptans on chelation of cadmium from metallothionein, J. Toxicol. Environ. Health, 9: 389.

Cikrt M, Basinger MA, Jones SG and Jones MM (1986) Structural effects in the dithiocarbamate enhanced biliary excretion of cadmium. J. Toxicol. Environ. Health 17: 429.

Cikrt M, Bláha K, Nerudová J, Bittnerová D, Jehlicková H and Jones MM (1992) Distribution and excretion of cadmium and nickel after simultaneous exposure and the effect of N-Benzyl-D-Glucamine Dithiocarbamate od Their Biliary and urinary excretion. J.Tox. Environ. Health, 35: 211-220.

Cikrt M, Lepsi P, Bittnerová D and Jones MM (1984) The effect of sodium diethaolamine dithiocarbamation distribution and excretion of cadmium in rats drinking cadmium-containing water. Res. Commun. Chem. Pathol. Pharmacol., 43: 497.

Cikrt M, Lepši P, Horáková L, Bláha K, Jr, Bittnerová D and Jones MM (1986) Effects of dithiocarbamates on cadmium distribution and excretion in chronically exposed rats., J.Tox.Environ Health, 17: 419

Cherian MG (1980) Chelation of cadmium with BAL and DTPA. Nature 287: 871

Cherian MG, Onosaka S, Carson GK and Dean PAW (1982) Biliary excretion of cadmium in rat V. Effects of structurally related mercaptans on chelation of cadmium from metallothionein, J. Toxicol. Environ. Health, 9: 389.

Dalhamn T and Friberg L (1955) Dimercaprol (2,3-dimercaptopropanol) in chronic cadmium poisoning. Acta Pharmacol. Toxicol.: 1168

Eybl V, Sykora J and Mertl F (1966) Einfluss der Chelatbildner auf die Ausscheidung des Cadmiums bei Cadmiumvergiftung. Nauyn Schmids Arch. Exp. Pathol. Pharmakol.: 252

Gilman A, Philips FS, Allen RP and Koelle ES (1946) The treatment of acute cadmium intoxication in rabbits with 2,3-dimercaptopropanol (BAL) and other mercaptans. J. Pharmacol.Exp. Ther. 87, Suppl. No.4: 85.

Jones MM and Cherian MG (1990) The search for chelate antagonists for chronic cadmium intoxication. Toxicology 62: 1

Jones SG and Jones MM (1984) Structure activity relationship among dithiocarbamate antidotes for acute cadmium chloride intoxications. Environ. Hlth. Persp. 54: 285

Jones SG, Jones MM, Basinger MA, Burka LT and Shinobu LA (1983) Polarity and antidotal efficacy of dithiocarbamates in acute cadmium intoxication. Res. Comm. Chem. Pathol and Pharmacol. 40: 1

Jones MM, Singh PM, Jones SG (1990) Optimization of chelatating agent structure for the mobilization of aged renal and hepatic cadmium deposits: N-benzyl-4-O-(Ñ-O-galactopyranosyl)-D-glucamine-N-carbodithioate. Chem. Res. Toxicol 3: 248

Jones SG, Singh PK and Jones MM (1988) Use of the Topliss scheme for the design of more effective chelating agents for cadmium decorporation. Chem. Res. Toxicol. 1: 234.

Kojima S, Kiyozumi M and Saito K (1981) Studies on poisonous metals. II Effect of chelating agents on excretion of cadmium through bile and gastrointestinal mucosa in rats. Chem. Pharm. Bull. 21: 16

Nordberg GF (1984) Chelating agents and cadmium toxicity, problems and prospects. Environ. Health Persp. 54: 213.

Shimada H, Kamenosono T, Kawagoe M, Funakoshi T, Kiyozumi M, Takadate A and Kojima S (1991) Mobilization of Renal and Hepatic Cadmium by Dithiocarbamates in Rats. J. Pharmacobio-Dyn., 14: 555

Shinobu LA, Jones SG and Jones MM (1984) Sodium N-methyl-D-glucamine dithiocarbamate and cadmium intoxication. Acta Pharmacol. Toxicol. 54: 189#

Chemical, Pharmacological, Toxicological and Therapeutic Advances of Deferiprone (L1) and Other Iron and Aluminium Chelators

G. J. Kontoghiorghes
Department of Haematology, Royal Free Hospital School of Medicine, London NW3 2QG, UK

A global breakthrough in oral chelation therapy with the α-ketohydroxypyridine chelating drug 1,2-dimethyl-3-hydroxypyrid-4-one (L1, INN/BAN: Deferiprone) (Kontoghiorghes 1982) in iron and aluminium overloaded patients has recently been achieved. Since the initiation of clinical trials with L1 in 1987 and prior to its recent registration in India, over 800 patients with ten different conditions in 16 countries have received it, in some cases daily for over 6 years. The major toxic side effects, which all appear to be reversible, include agranulocytosis, musculoskeletal and joint pains, zinc deficiency and gastric intolerance. Oral L1 appears to be as effective as subcutaneous desferrioxamine (DF) in iron and aluminium removal and has low toxicity. Other α-ketohydroxypyridines are also currently being developed.

The role of metal ions in biological systems varies widely, with some being essential for life while others are very toxic. Some metals for example, are used for medicinal purposes eg as therapeutics, in nutrition, as radiopharmaceuticals and also in nuclear medicine. In a different context, the epidemiology of major metal imbalance and toxicity could be classified into three major categories, namely metal deficiency, metal overload and metal toxicity. Exposure to toxic environmental pollutant metals, including the actinides used in the nuclear industry, is currently the subject of a major debate, especially between environmentalists and politicians. However, iron overload is generally being considered as the metal related toxic condition with the highest mortality rate worldwide.

Iron is an essential element for almost all living organisms. In humans, iron regulation is governed mainly by gastrointestinal iron absorption and the rate of iron turnover during the erythropoietic activity of the bone marrow, with iron excretion playing a smaller role. Iron metabolic disorders are common, with iron deficiency affecting about a quarter of the world's population at different stages of life. Iron overloading disorders are less common but have higher morbidity and mortality rates. At least 100 000 children are believed to be born with thalassaemia every year, the vast majority of whom require red blood cell transfusions for survival (WHO Bulletin 1983). Excess body iron levels cause tissue damage, which could become irreversible if untreated. This occurs mainly in refractory anaemias following the transfusion of about 50-100 units of red blood cells and in the absence of chelation therapy. Such excess iron is deposited in the liver and spleen, causing their enlargement, as well as in many other organs. Excess iron in the pancreas causes diabetes mellitus, whereas in the endocrine organs it causes growth failure and delayed or absent puberty. Deposits of iron in the cells of such tissues is mainly in the form of the proteins haemosiderin (insoluble in water) and to a lesser extent ferritin (soluble form). Transferrin iron saturation in such patients usually exceeds 100% and variable levels of low molecular weight iron, originating from the breakdown of effete red blood cells in the reticuloendothelial system, can also be detected in the serum.

Iron toxicity at the molecular level arises mainly from the catalytic production of oxygen activated products such as hydrogen peroxide and free radicals, e.g. superoxide and hydroxyl radicals, which are capable of damaging all biomolecular components such as sugars, fatty acids, amino acids, DNA etc (Halliwell et al 1992). Iron chelating drugs could in principle be used to reduce iron overload and toxicity in transfusional iron overload in thalassaemia and other refractory anaemias.

Desferrioxamine Therapy, its Complications and Efforts for its Replacement with Better Chelators

Desferrioxamine (DF), a natural product isolated from *Streptomyces pilosus* has been the mainstay of iron chelation therapy in the last 30 years. DF is effective in most cases in reducing iron overload at doses of 50-80 mg/kg, administered subcutaneously over 8-12 hours at least 5 days per week (Zurlo et al 1989). There are several thalassaemia major patients using DF who have now exceeded the age of 40, while others not using chelation therapy usually die from iron overload toxicity before the age of 20 years. However, despite the good prognosis with DF therapy, the general use of DF is restricted in over 90% of patients that need it because of its high cost (£5000/patient/year) and the low compliance with its 8-12 hours per day prolonged subcutaneous administration. Further limitation of its use is also observed because of a number of toxic side effects such as ocular, auditory and neurotoxic abnormalities, pulmonary and renal complications, thrombocytopenia, leukopenia, aplastic anaemia, yersiniasis, mucormycosis, allergic and anaphylactic reactions, etc (Kontoghiorghes 1995, in press). The most common, less serious toxic side effects observed in over 80% of patients are hardness, swelling and soreness at the site of the subcutaneous injection of DF (Politis et al 1983).

Animal studies using DF in mice, rats and rabbits have shown embryo, maternal and teratogenic toxicity at 50-200 mg/kg doses, while in other species, for example dogs, 70% mortalities were observed at doses of over 150 mg/kg within 4 months, as well as systemic toxicity at doses as low as 25 mg/kg. The metabolic and pharmacological properties of DF have not yet been fully investigated. However, DF is known to be metabolised to several metabolic byproducts some of which have iron binding properties. It is also known to be rapidly eliminated from blood with $T_{1/2}$ = 5-10 mins, while its iron complex (ferrioxamine) is cleared much more slowly with $T_{1/2}$ = 90 mins. In chronic transfusional iron overload DF is usually administered subcutaneously but it can also be administered intravenously and sometimes intramuscularly. In the case of acute iron poisoning, DF could also be administered orally in order to prevent iron absorption. Overall, despite the potentially satisfactory therapeutic effects of DF, the majority of iron loaded patients worldwide cannot use it either because of its high cost, oral inactivity or toxicity. An inexpensive, orally active and non toxic chelator is therefore needed for more general use.

In the last 20 years, hundreds of chelators have been tested in vitro and in animals for the replacement of DF with a better chelating drug. Homologues of microbial siderophores with a catechol and a hydroxamate iron binding site and also polyaminocarboxylic acids such as EDTA and DTPA have generally been tested (Anderson and Hiller 1975, Martell et al 1980). From the hundreds of these experimental new chelators, only fourteen reached the stage of clinical trials and most of these were also later abandoned because they were either toxic or ineffective in iron removal (Kontoghiorghes 1992, 1993). For example, rhodotoroulic acid, cholylhydroxamic acid, and 5-hydroxy-2-formyl-pyridinethiosemicarbazone were abandoned because of local or systemictoxicity, whereas other chelators such as 2,3-dihydroxybezoic acid, ethylenediamine-hydroxyphenylacetic acid, ethylenediamine-N, N'-diacetic acid and pyridoxal-isonicotinolylhydrazone were abandined because of low efficacy in iron loaded patients, despite that in earlier animal studies these were reported to be effective in iron removal. The polyaminocarboxylic acids EDTA and to a greater extent DTPA were shown to be effective in removing iron from iron loaded patients but were also non-selective causing substantial increases of other metal excretions such as Zn, Mg, Mn, etc. (Pippard et al 1986). Salicyl hydroxamic acid, which is primarily used as a urolithiostatic drug, has also been used as a chelating drug for iron removal in iron loaded patients in Egypt because of its low cost, but again as with the above other chelators combination of toxicity and low efficacy resulted in minimal use.

The Design and Development of Deferiprone (L1) and Other α-Ketohydroxypyridines

Investigations in the early 80s in the UK for the identification of new, acid stable, orally absorbable and effective iron chelators have resulted in the design and development of a new class of iron specific chelators, namely the α-keto-hydroxypyridines (KHP) (Kontoghiorghes 1982, 1990a). The KHP were in principle designed to mimic some of the properties of the naturally occurring α-ketohydroxy aromatic ring chelators mimosine, tropolone and maltol, which were previously shown to be absorbed orally, have high affinity for iron and in the case of mimosine, an amino acid, to increase iron excretion in animals (Lin and Ling 1961, Tsai and Ling 1971, 1773). More than 100 bi-, tetra-, and hexadentate KHPs have been synthesised and tested for the binding and removal of iron in vitro and in vivo (Kontoghiorghes 1982,1986, Kontoghiorghes and Sheppard 1987, 1989, Kontoghiorghes et al 1993, Molenda et al 1994, Sheppard and Kontoghiorghes 1992). Further development of most of these chelators was abandoned because of low water solubility, ineffectiveness in iron removal from iron containing proteins such as transferrin, ferritin and haemosiderin and also because of low efficacy in iron removal or high toxicity in animals. The most promising KHP in iron removal in animals appear so far to be the 1-substituted-2-alkyl-3-hydroxypyrid-4-ones and in particular the 1-methyl-, 1-ethyl-, 1-allyl- and 1-cyclopropyl-2-methyl-3-hydroxypyrid-4-one derivatives (Kontoghiorghes et al 1993). In particular, 1,2-dimethyl-3-hydroxypyrid-4-one (L1 or BAN/INN: Deferiprone) was shown in animal studies to have the highest therapeutic index in the KHP series and was then selected for further, more extensive preclinical and later clinical investigations. Permission for short term clinical trials with L1 was obtained in 1986 following the submission of the in vitro and in vivo studies to the Ethical Committee of the Royal Free Hospital and the Department of Health and Social Security in the UK. The first clinical trials with oral L1 in iron loaded patients were reported in 1987 (Kontoghiorghes et al 1987 a,b). Permission from the same authorities had subsequently been granted for prolonged clinical trials involving many more patients.

The promising clinical results in the UK prompted investigations with L1 in many other countries/centres and with several other categories of patients which are currently ongoing and expanding. Recently, L1 has been approved by the government of India and registered as a drug by one of the top three Indian pharmaceutical companies, namely CIPLA Ltd. Registration is pending in many other countries. Three other KHP chelators namely the 1-ethyl-2-methyl-, 1-allyl-2-methyl- and 1,2-diethyl-3-hydroxypyrid-4-ones have also been tested in preliminary clinical trials, but their development is no longer pursued. 1,2-Diethyl-3-hydroxypyrid-4-one (EL1NEt) was tested in humans in the UK following controversial findings in a number of animal toxicity and metabolic studies. Clinical studies in iron loaded patients using different doses of maximum 50 mg/kg EL1NEt have shown that this chelator was not sufficiently effective in increasing iron excretion (mean 7±2 mg/day) and was therefore abandoned (Porter et al 1993). In contrast, 1-ethyl-2-methyl-3-hydroxypyrid-4-one (L1NEt) was shown in a cross over study in ten iron loaded patients to be equally active to DF and L1 in increasing urinary iron excretion (22-84 mg iron per 50-100 mg/kg) (Kontoghiorghes et al 1990 a) but further studies were abandoned because of the widely successful use of L1. 1-Allyl-2-methyl-3-hydroxypyrid-4-one (L1NAll) has also been tested in preliminary clinical studies in Russia with similar results in iron excretion to those obtained with L1 and with no major toxicity (unpublished).

Chemical, Pharmacological and Metabolic Transformation Properties of L1

L1 (MW = 139) is a white crystalline solid which is water soluble at acidic and physiological pH, stable at room temperature for up to at least 3 years and has a very bitter taste. It forms red coloured complexes with iron in solutions at pH 7.4 in a 3 L1 : 1 Fe molar ratio (MW = 470) (Kontoghiorghes et al 1987c) and also complexes with other metals such as Cu,Al, Zn, Ga, In, Pu etc with a lower affinity than iron (Eybl et al 1992, Sheppard and Kontoghiorghes 1993, Taylor and Kontoghiorghes 1986). Its iron binding constant (log = 36) is higher than those of DF (log = 31) and DTPA (log = 29), the two other chelators currently used clinically for the treatment of iron overload (Kontoghiorghes 1993).

The chemical synthesis of L1 is simple and inexpensive, with current sale price ten times cheaper than DF and with a potential to be reduced even further, thus becoming more accesible to patients in poor countries. L1 is synthesised from the natural product and food additive maltol and methylamine in a one-step exothermic reflux reaction involving only water. Water is also used for the purification/recrystallization of L1 and in the abscence of any other potentially toxic chemicals. Both L1 and its 3 L1 : 1 Fe complex have neutral charge at physiological pH, but L1 is positively charged at the acidic pH of the stomach, where it forms 1:1 and 2:1 L1:Fe complexes which are also positively charged (Kontoghiorghes et al 1987c). The 3 L1 : 1 Fe complex of L1 is not stable at high dilutions (10^{-6}M), where 1:1 and 2:1 L1:Fe complexes could also be formed at neutral pH (Motekaitis and Martell 1991) and also when the concentration of iron in a soluble form is higher than that of L1. Similarly, complexes of a 3 L1 : 1 metal stoichiometry are being formed with Al and Ga, and 2 L1 : 1 metal complexes with Cu and Zn at pH 7.4 (Sheppard and Kontoghiorghes 1993).

The metabolism and pharmacological properties of L1 have been reported in a number of studies involving iron loaded patients and normal volunteers (Kontoghiorghes et al 1990b, Sheppard and Kontoghiorghes 1992) The oral administration of gelatin capsules of 35-60 mg/kg single doses of L1 have shown that L1 is absorbed and detected in blood within a few minutes, suggesting rapid uptake from the stomach. In almost all the cases L1 reaches maximum blood level concentrations within 1 h. It is eliminated from blood with a half-life of 1-2 h and it is almost completely undetectable in blood within 5-7 h (Kontoghiorghes et al 1990b). Orally administered L1 is mostly metabolised to a glucuronide conjugate which reaches maximum blood levels within 1-1.5 h, following peaked L1 levels. The glucuronide metabolite of L1 has no iron chelation potential. L1, its iron complex and its glucuronide conjugate are detectable in serum and urine and in most cases the total amount of all three components accounts for almost 100 % recovery of the administered dose in the urine (Kontoghiorghes et al 1990b). The finding that L1 and its metabolic products are eliminated from the kidneys is also supported from the abscence of L1, L1 glucuronide conjugate or indeed increased iron excretion in faecal samples of patients taking L1. Further evidence suggesting that the urinary route is the site of L1 and metabolic product elimination comes from human iron metabolic studies using ^{59}Fe labelling where it was shown that increased ^{59}Fe excretion is only apparent in the urine but not the faeces (Nielsen et al 1992).

Despite the above general observations there appears to be many variations in the pharmacokinetic properties and metabolism of L1 amongst patients. So far there are no clear indications that such variations are due to any individual factors. In particular, there have been studies recently showing that less than 10% of patients may excrete L1 in the urine unchanged with little or no evidence of glucuronidation (Kontoghiorghes et al 1995). Another observed variation in some patients taking repeated doses of L1, is the presence of a lag period of 0.5 - 1.0 h of L1 detection in blood during a second oral dose administration. This is probably an indication of a delayed dissolution of the gelatine capsules and powder of L1 in the gastrointestinal tract or a delayed absorption pattern of L1, which may probably take place from the small intestine, rather than the stomach. Additional supporting evidence come from a patient with partial gastrectomy, where a liquid formulation of L1 resulted in a tenfold increase of iron excretion by comparison to receiving the same dose in gelatine capsules. It will appear that the acidity of the stomach and the overall solubility properties of L1 in the stomach or the rest of the gastrointestinal tract could play an important role in its absorption properties and efficacy in iron excretion.

Toxicological Aspects of L1 Therapy

Many toxicity aspects of the use of L1 have been reported so far, which are mostly variable and generally non predictable for the protection of patients at risk. The relevance of such reports is complicated by many factors including differences between human and animal species metabolism, dose protocols, iron and other metal body loading and idiosyncratic considerations. It is estimated from published data and conference proceedings that since the clinical use of L1 in 1987, more than 800 patients in 16 countries and 31 centres worldwide have received L1 at different doses and variable time periods. This estimate excludes those taking L1 in India after registration. Some thalassaemia patients in Berne, Switzerland have received L1 daily for periods exceeding 6 years. The toxicity reported from the different centres could be summarised as follows: a) transient agranulocytosis 0-2%, b) transient musculoskeletal and joint pains 0-30%, c) gastric intolerance 0-6% and d) zinc deficiency 0-2%. Almost all the toxic side effects were reported to be reversible and most to be dose related, with their incidence reduced by using doses of 50-75 mg/kg/day. The causes of agranulocytosis, musculoskeletal and joint pains as well as gastric intolerance have not yet been fully identified. However, there are preliminary evidence that Cu deficiency may play a role in the cause of agranulocytosis and weak L1 - iron complexes at low (μM range) concentrations may play a role in the excacerbation of toxicity in the joints.

Prophylactic measures are currently being introduced to reduce further the incidence of the toxic side effects of L1. Of particular concern is agranulocytosis where monitoring of white blood cell count is essential and drug withdrawal during leukopenia or severe infections or inflammation is recommended. Similarly, in a few persistent cases of joint pains, temporary L1 withdrawal and use of anti-inflammatory drugs is also being recommended. The mild Zn deficiency observed in a few patients treated with L1 has easily been corrected using Zn supplements, whereas gastric intolerance could also be overcome in most cases using divided low doses of L1 and/or L1 administration during meals. The overall toxicity profile of L1 in chronically treated iron loaded patients is not different from that of DF and indeed is much better by comparison to many other drugs used in other chronic conditions (Kontoghiorghes 1995, in press).

In addition to clinical investigations and toxicity monitoring, a large number of studies with L1 have been carried out in vitro and in vivo for the determination of the mechanisms of its toxicity. Free radical damage catalysed by iron has been shown to be inhibited by L1 in a variety of in vitro and in vivo systems. The inhibition was concentration dependent but at low L1 concentrations (1 - 50 μM) where weak L1 - iron complexes are formed, no inhibition or in some cases activation of the iron induced free radical damage was observed. Similar effects on free radical damage caused by Fe could also be envisaged in the presence of Cu, which also appears to have high affinity to L1 (Sheppard and Kontoghiorghes 1993). The ability of L1 to bind, remove and redistribute Fe, Cu and Zn from essential pathways involving many enzymes containing these metals, is another area which has been considered in relation to possible adverse effects (Kontoghiorghes and Weinberg 1995). L1 and other KHP were shown for example to inhibit the two iron containing proteins involved in the arachidonic acid metabolism, namely lipoxygenase and cyclooxygenase at concentrations in the 1-4 mM range but not at concentrations usually obtained in humans ie about 0.6 mM maximum (Jeremy et al 1988). Similarly, mM range concentrations of L1 were also shown in cell studies to inhibit ribonucleotide reductase, an iron containing enzyme involved in DNA synthesis (Ganeshaguru et al 1992). In cell studies it was also shown that white cell progenitors are more susceptible to L1 toxicity than other cells or cell lines, with IC_{50} of about100 μM, a concentration easily achieved during chelation therapy with L1 (Cunningham et al 1991, Kontoghiorghes et al 1990b). Mutagenicity studies using L1 and its iron complex have indicated that neither of these are mutagenic.

Acute, subacute and chronic toxicity of oral and parenteral L1 is comparable to that of parenteral DF in various animal species, as recently reviewed (Kontoghiorghes 1995, in press). In summary, the LD_{50} of oral L1 is estimated to be greater than 1g/kg in several animal species and at higher doses, e.g. 2g/kg in rats, the cause of death is thought to be heart failure. Toxicity reported in rodents includes fatigue, loss of appetite and diarrhoea in mice at 3 x 250 mg/kg/day and hypersalivation, lowering of the white blood cell counts, hypocellularity of the bone marrow, increase in the size of the adrenals, decrease in the weight of thymus and testes in mice and rats at 200 mg/Kg. Similar pathological changes have been observed in other animal species such as dogs, where doses of 100-400 mg/kg/day, have been administered over a two week period. Mortalities at 600 mg/kg/day over the same period have also been observed.

Maternal, embryo and teratogenic toxicity caused by L1 has been suggested by the pharmaceutical company Ciba-Geigy but detailed data have not yet been published. Similar toxic side effects were also previously observed in maternal, embryo and teratogenic toxicity studies using DF. It would appear that overall, L1 has a similar profile of toxicity to DF in both cell and in vivo systems (Kontoghiorghes 1995, in press).

Therapeutic Implications of the Introduction of L1 in Iron Overloading Conditions

Clinical trials and compassionate use of L1 were initiated in several countries following the first reports on the clinical use of L1 in transfusionally iron loaded myelodysplasia and thalassaemia patients in 1987 (Kontoghiorghes et al 1987a,b) The main countries where L1 has been used are India (Agarwal et al 1992a), Italy (Carnelli et al 1992), Germany (Jaeger et al 1992), Netherlands (Goudsmit and Kersten 1992, Vreugdenhil et al 1989), Canada (Olivieri et al, 1995), Switzerland (Tondury et al, 1990) and Russia (Korkina et al, 1992). Ten different categories of patients have received L1 mainly for the removal of iron in cases of thalassaemia and myelodysplasia, for the removal of aluminium in cases of renal dialysis and for antiinflammatory response in cases of rheumatoid arthritis. L1 is currently being widely used for the removal of iron, whereas its use in other conditions is still under investigation.

In general, there are slight variations in the reported response to L1 from the various clinical centres in relation to iron excretion, but overall its use in transfusional iron loaded patients at 55-100 mg/kg/day was identified to be sufficient in bringing most patients to near normal iron balance within two years. Daily iron excretion in response to L1 varies but generally depends on the iron load of patients as well as the dose and frequency of its administration, and other not yet determined factors. While daily intake of iron from transfusions may range from 15-35 mg, iron excretion in response to the above doses may reach 100 mg/day, especially in some cases of heavily iron loaded patients. For example, maximum iron excretion of 84 mg per 50 mg/kg and 325 mg per 250 mg/kg dose of L1 was observed during chronic and intensive iron chelation studies respectively (Kontoghiorghes et al 1990c). In many patients the long term use (6 to 18 months) of L1 at doses of 55-100 mg/kg/day resulted in reduction of serum ferritin to near normal levels and an overall substantial decrease in liver iron (Agarwal et al 1992 b and Olivieri et al 1995).

Iron mobilisation by L1 from regularly transfused patients appears to be mainly from a) the low molecular weight serum iron in excess of transferrin saturation, b) transferrin bound iron, when transferrin iron saturation is much higher than normal and c) tissue iron mainly from the liver but also from other organs. The progressive removal of iron from all other tissues could be achieved through direct chelation from the tissues or mainly through equilibrium with serum transferrin, provided that the overall iron excretion by L1 would be higher than the iron intake from transfusions and the gastrointestinal tract.

The efficacy of L1 in increasing iron excretion is similar to that of DF. However, its low cost and improved compliance with its oral administration would be of benefit to many thousands of patients who are not currently chelated because of the drawbacks of the DF therapy, namely its high cost, low compliance and in many cases toxicity. It should be noted that for therapeutic purposes, several other factors may affect the choice for clinical use of either of these two chelators. For example, because of the many years of experience obtained with DF therapy, the use of L1 may be delayed, especially in countries where DF has long been used and is currently freely available. Similarly, the use of oral L1 may have an advantage in many other categories of patients with iron or aluminium overload, where bleeding due to subcutaneous infusion of DF could be a complication (Kontoghiorghes and Weinberg 1995).

Other Uses of L1 and Related Chelators

The therapeutic uses of L1 could broadly resemble those of DF with some exceptions. For example, L1 is less effective in preventing iron absorption than DF (Kontoghiorghes 1990b) and could not therefore have advantages over DF in the treatment of acute iron poisoning. However, in other areas the use of L1 may be more appropriate than DF. The benefits of the use of L1 may be highly significant in renal dialysis patients, who have high toxic body levels of Al and cannot take DF because of toxicity, non compliance or cost. Clinical trials in moderately Al loaded renal dialysis patients have shown that administration of L1 at 40 mg/kg or over, could result in an increase in serum Al levels within 1 h. Serum Al levels return to normal in about 6 h, which is about the time period required for the clearance of L1 from blood (Kontoghiorghes et al 1992, 1994). Administration of L1 in Al loaded peritoneal dialysis patients was similarly effective, resulting in a manyfold increase of Al levels in the peritoneal dialysis fluid collected within 12 hrs of the L1 administration. Despite that the levels of Al mobilisation and removal by L1 and DF appear to be similar, their toxicity and route of clearance is different. Different protocols for Al removal by L1 and DF are therefore required in order to achieve maximum efficacy and low toxicity (Kontoghiorghes et al 1994). Al removal by L1 may also be beneficial in Alzheimer's disease, where Al is thought to be accumulated in the senile plaques of the brain (Edwardson et al 1992), or in other conditions of environmental, pharmacogenic or acute Al toxicity.

Bacterial or parasitic growth could be inhibited by withholding iron from these microorganisms as previously shown to occur with DF and L1 (Weinberg 1984, Brock et al 1988, Heppner et al 1988). Thus, chelators may have a use in antimicrobial therapy and parasitic infections, especially in cases of resistance which is expressed by the microbe in relation to existing drugs.

Another area of developing interest on iron chelators, is the correction of the anaemia of chronic disease where iron is stored in the reticuloendothelial system and is noi made available to the erythron for the production of haemoglobin. Redistribution of iron in rheumatoid arthritis parients with chronic anaemia has already been shown to occur by both L1 and DF. In separate studies both chelators caused an increase in the production of haemoglobin and their effects were thought to be further potentiated by combination therapy with erythropoietin (Giordano et al 1986, Vreugdenhil et al 1991). Similarly, the combination of chelators with cytotoxic drugs has also been shown to cause reduction of malignant cells in leukaemia and neuroblastoma patients (Estrow et al 1987, Donfrancensco et al 1990).

The possible therapeutic uses of iron chelators in free radical toxicity such as ischaemia/reperfusion injury, atherosclerosis, cancer, ageing, etc, is also being considered. The prevention or reduction of free radical toxicity is based on the binding and removal by chelators of metal ions such as iron and copper, or inhibition of metal containing enzymes all of which play a major catalytic role in oxidative damage to tissues (Halliwell et al 1992). There are many iron and other related metal metabolic disorders where chelators may also play a therapeutic role, some of which are currently under investigation by both DF and L1. The design of more specific chelators for each of the above conditions will depend on the succesful targetting of iron or other metal pools or organs. Similarly, the identification of the causes of toxicity of L1, DF and other chelators and the use of appropriate safety interventions may increase the prospect of the therapeutic uses of chelators in a large number of other conditions.

In conclusion, progress in the chemical, pharmacological, toxicological and therapeutic aspects of oral L1 in the treatment of transfusional iron overload, has stimulated a new interest in the development of chelators for clinical use in many other conditions of metal imbalance and toxicity. It is envisaged that with further improvements on safety, the number of applications of L1 and other chelating drugs in medicine will be substiantially increased.

References

Agarwal MB, Gupta SS, Viswanathan C, et al (1992a) Long term assessment of efficacy and safety of L1, an oral iron chelator in transfusion dependent thalassaemia. Indian trial. Drugs Today 28 (Suppl A):107-114

Anderson WF, Hiller MC (eds) (1975) Development of iron chelators for clinical use. National Institute of Health. Bethesda USA:1-277

Brock J, Liceaga J, Kontoghiorghes GJ (1988) The effect of synthetic iron chelators on bacterial growth in human serum. FEMS Microbiol Immunol 47:55-60

Carnelli V, Spadaro C, Stefano V, Battaglioli M, Fossati G, Perri M (1992) L1 efficacy and toxicity in poorly compliant and for refractory to desferrioxamine thalassaemia patients: Interim report. Drugs Today. 28 (Suppl A):119-121

Cunningham JM, Hunter A, Francis GE, Kontoghiorghes GJ, Hoffbrand AV (1991) Differrential cytotoxicity of oral iron chelators and desferrioxamine to human bone marrow haemopoietic progenitor cells in vitro. Brit J Haematol 77 (suppl 1): 381

Donfrancesco A, Deb G, Dominici C, Castello MA, Helson L (1990) Effects of a single course of deferoxamine in Neuroblastoma patients. Cancer Res 50:4929- 4930

Edwardson JA, Ferrier IN, McArthur FK et al (1992) Alzheimer's disease and the aluminium hypothesis. In: Aluminium in Chemistry, Biology and Medicine. Nicolini, Zatta PF, Corain B (eds). Raven Press, New York 1:85-96

Estrow ZA, Tawa XH, Dube ID et al (1987) In vitro and in vivo effects of deferioxamine in neonatal acute leukaemia. Blood 69:757-761

Eybl V, Svihovcov P, Kontensky J, Kontoghiorghes GJ (1992) Interaction of L1, L1NAll and deferoxamine with Gallium in vivo. Drugs Today 28 (Suppl A):173-175

Ganeshaguru K, Lally KM, Piga A, Hoffbrand AV, Kontoghiorghes GJ (1992) Cytotoxic mechanisms of iron chelators. Drugs Today 28 (Suppl A):29-34

Giordano N, Sancasciani S, Borghi C, Fiorauenti A, Marcolongo P (1986) Antianemic and potential anti-inflammatory activity of desferrioxamine: Possible usefulness in rheumatoid arthritis. Clin Exp Rheumatol 4:25-29

Goudsmit R, Kersten MJ (1992) Long term treatment of transfusion hemosiderosis with the oral iron chelator L1. Drugs Today 28 (Suppl A):133-135

Halliwell B, Gutteridge JMC, Cross CE (1992). Free radicals, antioxidants and human disease: Where are we now? J Lab Clin Med 119:598-620

Heppner DG, Hallaway PE, Kontoghiorghes GJ, Eaton JW (1988) Antimalarial properties of orally active iron chelators. Blood 72:358-361

Jaeger M, Aul C, Sohngen D, Germing U, Schneider W (1992) Iron overload in polytransfused patients with MDS; Use of L1 for oral iron chelation. Drugs Today 28 (Suppl A):143-147

Jeremy JY, Kontoghiorghes GJ, Hoffbrand AV, Dandona P (1988) The iron chelators desferrioxamine and 1-alkyl-2-methyl-3-hydroxypyrid-4-ones inhibit vascular prostacylin synthesis in vitro. Biochem J 254:239-244

Kontoghiorghes GJ (1982) The design of orally active iron chelators for the treatment of thalassaemia. PhD thesis, University of Essex, Colchester UK. British Library Microfilm No D66194/86

Kontoghiorghes GJ (1986) Orally active α-ketohydroxypyridine iron chelators: studies in mice. Mol Pharmacol 30:670-673

Kontoghiorghes GJ (1990a) Design, properties and effective use of the oral chelator L1 and other α-ketohydroxypyridines in the treatment of transfusional iron overload in thalassaemia Ann NY Acad Sci 612:339-350

Kontoghiorghes GJ (1990 b) Chelators affecting iron absorption in mice. Arzneim.Forsch/Drug Res. 40:1332-1335

Kontoghiorghes GJ (1992) Advances in oral iron chelation in man. Intern J Haematol 55:27-38

Kontoghiorghes GJ (1993) Present status and future prospects of oral iron chelation therapy in thalassaemia and other diseases. Indian J Paediatr 60:485-507

Kontoghiorghes GJ (1995) Comparative efficacy and toxicity of desferrioxamine, deferiprone and other iron and aluminium chelating drugs. Toxicol Letters (in press)

Kontoghiorghes GJ, Aldouri MA, Hoffbrand AV et al (1987a) Effective chelation of iron in β–thalassaemia with the oral chelator 1,2-dimethyl-3-hydroxypyrimid-4-one. Br Med J 295:1509-1512

Kontoghiorghes GJ, Aldouri MA, Sheppard LN, Hoffbrand AV (1987b) 1,2-Dimethyl-3-hydroxypyrid-4-one, an orally active chelator for the treatment of transfusional iron overload. Lancet i:1294-1295

Kontoghiorghes GJ, Barr J, Baillot R (1992) Aluminium mobilisation in renal dialysis patients using the oral chelator 1,2,,dimethyl,,3,,hydroxypyrid,,4,,one (L1). Drugs Today 28 (Suppl A):183-187

Kontoghiorghes GJ, Barr J, Baillot R (1994) Studies of aluminium mobilisation in renal dialysis patients using the oral chelator 1,2-dimethyl-3-hydroxypyrid-4-one. Arzneim Forsch/Drug Res 44:522-526

Kontoghiorghes GJ, Barr J, Nortey P, Sheppard L.(1993) Selection of a newgeneration of new orally active α-ketohydroxypyridine iron chelators intended for use in the treatment of iron overload. Amer J Haematol 42:440-444
Kontoghiorghes GJ, Bartlett AN, Hoffbrand AV et al (1990c) Long term trial with the oral iron chelator 1,2-dimethyl-3-hydroxypyrid-4-one (L1). (1) Iron chelation and metabolic studies. Br. J Haematol 76:295-300
Kontoghiorghes GJ, Bartlett AN, Sheppard L, Barr J, Nortey P (1995) Oral iron chelation therapy with deferiprone. Monitoring of biochemical, drug and iron excretion changes. Arzneim Forsch/Drug Res 45:65-69
Kontoghiorghes GJ, Goddard JG, Bartlett AN, Sheppard L (1990 b) Pharmacokinetic studies in humans with the oral iron chelator 1,2-dimethyl-3-hydroxypyrid-4-one. Clin Pharm Ther 48:255-261
Kontoghiorghes GJ, Sheppard LN (1987) Simple synthesis of the potent iron chelators 1-hydroxy-2-methylpyrid-4-ones. Inorg Chim Acta 136:L11-L12
Kontoghiorghes GJ and Sheppard LN (1989) Process for producing pyrid-4-ones. European Patent Application No:0335745
Kontoghiorghes GJ, Sheppard LN, Hoffbrand AV, Bartlett AN, Barr J (1990a) 1-Ethyl-2-methyl-3-hydroxypyrid-4-one (L1NEt) A new orally effective iron chelator for the treatment of transfusional iron overload. Blood 76:66a
Kontoghiorghes GJ, Weinberg ED (1995) Iron: mammalian defense systems, mechanisms of disease, and chelation therapy approaches. Blood Rev 9:33-45
Korkina LG, Afanas'ev IB, Deeva IB et al (1992) Free radical status of blood of patients with iron overload: the effect of chelating treatment. Drugs Today 28 (Suppl A):137-141
Lin JY, Ling KH (1973) Studies on free amino acids in the seeds of Leucaena Glauca Benth:III, biological study of mimosine. J Formosan Med Assoc 60:657-664
Martell AE, Anderson WF, Badman DG (eds) (1980) Development of iron chelators for clinical use. Amsterdam: Elsevier:1-311
Molenda JJ, Jones MM, Basinger MA (1994) Enchancement of iron excretion via monoanionic 3-hydroxypyrid-4-ones. J Med Chem 37:93-98
Motekaitis RJ, Martell AE (1991) Stabilities of the iron (III) chelates of 1,2-dimethyl-3-hydroxy-4-pyridinone and related ligands. Inorg Chim Acta 183:71-80
Nielsen P, Frtjes M, Drescow B, Fischer R, Engelhardt R, Heinrich HC (1992) The iron-decorporating effect of L1 in normal and TMH-ferrocene iron loaded rats and in one patient with post-transfusional siderosis as judged by ^{59}Fe-labelling technique. Drugs Today 28 (Suppl A):45-53
Olivieri NF, Brittenham GM, Matsui D et al (1995) Iron-chelation therapy with oral deferiprone in patients with thalassaemia major. N Engl J Med 332:918-922
Pippard MJ, Jackson MJ, Hoffman K, Petrou M, Modell CB (1986) Iron chelation using subcutaneous infusion of diethylene triamine penta-acetic acid (DTPA). Scand J Haematol 36:466-472
Politis C, Vrettou H, Fragatou S et al (1993) Reactions to subcutaneous application of iron chelation therapy with desferrioxamine. An urgent need for oral chelators. Abstracts, 4th ICOC Limassol, Cyprus 4
Porter JB, Weir TB, Marshall L et al (1993) Clinical trials with the orally active hydroxypyridone CP94: Implications for future strategies in iron chelation therapy. Abstracts, 4th ICOC Limassol, Cyprus 18
Sheppard L, Kontoghiorghes GJ (1992) Synthesis and metabolism of L1 and other novel alpha-ketohydroxypyridine iron chelators and their metal complexes. Drugs Today. 28(Suppl A):3-10

Sheppard LN and Kontoghiorghes GJ (1993) Competition between deferiprone, desferrioxamine and other chelators for iron and the effect of other metals. Arzneim. Forsch/Drug Res 43:659-663

Taylor D, Kontoghiorghes GJ (1986) Mobilisation of plutonium and iron from transferrin and ferritin by α-oxohydroxypyridine chelators. Inorg Chim Acta 125:L35-L38

Tsai WC, Ling KH (1971) Effect of metals on the absorption and excretion of mimosine and 3,4-dihydroxypyridine in rat in vivo. J Formosan Med Assoc 73:543-549

Tsai WC, Ling KH (1973) Study of the stability constant of some metal ion chelates of mimosine and 3,4-dihydroxypyridine. J Chin Biol Soc 2:70-86

Tondury P, Kontoghiorghes GJ, Ridolfi,,Luthy R et al (1990) L1 (1,2-dimethyl-3-hydroxypyrid-4-one) for oral iron chelation in patients with β-thalassaemia major. Br J Haematol 76:550-553

Vreugdenhil G, Kontoghiorghes GJ, Van Eijk HG, Swaak AJG (1991) Impaired erythropoietin responsiveness to the anemia in rheumatoid arthritis. A possible inverse relationship with iron stores and effects of the oral iron chelator 1,2- dimethyl-3-hydroxypyrid-4-one. Clin Exp Rheumatol 9:35-40

Weinberg ED (1984) Iron withholding: a defense against infection and neoplasia. Physiol. Revs 64:65-102

World Health Organisation Bulletin (1983) Community control of hereditary anaemias 61:63-80

Zurlo MG, De Stefano P, Borgna-Pignatti C et al (1989) Survival and causes of death in thalassaemia major. Lancet ii:27-29

Free Radicals

(Chairs: H. Sies, Germany, and
L. Beneš, Slovak Republic)

Free Radicals in Toxicology: Redox Cycling and NAD(P)H:Quinone Oxidoreductase

Wolfgang A. Schulz, Peter Eickelmann, and Helmut Sies
Institut für Physiologische Chemie I, Heinrich-Heine-Universität Düsseldorf, Postfach 101007, D-40001 Düsseldorf, Germany

Introduction

Redox cycling compounds are among the multiple sources of free radicals generated in situations of toxicological interest. These compounds can undergo one-electron reduction to autoxidizable radical intermediates. The radicals react with oxygen to form the superoxide anion radical which can disproportionate to hydrogen peroxide and give rise to further oxidants. The other product of the autoxidation reaction is the parent compound, ready to go through further reaction cycles. One-electron reductions are frequently catalyzed by the electron-transfer chain of the endoplasmatic reticulum, the electron being supplied by NADPH. Detoxication of hydrogen peroxide utilizes NADPH as well in the GSSG reductase reaction. In this fashion, the cellular pool of NADPH may become severely depleted during redox cycling, enhancing toxicity. Toxic drugs of this type include quinones and nitroaromatics (for a review, see Kappus and Sies, 1981).

Defense against quinone redox cyclers is provided by the NAD(P)H:quinone oxidoreductase (NQOR) (Ernster & Navazio, 1958) which catalyzes an obligatory two-electron transfer from either NADPH or NADH to quinones (yielding hydroquinones) or nitroaromatic compounds (yielding nitrosoaromatic compounds). This reaction circumvents the reactive radical intermediates and allows Phase II reactions such as glucuronidation of hydroquinones or glutathione conjugation to take place. It is important to consider that the hydroquinones resulting from the NQOR reaction can also be subject to autoxidation. If the hydroquinone is itself highly susceptible to autoxidation, the two-electron reduction of the quinone may still present a prerequisite for its conjugation and excretion, but redox cycling will not become efficiently inhibited. It therefore depends on the redox potential of the hydroquinone/semiquinone pair, to what extent the NQOR reaction contributes to detoxification (see (Cadenas, 1995) for a detailed discussion). Some cytostatic drugs such as mitomycin C (Riley and Workman, 1992) are even activated by the NQOR reaction towards the ultimately reactive compounds alkylating and crosslinking DNA. In the case of the newly developed indoloquinone EO9, it may depend on the oxygen concentration whether NQOR is protective (Plumb et al, 1994).

Therefore, a lack of NQOR activity will on one hand enhance the toxicity and carcinogenicity of many quinones and nitroaromatic compounds, but on the other hand cause resistance to cytostatic drugs such as mitomycin C. In this article, we will illustrate these two aspects of NQOR action by examples from our own work, and finally provide evidence that they are reflected in the occurence of a functional polymorphism in the human population.

Results and Discussion

Decrease of NQOR expression in renal cell carcinoma: In humans the kidney is one of the organs expressing the highest enzyme levels of NQOR (Riley and Workman, 1992). Glutathione transferase (GST) activity is also high; in particular the isoenzymes of the alpha class (Morel et al, 1994) are highly expressed. GST overall activity is diminished in renal cell carcinoma due to a drastic decrease of GST alpha expression (Klöne et al, 1990, di Ilio et al, 1991, Eickelmann et al, 1994a). On average, a 17-fold decrease in GST alpha mRNA levels was found in renal cell carcinoma, with very few patients retaining close to normal levels of expression. The average expression of NQOR mRNA in 23 samples of renal cell carcinoma was only 28% of that in adjacent normal tissues, although the decrease in expression was more variable than that of GST alpha (Eickelmann et al, 1994). Accordingly, the average NQOR enzyme activity, measured as the reduction of menadione, decreased from 41 ± 39 to 18 ± 27 mU/mg protein. However, as can be inferred from the standard deviations, even in normal tissues the enzyme activity displayed a high interindividual variation (Fig. 1). In three patients, no specific enzyme activity was detected at all, although GST activity (Fig. 1) and lactate dehydrogenase activity (not shown) were within the normal range of the samples. This high variability can probably be explained to some extent by the inducibility of the enzyme by drugs and food components (Benson et al, 1980; Li and Jaiswal, 1992). However, as will be discussed below, it also reflects a genetic polymorphism in the human NQOR gene. Taken together, the decrease in the activity of two enzymes protective against oxidative stress, GST alpha and NQOR, in renal cell carcinoma suggests a relation to renal carcinogenesis.

NQOR and mitomycin C resistance in bladder carcinoma cells: A decrease in NQOR activity is frequently observed in tumor cell lines during development of resistance to cytostatic drugs such as mitomycin C. We have investigated a pair of human bladder carcinoma cell lines, RT112 and RT112MMC. RT112 cells are

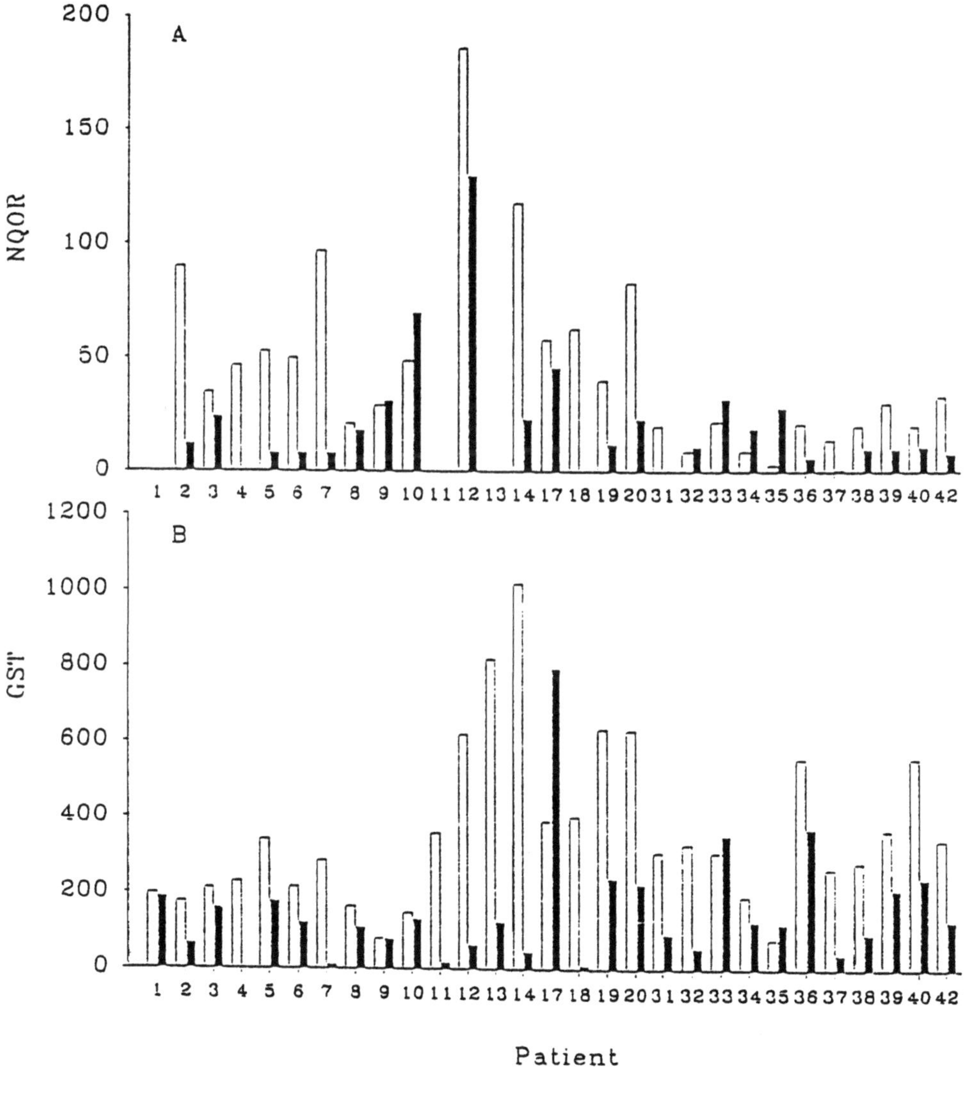

Fig. 1. Enzyme activities of NQOR (top) and total GST (bottom) in 23 renal cell carcinomas (solid bars) and corresponding control tissues (open bars). Activities were not determined in patient 16.

Cell line:	RT112	RT112MMC
Dose of mitomycin C inhibiting 50% of cell proliferation	60 μmol/l	2.5 mmol/l
NQOR enzyme activity (mU/mg protein)	391 ± 140	0 ± 0
GST enzyme activity (mU/mg protein)	95 ± 62	26 ± 6
NQOR mRNA (relative to GAPDH mRNA)	≡100%	15%
GST π mRNA (relative to GAPDH mRNA)	≡100%	70%
Genotype at NQOR locus (wt: wild-type; 0: null-allele)	wt/0	0/0

Table 1: Selected properties of the bladder carcinoma cell lines RT112 and RT112MMC

approximately 40-fold more sensitive towards mitomycin C than the daughter cell line RT112MMC (Table I). NQOR enzyme activity was 391 ± 140 mU/mg protein in RT112, but was nondetectable in RT112MMC cells. However, NQOR mRNA could be detected in both cell lines, although the level was lower in RT112MMC. GST enzyme activity, determined with 1-chloro-2,4-dinitrobenzene as a substrate, was also decreased in RT112MMC. Accordingly, the level of mRNA for GST Pi, which is the major isoenzyme in urothelial carcinoma cells, was also diminished (Table I). Thus, in this pair of cells several changes are associated with increased resistance to mitomycin C, a particularly striking one being a complete loss of NQOR activity.

In a similar investigation, Traver et al (1992) identified a C to T point mutation at base 609 of NQOR mRNA in the human BE colon carcinoma cell line, resulting in a proline to serine amino acid change at position 187 of the protein, as a possible cause for NQOR deficiency. In addition, the NQOR mRNA sequence in BE cells deviated from the 'wild-type' sequence at another position, base 98, which does not affect the amino acid sequence. Following amplification by RT-PCR, we sequenced the mRNA from both bladder carcinoma cell lines. In RT112MMC, we found the same two base changes described by Traver et al

(1992), whereas in RT112 both the 'wild-type' and the 'mutated' sequence were found (Eickelmann et al, 1994b). The transition at base 609 creates a polymorphism for the restriction enzyme HinfI which can be used to distinguish the two sequences following RT-PCR or PCR. This assay confirmed that RT112 is heterozygous at the NQOR locus, whereas RT112 is homozygous. The loss of NQOR enzyme activity in RT112MMC is therefore most probably due to a loss of heterozygosity at the NQOR locus.

A functional polymorphism in the human NQOR gene: The finding of identical mutations in the NQOR gene in two independently derived cell lines, i.e. BE and RT112MMC, points to the existence of two different NQOR alleles. We therefore used the PCR-RFLP assay described above to analyze DNA from several renal cell carcinoma patients. Indeed, all three patients that had been found to lack NQOR enzyme activity turned out to be homozygous for the mutant allele, whereas those with detectable enzyme activity were either heterozygous or homozygous for the wild-type allele. This suggests that the form of the gene with the mutation at base 609 represents a 'null-allele'. From an investigation of 270 unselected persons, we estimate the frequency of homozygotes for the null-allele in the Central European population (Germany) as 1.4%. Given the role of NQOR in protection against redox cycling, it is conceivable that homozygotes for the null-allele are more susceptible to the toxic and carcinogenic effects of certain quinones and nitroaromatic compounds. On the other hand, the activation of cytostatic drugs such as mitomycin C or E09 may be compromised in such patients. Thus, the occurence of three homozygotes for the NQOR null-allele among 23 renal carcinoma patients has prompted us to begin a larger study of the relationship between NQOR genotype and urological malignancies.

References

Benson AM, Hunkeler MJ and Talalay P (1980) Increase of NAD(P)H:quinone reductase by dietary antioxidants: possible role in protection against carcinogenesis and toxicity. Proc. Natl. Acad. Sci. USA 77, 5216-5220.

Cadenas E (1995) Antioxidant and prooxidant funtions of DT-diaphorase in quinone metabolism. Biochem. Pharmacol. 49, 127-140.

Di Ilio C, Aceto A, Bucciarelli T, Angelucci S, Felaco M, Grilli A, Zezza A, Tenaglia R and Federici G (1991) Glutathione transferase isoenzymes in normal and neoplastic human kidney tissue. Carcinogenesis 12, 1471-1475

Eickelmann P, Ebert T, Warskulat U, Schulz WA and Sies H (1994) Expression of NAD(P)H:quinone oxidoreductase and glutathione S-transferases α and π in human renal cell carcinoma and in kidney cancer-derived cell lines. Carcinogenesis 15, 219-225.

Eickelmann P, Schulz WA, Rohde D, Schmitz-Dräger B and Sies H (1994) Loss of heterozygosity at the NAD(P)H:quinone oxidoreductase locus associated with increased resistance against mitomycin C in a human bladder carcinoma cell line. Biol. Chem. Hoppe-Seyler 375, 439-445.

Ernster L and Navazio F (1958) Soluble diaphorases in animal tissues. Acta Chem. Scand. 12, 595.

Kappus H, Sies H (1981) Toxic drug effects associated with oxygen metabolism: redox cycling and lipid peroxidation. Experientia 37, 1233-1241

Klöne A, Weidner U, Hussnätter R, Harris J, Meyer D, Peter S, Ketterer B and Sies H (1990) Decrease expression of the glutathione S-transferases alpha and pi genes in human renal cell carcinoma. Carcinogenesis 11, 2179-2183

Li Y and Jaiswal AK (1992) Regulation of human NAD(P)H:quinone oxidoreductase gene. J. Biol. Chem. 267, 15097-15104.

Morel F, Schulz WA and Sies H (1994) Gene structure and regulation of expression of human glutathione S-transferases alpha. Biol. Chem. Hoppe-Seyler 375, 641-649.

Plumb JA, Gerritsen M and Workman P (1994) DT-diaphorase protects cells from the hypoxic cytotoxicity of indoloquinone EO9. Br. J. Cancer 70, 1136-1143.

Riley RJ and Workman P (1992) DT-diaphorase and cancer chemotherapy. Biochem. Pharmacol. 43, 1657-1669

Traver RD, Horikoshi T, Danenberg KD, Stadlbauer THW, Danenberg PV, Ross D and Gibson NW (1992) NAD(P)H:quinone oxidoreductase gene expression in human colon carcinoma cells: characterization of a mutation which modulates DT-diaphorase activity and mitomycin C sensitivity. Cancer Res. 52, 797-802.

Cytotoxic Impact of DNA Single vs Double Strand Breaks in Oxidatively Injured Cells

O. Cantoni, P. Sestili, A. Guidarelli, L. Palomba, L. Brambilla and F. Cattabeni.
Istituto di Farmacologia e Farmacognosia and Centro di Farmacologia Oncologica Sperimentale, Università degli Studi di Urbino, Via S. Chiara, 27, 61029, Urbino, Italy.

Summary

Hydrogen peroxide is a potent inducer of DNA single strand breaks (SSBs) in cultured mammalian cells. These lesions, however, are efficiently repaired and do not appear to mediate the cytotoxic response. This inference is based on the observations that a) inhibiting the rate of SSB-removal does not result in an increased cytotoxicity; b) using different experimental conditions it is possible to dissociate the formation of DNA SSBs from the cytotoxic response; c) the induction/loss of the oxidant-resistant phenotype in cell variants characterized by different levels of resistance to the lethal effect of the oxidant does not correlate with resistance to DNA SSB-induction; d) a much larger accumulation of DNA SSBs can be observed following treatment with H_2O_2 at 4° C, as compared to 37° C, although the opposite is true in terms of cytotoxicity. In the presence of micromolar levels of L-Histidine, H_2O_2 also induces DNA double strand breaks (DSBs), a type of lesion which we suggest may mediate the lethal event. This conclusion finds experimental support in the following observations: a) DNA DSBs are generated at survival-range concentrations, and a linear correlation exists between the level of this lesion and cytotoxicity; b) this correlation curve overlaps with the curves generated under similar experimental conditions using different cell lines with different sensitivity to the oxidant alone, or different clones derived from the same cell line, some of which showed a high degree of resistance to H_2O_2. Finally, the formation of DNA DSBs appears to enhance both apoptotic and necrotic cell death.

Introduction

In aerobic organisms hydrogen peroxide is a normal metabolite, the production of which can be stimulated by the metabolism of certain chemicals and chemotherapeutic agents (Chance, 1979; Subrahmaniam et al., 1987). The toxicity of H_2O_2 seems to depend on the formation of hydroxyl radicals ($OH^{\cdot}$) since scavengers of this reactive oxygen species efficiently prevent the lethal

$$O_2^{\cdot -} + Fe^{3+} \longrightarrow Fe^{2+} + O_2$$

$$H_2O_2 + Fe^{2+} \longrightarrow HO^{\cdot} + HO^- + Fe^{3+}$$

$$H_2O_2 + O_2^{\cdot -} \longrightarrow HO^{\cdot} + HO^- + O_2$$

Fig. 1. The Haber-Weiss reaction.

action of H_2O_2 (Meneghini and Hoffmann, 1980; Mello Filho and Meneghini, 1984; Mello Filho et al., 1984). In addition, iron chelators are also capable of preventing the cell inactivation induced by the oxidant (Mello Filho and Meneghini, 1984; Mello Filho et al., 1984; Starke and Farber, 1985) indicating that the formation of $OH^{\cdot}$ occurs via an iron-catalyzed reduction of H_2O_2. The Haber-Weiss reaction is illustrated in Fig. 1. Hydrogen peroxide, in the presence of traces of divalent iron, produces the highly reactive $OH^{\cdot}$ Superoxides participate in this reaction by providing the means for recycling of trivalent iron to the divalent form. In principle, other intracellular reducing agents such as ascorbate may also participate in this reaction. It is important to point out, however, that the putative reductant has to reach the chromatin since chromatin-bound iron is likely to participate in those reactions leading to DNA damage (Meneghini and Hoffmann, 1980). The supposition that these reactions take place in mammalian cells and mediate the formation of DNA SSBs finds experimental support in the observation that this type of DNA lesion can be prevented by both $OH^{\cdot}$ scavengers and iron chelators (Mello Filho and Meneghini, 1984; Mello Filho et al., 1984).

This paper is designed to provide an overview of the current knowledge on the relationships between DNA SSBs/DSBs and cytotoxicity in mammalian cells exposed to hydrogen peroxide.

DNA Single Strand Breaks and Cytotoxicity: Hydrogen peroxide is a very potent inducer of DNA SSBs. In a previous study (Cantoni et al., 1988) we compared the effects of X-rays and H_2O_2 on SSB-induction in CHO cells and found that 450 rads produced as many breaks as 30 μM H_2O_2 (treatment was performed at ice temperature to prevent repair). This was somewhat surprising, since most of the cells treated with 450 rads did not survive whereas cells exposed to 30 μM H_2O_2 at 4°C remained viable and, in order to get a 50 % decrease in the surviving fraction, it was necessary to increase the levels of the oxidant by about two orders of magnitude. If we postulate that the level of DNA SSBs is a linear function of the peroxide dose, a concentration of H_2O_2 resulting in 50% killing should produce the same level of SSBs that is generated by 6000 rads. This is difficult to test since no available technique has a sensitivity low enough to allow the appreciation of differences with these high levels of DNA lesions. Nevertheless, we estimated that the production of SSBs remained linear at least up to 500-700 μM H_2O_2. We may also add that these breaks were repaired very rapidly (t1/2 of approx. 3-5 min) and probably activated the enzyme poly(ADP-ribose)polymerase (ADPRP) since 3-aminobenzamide (3-AB), a competitive inhibitor of this enzyme, delayed the kinetics of SSB-removal (Cantoni et al., 1986).

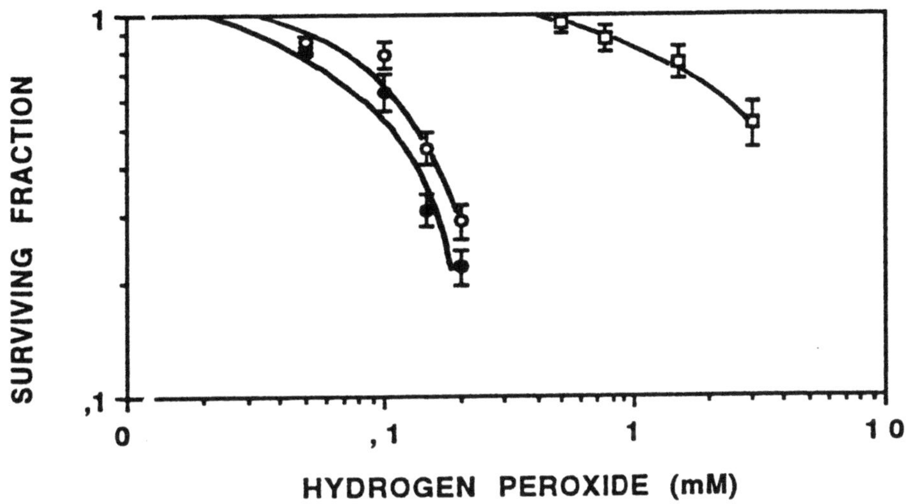

Fig. 2. Hydrogen peroxide cytotoxicity as a function of exposure temperature. Chinese Hamster Ovary (CHO) cells were challenged for 30 min in a Saline A buffer with increasing concentrations of hydrogen peroxide. Treatments with the oxidant were performed either at physiological (o) or at ice-bath temperature (□). Cytotoxicity was assessed using a clonogenic assay. Also shown is the effect of 5 mM 3-aminobenzamide on the survival of cells exposed to H_2O_2 at 37°C (•). Results represent the mean ± S.E.M. calculated from 3 - 4 separate experiments, each performed in duplicate (adapted from Cantoni et al., 1986 and 1989).

Thus, under the above experimental conditions, H_2O_2 efficiently generates DNA SSBs but is a very poor inducer of cytotoxicity, suggesting that DNA SSBs are not a lethal event. A similar conclusion was reached by Ward and coworkers (Ward et al., 1985 and 1987). They calculated that a cell dying from hydrogen peroxide exposure at 4° accumulates about 0.4 million DNA SSBs, but of course this does not mean that the cell dies as a result of these SSBs. The results thus far discussed would indicate that DNA SSBs produced by hydrogen peroxide are not toxic, although we should emphasize that this conclusion is based upon observations that were made studying the effects of hydrogen peroxide at ice bath temperature. We should therefore pose the question of whether the same conclusion holds true in cells treated with the oxidant at physiological temperature.

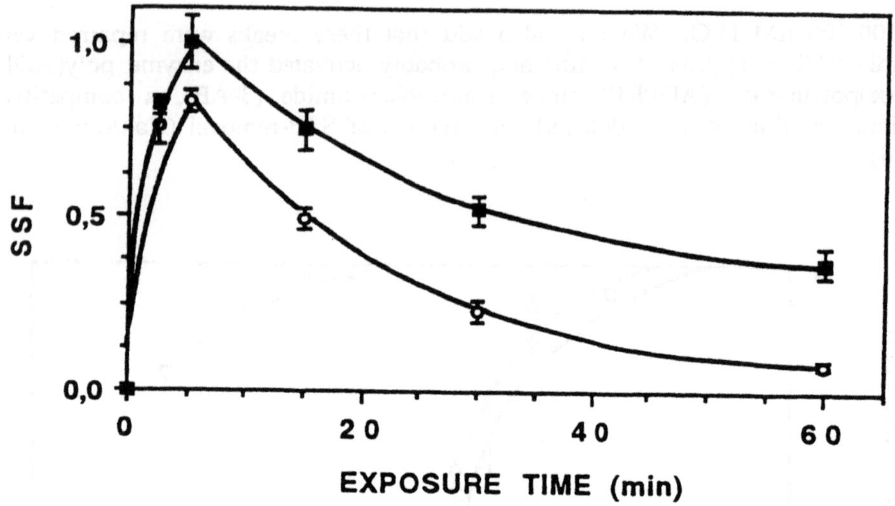

Fig. 3. Induction/removal of DNA SSBs in cells treated with H_2O_2 at physiological or ice-bath temperature.
CHO cells were treated for increasing legths of time with 100 µM H_2O_2 at 37°C in the absence (o) or presence (■) of 5 mM 3-aminobenzamide and analyzed for DNA damage with the alkaline elution assay. Results represent the mean ± S.E.M. calculated from 3 - 4 separate experiments, each performed in duplicate (adapted from Cantoni et al., 1989).

As previously mentioned, H_2O_2 displays a very low toxicity in cells when given at 4° but, as depicted in Fig. 2, a much greater cytotoxic response is observed following exposure to the oxidant at 37° C. We previously reported that DNA SSBs produced by the oxidant at 4° are removed within minutes, upon post-incubation at physiological temperature (Cantoni et al., 1986). Therefore, under conditions in which the treatments are given at 37°, the level of SSBs present at a fixed time is not simply a function of the net amount of SSBs that have been produced but also depends on the proportion of those breaks that have been rejoined during the treatment. The influence of SSB-rejoining on the steady-state level of SSBs can be appreciated from the data illustrated in Fig. 3. The SSF value is not a linear function of the time of exposure; rather, there is a rapid accumulation of DNA breaks during the first 5-10 min, followed by a marked decrease in the level of these lesions at longer times. Coincubation with the ADPRP inhibitor 3-AB increases the level of SSBs at all times after addition of the oxidant. This would suggest that the enzyme ADPRP is induced under these experimental conditions and the results obtained in our (Cantoni et al., 1989) and other laboratories (Schraufstatter et al., 1986a; Junod et al., 1989; Kirkland, 1991) confirm that this is indeed what happens. We observed maximal polymer accumulation at about 20 min followed by a rapid return to control levels. The levels of NAD^+, the substrate of this enzyme, concomitantly declined and thereafter returned slowly to control levels. Addition of 3-AB blocked the stimulation of ADPRP as well as the drop in NAD^+. In conclusion, DNA SSBs produced by H_2O_2 activate ADPRP, an enzyme postulated to regulate a number of cell functions including DNA repair, but also gene expression and differentiation. The question is then: does the activation of ADPRP trigger a machinery involved in the recovery from oxidative damage? Cochcrane and coworkers (Schraufstatter et al., 1986b) reported that inhibition of ADPRP markedly reduces the toxicity of H_2O_2 and this is in contrast with our results, since we did not find any significant effect (Fig. 2; Cantoni et al., 1989). It should be noted, however, that they used millimolar levels of the oxidant and assessed toxicity by trypan blue exclusion. This is an extremely important difference which emphasizes the importance of being extremely cautious when comparing results from different papers dealing with oxidative stress. There are many parameters that should be considered and, among these, the concentration of the oxidant utilized is extremely important. It would now appear that low and high concentrations of hydrogen peroxide kill cells via different mechanisms. High levels of H_2O_2 induce massive DNA damage which activates ADPRP resulting in NAD^+ depletion, which in turn will also lower cellular ATP content. Thus, the finding of Cochrane and coworkers (Schraufstatter et al., 1986b) that by preventing ADPRP induction the decrease in NAD^+ and ATP will also be reduced, as will be toxicity, is not surprising. Another important parameter to be considered when comparing the results of different investigators is the cell type utilized, since different cell lines may respond with different modes of cell death. Indeed, we also find that 3-AB reduces the level of cell death detected after a short period of incubation of U937 cells in the presence of millimolar levels of H_2O_2 (Palomba et al., manuscript submitted). On the other hand, cells are not rescued and the death process is simply delayed. The difference lies in the mode of cell death which is necrotic in cells receiving the oxidant alone and apoptotic in cells exposed to the oxidant in the presence of 3-AB. Preserving ATP levels via inhibition of ADPRP might therefore allow the cells to undergo the complex series of events involved in apoptosis. In conclusion, high concentrations of H_2O_2 elicit necrotic cell death which appears to be triggered by energy depletion.

What then is the mechanism of cell lethality triggered by low concentrations of hydrogen peroxide? This has not as yet been established. As previously mentioned, the toxicity of H_2O_2 is mediated by the highly reactive hydroxyl radical which is produced via interaction of the oxidant with divalent iron. The extreme reactivity of this radical species makes its interaction with the biomolecule closest to the site in which the same radical is generated highly probable. Thus it is clear that this type of oxidative injury will result in a wide spectrum of lesions, and therefore establishing the cytotoxic potential of each of these specific lesions is not an easy task. We can safely predict, however, that cell killing elicited by low concentrations of hydrogen peroxide is likely to depend on a multitarget attack.

On the other hand, since we are now dealing with DNA damage we must examine whether the SSBs brought about by H_2O_2 represent a potentially lethal event. In the past, a number of reports (Mello Filho et al., 1984; Hoffmann et al., 1984) suggested that these lesions can indeed promote cell killing, a conclusion mainly based on the observation that by inhibiting the production of DNA SSBs with iron chelators or $OH^·$ scavengers, toxicity can also be reduced. On the other hand, the production of SSBs could be an epiphenomenon and an alternative interpretation of the above results could be that both events are generated by the same radical species. Therefore, by inhibiting their production with iron chelators or their effects with OH· scavengers, both DNA damage and cytotoxicity can be suppressed. Experimental results obtained in our laboratory during the last few years support the notion that SSBs generated by low concentrations of H_2O_2 are in fact not lethal. For example, we previously discussed data (see also Cantoni et al., 1986 and 1989) indicating that the toxicity elicited by the oxidant did not increase when the repair of DNA SSBs was slowed down by incubating the cells in the presence of 3-AB (Fig. 3), making it unlikely that these lesions have anything to do with cytotoxicity. More recently, we demonstrated that the induction/loss of the oxidant-resistant phenotype is not correlated to the cytotoxicity elicited by hydrogen peroxide (Sestili et al., 1994). Additional evidence suggesting that DNA SSBs generated under conditions of oxidative stress are not cytotoxic comes from recent work on the effects of L-Histidine in cells exposed to hydrogen peroxide. A number of laboratories have reported that this amino acid increases the toxicity of H_2O_2 (Oya and Yamamoto, 1988; Link, 1988, Tachon and Giacomoni, 1989; Tachon 1990; Cantoni et al., 1992). The L-Histidine-mediated enhancement of hydrogen peroxide-induced cytotoxicity, however, is not always associated with increased DNA single strand breakage. Three lines of evidence support this inference: 1) the addition of an excess of L-Glutamine, while virtually inhibiting L-Histidine uptake, abolishes the L-Histidine-mediated enhancement of H_2O_2-induced cytotoxicity but does not prevent the increased formation of DNA SSBs (Sestili et al., 1992); 2) treatment with the oxidant following pre-loading with the amino acid leads to an enhanced cytotoxic response with no increased formation of DNA SSBs (Cantoni et al., 1994a); 3) DNA single strand breakage - but not lethality - is increased by treatment with L-Histidine and hydrogen peroxide at ice temperature (Cantoni et al., 1992). These observations are in keeping with recently published data from Coleman et al., (1989) indicating a dissociation of the accumulation of DNA single strand breakage from the killing of hepatocytes treated with various agents producing oxidative stress and therefore suggesting that DNA single strand breaks caused by oxygen-derived radicals are not a lethal lesion.

In conclusion, a large body of evidence strongly suggests that DNA SSBs generated by low levels of H_2O_2 are not causally related to the cytotoxic response in oxidatively injured cells. In cells treated with very high concentrations of H_2O_2, the massive induction of DNA SSBs might indirectly affect the lethal response via ADPRP activation and consequent depletion of NAD^+ and ATP.

DNA Double Strand Breaks and Cytotoxicity: The results reported above indicate that hydrogen peroxide is a potent inducer of DNA SSBs and one should therefore expect that DNA DSBs are also being generated, at least under conditions in which the oxidant produces a large number of DNA SSBs. This assumption, however, has not yet been experimentally proven. Indeed, using the neutral elution technique and more recently the pulsed field gel electrophoresis (PFGE) assay, a number of laboratories, have failed to detect DNA DSBs, at least at survival-range concentrations (Ward et al., 1985 and 1987; Cantoni et al., 1989; Iliakis et al., 1992; Sestili et al., in press). It would therefore appear that this type of DNA lesions does not represent a harmful event in oxidatively injured cells for the simple reason that DNA DSBs are not being produced. On the other hand, we recently demonstrated that survival-range concentrations of hydrogen peroxide effectively induce DSBs in the presence of L-Histidine (Cantoni et al., 1994b; Sestili et al. in press). As shown in Fig. 4, H_2O_2 alone (up to 300 μM) does not produce DSBs, but in the presence of L-Histidine the oxidant produces DNA DSBs in a concentration-dependent fashion and this lesion becomes apparent at levels as low as 20 μM. Thus, this system represents a condition of oxidative stress in which DNA DSBs are effectively produced and therefore we

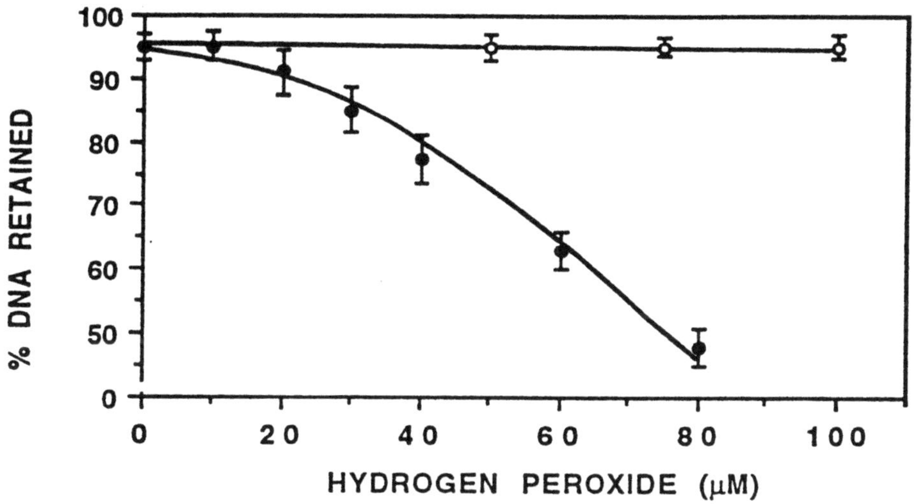

Fig. 4. Induction of DNA DSBs by the cocktail H_2O_2/L-Histidine.
CHO cells were treated with increasing concentrations of H_2O_2 in the absence (o) or presence (•) of 300 μM L-Histidine and then analyzed for DNA double strand breakage by the PFGE assay. Results are the mean ± S.E.M. calculated from three separate experiments, each performed in duplicate (adapted from Sestili et al., Biochim. Biophys. Acta, in press)

can investigate whether DNA DSBs mediate cell death. DNA double strand breakage in general is considered an extremely cytotoxic event and a number of agents such as X-rays kill cells via this mechanism (Lett, 1992; Frankenberg-Schwager, 1990; Elia et al., 1991). It is interesting to observe that an excellent correlation exists between DNA DSBs and toxicity in cells treated with H_2O_2 in the presence of L-Histidine, indicating that the cytotoxic response is a direct function of the level of DNA double strand breakage, and suggesting a cause-effect relationship (Fig. 5). Obviously, these results do not provide conclusive evidence that DNA DSBs are cytotoxic and therefore we attempted to collect additional data in support of this theory. Although we realize that the available evidence is still circumstantial, we nevertheless strongly believe that the DNA DSBs produced by H_2O_2/L-Histidine are toxic and are in fact the cause of the cytotoxicity elicited by this cocktail. The first evidence comes from a study in which we investigated whether metabolic processes were involved in the formation of DNA DSBs (Sestili et al., in press). CHO cells were incubated for 30 min in a saline with or without L-Histidine at physiological temperature, and then treated with H_2O_2 at either 37 or 4° C. The reason for studying the effect of L-Histidine in cells injured by the oxidant at 4° C is related to the fact that this temperature is not permissive for enzyme and/or energy-dependent processes; comparing the effects of the oxidant at the two temperatures in cells pre-loaded with the amino acid can therefore help us to understand whether the enhancing effects of the amino acid require active cellular metabolism. As previously discussed, the oxidant is more toxic at 37° than at 4°. Addition of L-Histidine increased the toxicity elicited by the oxidant at both temperatures, although the increase was only two-fold at 37° and more than one order of magnitude at 4°. DNA DSBs are not produced at survival-range concentrations after challenge with the oxidant at both temperatures. In the presence of L-Histidine, DNA DSBs were observed not only in cells challenged with the oxidant at 37° but also following exposure at 4°C. It was of interest that there was exactly the same relationship between the formation of DNA DSBs and toxicity in L-Histidine pre-loaded cells challenged with the oxidant at either 4 or 37°. All the data points fit in a single curve suggesting that 1) metabolism is not required for the production of DNA DSBs and for the increased cytotoxic response and 2) a cause-effect relationship is likely to exist between these two events. These data also suggest that similar molecular mechanisms mediate the formation of DNA DSBs under both experimental conditions. Two lines of evidence provide experimental support for this inference: 1) the kinetics of repair of DNA DSBs generated at 37° or 4°C were virtually superimposable (Sestili et al., in press); this would suggest that the same repair pathway(s) is/are responsible for the removal of the DNA DSBs generated at the two temperatures; 2) the size distribution of double-stranded DNA fragments produced under the two treatment conditions, resulting in a similar cytotoxic response, was basically identical (Sestili et al., in press); this is indicative of remarkable similarities in the topology of the chromosomal domains where DSBs are generated. These results were obtained by using a three-block assay -a modification of the PFGE assay used for measuring DNA DSBs - recently developed in our laboratory (Sestili et al., 1995).

Overall, the above results provide further experimental evidence in support of the notion that DNA DSBs are responsible for the L-Histidine-mediated enhancement of hydrogen peroxide-induced cytotoxicity, and demonstrate that the mechanism whereby the amino acid enhances the ability of hydrogen peroxide to produce DNA double strand breakage and cell killing does not depend on cellular metabolism and/or energy-dependent reactions.

Further evidence that the DNA DSBs produced by H_2O_2/L-Histidine are toxic was obtained in a study performed on a hydrogen peroxide-resistant CHO cell line (AG8). These cells display a 20 % survival after challenge with 20 mM H_2O_2 and the same toxicity is produced by 100 μM H_2O_2 in wild-type CHO cells (Cantoni et al., in press). L-Histidine increases the toxicity promoted by the oxidant in both cell types, although the level of enhancement was higher in AG8 cells (dose-modifying factor of about 5.5) compared to the parental cell line (dose modifying factor of about 2). AG8 cells have 4 times more catalase than wild-type cells. By treating AG8 cells with 10 mM aminotriazole for 6 hr we were able to lower catalase content of AG8 cells to the same levels detectable in wild-type cells. We next treated these cells with the oxidant in the absence or presence of L-Histidine. It is interesting to note that when these cells are treated with the oxidant alone, they still display a signifcant degree of resistance to H_2O_2 but, when challenged

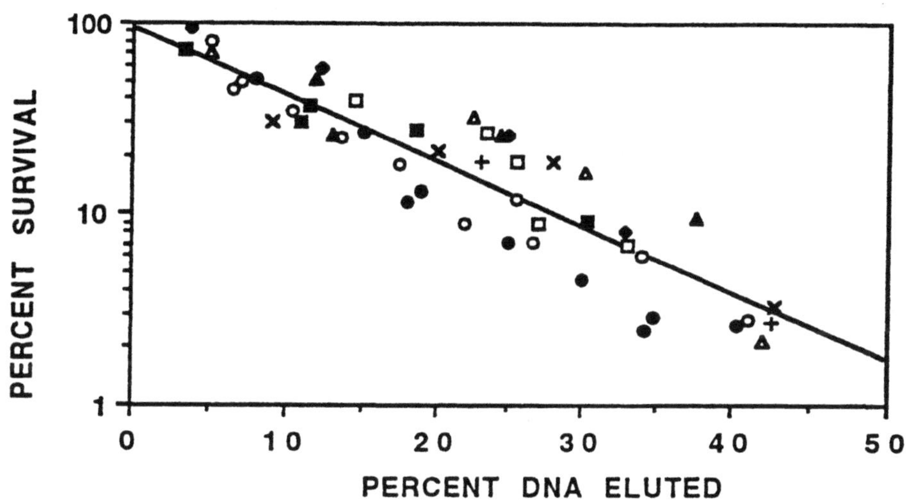

Fig. 5. Correlation analysis for DNA DSBs and cytotoxicity in various cell lines treated with hydrogen peroxide and L-Histidine under different experimental conditions.

Each point represents the results obtained by measuring cytotoxicity and DNA strand breakage in experiments using the following treatment protocols:

(o) CHO cells treated for 30 min with increasing concentrations of H_2O_2 in the presence of L-Histidine; CHO cells treated for 30 min with 100 μM H_2O_2 in the presence of increasing concentrations (0.01-1 mM) of L-Histidine; CHO cells treated for 30 min with increasing concentrations of H_2O_2 following a 20 min pre-incubation with 300 μM L-Histidine.

(•) 20 min exposure to L-Histidine (300 μM) at 37°C followed by treatment (30 min) with increasing concentrations of H_2O_2 at 4°C; 20 min exposure to increasing concentrations (0.01-1 mM) of L-Histidine at 37°C followed by treatment (30 min) with 300 μM H_2O_2 at 4°C.

(X) AG8 cells or catalase-depleted AG8 cells treated with increasing concentrations of H_2O_2 in the presence of L-Histidine (300 μM). Catalase depletion was achieved by pre-treating AG8 cells with 10 mM aminotriazole for 6 hr.

CCRF-CEM (▲), EAHY(△), McCoy's (◆), SCS1 (+), Vero (■), U937 (□) cells treated with increasing concentrations of H_2O_2 in the presence of L-Histidine (300 μM).

(adapted from Cantoni et al., 1994b; Sestili et al., Biochim. Biophys. Acta, in press; Sestili et al., Biochem. Pharm., in press) with the oxidant and L-Histidine, the survival is dramatically reduced, being superimposable to the survival of the parental cell line which was also treated with H_2O_2/L-Histidine. We then estimated the yield of DNA DSBs as a function of the H_2O_2 concentration in the presence of L-Histidine under conditions of normal or reduced catalase activity. In AG8 cells with normal catalase, DSBs became apparent at 200 μM H_2O_2 and then linearly increased in a concentration-dependent fashion. In cells with reduced catalase activity we obtained a curve which was similar although markedly shifted to the left. Once again, this curve was found to overlap with the dose-response curve obtained in the parental cell line following treatment with the oxidant and L-Histidine. A remarkable correlation was found to exist between cytotoxicity and DNA double strand breakage in wild type and AG8 cells under conditions of normal or reduced catalase content, following treatment with H_2O_2/L-Histidine. The experimental points fit once again in the curve illustrated in Fig. 5. These results led us to make two main considerations: firstly, a cell line which is resistant to H_2O_2, is not resistant to the oxidant in the presence of L-Histidine. Thus assuming that AG8 cells developed defense mechanisms which specifically protect these cells against the H_2O_2-induced injury, we can conclude that this machinery does not recognize the H_2O_2/L-Histidine-induced injury because different modes of action are responsible for the toxic events elicited by the two treatments. Secondly, the level of DNA DSBs produced on a per cell basis always results in an equal level of toxicity, regardless of cell type (resistant-vs sensitive cell line) since the lethality of each of these cell lines is dependent on the number of induced DSBs.

The third line of evidence that DNA DSBs produced by H_2O_2/L-Histidine are toxic was obtained by investigating the relationships existing between cytotoxicity and the formation of DNA DSBs in different cell lines derived from different animal species and characterized by different levels of sensitivity to the oxidant (Cantoni et al., 1994b). To make a long story short, we were able to demonstrate the existence of an excellent level of correlation by relating the cytotoxicity and the level of DNA DSBs obtained in all these cell lines challenged with the cocktail H_2O_2 /L-Histidine. These results suggest that, regardless of the concentration of the oxidant or the cell type; a fixed level of DNA DSBs is always paralleled by the same cytotoxic response. One can basically predict the toxicity of a given dose of H_2O_2 when it is associated with L-Histidine, in whatever cell line, by simply assessing the level of DSBs or vice versa.

The above experimental results strongly support the notion that DNA DSBs generated by an oxidative stress are lethal and, more specifically, that DNA double strand breakage mediates the cell killing elicited by H_2O_2 /L-Histidine.

In these studies, however, cytotoxicity was measured using a conventional cloning efficiency assay, which has the great advantage of being sensitive but also the important and, to some extent, critical limitation of providing information only about the number of cells still able to proliferate after exposure to a toxin. In other words, these toxicity measurements are based on the assumption that a cell which loses the ability to form colonies has also lost its viability. Although this might be true, this approach does not allow to distinguish between apoptotic and necrotic cell death. Additional studies were therefore undertaken to clarify the mechanism whereby L-Histidine enhances the toxicity of low concentrations of hydrogen peroxide using different cell lines (Palomba et al., manuscript submitted). It was found that in U937 cells exposed to non-toxic, albeit growth-inhibitory concentrations of the oxidant, the amino acid elicits an apoptotic response and this event is preceded by the formation of a low level of DNA DSBs. Using higher concentrations of H_2O_2, L-Histidine still remarkably increased cell killing but cell death occurred via a necrotic process. Thus, we suggest that a low level of DNA DSBs might be a triggering event for apoptosis whereas a high level of this lesions promptly promote necrotic cell death.

Conclusions

The results summarized in this paper emphasize that the cellular responses to H_2O_2 depend on the cell type and the oxidant concentration used. In particular, we have examined the relative impact of DNA SSBs vs DSBs and found that the former lesion, unlike the latter, does not appear to be able to directly promote the lethal response. We suggest, however, a potential role for DNA SSBs in triggering necrotic cell death -in concert with other lesions to different subcellular targets- following challenge with very high concentrations of H_2O_2. This effect might be the consequence of the ability of DNA SSBs to activate the enzyme ADPRP, an event which is parallelled by depletion of NAD^+ and ATP. Necrotic cell death would then be caused by energy depletion.

DNA DSBs caused under conditions of oxidative stress appear to be highly toxic and capable of triggering necrosis. The fact that, in the presence of L-Histidine, non toxic levels of H_2O_2 generate a low level of DNA DSBs and that this event is followed by apoptotic cell death, suggests that this type of DNA lesion might also be capable of triggering apoptosis.

References

Cantoni O, Murray D, Meyn RE (1986) Effect of 3-aminobenzamide on DNA strand-break rejoining and cytotoxicity in CHO cells treated with hydrogen peroxide. Biochim Biophys Acta 867: 135 - 143

Cantoni O, Sestili P and Cattabeni F (1988) Randomly distributed DNA breaks are not lethal for mammalian cells. Xenobiotica 18: 1481-1487

Cantoni O, Cattabeni F, Stocchi V, Meyn RE, Cerutti P, Murray D (1989) Hydrogen peroxide insult in cultured mammalian cells: relationships between DNA single strand breakage, poly(ADP-ribose) metabolism and cell killing. Biochim Biophys Acta 1014: 1-7

Cantoni O, Sestili P, Guidarelli A, Giacomoni PU and Cattabeni F (1992) Effects of L-histidine on hydrogen peroxide-induced DNA damage and cytotoxicity in cultured mammalian cells. Mol Pharm 41: 969-974

Cantoni O, Guidarelli A, Sestili P, Giacomoni PU, Cattabeni F (1994a) L-histidine-mediated enhancement of hydrogen peroxide-induced cytotoxicity: relationships between DNA single/double strand breakage and cell killing. Pharm Res 29: 169-177

Cantoni O, Sestili P, Brandi G, Cattabeni F (1994b) The L-histidine mediated enhancement of hydrogen peroxide-induced cytotoxicity is a general response in cultured mammalian cell lines and is always associated with the formation of DNA double strand breaks. FEBS Lett 353: 75-78

Cantoni O, Sestili P, Palomba L, Guidarelli A, Cattabeni F, Murray D Isolation and characterization of a chinese hamster ovary cell line with high-degree resistance to hydrogen peroxide. Biochem Pharm , in press

Chance B, Sies H and Boveris A (1979) Hydroperoxide metabolism in mammalian organs. Physiol Rev 59: 527-605

Coleman JB, Gilfor D, Farber JL (1989) Dissociation of the accumulation of single-strand breaks in DNA from killing of cultured hepatocytes by an oxidative stress. Mol Pharm 36: 193-200

Elia MC, Deluca JG, Bradley MO (1991) Significance and measurement of DNA double strand breaks in mammalian cells. Pharmacol and Therap 51: 291-327

Frankenberg-Schwager M (1990) Induction, repair and biological relevance of radiation-induced DNA lesions in eucaryotic cells. Radiat Environ Biophys 29: 272-292

Hoffmann ME, Mello Filho AC, Meneghini R (1984) Correlation between cytotoxic effects of hydrogen peroxide and the yeld of DNA strand breaks in cells of different species. Biochim Biophys Acta 781: 234-238

Iliakis G E, Pantelias GE, Okayasu R, Blakely WF (1992) Induction by H_2O_2 of DNA and interphase chromosome damage in plateau-phase Chinese hamster ovary cells. Radiat Res 131: 192-203

Junod AF, Jornod L, Petersen R (1989) Differential effects of hyperoxia and hydrogen peroxide on DNA damage, polyadenosine diphosphate-ribose polymerase activity, and nicotinamide adenine dinucleotide and adenosine triphosphate contents in cultured endothelial cells and fibroblasts. J Cell Physiol 140: 177-182

Kirkland JB (1991) Lipid peroxidation, protein thiol oxidation and DNA damage in hydrogen peroxide-induced injury to endothelial cells: role of activation of poly(ADP-ribose)polymerase. Biochim Biophys Acta 1092: 319-323

Lett J T (1992) Damage to cellular DNA from particulate radiations, the efficacy of its processing and the radiosensitivity of mammalian cells. Emphasis on DNA double-strand breaks and chromatin breaks. Radiat Environ Biophys 31: 257-277

Link, EM (1988) The mechanism of pH-dependent hydrogen peroxide cytotoxicity in vitro. Arch Biochem Biophys 265: 362-372

Mello Filho AC, Hoffmann E, Meneghini R (1984) Cell killing and DNA damage by hydrogen peroxide are mediated by intracellular iron. Biochem J 218: 273-275

Mello Filho AC, Meneghini R (1984) *In vivo* formation of single strand breaks in DNA by hydrogen peroxide is mediated by the Haber-Weiss reaction. Biochim Biophys Acta 781: 56-63

Meneghini R, Hoffmann E (1980) The damaging action of hydrogen peroxide on DNA of human fibroblasts is mediated by a non-dialyzable compound. Biochim Biophys Acta 608: 167-173.

Oya Y, Yamamoto K (1988) The biological activity of hydrogen peroxide IV. Enhancement of its clastogenic actions by coadministration of L-histidine. Mutat Res 198: 233-240

Schraufstatter IU, Hinshaw DB, Hyslop PA, Spragg RG, Cochrane CG (1986a) Oxidant injury of cells. DNA strand breaks activate polyadenosine diphosphate ribose polymerase and lead to depletion of nicotinamide adenine dinucleotide. J Clin Invest 76: 1131-1139

Schraufstatter IU, Hyslop PA, Hinshaw DB, Spragg RG, Sklar LA, Cochrane CG (1986b) Hydrogen peroxide-induced injury of cells and its prevention by inhibitors of poly(ADP-ribose)polymerase. PNAS 83: 4908-4912

Sestili P, Giacomoni PU, Cattabeni F, Cantoni O (1992) L-glutamine prevents the L-histidine-mediated enhancement of hydrogen peroxide-induced cytotoxicity. Biochem Pharm 44: 2418-2421

Sestili P, Cattabeni F, Cantoni O (1994) The induction/loss of the oxidant-resistant phenotype of Chinese Hamster Ovary (CHO) cell variants does not correlate with sensitivity to DNA single strand breakage by hydrogen peroxide. Biochem Pharm 48: 1701-1706

Sestili P, Cattabeni F, Cantoni O (1995) Simultaneous determination of DNA double strand breaks and DNA fragment size in cultured mammalian cells exposed to hydrogen peroxide/histidine or etoposide with CHEF electrophoresis. Carcinogenesis 16: 703-706

Sestili P, Cantoni O, Murray D Evidence for separate mechanisms of cytotoxicity in mammalian cells treated with hydrogen peroxide in the absence or presence of L-histidine. Biochim Biophys Acta , in press

Sestili P, Cattabeni F, Cantoni O The L-histidine-mediated enhancement of hydrogen peroxide-induced DNA double strand breakage and cytotoxicity does not involve metabolic processes. Biochem Pharmacol, in press

Starke PE, and Farber JE (1985) Ferric iron and superoxide ions are required for the killing of cultured hepatocytes by hydrogen peroxide. J Biol Chem 260: 10099-10104

Subrahmaniam VV, McGirr LG, O'Brien PJ (1987) Oxygen activation during drug metabolism. Pharmacol and Therap 33: 63-77

Tachon, P (1990) Intracellular iron mediates the enhancing effect of histidine on the cellular killing and clastogenicity induced by H_2O_2. Mutat Res 228: 221-228

Tachon P, Giacomoni PU (1989) Histidine modulates the clastogenic effect of oxidative stress. Mutat Res 211: 103 - 109

Ward JF, Blakely WF, Joner EI (1985) Mammalian cells are not killed by DNA single-strand breaks caused by hydroxyl radicals from hydrogen peroxide. Radiat Res 103: 383-392

Ward JF, Evans CL, Limoli CL, Calabro-Jones PM (1987) Radiation and hydrogen peroxide induced free radical damage to DNA. Brit J Cancer 55 (Suppl. VIII): 105-112

Mineral Fiber-Induced Oxidative Stress in Phagocytes

Kai M. Savolainen[1,2,*] and Marjo Ruotsalainen[2]
[1]University of Kuopio, Department of Pharmacology and Toxicology, P.O.B. 1627, FIN-70211 Kuopio
[2]National Public Health Institute, Department of Toxicology, P.O.Box 95, FIN-70701 Kuopio, Finland

Chronic exposure to mineral particles may cause development of pulmonary inflammation both in humans and experimental animals. Inhalation of mineral particles results in their uptake by alveolar macrophages. Moreover, a small part of the particles may be taken up also be the epithelial cells lining the pulmonary alveoli (see Doelman et al., 1990). The phagocytized mineral particles may induce inflammatory reaction during which chemotactic factors are released from the macrophages. Upon stimulation by macrophage-derived cytokines leukocytes infiltrate to the site of inflammation in the lung (Lugano et al., 1982). Prolonged exposure to mineral particles is associated with accumulation of activated phagocytic cells, i.e. neutrophils, eosinophils and basophils, in the lungs. Activation of alveolar macrophages by mineral fibers may be of particular importance in the pulmonary inflammation and fibrosis because they are the first line of defence against foreign intruders (Doelman et al., 1990). Furthermore, activation of macrophages and other phagocytic cells, monocytes and polymorphonuclear leukocytes (PMNL), results in a release of proteolytic enzymes and reactive oxygen metabolites (ROM) such as superoxide anion radical, hydrogen peroxide, hydroxyl radical and singlet oxygen (Adamson and Bowden, 1988) which contribute to oxidative stress in phagocytes (Halliwell and Gutteridge, 1984; Seifert and Schultz, 1991).

Even if the pulmonary pathology induced by quartz or asbestos has been well characterized (Mossman and Marsh, 1989), the mechanisms of this pathology are poorly known. A plethora of recent studies have provided, however, evidence that fibers stimulate phagocytic cells capable of producing ROM, and the hypothesis of oxidative stress-driven pathophysiology behind mineral fiber-induced lung diseases has gained much support (Mossman and Marsh, 1989).

ROM production during phagocytosis is a vital part of the normal function of phagocytic cells. The goal of ROM production, together with the release of proteases and other secretory products, is to destroy foreign particles such as bacteria, viruses and mineral particles in phagocytic cells (Doelman et al., 1990). It is the inability of phagocytic cells to destroy mineral particles that leads to a pathological persistent production of ROM, often called frustrating phagocytosis. When excessive amounts of ROM are produced, they damage, in addition to the phagocytic cell, also surrounding cells and tissue by attacking biological macromolecules, i.e. membrane proteins and lipids, cytosolic proteins and nucleic acids (Doelman et al., 1990; Adachi et al., 1992).

Recent findings have provided evidence that nitric oxide, also a radical species produced by phagocytic cells, may have a role in the activation, and in the deleterious effects of activation, of immunological cells, and in the resulting lung pathology (Moncada and Higgs, 1993). The role of NO may be especially remarkable in controlling of the functions of macrophages and leukocytes. However, the mechanisms of the effects of NO in phagocytic cells and in mineral particle-induced functional and morphological pulmonary changes remain to be elucidated.

Phagocytic Cell Activation and Lung Pathology

Inhaled mineral particles cause inflammation of the lung which at later stages of the irritation is often associated with fibrosis. Pulmonary inflammation is associated with phagocytic cell activation and the release of bioactive substances. Subsequent to release of chemotactic factors, leukocytes infiltrate from the blood stream into the pulmonary tissue. Among mineral particles, asbestos seems to be active in the production of chemotactic factors (Yano, 1988). Mineral particle-exposed phagocytic cells are activated during the inflammation to and are evoked to produce ROM. Also a number of soluble and particulate stimuli also activate alveolar macrophages and PMNL to produce reactive oxygen species (von Tscharner et al., 1986; Roney and Holian, 1989; Tuomala et al., 1992; 1993a; b). Sustained production of ROM by immunological cells upon stimulation by inhalable mineral particles may be involved in various pathological lung conditions (Mossman et al., 1983). In a recent study, Driscoll and coworkers (Driscoll et al., 1990) demonstrated an association between increased quartz particle retention, lung injury, activation of alveolar macrophage fibronectin release, and the development of fibrosis in rat lung. These and other experimental animal studies are in good agreement with the results of human studies. Prolonged occupational exposure to asbestos or quartz has been specifically implicated in pulmonary fibrosis (Mossman et al, 1983; Heppleston, 1991) and in the increased risk of lung cancer in humans (Troll and Wiesner, 1985; Koskela et al, 1987). Moreover, exposure to chrysotile asbestos may cause pulmonary fibrosis, bronchogenic carcinoma, and mesothelioma (Mossman and Gee, 1989).

The mechanisms whereby asbestos, quartz or other mineral fibers, or even organic particles (Korchak et al., 1988a; Tuomala et al., 1992; 1993a; b), activate PMNL and other phagocytic cells to produce ROM have remained unresolved, but there are data to suggest that these cellular events may at least partially be receptor-mediated phenomena. Chrysotile asbestos stimulates the production of superoxide anion in guinea pig alveolar macrophages by increasing the turnover of phosphatidic acid (PA) (Roney and Holian, 1989). The change in PA turnover expresses the activation of inositol lipid signalling system (Berridge and Irvine, 1989). These events are, in turn, associated with Ca^{2+}-mediated cellular signalling and the stimulation of protein kinase C (PKC) (Nishizuka, 1986; Berridge and Irvine, 1989; Seifert and Schultz, 1991). Moreover, Tuomala and coworkers (Tuomala et al., 1992) have shown that facilitated inositol lipid signalling and elevation of intracellular Ca^{2+} levels ($[Ca^{2+}]_i$) precede the production of ROM in PMNL by quartz. Thus, there are data to suggest that receptor-mediated cell signalling may be involved in particulate-induced stimulation of immunological cells to produce ROM and subsequent oxidative stress in phagocytes.

Activation Mechanisms of Phagocytic Cells

Even if the mechanisms whereby mineral particles activate phagocytes to produce ROM and oxidative stress are poorly known, those of various soluble agonists are much better understood. Therefore, the activation of phagocytic cells by soluble stimuli will be discussed first, and will compared with the mechanistic information from mineral particle-induced cell activation and production of ROM.

A chemotactic peptide, FMLP, is a bacterial chemoattractant released from bacterial membrane that causes the invasion of leukocytes to the site of the release of FMLP. Moreover, FMLP is also used as a standard stimulant to explore mechanisms of receptor-mediated activation of phagocytic cells. FMLP causes the aggregation of PMNL, depolarization of the cells, and stimulation of respiratory burst, i.e. increased O_2 consumption and increased activity of hexose monophosphate shunt (Holian and Stickle, 1985). A key-event in this process is the binding of FMLP to its cell surface receptor which through the activation of G-protein stimulates G-protein-coupled phospholipase C (PLC). This initiates a bifurcating cell signalling pathway, the production of inositol lipid-derived second messengers. Membrane phosphatidylinositol-4,5-bisphosphate is hydrolyzed by PLC to inositol-1,4,5-trisphosphate (Ins-1,4,5-P_3) and diacylglycerol (DG) (Korchak et al., 1988a; b; Berridge and Irvine, 1989; Baggiolini et al., 1993). Ins-1,4,5-P3 stimulates the release of calcium from intracellular stores (Berridge and Irvine 1989), and DG stimulates PKC (Nishizuka, 1986). PKC may phosphorylate plasma membrane NADPH oxidase which then catalyzes the reduction of oxygen to superoxide anion (Cox et al., 1985; Sadler and Badwey, 1988; Baggiolini and Wyman, 1990, Seifert and Schultz, 1991). This conclusion agrees well with the observation that in human PMNL oxidative burst correlates with the amount of PKC in the cells (Salamino et al., 1991).

PMA, a tumor promoter, is a direct activator of PKC that is able to by-pass receptor stimulation (Nishizuka, 1986). PKC has a role in receptor-mediated and G-protein coupled inositol lipid signalling pathway as one of the regulatory enzymes of cellular functions and activation (Lad et al., 1992). Because of the ability of PMA to directly stimulate PKC, PMA has been widely used to explore the roles of PKC in various types of activation of phagocytic cells. An important feature of DG is that it induces a transient translocation, and subsequent activation of PKC, into the membrane (Nishizuka, 1986) and concomitantly causes the activation of NADPH oxidase, and subsequently increases ROM production and oxidative stress in phagocytic cells (Seifert and Schultz, 1991). In contrast to DG, PMA induces a permanent translocation of PKC into the membrane. This difference in effects between DG and PMA may explain many of the different effects in phagocytic cells during their receptor-mediated or direct stimulation of PKC. For example, Chabot et al. (1992) suggest that FMLP stimulates DG production through phospholipase D (PLD), whereas PMA-induced DG generation may involve several metabolic pathways including PLD activation, and activation of PA phosphohydrolase (see figure 1).

Protein Kinase C in the Activation of Phagocytes

Recent observations indicate that changes in cellular phospholipid metabolism and subsequent activation of PKC are important in agonist-induced production of ROM in phagocytic cells. Umei et al. (1993) reported that priming of neutrophils with PMA, but not with FMLP, prior to their stimulation with the same compounds causes translocation of PKC into the membrane. PKC translocation to the membrane is, in turn, a prerequisite for the enzyme activity, subsequent phosphorylation of NADPH oxidase, ROM production, and oxidative stress. In another study, Ding and Badwey (1993a; b) showed in guinea pig neutrophils that these cells contain a battery of protein kinases that are rapidly and transiently activated upon stimulation of the cells by FMLP. In contrast, stimulation of these cells with PMA resulted in a decrease of these kinase activities. Even if the time-course of FMLP-induced activation of these kinases in guinea pig neutrophils resembles the time-course of FMLP-induced inositol lipid signalling (Korchak et al., 1988a; b), the elevation of $[Ca^{2+}]_i$, and the increase of the production of ROM (Tuomala et al., 1992; 1993a) in human PMNL it is not likely that they belong to closely related signalling pathways because of the different role of PKC in these cascades of events of cell activation. The differences may be due to differences between cell types. Alternatively, FMLP and PMA may stimulate at least partially different PKC isoenzymes (Nishizuka, 1986) with a different role in the production of ROM by phagocytes. Both FMLP and PMA are strong inducers of ROM production in phagocytic cells. Oxidative burst induced by FMLP and PMA is at least partially sensitive to PKC inhibitors even if the ROM production in these cases exhibits a clearly different time-course (Ruotsalainen and Savolainen, 1995b).

Kessels et al. (1993) suggested that PKC is not involved in FMLP-induced activation of phospholipase D (PLD), but is essential for concomitant activation of NADPH oxidase. Activation of PLD results, in turn, in the formation of PA which is further converted to DG by PA phosphohydrolase. DG plays an important role in the activation of PKC an the oxidative burst in PMNL. However, also PA, derived from phosphatidylcholine (PC) directly induces respiratory burst in electropermeabilized human PMNL by acting on a downstream step of PKC. This is most likely due to direct activation of NADPH oxidase which produces superoxide anion (Mitsuyama et al., 1993).

In most cases the activation of PKC seems to occur in concert with the activation of NADPH oxidase, responsible for FMLP-induced rapid and transient oxidative burst. After PMA exposure, a PKC inhibitor staurosporine inhibits NADPH-oxidase, most likely stimulated through increased activity of PKC in PMNL. PKC inhibition has, however, not an effect on the initiation or the maximal rate of superoxide anion secretion by FMLP, but results in a more rapid termination of oxidant production (Watson et al., 1991). Thus, the mechanisms which regulate the activity of PKC during different stimulations of phagocytic cells have not yet thoroughly clarified. However, mineral fibers seem to modulate the activation of phagocytic cells at several steps of communicating cell signalling pathways (Tuomala et al., 1992; 1993a; b; Ruotsalainen et al., 1995; Ruotsalainen and Savolainen 1995a; b) but many of the final regulatory pathways remain still to be elucidated. Figure 1. depicts some of the details of cell signalling pathways in human PMNL.

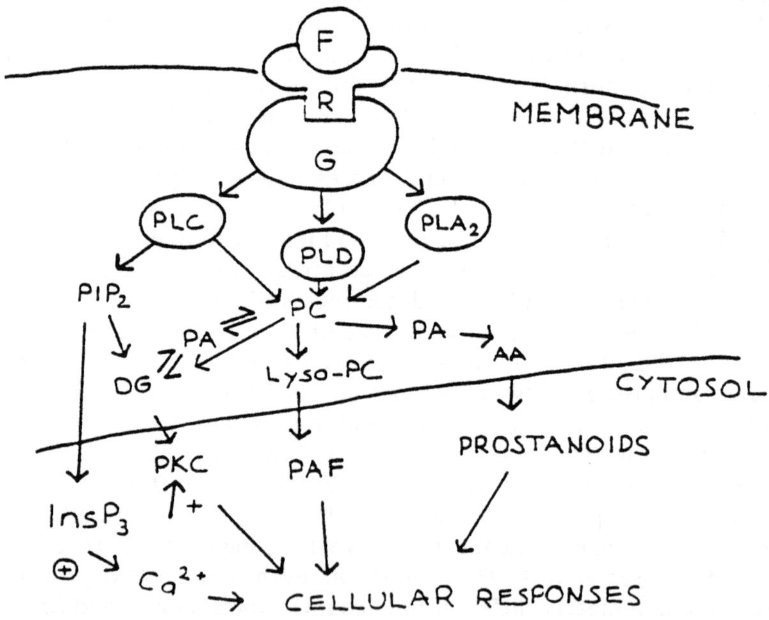

Fig. 1. A simplified summary of cellular lipid interactions. F indicates fiber bound to a recognition site (R) coupled to a G protein (G). The hydrolysis of phosphatidylinositol 4,5-bisphosphate (PIP_2) by phospholipase C (PLC) leads to generation of inositol 1,4,5-trisphosphate (IP_3) and diacylglycerol (DG). IP_3 mobilizes calcium from intracellular stores and DG activates protein kinase C (PKC). Hydrolysis of phosphatidylcholine (PC) by phospholipase A_2 (PLA_2) generates lysophosphatidylcholine (lyso-PC). Also PIP_2 may converted to lyso-PC, a source for platelet activating factor (PAF). Phospholipase D converts PC DG which can be metabolized to phosphaticidc acid (PA). DG-kinase activates DG to PA and PA-phosphohydrolase forms DG from PA. PLA_2 activates PA to arachidonic acid (AA), which is converted via the cyclo-oxygenase and lipo-oxygenase pathways to eicosanoids, e.g. prostaglandins and prostacyclins.

Effects of Mineral Fibers on the Activation of Phagocytic Cells

The ability of asbestos and quartz to increase the production of ROM in human PMNL and macrophages (Hedenborg and Klockars, 1987; 1989; Klockars et al., 1990; Nyberg and Klockars, 1990a; b; c) has been convincingly demonstrated. Perhaps the most striking new feature in fiber-induced phagocytic cell activation and ROM production is that the involvement of G-protein-coupled recognition sites for mineral particles, subsequent activation of inositol lipid signalling pathway, $[Ca^{2+}]_i$, and the activity of PKC may play an important role in mineral fiber-induced activation of immunological cells (Ruotsalainen and Savolainen, 1995a). The exact nature of the recognition sites, and their coupling mechanisms, of fibers on phagocyte surface has remained open. Opsonization of mineral particles seems to be important for the mineral particle-induced oxidative burst (Kobzik et al., 1993), and the role of Fc-receptors has been emphasized (Kobzik, 1995). However, Kobzik (1995) has observed that also scavenger-type receptors on phagocytes bind inert particles but without evoking oxidative burst in these cells. However, the role of these types of receptors remains to be clarified in mineral particle-induced responses in phagocytic cells.

Meshulam et al. (1988) showed that particulate stimuli, notably serum opsonized and non-opsonized zymosan, elicite oxidative burst in PMNL that preceded increased production of Ins-1,4,5-P_3 and elevation of $[Ca^{2+}]_i$. The kinetics of these events were such that they clearly proposed for the first time an association between the oxidative burst, and the production of inositol lipid second messengers. In the same paper, also Candida albicans hyphae induced similar effects in PMNL. For details of inositol lipid signalling, see figure 2.

Roney and Holian (1989) induced excessive production of ROM in guinea pig alveolar macrophages by exposing them to chrysotile asbestos. In these cells, chrysotile dose-dependently increased the production of ROM. Oxidative burst was preceded by increased turnover of PA, a metabolite of inositol lipid-derived DG (Berridge and Irvine, 1989). It is also possible that PA could have derived from phosphatidylcholine and had directly activated NADPH oxidase (Mitsuyama et al., 1993). This interpretation is, however, unlikely to be correct because increased PA turnover was followed by elevations of $[Ca^{2+}]_i$. Moreover, chrysotile-induced ROM production could be inhibited with a non-specific PKC inhibitor, staurosporine, and partially blocked with pertussis toxin that inhibits a stimulatory G-protein (Gp) coupled to PLC. Therefore, these results provide evidence that chrysotile induces oxidative burst in guinea pig alveolar macrophages by acting through a receptor-coupled and second messenger-mediated mechanism in which a G-protein and PKC may have at least a contributing role.

Another important piece of information suggesting a role for second messengers in mineral fiber-induced oxidative burst in phagocytic cells has derived from studies of Tuomala et al. (1992) who demonstrated in PMNL that quartz increases the production of Ins-1,4,5-P3 prior to the increased production of reactive oxygen species. In another study, Tuomala et al. (1993a) showed that quartz also dose-dependently elevated $[Ca^{2+}]_i$ in stimulated PMNL. The time-course of the quartz-induced $[Ca^{2+}]_i$ elevation was such that it was clear that also the changes in $[Ca^{2+}]_i$ preceded increased ROM production. This is consistent with the assumption that quartz, in PMNL, may induce the production of ROM through a mechanism which requires both the generation of inositol lipid second messengers and subsequent elevation of $[Ca^{2+}]_i$. It was notable that the quartz-induced ROM production, both in terms of time-course and dose-effect, resembled closely those induced by FMLP and PMA. This is consistent with the view that receptor-mediated mechanisms and PKC may be involved in quartz-induced PMNL activation. In the same study, chrysotile had effects on $[Ca^{2+}]_i$ and ROM production that closely resembled the effects of quartz. A putative mechanism through which mineral fibers may stimulate oxidative burst in human polymorphonuclear leukocytes is depicted in figure 2.

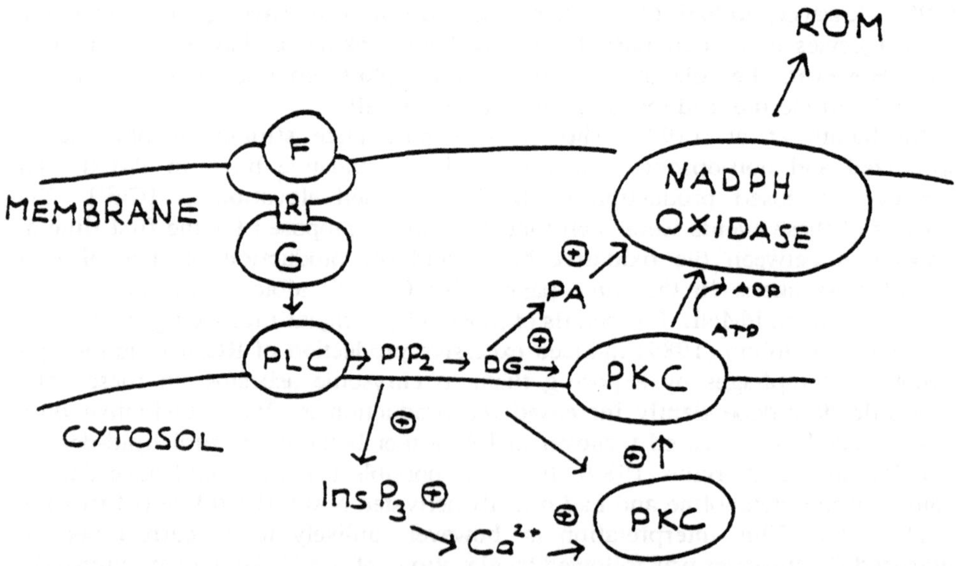

Fig. 2. A putative mechanism for mineral fiber-induced activation of leukocytes through inositol lipid signalling system. F indicates a fiber that is bound to a recognition site (R) coupled to a G-protein (G). The hydrolysis of phosphatidylinositol 4,5-bis-phosphate (PIP2) by phospholipase C (PLC) produces inositol 1,4,5-trisphosphate (Ins(1,4,5)P3) and diacylglycerol (DG). InsP3 evoked Ca++-mobilization enhances translocation of PKC from the cytosol to the membrane and stimulates PKC. DG also translocates and stimulates PKC which subsequently phosphorylates NADPH oxidase and evokes oxidative burst. Phosphatidic acid (PA), metabolized from DG or phosphatidylcholine (PC, see figure 1) may directly stimulate NADPH oxidase and induce oxidative burst in phagocytes.

The Role of Calcium in Mineral Fiber-Induced Cell Damage

Also other studies have shed light on the possible mechanisms whereby quartz or chrysotile asbestos activate human PMNL to increase the production of reactive oxygen species. Tuomala et al. (1993b) found that a single dose of quartz dose-dependently elevated $[Ca^{2+}]_i$ and ROM production, whereas such effect was not seen after a repeated dose of quartz. On the other hand, quartz administered prior to FMLP did not modify FMLP-induced $[Ca^{2+}]_i$, but attenuated FMLP-induced ROM production. Priming of PMNL with FMLP prior to quartz did not affect quartz-induced $[Ca^{2+}]_i$ changes but quartz-induced ROM production was greatly amplified. These studies provide evidence that both quartz and FMLP increase the production of ROM in PMNL by mechanisms in which $[Ca^{2+}]_i$ may have a key-role. It seems possible that both $[Ca^{2+}]_i$-dependent and -independent mechanisms may account to quartz-induced ROM production in PMNL.

In a series of studies Ruotsalainen and Savolainen (1995b) have shown that a novel PKC inhibitor, Ro 31-7549 exhibiting a high specificity and affinity toward PKC (Twomey et al., 1990), almost completely blocks both quartz- and chrysotile-induced oxidative burst in PMNL suggesting a role for PKC in mineral fiber-induced production of reactive oxygen species. The role of G-protein in mineral fiber-induced oxidative burst was explored by pretreating human PMNL with pertussis or chlolera toxin after permealization of the cells. These studies (Ruotsalainen and Savolainen, 1995a) showed that pretreatment of the cells with cholera toxin, but not with pertussis toxin, partially inhibited quartz- and chrysotile-induced production of ROM thereby providing evidence that a cholera, but not a pertussis toxin sensitive G-protein may be involved in quartz- and chrysotile-induced activation of PMNL to produce reactive oxygen species.

Extracellular calcium seems also play a role in the activation of phagocytes to produce reactive oxygen species. Ruotsalainen and coworkers (Ruotsalainen et al., 1995) demonstrated that a calcium channel blocker verapamil dose-dependently attenuated both quartz- and chrysotile-induced elevations of Ca^{2+} suggesting that the influx of Ca^{2+} through verapamil-sensitive Ca^{2+} channels may play a role in mineral fiber-induced activation of PMNL. Moreover, the steepness of quartz- and chrysotile-induced elevation of $[Ca^{2+}]_i$ decreased dose-dependently by verapamil indicating that verapamil inhibited the influx of extracellular Ca^{2+} into the cell through L-type Ca^{2+} channels. These findings were also confirmed by similar findings when extracellular Ca^{2+} was chelated with EGTA (Ruotsalainen et al., 1995). These data suggest that a release of Ca^{2+} from intracellular stores may be the initial event in mineral fiber-induced activation of PMNL. Thus, even if both intra- and extracellular Ca^{2+} are essential for PMNL activation by mineral fibers, the release of Ca^{2+} from intracellular stores seems to be essential for the initiation of the activation, then amplified by a slower influx of extracellular calcium, possibly due a co-operative action of elevated $[Ca^{2+}]_i$ on membrane calcium channels (von Tscharner et al., 1986). These findings are supported by observations of Chen et al. (1991) who showed that quartz induces a dose-dependent increase in $[Ca^{2+}]_i$ in rat alveolar macrophages. Moreover, chelation of extracellular calcium by EGTA partially inhibits quartz-induced $[Ca^{2+}]_i$ elevation, LDH release, and cellular damage. Chen and coworkers (1991) concluded that the elevation of $[Ca^{2+}]_i$ is an early event in quartz-induced macrophage damage subsequent to excessive cell stimulation by fibers. These results have been in essence confirmed by the findings of Rojanasakul et al. (1993) who showed that quartz produced damage in rat alveolar macrophages which was associated with increased $[Ca^{2+}]_i$ that could be partially blocked by a calcium channel blocker nifedipine or by chelation of extracellular calcium by EGTA.

Conclusions

The activation of phagocytic cells to produce reactive oxygen species may take place via mechanisms which at least partially are mediated through receptor-mediated mechanisms. Involvement of a cell membrane recognition site coupled to a pertussis or a cholera toxin sensitive G protein is a possibility that should be further explored. There are data from several studies utilizing different cell types that mineral particles, such as quartz or chrysotile, are able to increase the production of inositol lipid second messengers and concomitantly elevate $[Ca^{2+}]_i$ which seems to be a prerequisite for mineral fibre-induced activation of phagocytic cells to produce ROM. Both intra- and extracellular calcium seem to be important for phagocytic cell activation, intracellular calcium as an initiator, and extracellular calcium as an amplifier, of cellular responses. PKC seems to play a key-role in the activation of both PMNL and macrophages to produce ROM. Regulation of PKC activity is most likely significantly affected by changes in cellular calcium homeostasis, inositol lipid signalling, and activity of number of important enzymes such as PLD and PA phosphohydrolase. They are important in the metabolism of various cellular second messengers derived from membrane phospholipid components. For future studies, effects of mineral fiber-evoked alterations in gene expression may be important for understanding mineral fiber-induced harmful effects in the lungs.

Acknowledgements

The authors wish to thank Ms. Leena Heikkinen and Ms. Liisa Kujanpää for skillful technical assistance and for drawing of the figures, Ms. Heli Leinonen for careful typing of the manuscript, and the Finnish Work Environment Fund and The Academy of Finland for financial support.

References

Adachi S, Kawamura K, Yoshida S, Takemoto K (1992) Oxidative damage on DNA induced by asbestos and man-made fibers in vitro. Int Arch Occup Health 63:553-557

Adamson IY, Bowden DH (1988) Relationship of alveolar epithelial injury and repair to induction of pulmonary fibrosis. Am J Pathol 130:377-383

Baggiolini M, Boulay F, Badwey J.A, Curnutte JT (1993) Activation of neutrophil leukocytes: chemoattractant receptors and respiratory burst. FASEB J 7:1004-1010

Baggiolini M, Wymann P (1990) Turning on the respiratory burst. Trends Biol Sci 15:69-72

Berridge MJ, Irvine RF (1989) Inositol phosphates and cell signalling. Nature 341:197-205

Chabot MC, McPhail C, Wykle RL, Kennerly AA, McCall CE (1992) Comparison of diglyceride production from choline-containing phosphoglycerides in human neutrophils stimulated with N-formylmethionyl-leucylphenylalanine, ionophore A23187 or phorbol 12-myristate 13-acetate. Biochem J 286:693-699

Chen J, Armstrong LC, Liu S, Gerriets JE, Last JA (1991) Silica increases cytosolic free calcium ion concentration of alveolar macrophages in vitro. Toxicol Appl Pharmacol 111:211-220

Cox JA, Jeng AY, Sharkey NA, Blumberg PM, Tauber AI (1985) Activation of the human neutrophil nicotinamide adenine dinucleotide phosphate (NADPH)-oxidase by protein kinase C. J Clin Invest 76:1932- 1938

Ding J, Badwey JA (1993a) Neutrophils stimulated with a chemotactic peptide or a phorbol ester exhibit different alterations in the activities of a battery of protein kinases. J Biol Chem 268:5234-5240

Ding J, Badwey JA (1993b) Stimulation of neutrophils with a chemoattractant activates several novel protein kinases that can catalyze the phosphorylation of peptides derived from the 47-kDa protein component of the phagocyte oxidase and myristoylated alanine-rich C kinase substrate. J Biol Chem 268:17326-17333.

Doelman CJA, Leurs R, Oosterom WC, Bast A (1990) Mineral dust exposure and free radical-mediated lung damage. Exp Lund Res 16:41-55

Driscoll KE, Maurer JK, Lindenschmidt RC, Romberger D, Rennard SI, Crosby L (1990) Respiratory tract responses to dust: Relationship between dust burden, lung injury, alveolar macrophage fibronectin release, and the development of pulmonary fibrosis. Toxicol Appl Pharmacol 106:88-101

Halliwell B, Cutteridge JMC (1984) Oxygen toxicity, oxygen radicals, transition metals and diseases. Biochem J 219:1-14

Hedenborg M, Klockars M (1987) Production of reactive oxygen metabolites induced by asbestos fibers in human polymorphonuclear leucocytes. J Clin Pathol 40:1189-1193

Hedenborg M, Klockars M (1989) Quartz-dust-induced production of reactive oxygen metabolites by human granulocytes. J Lung 167:23

Heppleston AG (1991) Minerals, fibrosis, and the lung. Environ Health Persp 94:149-198

Holian A, Stickle D (1985) Calcium regulation of phosphatidyl inositol turnover in macrophage activation by formyl peptides. J Cell Physiol 123:39-45

Kessels GC, Krause K-H, Verhoeven AJ (1993) Protein kinase C activity is not involved in N-formylmethionyl-leucyl-phenylalanine-induced phospholipase D activation in human neurophils, but is essential for concomitant NADPH oxidase activation: studies with a staurosporine analogue with improved selectivity for protein kinase C. Biochem J 292:781-785

Klockars M, Hedenbort M, Vanhala E (1990) Effect of two particle surface- modifying agents, polyvinylpyridine-N-oxide and carboxymethylcellulose, on the quartz and asbestos mineral fiber-induced production of reactive oxygen metabolites by human polymorphonuclear leukocytes. Arch Environ Health 45: 8-14

Kobzik L (1995) Lung macrophage uptake of unopsonized environmental particulates. J Immunol 155:367-376

Kobzik L, Huang S, Paulauskis JD, Godleski JJ (1993) Particle opsonization and lung macrophage cytokine response. J Immunol 151:2753-2759

Korchak HM, Vosshall LB, Zagon G, Ljubich P, Rich AM, Weissmann G (1988a) Activation of the neutrophil by calcium-mobilizing ligands. J Biol Chem 263:11090-11097

Korchak HM, Vosshall LB, Zagon G, Ljubich P, Rich AM, Weissmann G (1988b) Activation of the neutrophil by calcium-mobilizing ligands II. Correlations of calcium, diacylglycerol, and phosphatidic acid generation with superoxide anion generation. J Biol Chem 263:11098-11105

Koskela R-S, Klockars M, Järvinen E, Kolari PJ, Rossi A (1987) Cancer mortality of granite workers. Scand J Work Environ Health 13:26-31

Lad PM, Kaptein JS, lin C-KE, Kalunta CI, Scott SJ, Gu DG (1992) G-proteins and the role of second messengers in the regulation of the human neutrophil. Immunol Ser 57:107-136

Lugano EM, Dauber JH, Daniele RP (1982) Acute experimental silicosis. Lung morphology, histology, and macrophage chemotaxin secretion. Am J Pathol 109:27-36

Meshulam T, Diamond RD, Lyman CA, Wysong DR, Melnick DA (1988) Temporal association of calcium mobilization, inositol trisphosphate generation, and superoxide anion release by human neutrophils activated by serum opsonized and nonopsonized particulate stimuli. Biochem Biophys Res Communications 150:532- 539

Mitsuyama T, Takeshige K, Minakami S (1993) Phosphatidic acids induces the respiratory burst of electropermeabilized human neurophils by acting on a downstream step of protein kinase C. Febs 328:67-70

Moncada S, Higgs A (1993) The L-argine-nitric oxide pathway. New England J Med 329:2002-2012

Mossman BT, Gee JBL (1989) Asbestos-related diseases. New Engl J Med 320:1921

Mossman BT, Light W, Wei E (1983) Asbestos: Mechanisms of toxicity and carcinogenicity in the respiratory tract. Ann Rev Pharmacol Toxicol 23:595-615

Mossman BT, March JP (1989) Evidence supporting a role for active oxygen species in asbestos induced toxicity and lung disease. Env Health Perspectives 81:91-94

Nishizuka Y (1986) The role of protein kinase C in cell surface signal transduction and tumor promotion. Science 233:305

Nyberg P, Klockars M (1990a) Measurement of reactive oxygen metabolites produced by human monocyte- derived macrophages exposed to mineral dusts. Int J Exp Path 71:537-544

Nyberg P, Klockars M (1990b) Quartz-induced production of reactive oxygen metabolites by activated human monocyte-derived macrophages. APMIS 98:823-827

Nyberg P, Klockars M (1990c) Effect of immunoglobulins on mineral dust- induced production of reactive oxygen metabolites by human macrophages. Inflammation 14:621-629

Rojanasakul Y, Wang L, Malanga CJ, Ma JYC, Banks DE, Ma JKH (1993) Altered calcium homeostasis and cell injury in silica-exposed alveolar macrophages. J Cell Phys 154:310-316

Roney PL, Holian A (1989) Possible mechanism of chrysotile asbestos-stimulated superoxide anion production in guinea pig alveolar macrophages. Toxicol Appl Pharmacol 100:132-144

Ruotsalainen M, Naarala J, Savolainen KM (1995) Mineral fiber-induced leukocyte activation: the role of intra- and extracellular calcium. Toxicol Lett 78:195-205

Ruotsalainen M, Savolainen KM (1995a) The role of G-proteins in the activation of human leukocytes by particulate stimuli to produce reactive oxygen metabolites. Toxicology 99:67-76

Ruotsalainen M, Savolainen KM (1995b) Effects of a protein kinase C inhibitor, Ro 31-7549, on the activation of human leukocytes by particulate stimuli. Human & Experimental Toxicology 14:266-272

Sadler KL, Badwey JA (1988) Second messengers involved in superoxide production by neutrophils. Hemat Oncol Clin North Am 2:185-200

Salamino F, Sparatore B, De Tullio, Mengotti P, Melloni E, Pontremoli S (1991) Respiratory burst in activated neutrophils is directly correlated to the intracellular level of protein kinase C. Eur J Biochem 200:573-577

Seifert R, Schultz G (1991) The superoxide-forming NADPH oxidase of phagocytes: an enzyme system regulated by multiple mechanism. Reviews of Physiology, Biochemistry, and Pharmacology 117, Springer-Verlag, Heidelberg, p. 19

Troll W, Wiesner R (1985) The Role of Oxygen Radicals as a Possible Mechanism of Tumor Promotion. Annu Rev Pharmacol Toxicol 25:509-528

von Tscharner V, Prod'hom B, Baggiolini M, Reuter H (1986) Ion channels in human neutrophils activated by a rise in free cytosolic calcium concentration. Nature 324:369-372

Tuomala M, Hirvonen M-R, Holopainen M, Savolainen K (1993a) Stimulation of human polymorphonuclear leukocytes by consecutive doses of quartz and interactions of quartz with fMLP. Toxicol Appl Pharmacol 118: 224-232

Tuomala MH, Hirvonen M-R, Savolainen KM (1992) Production of inositol phosphates and reactive oxygen metabolites in quartz-dust-stimulated human polymorphonuclear leukocytes. FEBS 296:57-60

Tuomala M, Hirvonen M-R, Savolainen KM (1993b) Changes in free intracellular calcium and production of reactive oxygen metabolites in human leukocytes by soluble and particulate stimuli. Toxicol 80:71-82

Twomey B, Muid RE, Nixon JS, Sedwick AD, Wilkonson SE, Dale MM (1990) The effect of new potent selective inhibitors of protein kinase C on the neutrophil respiratory burst. Biochem Biophys Res Communications 171:1087-1092

Umei T, Ohhara N, Okamura S, Harada M, Nakao M, Shirai T, Niho Y (1993) Activation of neutrophils NADPH oxidase by PMA: cytosol activity is translocated in phorbol-primed neutrophils. Int J Biochem 25:631-633

Watson F, Robinson J, Edwards SW (1991) Protein kinase C-dependent and -independent activation of the NADPH oxidase of human neutrophils. J Biol Chem 266:7432-7439

Yano E (1988) Mineral fiber-induced malondialdehyde formation and effect of oxidant scavengers in phagocytic cells. Int Arch Occup Environ Health 61:19-23

Toward Risk Assessment of Heterocyclic Amines in Food

(Chairs: E. Dybing, Norway, and
J. Ruprich, Czech Republic)

Chemical Methods for Assessing Systemic Exposure to Dietary Heterocyclic Amines in Man

Donald S. Davies[1], Nigel J. Gooderham[1], Stephen Murray[1], Anthony Lynch[1], Rafael de la Torre[2], Jorge Segura[2] and Alan R. Boobis[1].
[1]Department of Clinical Pharmacology, Royal Postgraduate Medical School, London W12 0NN
[2]Department of Clinical Pharmacology/Toxicology, Consori d'Hospitals de Barcelona, Institut Municipal d'Investigacio Medica, Passeig Maritim, Barcelona, Spain.

Abstract

A significant proportion of the mutagenic material present in cooked beef is accounted for by 2-Amino-3,8-dimethylimidazo[4,5-f]quinoxaline (MeIQx) and 2-amino-1-methyl-6-phenylimidazo[4,5-b]pyridine (PhIP) which are formed during the cooking of meat. N-hydroxylation catalyzed by CYP1A2 is the major pathway of metbolism of MeIQx and PhIP and is solely responsible for the generation of mutagenic species. Assays for MeIQx and PhIP in foods and body fluids were developed utilising gas chromatography/mass spectrometry with stable isotope labelled analogues as internal standards. Studies using these assays have demonstrated that both MeIQx and PhIP are well absorbed and extensively metabolised following ingestion of amine-containing beef by humans. Studies in which furafylline, a potent and selective inhibitor of human CYP1A2, was administered before ingestion of beef revealed that more than 90% of MeIQx and 70% of PhIP are N-hydroxylated *in vivo*, probably pre-systemically in the liver. The results demonstrate that unchanged MeIQx and PhIP in urine are accurate and sensitive measures of systemic exposure to the amines.

Introduction

A substantial proportion of neoplastic disease in humans is associated with diet (Doll & Peto 1981, Doll 1992) and the consumption of meat or meat products makes a prominent contribution (Gerhardsson de Verdier 1991). The mechanism whereby consumption of meat increases the risk of cancer is not known, but a number of potent genotoxic chemicals are formed from the natural constituents of meat during cooking (Sugimura & Sato 1983) including various heterocyclic amines (HCA) (Felton et al 1986, Felton & Knize 1990). Almost 60% of the total mutagenic activity of fried beef is accounted for by PhIP (10-50ng/g cooked meat), MeIQx and DiMeIQx (1-2ng/g cooked meat) (Felton et al 1986, Felton & Knize 1990, Murray et al 1988, Lynch et al 1992). PhIP and MeIQx have been found to be carcinogens in rodents (Ohgaki et al 1987, Kato et al 1988, Esumi et al 1989, Ochiai et al 1991) with PhIP producing tumours of the colon and mammary glands (Ito et al 1991).

In common with most genotoxic carcinogens, these amines do not interact directly with DNA but require enzymic activation to reactive metabolites (Yamazoe et al 1988, McManus et al 1989, Turesky et al 1990). Activation is via the formation of N-hydroxy intermediates, a reaction catalyzed by CYP1A2 in rodents (Snyderwine et al 1992, Boobis et al 1991) and this was assumed to be true for humans (Turesky et al 1991, Shimadu et al 1989, McManus et al 1990, Rich et al 1992b). Direct evidence for the exclusive involvement of CYP1A2 in the activation of MeIQx and PhIP by human liver has been obtained in our laboratory (Rich et al 1992a, Zhao et al 1992). There are numerous studies of the fate of HCAs in experimental animals administered very high doses but these are probably not helpful in assessing the absorption, metabolism and excretion of low doses of HCAs in man. In order to carry out studies in man we developed sensitive and specific analytical procedures using gas chromatography-mass spectrometry (GC-MS).

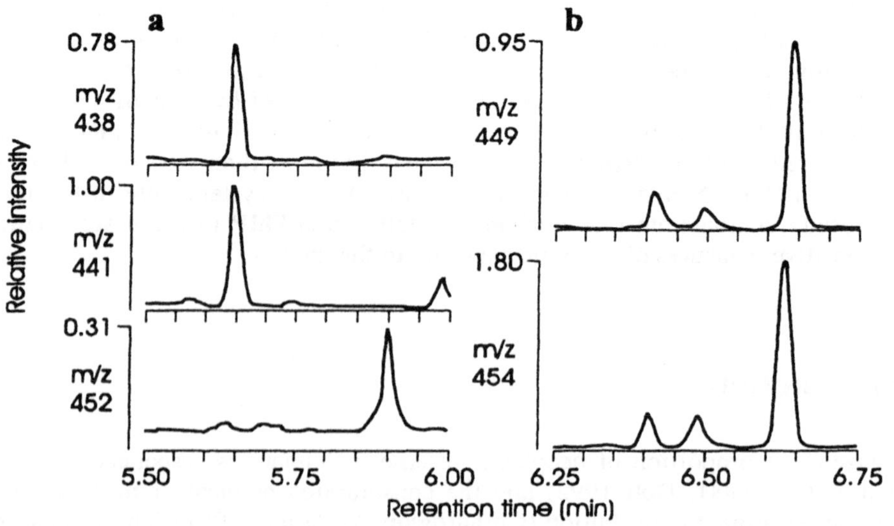

Figure 1 SIM of a. MeIQx (m/z 438), DiMeIQx (m/z 452) and [13C,15N2]MeIQx (m/z 441); b. PhIP (m/z 449) and [2H5] PhIP (m/z 454)

Measurement of MeIQx, DiMeIQx and PhIP in food and urine

Combined gas chromatography/mass spectrometry was considered to be the only analytical technique with the necessary specificity and sensitivity to measure HCAs in food and urine. The analytical methods have been described in detail (Murray et al 1988, Lynch et al 1992), and utilize stable isotope labelled analogues of MeIQx and PhIP as internal standards. The amines and internal standards are extracted from food or urine and chromatographed as their di-3,5-bistrifluoromethylbenzyl derivatives. Electron capture negative ion chemical ionization was used with ammonia as reagent gas. Under the conditions of the assay, MeIQx, DiMeIQx and PhIP were measured in a single chromatographic run by monitoring the appropriate ions for the amines and their internal standards (Figure 1). All three amines were detected in cooked meat but were absent from raw meat.

Systemic Exposure of Humans to MeIQx and PhIP

On four separate occasions, ten normal, healthy volunteers ate a cooked meal of lean beef (400-450g) prepared as patties and cooked on a griddle having refrained from eating meat or meat products for the previous 24 hours. A sample of urine was obtained at the end of the 24 hour period of abstention, just before the test meal. After the test meal, urine was collected for 8 hours as we have shown that this is sufficient time for complete excretion of unchanged amine. The ingested dose was determined by measuring the amines in portions of the cooked meat patties.

In this study the subjects ate 240 ± 9 (SE) g of cooked meat which contained 2.2 ± 0.2ng of MeIQx, 0.7 ± 0.1ng of DiMeIQx and 16.4 ± 2.1 ng of PhIP per g. The percentage of ingested amine excreted unchanged in urine was relatively constant within an individual over the dose range studied but there was considerable inter-subject variation (Table 1). In the ten subjects, a mean of 1.1% of the ingested dose of PhIP and 2.1% of MeIQx was excreted in the urine in 8 hours (Table 1). Unchanged DiMeIQx was not detected in any of the urine samples.

By analogy with studies in laboratory animals, the low percentage of dose excreted unchanged in urine probably represents almost complete absorption and extensive metabolism and can be used as a reproducible measure of systemic exposure to the amines within individuals.

Table 1. Excretion of unmetabolized MeIQx and PhIP in urine following a fried beef meal

Subject	MeIQx (% dose in 8h urine)	PhIP (% dose in 8h urine)
1	1.4 ± 0.5 (3)	0.6 ± 0.2 (3)
2	1.6 ± 0.3 (3)	0.9 ± 0.3 (3)
3	4.3 ± 1.0 (3)	2.3 ± 0.8 (3)
4	1.2 ± 0.1 (3)	0.8 ± 0.2 (3)
5	1.7 ± 0.6 (3)	0.7 ± 0.2 (3)
6	1.5 ± 0.3 (4)	0.8 (0.9,0.6) (2)*
7	1.7 ± 0.7 (4)	1.5 (0.9,2.0) (2)*
8	2.1 ± 1.1 (3)	1.0 (1.0,1.0) (2)*
9	3.7 (5.4, 2.0) (2)*	0.9 ± 0.2 (3)
10	1.5 ± 0.3 (4)	1.1 (1.0, 1.2) (2)*
Mean±SD (n=10)	2.1 ± 1.1	1.1 ± 0.5

*Results are means ± SD, except for subject 9 for MeIQx and subjects 6,7,8 and 10 for PhIP in whom only 2 results could be obtained due to assay interference on the other two test days. The mean and individual values are shown for these subjects.

CYP1A2 and the Metabolism of MeIQx and PhIP in Humans

N-hydroxylation, catalyzed exclusively by CYP1A2, is the major pathway of oxidation of MeIQx and PhIP (Rich et al 1992a, Rich et al 1992b, Zhao et al 1994) by human liver microsomes. The contribution of this pathway of metabolism to the elimination of the amines *in vivo* in man is not known but can be determined by measuring the effect of furafylline, a potent and selective inhibitor of CYP1A2 activity , on the excretion of unchanged amines in urine (Davies et al 1995, Boobis et al 1995).

The total body clearance of caffeine is greatly reduced by furafylline, probably by inhibition of CYP1A2 catalyzed N3-demethylation (Tarrus et al 1987). Potent and selective inhibition of CYP1A2 *in vitro* by furafylline was confirmed in studies of the O-deethylation of phenacetin to paracetamol by human liver microsomes (Sesardic et al 1990).

In a study (Boobis et al 1994) conducted in eight healthy volunteers, we have shown that a dose of 125mg of furafylline given 4 hours before a 100mg dose of phenacetin almost completely inhibited its oxidation. Plasma concentrations greatly in excess of its Ki value for CYP1A2 (0.18μg/ml) were maintained throughout the 24 hours of the study. Phenacetin plasma concentration was increased 100-fold, area under curve (AUC) 200-fold and total body clearance of phenacetin was reduced to 0.52% of control values. This dose of furafylline produced more than 99% inhibition of the CYP1A2 dependent oxidation of phenacetin *in vivo* in humans, confirming that the methylxanthine compound could be used to measure the contribution of this form of P450 to the metabolism of MeIQx and PhIP.

A double-blind, balanced cross-over study in 6 subjects who took 125mg of the methylxanthine or placebo 2 hours before eating a standard cooked beef meal with a known content of MeIQx and PhIP was conducted (Boobis et al 1994). Urine was collected for 28 hours after the meal and analyzed for unchanged MeIQx and PhIP by GC-MS (Lynch et al 1992). There was a very large increase in the excretion of MeIQx (14.3-fold) and a significant increase in PhIP (4.1 fold; see Table 2).

Table 2. Inhibition of the metabolism of MeIQx and PhIP by furafylline

	MeIQx*		PhIP*	
Subject	P	F	P	F
1	<4.2**	30.6	0.6	4.9
2	2.6	36.9	2.6	7.7
3	2.6	57.9	1.1	5.0
4	1.1	23.2	1.0	3.3
5	1.2	18.8	0.5	1.9
6	3.7	19.9	1.4	2.5
mean ± sd	2.6 ± 1.3	31.2 ± 14.8	1.2 ± 0.8	4.2 ± 2.1

* Results are expressed as a percentage of the ingested dose in 6 volunteers following administration of (P) placebo or (F) furafylline (125mg)

** The poor sensitivity observed in Subject 1 was due to the relatively low amount of MeIQx ingested and large volume of urine excreted by this individual.

Since the renal excretion of the amines is first order and is unaffected by furafylline or urine flow, the increase in the fraction of dose excreted unchanged in urine is inversely proportional to the decrease in total clearance. For MeIQx the 14.3 fold increase in the proportion of the dose excreted unchanged in urine indicates that 90% of its clearance is by the furafylline-inhibitable (CYP1A2) pathway. Seventy percent of the metabolism of PhIP is catalyzed by CYP1A2 (Table 3). Studies *in vitro* have shown that for both amines CYP1A2 produces a single product, the N-hydroxy metabolite, which is responsible for the mutagenicity of MeIQx and PhIP in the Ames salmonella assay (Riche et al 1992b, Zhao et al 1994).

Table 3. The contribution of CYP1A2-catalysed oxidation to the systemic elimination of MeIQx and PhIP

	Increase in AUC*	% Clearance		
		CYP1A2	Renal	Other
MeIQx	14.3 fold	90.8	2.6	6.6
PhIP	4.1 fold	69.9	1.2	28.9

Urinary excretion rates show first order elimination with no effect of urine flow or furafylline on renal clearance
* After furafylline

These results and our findings of the importance of CYP1A2 for the metabolism of MeIQx and PhIP *in vivo* in man confirm that these amines are a significant source of gentotoxic metabolites in humans eating meat and meat products. Additional genetic and epidemiologic studies are needed to assess their influence on the cancer burden in humans.

Acknowledgements

This work was supported by grants from the National Cancer Institute, DHHS (PHS grant number CA40895), USA; the Cancer Research Campaign UK; and the Ministry of Agriculture, Fisheries & Food, UK. We are grateful to Laboratorias Almirall, Barcelona, Spain, for supplying furafylline.

References

Boobis A R, Lynch, A M, Murray, S, De La Torre R, Solans A, Farré M, Segura J, Gooderham N J and Davies D S (1994) CYPIA2 catalyzed conversion of dietary heterocyclic amines to their proximate carcinogens is their major route of metabolism in humans. Cancer Res 54:89-94.

Boobis A R, Gooderham N J, Rich K J, Zhao K, Edwards R J, Murray B P, Lynch A M, Murray S and Davies D S (1995) Enzymatic studies of the activation of heterocyclic food mutagens in man. In: Heterocyclic Amines in Cooked Foods: Possible Human Carcinogens. Proc of the 23rd International Symposium of the Princess Takamatsu Cancer Research Fund. Adamson RH, Gustafsson J-A, Ito N, Nagao M, Sugimura T, Wakabayashi K and Yamazoe Y (eds). Princeton Scientific Publishing Co Inc, New Jersey, USA. pp134-144

Boobis A R, Sesardic D, Murray B P, Edwards R J and Davies D S (1991) Specificity and inducibility of cytochrome P-450 catalysing the activation of food-derived mutagenic heterocyclic amines. In: Hlavica P and Damani L A (eds) N-Oxidation of drugs; Biochemistry, Pharmacology, Toxicology. Chapman and Hall, London, pp345-355

Davies D S, Gooderham N J, Murray S, Lynch A, de la Torre R, Segura J and Boobis A R (1995) Systemic exposure to dietary heterocyclic amines in man. In: Heterocyclic Amines in Cooked Foods: Possible Human Carcinogens. Proc of the 23rd International Symposium of the Princess Takamatsu Cancer Research Fund. Adamson RH, Gustafsson J-A, Ito N, Nagao M, Sugimura T, Wakabayashi K and Yamazoe Y (eds). Princeton Scientific Publishing Co Inc, New Jersey. pp 190-196

Doll R and Peto R (1981) The causes of cancer. In: pp 1226-1235 Oxford Medical Publications, Oxford University Press, Oxford

Doll R (1992) The lessons of life: Keynote address to the nutrition and cancer conference. Cancer Res 52 (Suppl):2024s-2029s

Esumi H, Ohgaki H, Kohsen E, Takayama S and Sugimura T (1989) Induction of lymphoma in CDF1 mice by the food mutagen, 2-amino-1-methyl-6-phenylimidazo[4,5-b]pyridine. Jpn J Cancer Res 80:1176-1178

Felton J S and Knize M G (1990) Heterocyclic amine mutagens/carcinogens in foods. In: C S Cooper and P L Grover (eds), Handbook of Experimental Pharmacology, Vol. 94, pp. 471-502 Springer-Verlag, Berlin

Felton J S, Knize M G, Shen N H, Andresen B D, Bjeldanes L F and Hatch F T (1986) Identification of the mutagens in cooked beef. Environ Health Perspect 67:17-24

Gerhardsson de Verdier M, Hagman U, Peters R K, Stenineck G and Overvik E (1991) Meat, cooking methods and colorectal cancer: a case-reference study in Stockholm. Int J Cancer 49:520-525

Ito N, Hasegawa R, Sano M, Tamano S, Esumi H, Takayama S and Sugimura T(1991) A new colon and mammary carcinogen in cooked food, 2-amino-1-methyl-6-phenylimidazo[4,5-b]pyridine (PhIP). Carcinogenesis 12:1503-1506

Kato T, Ohgaki H, Hasegawa H, Sato S, Takayama S and Sugimura T (1988) Induction of tumours in the Zymbal gland, oral cavity, colon, skin and mammary gland of F344 rats by a mutagenic compound 2-amino-3,8-dimethylimidazo[4,5-f]quinoxaline. Carcinogenesis 9:71-73

Lynch A M, Knize M G, Boobis A R, Gooderham N J, Davies D S and Murray S (1992) Intra-and inter-individual variability in systemic exposure in man to MeIQx and PhIP, carcinogens present in cooked beef. Cancer Res 52:6216-6223

McManus, M. E., Felton, J. S., Knize, M. G., Burgess, W. M., Roberts Thomson, S., Pond, S. M., Stupans, I. and Veronese, M. E. (1989) Activation of the food-derived mutagen 2-amino-1-methyl-6-phenylimidazo[4,5-b]pyridine by rabbit and human liver microsomes and purified forms of cytochrome P-450. Carcinogenesis, 10: 357-363

McManus, M., Burgess, W. M., Veronese, M. E., Huggett, A., Quattrochi, L. C. and Tukey, R. H. (1990) Metabolism of 2-acetylaminofluorene and benzo(a)pyrene and activation of food-derived heterocyclic amine mutagens by human cytochrome P-450. Cancer Res. 50: 3367-3376

Murray S, Gooderham N J, Boobis A R and Davies D S (1988) Measurement of MeIQx and DiMeIQx in fried beef by capillary column gas chromatography electron capture negative ion chemical ionisation mass spectrometry. Carcinogenesis 9:321-325

Ochiai M, Ogaw K, Wakabayashi K, Sugimura T, Nagase S, Esumi H and Nagao M (1991) Induction of intestinal adenocarcinomas by 2-amino-1-methyl-6-phenylimidazo[4,5-b]pyridine in Nagase analbuminemic rats. Jpn J Cancer Res 82:363-366

Ohgaki H, Hasegawa H, Suenaga M, Sato S, Takayama S and Sugimura T (1987) Carcinogenicity in mice of a mutagenic compound, 2-amino-3,8-dimethylimidazo[4,5-f]quinoxaline (MeIQx) from cooked foods. Carcinogenesis, 8:665-668

Rich K, Murray B P, Gooderham N J, Lewis I, Davies D S and Boobis A R (1992a) CYPIA2 is responsible for the hepatic microsomal metabolism of MeIQx to N-hydroxy MeIQx in man. Br J Clin Pharmacol 33:213

Rich K, Murray B P, Lewis I, Rendell N, Davies D S, Gooderham N J and Boobis A R (1992b) N-Hydroxy-MeIQx is the major microsomal oxidation product of the dietary carcinogen MeIQx with human liver. Carcinogenesis 13:2221-2226

Sesardic D, Boobis A R, Murray B P, Murray S, Segura S, De La Torre R and Davies D S (1990) Furafylline is a potent and selective inhibitor of cytochrome P450IA2 in man. Br J Clin Pharmacol 29:651-663

Shimadu T, Iwasaki M, Martin M V and Guergerich F P (1989) Human liver microsomal cytochrome P-450 enzymes involved in the bioactivation of procarcinogens detected by umu gene response in Salmonella typhimurium TA 1535/pSK1002. Cancer Res 49:3218-3228

Snyderwine E G, Schut H A, Adamson R H, Thorgeirsson U P and Thorgeirsson S S (1992) Metabolic activation and genetoxicity of heterocyclic arylamines. Cancer Res 52: 2099s-2103s

Sugimura T and Sato S (1983) Mutagens-carcinogens in food. Cancer Res 43 (Suppl): 2315s-2421s

Tarrus E, Cami J, Roberts D J, Spickett R G W, Celdran E and Segura J (1987) Accumulation of caffeine in healthy volunteers treated with furafylline. Br J Clin Pharmacol 23:9-18

Turesky R J, Lang N P, Butler M A, Teitel C H and Kadlubar F F (1991) Metabolic activation of carcinogenic heterocyclic aromatic amines by human liver and colon. Carcinogenesis 12:1839-1845

Turesky R J, Bracco Hammer I, Markovic J, Richli U, Kappeler A M and Welti D H (1990) The contribution of N-oxidation to the metabolism of the food-borne carcinogen 2-amino-3,8-dimethylimidazo[4,5-f]quinoxaline in rat hepatocytes. Chem Res Toxicol 3: 524-535

Yamazoe Y, Abu-Zeid M, Manabe S, Toyama S and Kato R (1988) Metabolic activation of a protein pyrolysate promutagen 2-amino-3,8-dimethylimidazo[4,5-f]quin-oxaline by rat liver microsomes and purified cytochrome P-450. Carcinogenesis, 9:105 109

Zhao K, Murray S, Davies D S, Boobis A R and Gooderhan NJ (1994). Metabolic of the food-derived mutagen and carcinogen 2-amino-1-mnethyl-6-phenylimidazo [4,5-b]pyridine (PhIP) by human liver microsomes. Carcinogenesis 15:1285-1288.

Immunological Methods for Dosimetry of Heterocyclic Amines

L.O. Dragsted[1], S.E. Nielsen[1], B.L. Heitmann[2], S. Grivas[3] and H. Frandsen[1].
[1]Institute of Toxicology, National Food Agency of Denmark, and Glostrup population studies, Medical Department C, Glostrup Hospital, University of Copenhagen, Denmark
[2]Danish Epidemiology Science Center at the Institute of preventive Medicine, Copenhagen Hospital Corporation, Kommunehospitalet, Copenhagen, Denmark.
[3] Department of Chemistry, Swedish University of Agricultural Sciences, Uppsala, Sweden.

Carcinogenic heterocyclic amines are formed by common household cooking procedures, and it is therefore of interest to evaluate exposures in the general population. High-affinity monoclonal antibodies to 2-amino-1-methyl-6-phenylimidazo[4,5-*b*]pyridine (PhIP[1]) were produced and a sensitive fluorescence-based ELISA assay developed with a 50% inhibition by 150 fmoles PhIP or PhIP-metabolites modified in the 2-position. PhIP antigenic material was excreted mainly from 0-24 hours after a single meal containing fried meat in two volunteers and dropped to background levels after 48 hours. Urine samples (24 hours) were collected from 265 Danish males and females aged 40-70 years. The participants were also asked to fill in a questionnaire regarding their recent intake of fried and otherwise cooked meat. Excretion of PhIP-antigenic material was observed among those who had been eating fried meat with a median excretion of 1.5µg/24 hours. A higher median excretion among all samples of about 2.3µg/24 hours was observed when acid hydrolyzed samples were analyzed. This indicates that a large proportion of conjugated metabolites are present before hydrolysis, since these metabolites are not as efficient competitors for the antibody in ELISA. The excretion of PhIP antigenic materials showed a skewed normal distribution. These observations are close to predicted exposure rates in the general population, based on food consumption figures and analyses of PhIP in fried meat samples. PhIP antigenic material was always found to be present in urine samples from individuals who reported a recent intake of fried meat. Positive samples were also found among individuals who reported no recent intakes of fried or browned meat. The possible sources of this exposure are discussed.

[1]**Abbreviations**: ELISA, enzyme-linked immunosorbent assay; MAb, monoclonal antibody; MeIQ, 2-amino-3,4-dimethylimidazo[4,5-*f*]quinoline; 8-MeIQx, 2-amino-3,8-dimethylimidazo[4,5-*f*]quinoxaline; N^2-OH-PhIP, 2-hydroxylamino-1-methyl-6-phenylimidazo[4,5-*b*]pyridine; 4'-OH-PhIP, 2-amino-1-methyl-6-(4-hydroxyphenyl)imidazo[4,5-*b*]pyridine; PBS, phosphate buffered saline; PBS-T, PBS with 0.05% Tween 20; PhIP, 2-amino-1-methyl-6-phenylimidazo[4,5-*b*]pyridine.

Introduction

Heterocyclic amines are formed by condensation reactions between amino acids, products of Amadori rearrangements and creatinine at usual temperatures used for frying of meat or fish (Skog 1993). They are strong mutagens *in vitro* and carcinogens *in vivo* (Felton and Knize 1990). Several estimates of human daily intakes of heterocyclic amines have been reported, based on quantitative data from analyses of fried food samples (Felton et al. 1992, Wakabayashi et al. 1993, Eisenbrand and Tang 1993). Estimates are generally in the order of 0.5-15μg per day for individuals in Europe, Japan and the USA. Based on direct analysis of food pellets representing average consumption figures in the Netherlands, an average consumption of 15-20μg per day was found (Alink et al. 1988). Based on analyses of human foods these numbers have been largely confirmed (Ushiyama et al. 1991, Wakabayashi et al. 1992, 1993). The most abundant heterocyclic amine known from fried meat crust is PhIP which comprises about 80% of the estimated exposures.

PhIP is biotransformed by the microsomal monooxygenase CYPIA2 to its N^2-hydroxy derivative which is further activated by conjugation enzymes, most notably N-acetyl transferase (NAT2) to strongly electrophilic products which are able to form adducts with DNA (Frandsen et al. 1990, 1992). In the rat about half of the PhIP is biotransformed by CYPIA1 into 2-amino-1-methyl-6-(4'-hydroxy)phenylimidazo[4,5-*b*]pyridine (4'-OH-PhIP) which is a detoxication product (Dragsted et al. 1993). Very little 4'-OH-PhIP is formed in the human, if any (Zhao et al. 1994). Consequently, the phenylimidazopyridine moiety does not

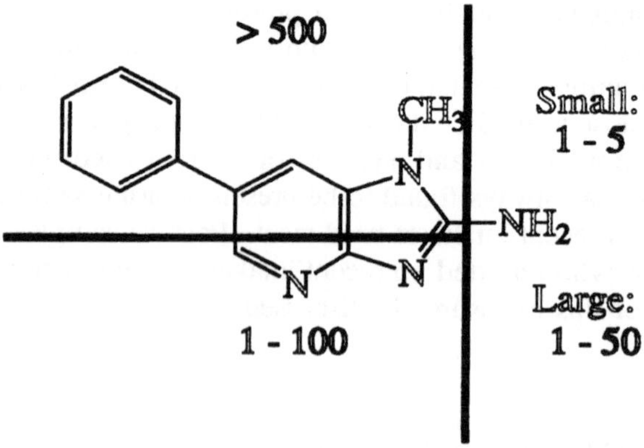

Fig. 1. The figure represents a map of the epitope recognized by the monoclonal anti-PhIP antibody, 7B7-1. Numbers represent the increase compared to PhIP in the concentration of a structural analog required to cause 50% inhibition in ELISA, when the analog is modified in the given area of the molecule. Small substituents modify MAb binding less than large substituents at the N^2-position.

seem to be changed during human biotransformation of PhIP, and antibodies recognizing exactly this structure could be very useful for human biomonitoring.

We have recently reported the production of such monoclonal antibodies with high affinity for PhIP and its N^2-modified metabolites (Dragsted et al. 1995a). A map showing the epitope of one high-affinity antibody is shown in Figure 1. The application of this antibody in a sensitive ELISA assay and its use for affinity chromatography of rat urine samples, rat faecal extracts and of PhIP-DNA adducts was also reported (Dragsted et al. 1995a, 1995b). In the present study we report the use of this antibody for analysis of human samples from a pilot study and from an ongoing study of a danish cohort of 366 middle-aged men and women from the western part of the Copenhagen County.

Materials and Methods

Chemicals: PhIP was a generous gift from Dr. Errol Zeiger, Research Triangle Park, North Carolina, USA. PhIP was labelled with tritium by Amersham International plc. (Little Chalfont, UK) using their TR7 catalytic tritiation protocol. The crude reaction mixture was purified by HPLC as previously described (Dragsted et al. 1993). Specific activities from 5-300 mCi/mmol were obtained. 2-amino-3-^{3}H-methyl-4-methylimidazo[4,5-*f*]quinoline (^{3}H-MeIQ, specific activity 4 Ci/mmol) was synthesized as described by Adolphsson and Olsson (1983). N^2-(vinylsulphonethyl)-2-amino-1-methyl-6-phenylimidazo[4,5-*b*]pyridine and its reaction products with egg albumin was prepared as previously described (Dragsted et al. 1995a). All other chemicals were of analytical grade and purchased from Sigma (St. Louis, Mo) or Merck (Darmstadt, Germany) unless otherwise mentioned in the text.

Study Subjects: Two of the authors (males), one aged 38 (smoker) and the other 41 (non-smoker) volunteered for a pilot study. Following 48 hours on a diet without pan-fried meat or fish, both subjects ingested a fried beef meal. Two 0.5 kg portions of beef in 1 cm slices, fried at 200°C for 10 min (temperature measured using thermo-element placed under the meat), were each split in two equal portions by weight. The pan residues were used for the preparation of two portions of brown gravy, which were also each split in two by weight. Each subject consumed one portion of beef and gravy, and the matching set was mixed (beef with gravy), extracted and analyzed for PhIP as described below. Following the intake of fried beef the two subjects avoided fried meat or fish for the following 72 hours. Boiled meat, fried vegetables and baked bread was ingested throughout this period, however. Urine samples were collected for the following time intervals after ingestion of the beef by both individuals: 0-4 hours, 4-24 hours, 24-48 hours, and 48-72 hours. Collection was done in 2 l polypropylene flasks with 12 cm wide opening, containing 20ml of 1 N HCl. The samples were kept refrigerated until the termination of each collection period as far as possible. After collection, two 50 ml aliquots were drawn from the samples, purged with Ar, and pH was adjusted to 2-3, if necessary by addition of 1 N HCl. The samples were stored at -80°C until further processing and analysis.

A randomly selected subcohort of 511 males and females, aged 40-70 years, drawn from the 1983 Monica-1 cohort of the Glostrup population study (Heitmann 1993) was asked to participate. Among these, 435 were alive and 366 were willing to participate. They were asked to collect a 24 hour urine sample, and to fill in a questionnaire regarding their intake of fried meat during the last 0-24 hours and 24-48 hours before the end of the collection period. Informed consent was obtained from each participant after accept of the study by the local ethical committee. The questionnaire was formed as a table, where the volunteers could fill in the amount of fried meat taken, the crust colour, the internal colour ("done"-ness), and the method of preparation (pan-frying, oven-baking, boiling or stewing). Apart from the intake of meat and gravy the questionnaire also contained questions regarding possible spills of urine during the full 24-hour period. Urine samples were collected as described above for the pilot study.

Calculation of PhIP intake from questionnaire: For each 100 g of well done whole pan-fried meat with dark crust, including gravy/stew, an exposure to 5 µg of PhIP was stipulated based on the review of heterocyclic amine analyses by Skog (1993). This number was divided by two if any of the conditions, 'medium done' or 'medium-brown crust' was reported, and consequently with four if both conditions were met. The number was divided with three when each of the conditions, 'rare' or 'light brown crust' was reported. The number was likewise divided by three if meat or fish patties or whole fish were fried in stead of whole meat. Grilled samples were set equal to fried samples for these calculations, except that grilled fish and grilled whole meat were treated equally. Browned and stewed meat or fish products may add to PhIP exposure, but the literature values do not allow simple evaluation of expected exposures from these cooking procedures. Boiled, oven baked, microwaved or otherwise treated meat and fish can not be evaluated based on literature data either but these methods of preparation are not expected to add significantly to the PhIP exposure. Consequently, an expected exposure could only be calculated for those participants who reported the ingestion of fried meat on both days, or who reported that no meat (or only minute amounts) had been ingested in the periods.

The PhIP-exposures during the last 0-24 and 24-48 hours before completion of the urine samples were calculated for this sub-group (55 individuals). The latter exposures (previous 24-48 hours) were divided by 10 and added to the former (previous 0-24 hours) to give the expected (or calculated) urinary excretion. The division by 10 was based the reported urinary excretion of PhIP in humans at different times after ingestion of fried meat (Lynch et al. 1992).

Preparation of Samples: All further handling of the urine samples were performed under Ar. The urine samples were thawed and an aliquot adjusted to pH

4-5, diluted two or four times with PBS-T (PBS containing 0.05% Tween 20) and analyzed by fluorescence ELISA. Some aliquots were stored again at -80°C under Ar and fluorescence ELISA performed later. The remaining sample (45 ml) was added with 10.000 DPM ^{3}H-MeIQ, treated with 1 N HCl at 95°C for 30 min, divided in two and each half adsorbed onto a C_{18}-cartridge as previously described (Dragsted et al. 1995a). The two eluates were pooled (1.2 ml) and analyzed at 10% and 4% dilutions in PBS-T by the competitive photometric ELISA. An aliquot was analyzed by scintillation counting in a TR2500 scintillation counter (Packard Instruments, Meriden, CT) to calculate recovery of heterocyclic amines. The remainder of each urine concentrate (70%) was stored under an Ar atmosphere at -80°C for affinity chromatography (see below).

The mixed meat and gravy samples were finely divided and homogenized in a Polytron homogenizer and 50 ml (15%) mixed with 200 DPM ^{3}H-MeIQ was used for further analysis. This sample was homogenized in an equal volume of 0.02 M HCl containing 50% ethanol and 5% dimethylformamide, centrifuged at 4.000 RPM, and the supernatant collected. The residue was homogenized and centrifuged two more times, and the supernatants pooled and evaporated to a volume of 10-15 ml. The samples were adjusted to pH 4 with dilute NaOH and applied to a preconditioned C_{18}-cartridge, which was subsequently washed with 10ml of 5% ethanol in PBS-T and finally eluted with 50% ethanol. The eluate was evaporated to dryness, redissolved in PBS-T containing 5% acetonitrile and an aliquot analyzed by scintillation counting to calculate recovery of heterocyclic amines. The rest was diluted 1000 and 10.000 times and analyzed by fluorescence ELISA.

ELISA assays: The production of monoclonal anti-PhIP antibodies and the conduction of the competitive photometric ELISA assay has previously been described (Dragsted et al. 1995a).

The competitive fluorescence ELISA was performed in white opaque Maxisorp FluoroNunc microtiter plates (NUNC, Roskilde, Denmark) which were coated overnight at room temperature with PhIP-egg albumin antigen conjugate (0.13 µg/well) in PBS, and washed twice with PBS and once with PBS-T. Part of each plate was coated with divinylsulphone modified egg albumin (500 ng/well) for assay blanks. Samples or standards (100 µl) were added in PBS-T immediately before the addition of Prosep-A (Biotech IgG, Durham, England) purified antibody, diluted 1:20.000 (approx. 10 ng IgG/well). Plates were incubated at 4°C for 16 hrs. The plates were washed four times with PBS-T and 100 µl of alkaline phosphatase-coupled rabbit anti mouse-Ig (E224, Dakopatts, Glostrup, Denmark), 1:500 in PBS-T with 1% egg albumin, was added to each well. After two hours of incubation, the plates were washed four times in PBS-T and three times in 0.1 M diethanolamine, pH 9.4, and 100 µl 1 mg/ml 4-methylumbelliferylphosphate in 1 M diethanolamine pH 9.4 was added to each well. The fluorescence in each well was read with excitation at 360nm (slit, 2.5 nm bandwidth) and emission at 450nm (5 nm bandwidth) in a fluorescence ELISA plate reader coupled to a Perkin-Elmer LS-50B fluorometer (Perkin-Elmer, Beaconsfield, UK) at intervals from one to three hours. For each sample on a plate the equivalent inhibition in ng/ml was calculated as the amount of PhIP that would accomplish a similar inhibition. The calculations were based on the inhibition values of three standards and two blanks which were included on each plate in quadruplicate together with appropriate assay blanks.

Affinity Chromatography: The production of sepharose-based and porous glass bead-based affinity columns containing the anti-PhIP MAb 7B7-1 and their use for affinity chromatography was reported previously (Dragsted et al. 1995a).

Results

The competitive fluorescence ELISA was found to be 5-10 times more sensitive than the photometric assay, showing 50% inhibition with only 150 fmol PhIP per well. This assay was used to analyze the urine samples and meat samples from the pilot study and the total amount of PhIP-equivalents in each sample was calculated, see Figure 2. The meat samples consumed by the non-smoker and the smoker were found to contain 21.5 and 28.2 µg of PhIP, respectively. In the following 72 hours,

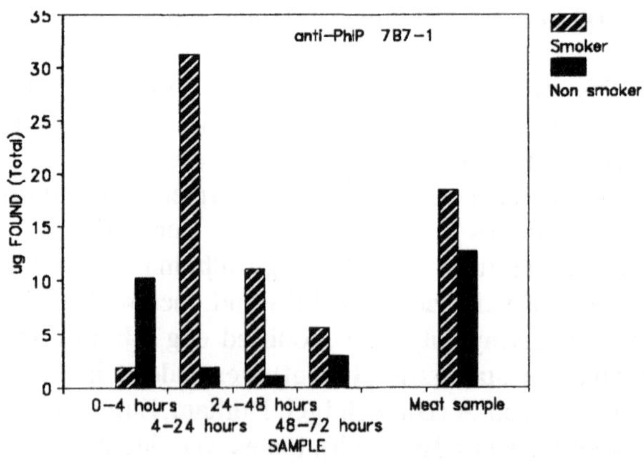

Fig. 2. The amounts of PhIP equivalents in the two meat samples (right) ingested by the smoker (hatched bars) and the non-smoker (filled bars), respectively, and the urinary excretion (left) of PhIP-equivalents in the periods 0-4, 4-24, 24-48, and 48-72 hours after a single meal. The values were calculated based on the competitive photometric ELISA analysis of each of the samples after acid hydrolysis and concentration on a C_{18}-cartridge. Samples were tested three times in quadruplicate.

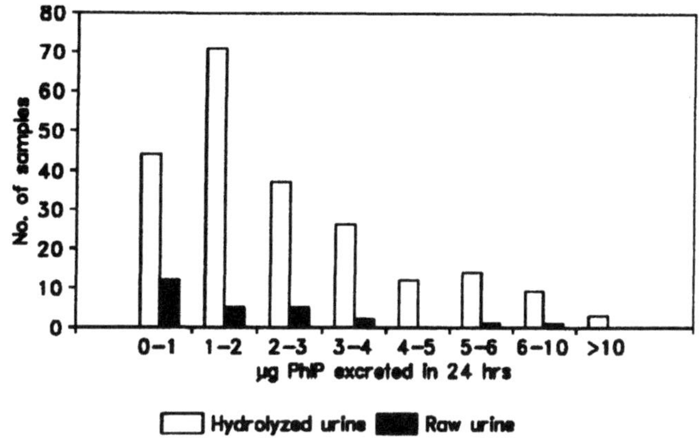

Fig. 3. The distribution of PhIP urinary excretion values as determined by ELISA. Filled bars, values obtained from fluorescence ELISA of raw urine samples. Only samples from subjects who reported intake of fried meat in both periods (0-24 hours and 24-48 hours before sample completion) were included (n=29). Open bars, values obtained with photometric ELISA of acid hydrolyzed, concentrated samples. All samples included (n=229). Analysis of each sample was performed two times in quadruplicate.

the non-smoker excreted 19.3 µg, and the smoker 46.1 µg of PhIP, respectively, as determined by ELISA. The majority was excreted within the first 24 hours, and more than 80% was excreted within 48 hours, but the amounts excreted between 48-72 hours unexpectedly did not reach zero level. The samples were also analyzed with the more stringent anti-PhIP antibody, 7-2 G10, with similar results.

A total of 265 (72%) out of the 366 individuals in the subcohort had completed a 24-hour urine sample, after two reminders. Of these, 229 (63%) were included into the study based on the criteria that a) they had completed the sample without reported spills, b) delivery of the sample had not been postponed for more than 12 hours, and c) the questionnaire was completed. Each of these samples were acid-hydrolyzed, concentrated on a C_{18}-cartridge, and analyzed by the photometric ELISA. A relatively small number of individuals, only 29, had fried meat on both days. The raw urine samples of this group were analyzed by the fluorescence-based ELISA. The distribution of PhIP-equivalents determined by the two methods are shown in Figure 3. The distributions are close to log-normal but both show some indication of bimodality. The analysis of the selected raw samples (n=29) show a median excretion of PhIP equivalents of 1.4 µg per 24 hours, whereas for the same acid-treated samples the median value is 4.17 µg per 24 hours. For all 229 samples the median excretion value was 2.53 µg per 24 hours as determined by ELISA of acid-hydrolyzed samples. The correlation of the two methods for the individual samples analyzed by both procedures are shown in Figure 4. The correlation coefficient indicates that only about 28% of the variation obtained by the analysis of raw samples can be explained by the results of analysis of acid hydrolyzed

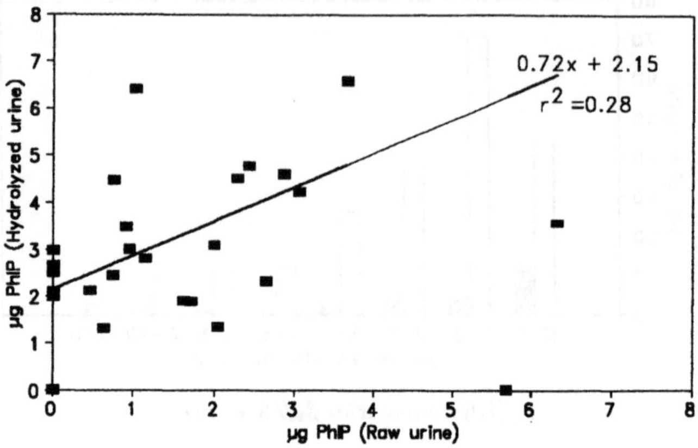

Fig. 4. Correlation between the amount of PhIP measured by ELISA before and after acid hydrolysis in each sample. Only samples from subjects who reported an intake of fried meat in both periods (0-24 hours and 24-48 hours before sample completion) were included.

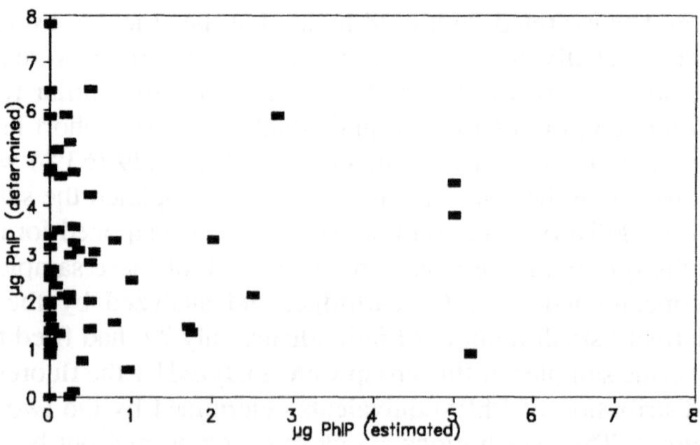

Fig. 5. Scatterplot of expected (calculated) total PhIP excretion based on questionnaire data as compared with total excretion measured by ELISA of acid hydrolyzed samples. See materials and methods for details.

samples. The calculated regression line of a = 0.72x + 2.15 is also shown in Figure 4.

Data from those individuals who reported ingestion of fried meat and those who reported no ingestion of meat at all were used to compare the PhIP intake values obtained from the questionnaire with the analytical urinary excretion data obtained by ELISA from acid hydrolyzed samples. This new subset comprised 55 individuals. A scatterplot of the results obtained are shown in Figure 5. As it can be seen, samples expected to contain significant amounts of PhIP based on the accompanying questionnaire were also generally found to contain expected levels by ELISA. On the other hand, samples expected to be devoid of PhIP based on the questionnaire were found to contain significant levels by ELISA in most instances.

Discussion

The formation of food mutagens is proportional to the pan temperature in the range from 150-300°C and increases linearly with duration of frying up to about 10 min (Bjeldanes et al. 1982, 1983; Nielsen, personal communication). Formation of mutagens has been observed at pan temperatures as low as 120°C (Tikkanen 1991). A considerable mutagenic activity has been observed in the pan residues (Felton et al., 1981, Övervik et al., 1987, Berg et al. 1990) and in oven residues (Laser Reuterswärd et al., 1989), but only little mutagenic activity was observed in the crust of oven-baked meats (Laser Reuterswärd et al., 1987), or in microwaved beef (Nader et al., 1981).

In fried meat patties, 2-amino-1-methyl-6-phenylimidazo[4,5-*b*]pyridine (PhIP) is the most abundant (1-15 µg/kg meat) of the known heterocyclic amine mutagens, constituting 83-93% by weight (Felton et al. 1992). The mutagens, 2-amino-3,8-dimethylimidazo[4,5-*f*]quinoxaline (8-MeIQx) and 2-amino-3,4,8-trimethylimidazo-[4,5-*f*]quinoxaline (4,8-DiMeIQx) are also commonly found in levels of about 0.3-12 and 0.1-4 µg/kg of fried ground beef, respectively, whereas 2-amino-3-methyl-imidazo[4,5-*f*]quinoline (IQ) is generally found in lower concentrations, 0.02-2 µg/kg (Skog 1993).

In a direct comparison between fried beef and pork patties, essentially similar profiles have been observed (Nielsen, personal communication). The formation of mutagenic compounds has also been observed in fried mutton, lamb and chicken (Skog 1993). The values reported in the literature vary considerably with the conditions of extraction, analysis and other factors. For example the content of PhIP varies from 0.56 ng/g in fried beef (Hayatsu et al. 1991) to 69.2 ng/g in fried whole fish (Zhang et al., 1988). Values are generally higher in fried whole meat than in ground meat, possibly due to a lower cooling effect from expelled meat juices (Turesky, personal communication).

In our pilot study we chose to use whole fried beef, fried at very high temperatures and the estimated content of PhIP formed in crust plus pan residues (gravy) were estimated by ELISA to be 80-120 ng/g, values similar to or a little higher than those reported by Turesky et al. (1988, 1989) and Johansson and Jägerstad (1994). The efficiency of our very simple extraction procedure was verified by the presence of a radiolabelled internal standard to be 70-90%. The use of ELISA for the analysis of PhIP in meat extracts has, however, not been validated. The numbers may therefore be influenced by other compounds in the extract which could competitively or non-competitively influence the analysis. The finding that 100-150% of the PhIP ingested in the pilot study was found to be excreted between 0-72 hours might reflect such analytical problems. On the other hand, the time-course of excretion was unexpected. Evidence from analysis of urine samples after a fried beef meal indicate that about 1% of the PhIP and about 2% of the 8-MeIQx ingested is excreted within 8 hours, whereas little was excreted between 8-12 hours (Lynch et al. 1992). In the present investigation we used acid hydrolysis to liberate PhIP, N^2-OH-PhIP and other possible metabolites from their conjugates and we consequently observed much higher amounts of PhiP equivalents. The majority of the excretion took place during the first 24 hours (4 hours for the non-smoker), as expected, but 30% was found at later times in the samples from both volunteers. One possibility to explain this discrepancy could be that other sources (food or environmental) might contribute to human PhIP exposures. Heterocyclic amines have also been observed in other food sources, including beef extracts (Takahashi et al., 1985a, 1985b; Turesky et al., 1988, 1989), smoked, dried fish (Kato et al., 1986), fried eggs (Grose et al., 1986), baked bread (Felton et al. 1993), roasted coffee beans (Takahashi et al., 1988; Kikugawa et al., 1989) and in beers and wines (Manabe et al. 1993), albeit at much lower levels than in fried meat. Non-food sources, including cigarette smoke condensates (Manabe et al. 1991) and diesel exhausts (Manabe et al. 1993) have also been found to contain PhIP. Curiously, the main DNA adduct observed in a bacterial tester strain used to determine the mutagenicity of urine from smokers was determined by ^{32}P-postlabelling to co-chromatograph with that of PhIP (Peluso et al. 1991). In our pilot study we observed the largest discrepancy between measured intake and excretion in the smoker, but the levels reported in cigarette smoke by Manabe et al. (1991) could hardly explain this.

Several estimates of the human daily intakes of heterocyclic amines have been reported. Based on the analytical data from fried foods, estimates of 0.8-8 µg (Eisenbrand and Tang 1993), 12.5 µg (Felton et al. 1992) and 0.4-16 µg (Wakabayashi et al. 1992) have been presented. Based on direct analyses of food pellets representing average consumption figures in the Netherlands, an average exposure of 15-20 µg/day was found (Alink et al., 1988). The intakes of meat and fish by Danish adults have been reported to be approximately 56 and 9 kg/capita/year, respectively (Larsen and Jensen 1985). About 70% of the consumed meat and fish is fried. Dragsted (1992) estimated the intake of individual heterocyclic aromatic amine mutagens in the Danish population to 0-16.5 µg/day depending on cooking habits. Still, the estimates are influenced by several analytical uncertainties. In the present study we observed excretion levels of 0-28 µg PhIP in 24 hour samples obtained from a group of middle-aged men and women. Only one individual excreted more than 14 μg PhIP, and the median excretion was 2.53 μg in the group. These results correspond well with the expected levels. Based on the presence of PhIP and 8-MeIQx in urines of 10 human volunteers, Wakabayashi et al. (1992, 1993) estimated the daily intakes of these two compounds to be in the ranges 0.1-13.8 μg and 0.2-2.6 μg, respectively. These estimates are close to those calculated from analyses of meat or total foods, and also agree with our results.

The very weak mutagen, 2-amino-1-methyl-6-(4'-hydroxyphenyl)imidazo[4,5-*b*] pyridine (4'-OH-PhIP), has also been found in broiled beef at levels of 21 μg/kg meat (Kurosaka et al. 1992). This observation has important implications for the evaluation of human exposures to PhIP by analysis of excretion products by immunoassay, since cross-reaction with 4'-OH-PhIP by the antibody used would lead to overestimations of PhIP exposures. The antibody used in this study does not cross-react with 4'-OH-PhIP (Dragsted et al. 1995a).

Both CYPIA1 and CYPIA2 can oxidize PhIP (Aoyama et al. 1990; Wallin et al. 1990). Ring oxidations lead to detoxication, whereas oxidation of the exocyclic amino group leads to activation (Frandsen et al. 1990). In a comparative study of the capacity of rat, monkey and human microsomes to activate PhIP to mutagenic products, or to products capable of binding to DNA, human liver microsomes had the highest capacity to activate PhIP followed by rat and monkey microsomes (Davis et al. 1993). Human microsomes have been found to form exclusively the activated metabolite, N^2-OH-PhIP, and not the ring-hydroxylated metabolite, 4'-OH-PhIP (Zhao et al. 1994). The antibody, 7B7-1, used in this study cross-reacts with N^2-OH-PhIP, and it is therefore capable of measuring the product of human microsomal PhIP metabolism as well as unmetabolized compound. N^2- and 3-N-glucuronides of N^2-OH-PhIP seem to be the major transport forms (Alexander et al. 1991) and lead to significant excretion of this activated metabolite in rats, mice, dogs and monkeys (Watkins et al. 1991; Bounarati et al. 1992; Kadlubar et al. 1992; Snyderwine et al. 1993). It is therefore conceivable that glucuronides of PhIP and N^2-OH-PhIP might also be excreted into the urine in humans. Such bulky conjugates decrease the affinity of our antibody to the PhIP epitope and might lead to underestimations of the total amount of PhIP-equivalents by ELISA of urine samples. We therefore compared the results of ELISA of raw samples with results obtained with acid treated samples, where the conjugates were hydrolyzed. Not unexpectedly, the values are almost consistently higher after acid hydrolysis, as seen in Figure 4. A similar result was previously obtained with rat urine samples (Dragsted et al. 1995a). For the present study, we had to develop a very sensitive fluorescence-based ELISA in order to quantitate PhIP in unconcentrated urines.

A comparison between expected PhIP intakes based on information in the questionnaires and measured PhIP excretion based on ELISA for a subset of the presently studied population indicates a very low concordance. Generally, data fall into two categories, those with similar values obtained by both procedures, and those with much higher 'measured' than 'expected' values as can be seen from Figure 5. One explanation might be that the method used to calculate the expected exposures from the questionnaires could lead to underestimations in many cases.

Since such calculations are always expected to be rather unprecise, it is remarkable, that almost no saples were found to be overestimated, as seen i fig. 5. This is therefore likely to be part of the explanation. Some individuals reported that they had no meat at all in the 48-hour period before the sample was completed, and for these individuals the expected excretion would be zero. Other explanations are necessary in order to account for the PhIP antigenic material found in their urine samples. One explanation might be that some unrelated substance in human urine might cross-react with the MAb used. The use of affinity chromatography to isolate the antigenic materials for further analysis is presently underway in order to evaluate this possibility. So far we have observed no PhIP or PhIP metabolites in affinity isolates from these samples, most probably due to problems with the washing procedure and with sensitivity of the subsequent HPLC analysis (data not shown). The existence of other possible sources of PhIP exposures in the present study will be analyzed by comparing personal and life-style factors for the individuals in our study with PhIP excretion patterns. Specifically smoking habits and sex will be analyzed as potential explanatory factors.

We conclude that PhIP antigenic materials in human urine samples could be determined by ELISA. The measured levels in human urine samples were in accordance with literature data on human PhIP intake. Information on the intake of fried meat during the last 48 hours from the presently used questionnaire can not explain the amounts of PhIP found in the urine samples. Consequently, more information and more exact data must be obtained by further analysis by affinity chromatography.

A controlled study on fried meat intake and excretion of PhIP and N^2-OH-PhIP in humans is presently under way in order to determine the fraction of ingested PhIP which is excreted into the urine as determined by the present method. Both ELISA analyses and affinity purification will be automated in order to increase the analytical capacity. Hopefully these analyses can aid in the evaluation of human PhIP exposures and result in a methodology that will be useful to evaluate human cancer risks from exposures to PhIP.

Acknowledgements: The authors want to thank Anni Schou, Vibeke Kegel Katrin Christiansen and Birgitte Gram for excellent technical assistance. This work was supported in part by The Danish Environmental Research Programme, The Danish Health Insurance Foundation, The Danish Ministry of Health, Research Centre for Environmental Medicine, and the Swedish Council for Agricultural Research (NKJ 82).

References

Adolfsson L and Olsson K (1983) A convenient synthesis of mutagenic 3*H*-imidazo[4,5*f*] quinolin-2-amines and their 2-^{14}C-labelled analogues. Acta Chem Scand Ser B 37:157-159

Alexander J, Wallin H, Rossland OJ, Solberg KE, Holme JA, Becher G Andersson R, Grivas S (1991) Formation of a glutathione conjugate and a semistable transportable glucuronide conjugate of N^2-oxidized species of 2-amino-1-methyl-6-phenylimidazo[4,5*b*]-pyridine (PhIP) in rat liver. Carcinogenesis 12:2239-2245

Alink GM, Knize MG, Shen NH, Hesse SP, Felton JS (1988) Mutagenicity of food pellets from human diets in The Netherlands. Mutat Res 206:387-393

Aoyama T, Gelboin HV, Gonzalez FJ (1990) Mutagenic activation of 2-amino-3-methylimidazo[4,5-*f*]quinoline by complementary DNA-expressed human liver P-450. Cancer Research 50:2060-2063

Berg I, Övervik E, Gustavsson J-Å (1990) Effect of cooking time on mutagen formation in smoke, crust and pan residue from pan-broiled pork. Food Chem Toxicol 28:421-426

Bjeldanes LF, Morris MM, Felton JS, Healy S, Stuermer D, Berry P, Timourian H, Hatch FT (1982) Mutagens from the cooking of food. II. Survey by Ames/Salmonella test of mutagen formation in the major protein-rich foods of the American diet. Food Chem Toxicol 20:357-363

Bjeldanes LF, Morris MM, Timourian H, Hatch FT (1983) Effects of meat composition and cooking conditions on mutagen formation in fried ground beef. J Agric Food Chem 31:18-21

Buonarati MH, Roper M, Morris CJ, Happe JA, Knize MG, Felton JS (1992) Metabolism of 2-amino-1-methyl-6-phenylimidazo[4,5-*b*]pyridine (PhIP) in mice. Carcinogenesis 13:621-627

Davis CD, Schut HAJ, Adamsson RH, Thorgeirsson UP, Thorgeirsson SS, Snyderwine EG (1993) Mutagenic activation of IQ, PhIP and MeIQx by hepatic microsomes from rat, monkey and man: low mutagenic activation of MeIQx in cynomolgus monkeys in vitro reflects low DNA adducts in vivo. Carcinogenesis 14:61-65

Dragsted LO (1992) Exposure and Carcinogenicity of Heterocyclic Amines. Report of the Toxicology Forum Annual European Meeting, Caset Associates Ltd., Fairfax, VA, pp. 141-148

Dragsted LO, Alexander J, Wallin H, Frandsen H, Vang O (1993) Bioactivation of food mutagens by liver microsomes from three different rat strains. Pharmacology and Toxicology 72:388-393

Dragsted LO, Grivas S, Frandsen H, Larsen JC (1995a) Antibodies to the food mutagens, 2-amino-1-methyl-6-phenylimidazo[4,5-*b*]pyridine and 2-amino-3,4,8-trimethylimidazo-[4,5-*f*]quinoxaline: ELISA and affinity chromatography of biological samples. Carcinogenesis 16:in press

Dragsted LO, Frandsen H, Reistad R, Alexander J, Larsen JC (1995b) DNA-binding and disposition of 2-amino-1-methyl-6-phenylimidazo[4,5-*b*]pyridine (PhIP) in the rat. Carcinogenesis 16:in press

Eisenbrand G and Tang W (1993) Food-borne heterocyclic amines. Chemistry, formation, occurrence and biological activities. A literature review. Toxicology 84:1-82

Felton JS, Knize MG (1990) Heterocyclic-amine mutagens/carcinogens in foods. In: Cooper CS, Grover PL (Eds) Handbook of experimental Pharmacology. Springer Verlag, Berlin, Heidelberg pp 471-502

Felton JS, Knize MG, Roper M, Fulz E, Shen NH, Turteltaub KW (1992) Chemical analysis, prevention, and low-level dosimetry of heterocyclic amines from cooked food. Cancer Research (Suppl), 52, 2103s-2107s

Felton JS, Knize MG, Cunningham PL, Dolbeare FA (1993) Sources and quantitation of heterocyclic amines from cooked foods. Proc Am Assoc Cancer Res 34:116

Frandsen H, Rasmussen ES, Nielsen PA, Farmer P, Dragsted LO, Larsen JC (1990) Metabolic formation, synthesis and genotoxicity of the N-hydroxy derivative of the food mutagen 2-amino-1-methyl-6-phenylimidazo[4,5-*b*]pyridine (PhIP). Mutagenesis 6:93-98

Frandsen H, Grivas S, Andersson R, Dragsted L, Larsen JC (1992) Reaction of the N-acetoxy derivative of 2-amino-1-methyl-6-phenylimidazo[4,5-*b*]pyridine (PhIP) with 2-deoxyguanosine and DNA. Synthesis and identification of N-(deoxyguanosin-8-yl)-PhIP. Carcinogenesis 13:629-635

Grose KR, Grant JL, Bjeldanes LF, Andersen BD, Healy SK, Lewis PR, Felton JS, Hatch FT (1986) Isolation of the carcinogen IQ from fried egg patties. J Agric Food Chem 34:201-202

Hayatsu H, Arimoto A, Wakabayashi K (1991) Methods for separation and detection of heterocyclic amines. In: Hayatsu H (ed) Mutagens in food: Detection and prevention. CRC Press, Boca Raton, Fl

Heitmann BL (1993) The influence of fatness, weight change, slimming history and other lifestyle variables on diet reporting in Danish men and women aged 35-65 years. Int J Obesity 17:329-336

Johansson MAE and Jägerstad M (1994) Occurrence of mutagenic/carcinogenic heterocyclic amines in meat and fish product, including pan residues, prepared under domestic conditions. Carcinogenesis 15:1511-1518

Kadlubar FF, Teitel CH, Lin D-X, Dooley KL, Friesen M, Bartsch H, Nagao M, Esumi H, Sugimura T (1992) Metabolism and DNA adduct formation of the food-borne carcinogen 2-amino-1-methyl-6-phenylimidazo[4,5-ß]pyridine (PhIP) in the dog. Proc Am Assoc Cancer Res 33:158

Kato T, Kikugawa K, Hayatsu H (1986) Occurrence of the mutagens, 2-amino,3,8-dimethyl-imidazo[4,5-*f*]quinoxaline (MeIQx) and 2-amino-3,4,8-trimethyl-imidazo[4,5-*f*]quinoxaline (4,8-DiMeIQx) in some Japanese smoked, dried fish products. J Agric Food Chem 34:810-814

Kikugawa K, Kato T, Takahashi S (1989) Possible presence of 2-amino-3,4-dimethyl-imidazo[4,5-*f*]quinoline and other heterocyclic amine-like mutagens in roasted coffee beans. J Agric Food Chem 37:881-886

Kurosaka R, Wakabayashi K, Ushiyama H, Nukaya H, Arakawa N, Sugimura T, Nahao M (1992) Detection of 2-amino-1-methyl-6-(4-hydroxyphenyl) imidazo[4,5-*b*] pyridine in boiled beef. Jpn J Cancer Res 83:919-922

Larsen PH, Jensen JH (1985) A Danish dietary survey, pilot study. National Food Agency of Denmark Publ 108:1-48

Laser Reutersward A, Skog K, Jägerstad M (1987) Effect of creatine and creatinine content on the mutagenic activity of meat extracts, bouillons and gravies from different meat sources. Food Chem Toxicol 25:747-754

Laser Reutersward A, Agerhem H, Skog K, Jägerstad M, Lidström A, Sundemar A (1989) The relationship between mutagenic activity and flavour development in fried beef steak. Vår Föda (suppl 2) 42:65-70

Lynch AM, Knize MG, Boobis AR, Gooderham NJ, Davies DS, Murray S (1992) Intra- and interindividual variability in systemic exposure in humans to 2-amino-3,8-dimethylimid-azo[4,5-*f*]quinoxaline and 2-amino-1-methyl-6-phenylimidazo[4,5-*b*]pyridine, carcinogens present in cooked beef. Cancer Res 52:6216-6223

Manabe S, Tohyama K, Wada O, Aramaki T (1991) Detection of a carcinogen 2-amino-1-methyl-6-phenylimidazo[4,5-*b*]pyridine (PhIP) in cigarette smoke condensata. Carcinogenesis 12:1945-1947

Manabe S, Suzuki H, Wada O, Ueki A (1993) Detection of the carcinogen 2-amino-1-methyl-6-phenylimidazo[4,5-*b*]pyridine (PhIP) in beer and wine. Carcinogenesis 14:899-901

Nader CJ, Spencer LK, Weller RA (1981) Mutagen production during pan-broiling compared with microwave irradiation of beef. Cancer Letters 13:147-152

Övervik E, Nilsson L, Fredholm L, Levin Ö, Nord CE, Gustavsson JÅ (1987) Mutagenicity of pan residues and gravy from fried meat. Mutat Res 187:47-53

Peluso M, Castegnaro M, Malaveille C, Friesen M, Garren L, Hautefeuille A, Veneis P, Kadlubar F, Bartsch H (1991) Phosphorus-32-postlabelling analysis of urinary mutagens from smokers of black tobacco implicates 2-amino-1-methyl-6-phenyl-imidazo[4,5-*b*]pyridine (PhIP) as a major DNA-damaging agent. Carcinogenesis 12:713-717

Skog K (1993) Cooking procedures and food mutagens: A literature review. Fd Chem Toxic 9:655-675

Snyderwine EG, Buonarati MH, Felton JS, Turteltaub KW (1993b) Metabolism of the food-derived mutagen/carcinogen 2-amino-1-methyl-6-phenylimidazo[4,5-*b*]pyridine (PhIP) in nonhuman primates. Carcinogenesis 14:

Takahashi M, Wakabayashi K, Nagao M, Yamamoto M, Masui T, Goto T, Kinae N, Tomita I, Sugimura T (1985a) Quantification of 2-amino-3,8-dimethyl-imidazo [4,5-*f*]quinoline (IQ) and 2-amino-3,8-dimethyl-imidazo[4,5-*f*]quinoxaline (MeIQx) in beef extracts by liquid chromatography with electrochemical detection (LCEC). Carcinogenesis 6:1195-1199

Takahashi M, Wakabayashi K, Nagao M, Yamaizumi, Sato S, Kinae N, Tomita I, Sugimura T (1985b) Identifikation and quantification of 2-amino-3,4,8-trimethyl-imidazo[4,5-*f*]quinoxaline (4,8-DiMeIQx) in beef extract. Carcinogenesis 6:1537-1539

Takahashi S, Kato T, Kikugawa K (1988) Formation of heterocyclic amine-like mutagens in roasted coffee beans. Mutat Res 203:390-394

Tiikkanen L (1991) Sources of mutagenicity in cooked finnish foods. Food Chem Toxicol 29:87-92

Turesky RJ, Bur H, Huynh-Ba T, Aeschbacher HU, Milon H (1988) Analysis of mutagenic heterocyclic amines in cooked beef products by high-performance liquid chromatography in combination with mass spectrometry. Food Chem Toxicol 26:501-509

Turesky RJ, Forster CM, Aeschbacher HU, Würzner HP, Skipper PL, Trudel LJ, Tannenbaum SR (1989) Purification of the food-borne carcinogens 2-amino-3-methyl imidazo[4,5-*f*]quinoline and 2-amino-3,8-dimethylimidazo[4,5-*f*]quinoxaline in heated meat products by immunoaffinity chromatography. Carcinogenesis 10:151-156

Ushiyama H, Wakabayashi K, Hirose M, Itoh M, Sugimura T, Nagao M. (1991) Presence of heterocyclic amines in the diet of healthy volunteers eating normal diet, but not of inpatients receiving parenteral alimentation. Carcinogenesis 12:1417-1422

Wakabayashi K, Nagao M, Esumi H, Sugimura T (1992) Food-derived mutagens and carcinogens. Cancer Research (Suppl) 52:2092s-2098s

Wakabayashi K, Uschiyama H, Takahashi M, Nukaya H, Kim S-B, Hirose M, Ochiai M, Sugimura T, Nagao M (1993) Exposure to heterocyclic amines. Environ Health Perspect 99: 129-133

Wallin H, Mikalsen A, Guengerich PF, Ingelman-Sundberg M, Solberg KE, Rossland OJ, Alexander J (1990) Differential rates of metabolic activation and detoxication of the food mutagen 2-amino-1-methyl-6-phenylimidazo[4,5-*b*]pyridine by different cytochrome P450 enzymes. Carcinogenesis 11:489-492

Watkins BE, Esumi H, Wakabayashi K, Nagao M, Sugimura T (1991a) Fate and distribution of 2-amino-1-methyl-6-phenylimidazo[4,5-*b*]pyridine (PhIP) in rats. Carcinogenesis 12:1073-1078

Zhang XM, Wakabayashi K, Liu ZC, Sugimura T, Nagao M (1988) Mutagenic and carcinogenic heterocyclic amines in Chinese cooked foods. Mutat Res 202:181-188

Zhao K, Murray S, Davies BS, Boobis AR, Gooderham NJ (1994) Metabolism of the food derived mutagen and carcinogen 2-amino-1-methyl-6-phenylimidazo[4,5-*b*]pyridine (PhIP) by human liver microsomes. Carcinogenesis 15:1285-1288

Formation and Processing of Reactive Metabolites of the Heterocyclic Amines

Jan Alexander
Department of Environmental Medicine, National Institute of Public Health, Geitmyrsveien 75, N-0462 Oslo, Norway

Introduction

More than twenty mutagenic and carcinogenic heterocyclic amines have hitherto been isolated from the crust of fried fish or meat (Sugimura, 1995). Since the heterocyclic amines, and particularly the group of amino-imidazoazaarenes (AIA), ("IQ-compounds") (Fig. 1), which are more commonly found in cooked food, are potential human carcinogens, the interest in these compounds has nearly exploded. A Med-line search gave more than 250 papers since 1990.

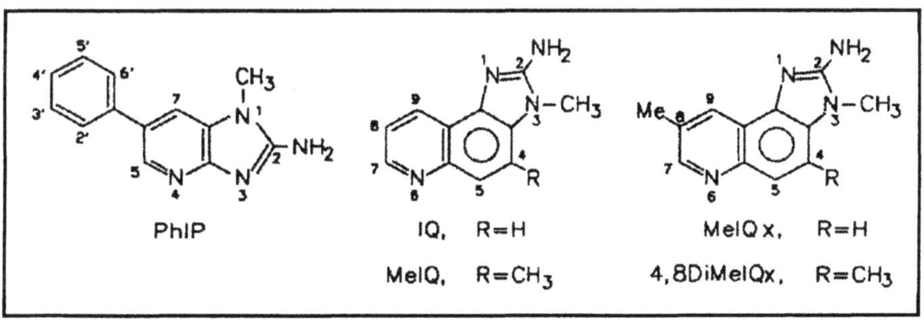

Fig. 1. Structures of some heterocyclic amines (AIA, "IQ-compounds)

The present paper review some recent results on metabolic activation of heterocyclic amines; further activation and deactivation of the proximate hydroxylamine metabolites as well as the formation of unstable thiol- and protein adducts of PhIP metabolites (Fig. 2). The amino-imidazo-azaarenes most common in cooked meat are PhIP and MeIQx (Alexander et al. 1989). It is focused on these and IQ which have been the most studied ones.

Activation of Heterocylic Amines

The dominating pathway of activation is via oxidation of the exocyclic followed by one or more steps ending up in a reactive nitrenium ion which reacts with DNA and forms several nucleotide adducts, the most common being the C^8-deoxyguanosine adduct of the exocyclic aminogroup, but several minor adducts are seen by ^{32}P-postlabelling, the N^2-deoxyguanosine-5-IQ/MeIQx has also been reported (Alexander and Wallin 1991, Frandsen et al. 1992 and 1994a, Turesky et al. 1992).

Although minor amounts are oxidized by prostaglandine synthetase or reactive oxygen during respiratory burst of white blood cells (Alexander and Wallin 1991, Banning et al. 1993, Yamazoe and Ozawa, 1990), P450 enzymes are the principle oxidation system for the amines to their corresponding hydroxylamines (Alexander and Wallin 1991). CYP1A2 appears, as for other aromatic amines, to be the key form in several species including humans, while other P450 enzymes, ia. CYP1A1, 2A3, 2C9, 2C11, 3A4 and 3A5 show intermediate activity depending on species and tissue. The CYP forms 2A1, 2B1, 2B2, 2B4, 2E1, 3A6, 4B1 and 2C4 have, however, very low or undetectable activity towards heterocyclic amines (Alexander and Wallin 1991, Boobis et al 1995, McKinnon et al. 1995, Övervik et al. 1995, Yamazoe and Ozawa 1990).

Depending on the conditions, nitroso- and nitro derivatives might be formed, particularly in *in vitro* experiments using microsomes (Alexander et al 1995, Alexander and Wallin 1991, Frandsen et al 1994b, Yamazoe and Ozawa 1990). Nitroso derivatives, particularly of non AIA heterocyclic amines (Glu-P1, Trp-P2), may form sulfinamides or sulfonamides with thiols (Alexander and Wallin 1991). Reaction products of nitrosos derivatives (IQ) have also been observed *in vivo* (Turesky et al 1987). For example it appears that nitro-PhIP can be formed *in vivo* (Alexander et al 1991, Alexander et al. 1995).

Although the hydroxylamine may probably form the nitrenium spontaneously and thereby bind to protein and DNA (Wallin et al. 1992, Dragstedt et al. 1993, Minchin et al. 1992), esterification appears to be an extremely important step to obtain binding for some of the amines, such as PhIP (Buonarati et al. 1990, Frandsen et al 1992, Minchin et al. 1992). This has also been observed in cells with and without esterification enzymes (Kerdar et al 1993). The esterification enzymes are acetyltransferases, sulfotransferases, phosphotransferases and amino acyl tRNA synthetase (Yamazoe and Ozawa, 1990). One of the reasons for the extremely high mutagenicity of heterocyclic amines in *Salmonella typhimurium* TA 98 and TA 1538 is that these bacteria contain an effective acetyl transferase enzyme (Kato 1986). Human enzymes have now been cloned into *Salmonella typhimurium* deficient in bacterial acetyl transferase (Grant et al. 1992).

The major oxidizing organ is the liver, but also extrahepatic tissues have ability to oxidize the heterocyclic amines. Their levels of P450 enzymes are, however, much lower. The hydroxylamines, provided they escape from the liver, are usually sufficiently stabile to be transported and further processed to ultimate DNA reactive metabolites in periferal tissues (Brooks et al. 1994, Holme et al. 1989, Kaderlik et al. 1994a).

The proximate hydroxylamine metabolites can be detoxified by formation of excretable conjugates, as they can be deactivated by reacting with thiols such as glutathione (Alexander et al. 1995, Alexander and Wallin 1991, Kaderlik et al. 1994b, Lin et al. 1995).

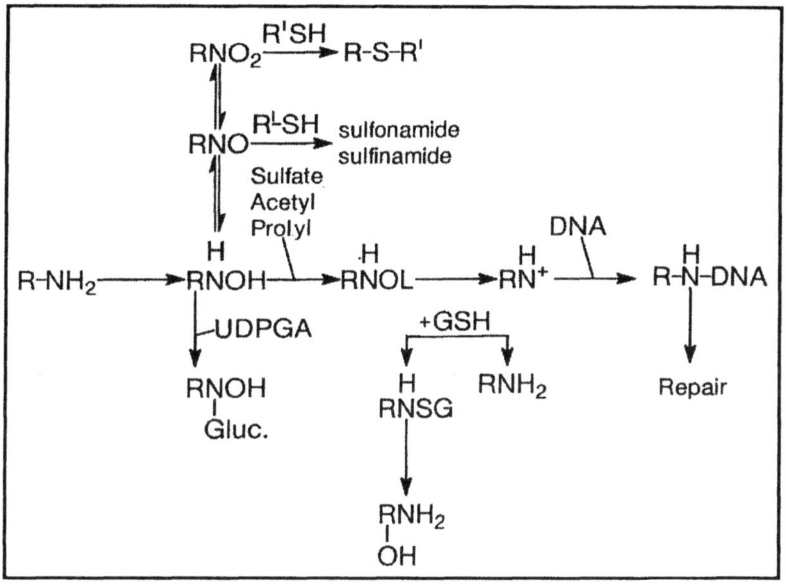

Fig. 2. Scheme of metabolic pathways of activated heterocyclic amines. R, denote the ring structure of the heterocyclic amines. The amines differ with respect to reported/preferred pathways.

N-acetyltransferases

Arylamine *N*-acetyltransferases are enzymes catalyzing central steps in the metabolic activation and/or detoxification of various arylamines including several heterocyclic amines (Hein 1988, Hein et al. 1993, Alexander and Wallin 1991, Yamazoe and Ozawa 1990). The enzymes are capable of *N*-,*O*, and *N,O*-acetylation and are present in the cytosol of many tissues (Hanna 1994). Human genetic variability in *N*-acetylation capacity may predispose toward a variety of drug-induced toxicities (Evans 1989) and in human epidemiological studies it appears that urinary bladder and colorectal cancers are associated with the slow and rapid acetylator phenotype, respectively (Hein 1988, Lang et al. 1994, Kadlubar et al. 1995). There are two isozymes, NAT1 and NAT2 of which NAT1 is monomorphic and NAT2 display genetic polymorphism with slow and rapid acetylation capacity (Blum et al. 1990, Hanna 1994). The enzymes' capability of detoxication via N-acetylation can, with regard to AIA, be considered of minor importance since the catalyzing activity toward particularly PhIP and MeIQx are very low (Alexander and Wallin 1991, Yamazoe et al. 1995). Thus, *O*-acetylation is the enzymatic activity of main concern.

Using recombinant human NAT1 and polymorphic NAT2 wild-type and mutant *N*-acetyltransferases Hein and coworkers (1994) found that both NAT1 and NAT2, the latter with greater specificity could catalyzed N-hydroxy-MeIQx and N-hydroxy-PhIP to products binding to DNA. Several mutant NAT2s had low activity. Similar results showing poor activity of NAT1 and high activity of NAT2 towards the *N*-hydroxy derivatives of Glu-P1, IQ, MeIQx and PhIP were found in cells expressing these enzymes (Minchin et al 1992). Yanagava et al. (1994) expressed CYP1A2 and NAT1 and NAT2 in cell lines. The cell line expressing CYP1A2 and NAT2 was approximately 370 and 100 times more susceptible to toxicity of IQ and MeIQx, respectively, than cells expressing NAT1 and NAT2 alone or CYP1A2 with NAT1. Kerdar et al. (1993) also expressing CYP1A1 or 1A2, NAT1 and NAT2 found greatest DNA binding of IQ with CYP1A2 and NAT2, less with CYP1A2 and NAT1 and only poorly binding with CYP1A2 alone. Thus, all four studies indicate greater activity of polymorphic NAT2 in catalyzing ultimate activation of AIA type of heterocyclic amines and indirectly support the importance of human genetic polymorphism in the susceptibility to heterocyclic amines.

In a recent study (Paulsen et al., submitted) we found that the number of aberrant crypt foci in colon (preneoplastic foci) (Paulsen et al. 1994, Pretlow et al. 1992) of congenic hamsters (Hein et al. 1994) treated with 3,2'-dimethyl-4-aminobiphenyl (resembling PhIP in the metabolism (Alexander et al. 1991, Kaderlik et al. 1994a)) were three times higher in the hamster line with rapid acetylator phenotype.

Arylsulfotransferases

Arylsulfotransferases play an important role both in the activation and detoxication of heterocyclic amines. Whereas heterocyclic amines are rather poor substrates for *N*-acetylation, their transformation to sulfamates is a major detoxication pathway (Alexander and Wallin 1991, Yamazoe et al. 1995). Except for IQ, the female rat liver appears to most effective in catlyzing *N*-sulfation. The activity is likely to be due to ST1A1 and a female form (Yamazoe et al. 1995). In addition, phenolic detoxication products are sulfonated (Alexander and Wallin 1991).

O-sulfation of the hydroxylamine derivatives of the heterocyclic amines are a major activating pathway (Yamazoe et al. 1995). This activity has been reported to be under control of pituitary growth hormone in rats and variation in enzyme activity among humans has been reported (Yamazoe et al. 1995).

Recently multiple forms of sulfotransferases in rats and humans have characterized at the molecular level (Yamazoe 1995, Ozawa et al 1994). HAST1 and HAST2 from livers of male rats were found to efficiently sulfate N-hydroxy-acetyl-aminofluorene (Yamazoe et al. 1995). Sulfation of N-hydroxy-PhIP was studied with several human and rat sulfotransferases expessed in cells (Ozawa et al. 1994). Highest specific catalytic activity was found for ST1A2 and ST1A3 and very low activity of ST1A1. Somewhat greater activity was found for ST1B1 and ST1C1. The activity of the most active forms were 5 to 6 times more active than ST1C1, a main activating sulfotransferase in rat liver (Ozawa et al. 1994).

Considerable species differences in the sulfation have been reported, eg. the relative activity in activation of N-hydroxy-PhIP by human liver cytosol was about 3 to 13 fold higher than that of male and female rats, respectively (Ozaw et al. 1994). The corresponding activity was about ten times higher in liver, kidney, colon and heart of the monkey in comparison with the rat (Davis, 1993).

Phenol-sulfating forms of sulfotransferases are also expressed in several human extrahepatic tissues, e.g. colon (Ilett et al. 1994, Turesky et al. 1991), the specific expression of different forms being an important determinant of ultimate activation of heterocyclic amines.

Other Activating Enzymes and their Relative Roles

Prolylation of the hydroxylamine derivatives of IQ, MeIQx and PhIP was considerable in the rat liver, kidney, colon and heart, whereas in the monkey this appears important in the kidney (Davis et al. 1993).

The relative role of the ultimate activation systems differs among species and between tissues depending on specific expression of enzymes and isoforms (Davis et al 1994) and also on the amine in question. Sulfation appears to be much more important in humans than in rats This makes extrapolation between species difficult. Polymorphism has been described for the N-acetyl transferase and it is likely that this occurs also for other important metabolic enzymes.

Glucuronidation

Glucuronidation is an important detoxication pathway for the parent amines and phenolic products (Alexander and Wallin 1991, Snyderwine et al. 1995). However, glucuronidation of the hydroxylamine AIA has only been described for MeIQx and PhIP displaying a major pathway for the latter (Alexander et al. 1991, Kaderlik et al. 1994b, Turesky et al. 1990). It has been suggested that the glucuronidated hydroxylamines which are excreted in the bile can be reactivated in the intestine by bacterial glucuronidases leaving the proximate metabolite free to act directly on the intestinal cells (Alexander et al. 1991). It appears, however, that further bacterial reduction of the hydroxylamine to the parent compound and excretion of this occurs to a large extent, particularly with PhIP (Alexander et al. 1995, Watkins et al. 1991). Glucuronidation of the hydroxylamine is probably of great importance for PhIP in protecting the liver from reactive metabolites. This is compatible with the fact that PhIP in contrast to other IQ-compounds does not induce liver cancer in rats and mice (Ohgaki et al. 1991, Sugimura 1995).

Role of Glutathione in Detoxication of Reactive metabolites

It has been reported for IQ and recently for PhIP that glutathione (GSH) apparently has a protective role against IQ and PhIP in hepatocytes and in rat liver since depletion of GSH resulted elevated levels of metabolites bound to DNA and DNA damage (Loretz et al. 1984, Kaderlik et al. 1994a,b). *In vitro* studies on the DNA binding of the *N*-acetoxy derivatives of PhIP, IQ and MeIQx in the presence of GSH alone inhibited the binding of IQ and MeIQx by about 40% while it only slightly affected the PhIP DNA binding (Lin et al. 1995). In the presence of glutathione transferases (GST) PhIP DNA binding was strongly inhibited. The resulting metabolic product found was PhIP (Lin et al. 1995). Among human GST the isozyme A1-1 was the most effective followed by A1-2 and negligible effect of A2-2 indicating activity in subunit A1. Also P1-1 had some activity as well as the rat forms 1-2 and 12-12. GST activity was expressed in human liver cytosol, but not in colon cytosols indicating organ specificity and protective effect in the liver. Regarding IQ and MeIQx GST appears not to be required for the effect of GSH (Lin et al. 1995).

Adducts of PhIP with Glutathione and Protein Thiols

PhIP may also form adducts with thiols including GSH. We previously characterized a PhIP-GSH adduct where the aminogroup had been substituted with GSH (Alexander et al., 1991). This metabolite was a major one in hepatocytes and bile of PhIP treated rats. A corresponding PhIP-*N*-acetylcystein adduct was found in urine (Alexander et al. 1995). This thiol substitution was *in vitro* found to occur spontaneously from nitro-PhIP in the presence of thiol either as GSH (Alexander et al. 1991) or with protein-SH such as albumine (Reistad et al. 1994).

In *in vitro* studies using N^2-acetoxy-PhIP we characterized unstable conjugates with GSH, cysteine and rat albumine (RSA). A structure of GS-N^2-PhIP was suggested on the basis of mass spectra and UV-spectra. The PhIP-RSA adduct decomposed rapidly by 30% and to 50% in 4 days. The degradation product was identified as a ring hydroxylated PhIP metabolite, 5-hydroxy-PhIP, which has also been identified in rat hepatocytes and bile (Reistad et al. 1994). We suggest that this metabolite might be a candidate for use as a biomarker of exposure to activated PhIP.

Similar reactions have to the best of our knowledge not been reported for other heterocyclic amines. Heterocyclic amines of the IQ type are known to form very low levels of protein adducts, formation of analogous unstable adducts could be one explanation.

Conclusion

Once oxidized in the amino group there are several options for further biotransfomation of heterocyclic amines (Fig 2.). Esterification pathways lead to the formation of DNA reactive metabolites, but it appears that there are also several pathways protecting from genotoxicity such as reactions with GSH and glucuronidation. Enzymes catalyzing the reactions are expressed differently among individuals, between species and in different tissues. Different isozymes may also dominate in activation vs. detoxication reactions, e.g. sulfation (Ozawa et al. 1994).

Acknowledgements

Support by a grant from the Norwegian Research Council as part of a Nordic collaborative project under the Scandinavian Contact Agency for Agricultural Research (NKJ prosjekt 82) is appreciated.

References

Alexander J, Becher G, Busk L (1989) Cooked food mutagens - a general overview. Vår Föda (Natl. Food Administration, Sweden) 42(Suppl. 2):9-59

Alexander J, Wallin H (1991) Metabolic fate of heterocyclic amines in cooked food. In Hayatsu H (ed) Mutagens in Food: Detection and Prevention, CRC Press, Boca Raton, FL, USA, pp 143-156

Alexander J, Wallin H, Rossland OJ, Solberg KE, Holme JA, Becher G, Andersson R, Grivas S (1991) Formation of a glutathione conjugate and a semi-stable transportable glucuronide conjugate of $\underline{N}^2$-oxidized species of 2-amino-1-methyl-6-phenylimidazo[4,5-*b*]pyridine (PhIP) in rat liver. Carcinogenesis 12:2239-2245

Alexander J, Fossum BH, Reistad R, Holme JA (1995) Metabolism of the Food Carcinogen 2-Amino-1-methyl-6-phenylimidazo[4,5-*b*]pyridine (PhIP) in the Rat and Other Rodents. In Adamson RH, Gustafsson J-Å, Ito N, Nagao M, Sugimura T, Wakabayashi K, YamazoeY (eds) Heterocyclic Amines in Cooked Foods: Possible Human Carcinogens, Proc. 23rd Int Symp Princess Takamatsu Cancer Research Fund, Princeton Scientific Publishing Co., Inc., Princeton, NJ, USA, pp 113-122

Banning DP, O'Farrell F, Gooderham NJ (1993) Activation of the food derived carcinogen 2-amino-3,8-dimethylimidazo(4,5-*f*)quinoxaline by rat pleural cavity inflammatory cells. Carcinogenesis 14(12):2457-2462.

Blum M, Grant DM, McBride W, Heim M, MeyerUA (1990) Human arylamine N-acetyltransferase genes: isolation, chromosomal location, and functional expression DNA Cell Biol 9:193-203

Boobis AR, Gooderham NJ, Rich KJ, Zhao K, Edwards RJ, Murray BP, Lynch AM, Murray S, Davis DS. Enzymatic studies of the activation of heterocyclic food mutagens in man

Boobis (1995). In Adamson RH, Gustafsson J-Å, Ito N, Nagao M, Sugimura T, Wakabayashi K, Yamazoe Y (eds) Heterocyclic Amines in Cooked Foods: Possible Human Carcinogens, Proc. 23rd Int. Symp. Princess Takamatsu Cancer Research Fund, Princeton Scientific Publishing Co., Inc., Princeton, NJ, USA, pp 134-144

Brookes RA, Gooderham NJ, Zhao K, Edwards RJ, Howard LA, Boobis AR, Winton DJ (1994) 2-Amino-1-methyl-6-phenylimidazo(4,5-*f*)pyridine is a potent mutagen in the mouse small intestine. Cancer Res 54(7):1665-1671

Buonarati MH, Turteltaub KW, Shen NH, Felton JS (1990) Role of sulfation and acetylation in the activation of 2-hydroxamino-1-methyl-6-phenylimidazo[4,5-*b*]pyridine to intermediates which bind to DNA. Carcinogenesis 11:1133-1138

Davis CD, Schut HA, Snyderwine EG (1993) Enzymatic phase II activation of the N-hydroxylamines of IQ, MeIQx and PhIP by various organs of monkeys and rats. Carcinogenesis 14:2091-2096

Davis CD, Ghoshal A, Schut HA, Snyderwine EG (1994) Metabolic activation of heterocyclic amine food mutagens in the mammary gland of lactating Fischer 344 rats. Cancer Lett 84(1):67-73

Dragsted L, Alexander J, Wallin H, Frandsen H, Vang O (1993) Bioactivation of food mutagens by liver microsomes from three different rat strains. Pharmacol Toxicol 72:388-393

Evans DAP (1989) N-acetyltransferase. Pharmacol Ther 42:157-234

Frandsen H, Grivas S, Andersson R, Dragsted L, Larsen JC (1992) Reaction of the N^2-acetoxy derivative of 2-amino-1-methyl-6-phenylimidazo[4,5-*b*]pyridine (PhIP) with 2'-deoxyguanosine and DNA. Synthesis and identification of N^2-(2'-deoxyguanosin-8-yl)-PhIP. Carcinogenesis 13:629-635

Frandsen H, Grivas S, Turesky RJ, Andersson R, Dragsted LO, Larsen JC (1994a) Formation of DNA adducts by the food mutagen 2-amino-3,4,8-trimethyl-3H-imidazo(4,5-*f*)quinoxaline (4,8-DiMeIQx) in vitro and in vivo. Identification of a N2-(2'-deoxyguanosin-8-yl)-4,8-DiMeIQx adduct. Carcinogenesis 15(11):2553-2558

Frandsen H, Nielsen PA, Grivas S, Larsen JC (1994b) Microsomal metabolism of the food mutagen 2-amino-3,4,8-trimethyl-3H-imidazo(4,5-*f*)quinoxaline to mutagenic metabolites. Mutagenesis 9(1):59-65

Grant DM, Josephy PD, Lord HL, Morrison LD (1992) Salmonella typhimurium strains expressing human arylamine N-acetyltransferases: metabolism and mutagenic activation of aromatic amines. Cancer Res 52:3961-3964

Hanna PE (1994) N-Acetyltransferases, O-acetyltransferases, and N,O-acetyltransferases: enzymology and bioactivation. Adv Pharmacol 27:401-430

Hein DW (1988) Acetylator genotype and arylamine-induced carcinogenesis. Biochim Biophys Acta 948:37-66

Hein DW, Doll MA, Rustan TD, Gray K, Ferguson RJ, Feng Y (1994) Construction of Syrian hamster lines congenic at the polymorphic acetyltransferase locus (*NAT2*): acetylator genotype-dependent N- and O-acetylation of arylamine carcinogens. Toxicol Appl Pharmacol 124:16-24

Hein DW, Rustan TD, Ferguson RJ, Doll MA, Gray K (1994) Metabolic activation of aromatic and heterocyclic N-hydroxyarylamines by wild-type and mutant recombinant human NATI and NAT2 acetyltransferases. Arch Toxicol 68(2):129-33

Holme JA, Wallin H, Brunborg G, Søderlund EJ, Hongslo JK, Alexander J (1989) Genotoxicity of the food mutagen 2-amino-1-methyl-6-phenylimidazo(4,5-*b*)-pyridine (PhIP): Formation of 2-hydroxamino-PhIP, a directly acting genotoxic metabolite. Carcinogenesis 10:1389-1396

Kaderlik KR, Mulder GJ, Shaddock J, Casciano DA, Teitel CH, Kadlubar FF (1994a) Effect of glutathione-depletion and inhibtion of glucuronidation and sulfation on 2-amino-1-methyl-6-phenylimidazo(4,5-b)pyridine (PhIP) metabolism, PhIP-DNA adduct formation, and unscheduled DNA synthesis in primary rat hepatocytes. Carcinogenesis 15:1711-1716

Kaderlik KR, Minchin RF, Mulder GJ, Ilett KF, Daugaard-Jenson M, Teitel CH, Kadlubar FF (1994b) Metabolic activation pathway for the formation of DNA adducts of the carcinogen 2-amino-1-methyl-6-phenylimidazo(4,5-*b*)pyridine (PhIP), in rat extrahepatic tissues. Carcinogenesis 15:1703-1709.

Kadlubar FF, Kaderlik KR, Mulder GM, Lin D, Butler MA, Teitel CH, Minchin RF, Ilett KF, Friesen MD, Bartsch H, Nagao M, Esumi H, Sugimura T, Lang NP (1995) Metabolic activation and DNA adduct detection of PhIP in dogs, rats and humans in relation to urinary bladder and colon carcinogenesis. In Adamson RH, Gustafsson J-Å, Ito N, Nagao M, Sugimura T, Wakabayashi K, Yamazoe Y (eds) Heterocyclic Amines in Cooked Foods: Possible Human Carcinogens, Proc. 23rd Int. Symp. Princess Takamatsu Cancer Research Fund, Princeton Scientific Publishing Co., Inc., Princeton, NJ, USA, pp 207-213

Kato R (1986) Metabolic activation of mutagenic heterocyclic aromatic amines from protein pyrolysates. CRC Crit Rev Toxicol 16:307

Kerdar RS, Fasshauer I, Probst M, Blum M, Meyer UA (1993) ^{32}P-postlabelling studies on the DNA adducts of the food mutagens/carcinogens IQ and PhIP-adduct formation in a chemical system, and by rat and human metabolism. IARC Sci Publ 124:179-179

Lang NP, Butler MA, Massengell J, Lawson M, Stotts RC, Hauer-Jensen M, Kadlubar FF (1994) Rapid metabolic phenotypes for acetyltransferase and cytochrome P4501A2 and putative exposure to food-borne heterocyclic amines increases the risk for colorectal cancer or polyps. Cancer Epidemiol Biomarkers Prev 3:675-682

Lin D, Meyer DJ, Ketterer B, Lang NP, Kadlubar FF (1995) Effects of human and rat glutathione S-transferases on the covalent DNA binding of the N-acetoxy derivatives of heterocyclic amine carcinogens in vitro: A possible mechanism of organ specificity in their carcinogenesis. Cancer Res 54:4920-4926

Loretz LJ, Pariza MW (1983) Effect of glutathione levels, sulfate levels, and metabolic inhibition on covalent binding of 2-amino-3-methylimidazo(4,5-*f*)quinoline and 2-acetylaminofluorene to cell macromolecules in primary monolayer cultures of adult rat hepatocytes. Carcinogenesis 5:895-899

McKinnon RA, Burgess WM, Hall PM, Roberts-Thomson SJ, Gonzalez FJ, McManus ME (1995) Characterization of CYP3A gene subfamily expression in human gastrointestinal tissues. Gut 36(2):259-267

Minchin RF, Reeves PT, Teitel CH, McManus ME, Mojarrabi B, Ilett KF, Kadlubar FF (1992) N- and O-Acetylation of aromatic and heterocyclic amine carcinogens by human monomorphic and polymorphic acetyltransferases expressed in Cos-1 cells. Biochem Biophys Res Commun 185:839-844

Ilett KF, Ingram DM, Carpenter DS, Teitel CH, Lang NP, Kadlubar FF, Minchin RF (1994) Expression of monomorphic and polymorphic N-acetyltransferases in human colon. Biochem Pharmacol 47:914-917

Ohgaki H, Takayama S, Sugimura T (1990) Carcinogenicity of heterocyclic amines in cooked food. Mutat Res 259:399-410

Ozawa S, Chou H-C, Kadlubar FF, Nagata K, Yamazoe Y, Kato R (1994) Activation of 2-Hydroxyamino-1-methyl-6-phenylimidazo(4,5-*b*)pyridine by cDNA-expressed human and rat arylsulfotransferases. Jpn J Cancer Res 85:1220-1228

Paulsen JE, Steffensen I-L, Namork E, Alexander J (1994) Scanning electron microscopy of aberrant crypt foci in rat colon. Carcinogenesis 15:2371-2373

Pretlow TP, O'Riordan MA, Pretlow TG, Stellato TA (1992) Aberrant crypts in human colonic mucosa: putative preneoplastic lesions. J Cell Biochem Suppl 16G:55-62

Reistad R, Frandsen HH, Grivas S, Alexander J (1994) *In vitro* formation and degradation of 2-amino-1-methyl-6-phenylimidazo(4,5-*b*)pyridine (PhIP) protein adducts. Carcinogenesis 15:2547-2552

Saito K, Yamazoe Y, Kamataki T, Kato R. Activation and detoxication of N-hydroxy-Trp-P-2 by glutathione and glutathione transferases. Carcinogenesis 4:1551-1557

Snyderwine EG, Turesky RJ, Buonarati MH, Turtletaub KW, Adamson RH (1995) Metabolic processing and disposition of 2-amino-3-methylimidazo(4,5-*f*)quinoline (IQ), 2-amino-3,8-dimethylimidazo(4,5-*f*)quinoxaline (MeIQx), and 2-amino-1-methyl-6-phenylimidazo(4,5-*b*)pyridine (PhIP) in nonhuman primates. In Adamson RH, Gustafsson J-Å, Ito N, Nagao M, Sugimura T, Wakabayashi K, Yamazoe Y (eds) Heterocyclic Amines in Cooked Foods: Possible Human Carcinogens, Proc. 23rd Int. Symp. Princess Takamatsu Cancer Research Fund, Princeton Scientific Publishing Co., Inc., Princeton, NJ, USA, pp 69-77

Sugimura T (1995) History, present and future, of heterocyclic amines, cooked food mutagens. In Adamson RH, Gustafsson J-Å, Ito N, Nagao M, Sugimura T, Wakabayashi K, Yamazoe Y (eds) Heterocyclic Amines in Cooked Foods: Possible Human Carcinogens, Proc. 23rd Int. Symp. Princess Takamatsu Cancer Research Fund, Princeton Scientific Publishing Co., Inc., Princeton, NJ, USA, pp 214-231

Turesky RJ, Skipper PL, Tannenbaum SR (1987) Binding of 2-amino-3-methylimidazo(4,5-*f*)quinoline to hemoglobin and albumin in vivo in the rat. Identification of an adduct suitable for dosimetry. Carcinogenesis 8:1537-1542

Turesky RJ, Bracco-Hammer I, Markovic J, Richli U, Kappeler A-M, Welti DH (1990) The contribution of N-oxidation to the metabolism of the food-borne carcinogen 2-amino-3,8-dimethylimadazo[4,5-*f*]quinoxaline in rat hepatocytes. Chem Res Toxicol 3:524-535

Turesky RJ, Lang NP, Butler MA, Teitel CH, Kadlubar FF (1991) Metabolic activation of carcinogenic heterocyclic aromatic amines by human liver and colon. Carcinogenesis 12:1839-1845

Turesky RJ, Rossi SC, Welti DH, Lay JO Jr, Kadlubar FF (1992) Characterization of DNA adducts formed in vitro by reaction of N-hydroxy-2-amino-3-methylimidazo(4,5-f)quinone and N-hydroxy-2-amino-3,8-dimethylimidazo(4,5-*f*)quinoxaline at the C-8 and N^2 atoms of guanine. Chem Res Toxicol 5:479-490

Wallin H, Holme JA, Alexander J (1992) Covalent binding of the food carcinogens MeIQx, MeIQ and IQ to DNA and protein in microsomal incubations and isolated rat hepatocytes. Pharmacol Toxicol 70:220-225

Watkins BE, Suzuki M, Wallin H, Wakabayashi K, Alexander J, Vanderlaan M, Sugimura T, Esumi H (1991) The effect of dose and enzyme inducers on the metabolism of 2-amino-l-methyl-6-phenylimidazo (4,5-*b*) pyridine (PhIP) in rats. Carcinogenesis 12:2291-2295

Yamazoe Y, Ozawa S (1990) Activation of food mutagens. In Hayatsu H (ed) Mutagens in Food: Detection and Prevention. CRC Press, Boca Raton, FL, USA, pp 143-156

Yamazoe (1995) In Adamson RH, Gustafsson J-Å, Ito N, Nagao M, Sugimura T, Wakabayashi K, Yamazoe Y (eds) Heterocyclic Amines in Cooked Foods: Possible Human Carcinogens, Proc. 23rd Int. Symp. Princess Takamatsu Cancer Research Fund, Princeton Scientific Publishing Co., Inc., Princeton, NJ, USA, pp

Yanagawa Y, Sawada M, Deguchi T, Gonzalez FJ, Kamataki T (1994) Stable expression of human CYP1A2 and N-acetyltransferases in chinese hamster CHL cells: Mutagenic activation of 2-amino-3-methylimidazo(4,5-*f*)quinoline and 2-amino-3,8-dimethylimidazo(4,5-f)quinoxaline. Cancer Res 54:3422-3427

Övervik E, Hellmold H, Branting C, Gustafsson J-Å (1995) Activation and effects of the food-derived heterocyclic amines in extrahepatic tissues. In Adamson RH, Gustafsson J-Å, Ito N, Nagao M, Sugimura T, Wakabayashi K, Yamazoe Y (eds) Heterocyclic Amines in Cooked Foods: Possible Human Carcinogens, Proc. 23rd Int. Symp. Princess Takamatsu Cancer Research Fund, Princeton Scientific Publishing Co., Inc., Princeton, NJ, USA, pp 123-133

Enzymic and Interindividual Differences in the Human Metabolism of Heterocyclic Amines

Alan R. Boobis, Nigel J. Gooderham, Robert J. Edwards, Stephen Murray, Anthony M. Lynch, Masoud Yadollahi-Farsani and Donald S. Davies.
Department of Clinical Pharmacology, Royal Postgraduate Medical School, London W12 0NN, United Kingdom.

Abstract

Heterocyclic amines (HAs) present in cooked meat (PhIP and MeIQx) are activated only by CYP1A2 in the liver of most species, including man. This enzyme exhibits marked interindividual differences in its expression, due to induction and possibly also genetically. The absence of CYP1A2 appears to protect from HA- (PhIP and MeIQx) induced cancer, as exemplified by results in the cynomolgus monkey. Differences in the potency of these HAs are not due to differences in the kinetics of their activation. The catalytic efficiency of CYP1A2 towards HAs and their oxidative fate varies amongst species, in both cases increasing the susceptibility of humans compared to that of the rat. Interindividual and inter-organ differences in the further metabolism of N-hydroxy-HAs appear to be important determinants of cancer susceptibility, as does the glutathione S-transferase catalysed detoxication of esters of N-hydroxy-PhIP. There is a need for an effective means of quantifying the *in vivo* activation of HAs in man to enable the possible risk posed by these compounds to be assessed effectively.

Introduction

Heterocyclic amines (HAs) are amongst the most potent bacterial mutagens yet described. Several members of this class of compounds, including 2-amino-3,8-dimethylimidazo[4,5-f]quinoxaline (MeIQx) and 2-amino-1-methyl-6-phenylimid-azo[4,5-b]pyridine (PhIP), are formed during the normal cooking of red meat at moderately high temperatures such as occurs during grilling or frying (Felton et al. 1986). All of the HAs tested to date are carcinogenic in rodents and of particular interest is the fact that PhIP causes tumours of the colon and breast in the rat (Sugimura 1985, Ito et al. 1991, Wakabayashi et al. 1992). In view of the strong epidemiological evidence for a positive association between the consumption of red meat (Potter and McMichael 1986), particularly when well cooked (Gerhardsson de Verdier et al. 1991, Schiffman and Felton 1990, Lang et al. 1994), and a number of tumour types, particularly of the colon, the possible risk to humans of the HAs is a major concern.

In common with many genotoxic carcinogens, the HAs are inactive until metabolised via the P450 system to the corresponding N-hydroxylamines (Felton and Knize 1991). The resultant N-hydroxy metabolite can be further activated by esterification leading to a potent electrophile (Buonarati et al. 1990, Turesky et al. 1991, Davis et al. 1993) which binds covalently to both DNA and protein and leads to mutations in critical gene targets, the identity of which has yet to be determined.

Given the pivotal role of N-hydroxylation in the carcinogenicity of HAs, the level of expression of the enzymes involved in the formation and disposition of this metabolite is of paramount importance. There is evidence that this can explain the difference in carcinogenicity profile of HAs, their target organ specificity, and inter-species and interindividual differences in susceptibility. Hence, the identification of the enzymes involved in the activation and detoxication of HAs and how these are regulated in animals and humans has profound implications for risk assessment of HAs and on strategies for reducing any such risk.

Hepatic Oxidation of HAs

HAs are extensively oxidised in the liver of most species. In rodents, this involves both ring and N-hydroxylation, the latter serving as a detoxication reaction (Turesky et al. 1988, Alexander et al. 1994). Hence, at non-saturating substrate concentrations, the balance between these two pathways could be an important determinant of the carcinogenic potential of HAs.

Human liver microsomes were incubated with either MeIQx or PhIP and the products separated by HPLC (Rich et al. 1992, Zhao et al. 1994). The elution positions of the respective N-hydroxylamines were determined by injecting samples of the authentic compounds, synthesised for that purpose. In both cases, the oxidation of HA by human liver resulted in the formation of a major (>90%) product which co-eluted with the N-hydroxylamine (Fig. 1). The identity of the respective peak was confirmed by mass spectrometry and UV spectroscopy. The formation of the N-hydroxylamine was absolutely dependent upon the presence of the co-factor NADPH. The oxidation of HAs by human liver is thus monospecific, in contrast to the situation in other species, such as rat, mouse and cynomolgus monkey, in which several products are observed, including ring hydroxylation derivatives (Snyderwine et al. 1995).

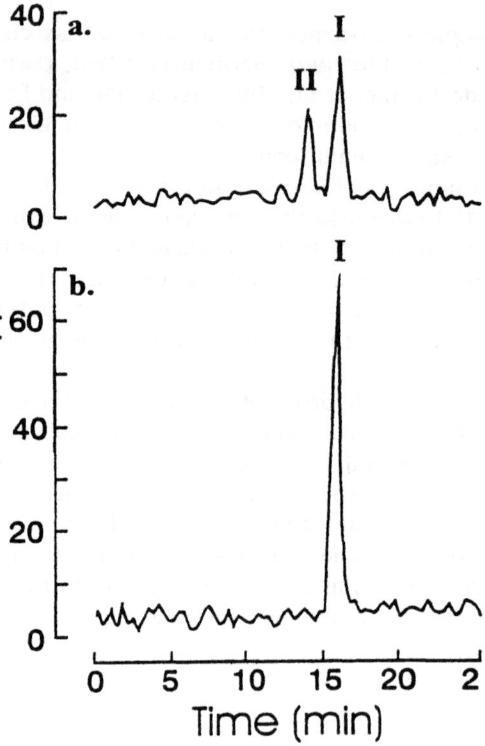

Fig. 1. HPLC chromatograms of reaction mixture following incubation of human hepatic microsomal fraction with MeIQx. Incubations were performed **a.** in the absence and **b.** in the presence of furafylline (5 μM). Peak I is unchanged MeIQx and peak II is N-hydroxy-MeIQx, as shown by LC-MS (Rich et al. 1992).

Identity of the Form of P450 Involved in the N-Hydroxylation of HAs

Studies in animals have strongly implicated members of the CYP1A subfamily in the N-hydroxylation of HAs (Kato and Yamazoe 1987). Like rodent liver, human liver contains numerous forms of P450 expressed at varying levels. These include CYP1A2,CYP2A6, CYP2C8, CYP2C9, CYPC19, CYP2D6, CYP2E1 and CYP3A4 (Shimada et al. 1994). Whilst many of these forms show considerable interindividual variability, they are expressed to some extent in most individuals.

The contribution of CYP1A2 to the N-hydroxylation of HAs was investigated using the highly selective inhibitor furafylline (Sesardic et al. 1990). Studies with representative oxidative reactions and with recombinant P450 enzymes have established that this compound inhibits human CYP1A2 at least two orders of magnitude more potently than any other form of P450 (Sesardic et al. 1990, Clarke et al. 1994). Hence, at concentrations of < 5 μM it can be used as a specific inhibitor of that enzyme. Following preincubation with furafylline, N-hydroxylation of both PhIP and MeIQx (Fig. 1) was inhibited >90% (Rich et al. 1992, Zhao et al. 1994), implicating the almost exclusive involvement of CYP1A2 in the formation of this metabolite by human liver microsomes. As indicated above, there are numerous forms of P450 expressed in human liver, and hence the ability to inhibit the N-hydroxylation of these amines with a selective inhibitor of a single form essentially precludes any important contribution of the other forms. However, one form for which this may not be the case is CYP1A1. There is evidence that this enzyme is not normally expressed in human liver (McKinnon et al. 1991). Using potent and highly selective anti-peptide antibodies directed against CYP1A1 and CYP1A2 (Murray et al. 1993, Edwards et al. 1995), we have screened over 50 human liver samples for the expression of these two proteins. Samples were obtained from a variety of donors, including a number of subjects who were actively smoking up to 20 cigarettes/day immediately prior to biopsy. In none of the samples was any CYP1A1 detected, whilst CYP1A2 was readily discernible in almost all of the samples. The antibodies did recognise human CYP1A1, as evidenced by their reactivity with the recombinant protein (Murray et al. 1993) and also the ability to detect the expression of this enzyme in placental samples from women who had smoked throughout pregnancy (Hakkola et al. in press). This would suggest that unlike in the rat and mouse, in humans CYP1A1 is less inducible in liver than in extrahepatic tissues. This is an area of work that clearly requires further investigation.

Hence, it appears that, in general, any role that CYP1A1 plays in the N-hydroxylation of HAs could occur only in extrahepatic tissues and even then only in those who smoke cigarettes or are exposed to other inducers of this enzyme. Normal levels of expression of CYP1A1 in the majority of tissues are below detection. However, in placenta (Hakkola et al. in press), lung (Yoshikawa et al. 1994, Nakajima et al. 1995) and possibly in intestine (McKinnon et al. 1992, Buchthal et al. 1995) and kidney (Lodovici et al. 1983) of cigarette smokers there is evidence for induced expression of CYP1A1.

McKinnon and McManus (1995) have compared the ability of recombinant human CYP1A1 and CYP1A2 to activate a number of HAs to mutagens, which will reflect the relative N-hydroxylation of the HAs (Table 1). For most HAs, including MeIQx, CYP1A2 was much more potent than CYP1A1. However, PhIP was an exception, and CYP1A2 was only approximately 2-fold more active than CYP1A1. Interpretation of these data, however, is somewhat complicated by the absence of detailed kinetic information and there is certainly circumstantial evidence that CYP1A1 does not play an important role in HA activation in humans (see below).

Catalytic Efficiency of HA Metabolism

At substrate concentrations below those saturating the enzyme, both the maximum velocity and the affinity of the enzyme are important in determining its relative catalytic efficiency. This will certainly be the situation in humans; it is possible

Table 1. Metabolic activation of HAs by human recombinant CYP1As

HA	Mutagenicity (rev/mg cell protein)	
	rCYP1A1	rCYP1A2
PhIP	148 ± 12	275 ± 61
IQ	1470 ± 40	29400 ± 4600
MeIQ	3130 ± 180	76700 ± 3700
MeIQx	585 ± 90	9100 ± 800
DiMeIQx	1310 ± 130	30000 ± 2200

Enzymes were expressed in COS cells.
Values are mean ± SD (n=3).
Data are modified from McKinnon and McManus (1995).

that the concentrations achieved at, or even below, the MTD may begin to saturate the enzyme (Turesky et al. 1991). Whilst all of the HAs are potent mutagens, there is still a wide range of activity, with compounds such as MeIQx being much more potent in the Ames test than PhIP (Felton and Knize 1991). There are several possible explanations for this, including the catalytic activity of the human enzymes for the different HAs. Hence, knowledge of the kinetics of the activation of HAs is important for a comparison between compounds, and also between enzymes and species.

The N-hydroxylation of MeIQx and PhIP has been determined using human liver microsomes (Boobis et al. 1995). Whilst there is a 2-fold difference in the V_{max} values, the K_m values are very similar for the two amines. Hence, the relative rate of activation, via N-hydroxylation of these two compounds by human liver microsomes will be very similar. As a consequence, the difference in mutagenic potency of these amines cannot be due to differences in the rates of formation of the proximate mutagens, and is more likely due to differences either in the intrinsic reactivity of the genotoxic metabolites or the further metabolism of the respective electrophiles. This was investigated using strains of bacteria competent and deficient in N-hydroxylamine esterification (via N-O-acetylation). Whilst the N-hydroxy metabolite of MeIQx is absolutely dependent upon esterification for mutagenicity, the corresponding metabolite of PhIP possesses substantial mutagenic activity without further metabolism (Ngao et al. 1983, Buonarati et al. 1990). In marked contrast, when cells competent in acetyltransferase activity were used, there were over two orders of magnitude difference in the potency of MeIQx compared to that of PhIP. One interpretation of these data is that the ester of N-hydroxy MeIQx is much more reactive than that of N-hydroxy PhIP. However, an alternative explanation is that the difference is due to the specificity of acetyltransferase for the respective N-hydroxylamines. In support of the latter explanation is the observation that whilst the mutagenicity of MeIQx was 25-fold greater with the acetyltransferase over-expressing strain YG1024, compared with TA98, that of PhIP was less than 2-fold greater. This is an area which is currently under investigation in a number of laboratories.

Table 2. Mutagenicity of N-hydroxy-HAs

Compound	Mutagenicity (rev/nmol)		
	TA98	TA98/DNP	YG1024
MeIQ[a]	43000	<30	11000000
N-OH-MeIQx[b]	+ + +	<30	+ + + + +
PhIP[a]	82	40	135
N-OH-PhIP	1130	440	1850

Both MeIQx and PhIP were non-mutagenic in the absence of an hepatic microsomal fraction.
[a]Activated by hepatic microsomal fraction.
[b]This compound was very mutagenic in both TA98 and YG1024 (more so in YG1024), but the specific reversion rate was not accurately quantifiable due to its instability.

Both rat (Sesardic et al. 1990) and human (Murray et al. 1993) liver express only CYP1A2 amongst members of the CYP1A subfamily. Hence, this is the only enzyme in either species responsible for the activation of the HAs (PhIP and MeIQx) by microsomal fractions from liver. However, there is a marked species difference in the capacity of hepatic microsomal fractions to activate MeIQx and PhIP. This is not due to the relative levels of expression of CYP1A2, which tends to be similar in uninduced humans and control rats (Table. 3). Normalising mutagenicity to the specific content of CYP1A2 would suggest that the rat enzyme is catalytically less efficient than the human enzyme at activating either MeIQx or PhIP, via N-hydroxylation. The difference is some 16-fold for MeIQx and almost 4-fold for PhIP (Table 3). This, combined with the almost exclusive N-hydroxylation of the amines by human liver, in contrast to the competing detoxicating hydroxylation reactions occurring in rat liver (Turesky et al. 1988, Alexander et al. 1989) will tend to magnify the relative susceptibility of humans compared to rats to the genotoxic effects of these HAs at dietary levels of exposure.

Table 3. Specific mutagenicity of HAs in rat and human liver

Parameter	HA	Species	
		Rat	Human
CYP1A2 (pmol/mg)		9	5
Mutagenicity (rev/ng)	MeIQx	25	200
	PhIP	0.12	0.23
Mutagenicity (rev/nmol)	MeIQx	5560	80000
	PhIP	27	92

Inter-species and Inter-individual Differences the in Activation of HAs

As uninduced animals do not express significant amounts of CYP1A1, and this is the only enzyme other than CYP1A2 that catalyses the activation of HAs such as MeIQx and PhIP, found in fried beef, the level of expression of CYP1A2 is likely to be an important determinant of the genotoxic activation of these compounds. This can be investigated by comparing the level of expression of CYP1A2, determined either immunologically with specific antibodies or by using a representative metabolic activity such as phenacetin O-deethylase, with the activation of the HAs. With both PhIP and MeIQx there is a highly significant correlation with CYP1A2 content or activity and their activation to mutagens (Murray et al. 1993). This supports the suggestion that variation in CYP1A2 content contributes substantially to differences in mutagenic activation of the amines.

Further evidence for the critical importance of CYP1A2 expression in the activation of some HAs can be obtained by comparison of HA activation in human, cynomolgus monkey and marmoset. The cynomolgus monkey does not express CYP1A2 at detectable levels in its liver, whereas marmoset expresses levels which are similar to those in human liver (Edwards et al. 1994). Hence, the cynomolgus monkey is in effect a gene knockout for constitutive expression of CYP1A2. In parallel with this lack of CYP1A2 expression, hepatic microsomes from cynomolgus monkey have a markedly reduced capacity to activate MeIQx or PhIP to mutagens in the Ames test whilst marmoset liver has an activity that is not dissimilar to that of human liver (Edwards et al. 1994). Interestingly, neither MeIQx not PhIP is carcinogenic in the cynomolgus monkey (Adamson et al. these Proceedings), whereas both amines are carcinogenic in all the other species tested (Sugimura 1995), in which CYP1A2 is constitutively expressed. The carcinogenicity of the HAs in marmoset has not yet been determined.

Interestingly, induction of the cynomolgus monkey with TCDD does result in detectable expression of CYP1A2 in the liver, to levels comparable to those in normal human liver. This is associated with a corresponding increase in the mutagenic activation of both HAs, to levels similar to those with human liver (Edwards et al. 1994).

Factors Contributing to the Variability in CYP1A2 Expression in Humans

CYP1A2 selectively catalyses several metabolic reactions that can be utilised to assess the activity of the enzyme *in vivo* in humans. These include caffeine N3-demethylation (Butler et al. 1989), phenacetin O-deethylation (Sesardic et al. 1988) and theophylline oxidation (Fuhr et al. 1992). A number of population studies have been performed *in vivo* and *in vitro* using these and other substrates.

These have generally failed to demonstrate any clear evidence for a polymorphism in CYP1A2 activity in humans. However, recently Nakajima et al. (1994) have reported that the distribution of CYP1A2-dependent caffeine oxidation is not normally distributed within the population, suggestive of a possible genetic polymorphism. However, sequencing of the CYP1A2 coding region from a number of individuals failed to identify any genetic explanation for this variation (Nakajima et al. 1994).

In our own studies we have demonstrated a wide variation in CYP1A2 activity, both *in vivo* and *in vitro*. In studies *in vitro* we have seen a single human liver sample amongst over 50 investigated in which there was no detectable CYP1A2 expressed nor any associated phenacetin O-deethylase activity, despite the presence of other forms of P450 and their associated activities (Sesardic et al. 1988, Murray et al. 1993). Unfortunately, it was not possible to pursue a possible genetic basis for this in this subject. At the other extreme, there is good evidence that a number of individuals have elevated activity for CYP1A2 and its associated reactions. This is due, in part, to the involvement of environmental inducers. It has been established that cigarette smoking and other hydrocarbon-containing mixtures induce CYP1A2 in human liver (Sesardic et al. 1988). However, as indicated above, at the levels of exposure encountered during cigarette smoking there is no corresponding expression of CYP1A1 in liver (Murray et al. 1993) although there is in other tissues, including the colon (Buchthal et al. 1995) and lung (Nakajima et al. 1995). The inducibility of CYP1A2 is mediated by the Ah receptor, and there is increasing evidence for genetic heterogeneity in the components of this system, the receptor itself and its nuclear translocator partner (Arnt) (Hayashi et al. 1994). In addition, the non-coding elements of the CYP1A2 gene responsible for regulating its expression have not yet all been identified and hence it is still possible that variability, with either increased or decreased expression, could be due to genetic heterogeneity within the population. Studies are currently in progress to elucidate this.

Importance of Further Metabolism of N-Hydroxylamines

As indicated above, studies with ester-conjugating deficient bacteria have provided strong evidence for the importance of further esterification of the N-hydroxylamines of HAs in their genotoxic activation. A number of enzymes can catalyse this reaction, including N-acetyltransferases and sulphotransferases (Turesky et al. 1991, Davis et al. 1993). The members of these two families of drug metabolising enzymes are being cloned and their relative activity towards HAs investigated. These studies have shown that there is marked variation in the specificity of these enzymes for the HAs. There is evidence that acetylation of the N-hydroxylamines is catalysed preferentially by the polymorphic form NAT2 (Kerdar et al. 1993). Similarly, two of the sulphotransferases (ST1A2 and ST1A3) cloned to date are particularly active in the esterification of N-hydroxy-PhIP (Ozawa et al. 1994).

Possible Protective Effects of Induction on HA-Mediated Genotoxicity

Whilst polycyclic aromatic hydrocarbons induce the hepatic content of CYP1A2 in many species including humans and mice, treatment of mice with the CYP1A inducer BNF has a paradoxical effect on the *in vivo* mutagenicity of PhIP. This was investigated in the *Dlb*-1 locus assay, in which mice heterozygous at a locus coding for a lectin binding protein expressed on the surface of enterocytes can be used to quantify *in vivo* stem cell mutations in the intestinal crypts (Winton et al. 1988). Such animals are very sensitive to the *in vivo* mutagenicity of PhIP (Brooks et al. 1994), whilst they are completely resistant to the effects of MeIQx, even after pretreatment of the animals with CYP1A inducers (Winton et al. 1990, Brooks et al. 1994). These data support the suggestion that only CYP1A2 is involved in heterocyclic amine activation and further that in the case of MeIQx either the intrinsic reactivity of the hydroxylamine ester or its failure to serve as a substrate for further activation in the intestine serves to protect this tissue from the carcinogenic effects of this HA. Treatment of mice with BNF results in a decrease in the frequency of stem cell mutations of the small intestine following administration of PhIP compared to that in corn oil treated controls. This is despite induction of both CYP1A1 and CYP1A2 in the liver of these animals and an induction of CYP1A1 in the intestine (Brooks et al. 1994). A possible explanation for this finding would be the induction of alternate routes of HA metabolism. Certainly, CYP1A1 selectively catalyses the detoxifying ring hydroxylation of some HAs including PhIP and this reaction is increased following treatment of animals with polycyclic hydrocarbons (Wallin et al. 1990). However, the increase in N-hydroxylation appears to be of more importance, as there is increased covalent binding of PhIP in the liver following CYP1A induction of mice (Buonarati et al. 1992). Hence, a more likely explanation for this finding comes from the discovery that the genotoxic esters of only N-hydroxy PhIP are selectively detoxified by certain forms of glutathione S-transferase (Kaderlik et al. 1994, Lin et al. 1994), GST A1-1 in man (Lin et al. 1994). The alpha class GSTs are inducible by a variety of agents in both animals and in humans. These include natural constituents of food (Bogaards et al. 1994, Nijhoff et al. 1995). To our knowledge, the effect of polycyclic aromatic hydrocarbons on the induction of this form of GST in the mouse has not yet been determined.

Whilst the effects of cigarette smoking on the activity of this enzyme have not yet been determined, it is intriguing that epidemiological studies suggest that smokers eating meat are at reduced risk of colorectal cancer compared with non-smokers who eat meat although smoking itself is a positive risk factor (Giovannucci et al. 1994). Interestingly, Morel et al. (1993) have demonstrated that 3-methylcholanthrene is capable of inducing GST A1-1 transcripts in cultured human hepatocytes. Hence, it is possible that cigarette smoking, although increasing hepatic CYP1A2 levels, which might be expected to increase the activation of HAs including PhIP, leads to reduced genotoxicity of these compounds through parallel induction of the protective enzyme GST A1-1 which serves to detoxify the esterified form of this HA, possibly in the colon. In their study, Morel et al. (1994) found that only two out of six liver samples responded to induction of GSTα by 3MC, despite induction of CYP1A2 in all of the preparations. Whether this reflects important interindividual variation in protection against PhIP has yet to be determined.

Assessment of Activation of HAs in Humans

Whilst the excretion of unchanged HA in urine can provide a measure of relative bioavailability of HAs and, as described by Davies et al. (these Proceedings) the use of furafylline, or a similar specific inhibitor of CYP1A2, permits the contribution of this enzyme to the metabolism of HAs *in vivo* in man to be determined (Boobis et al. 1994), these are neither sufficiently specific nor convenient to enable their widespread use in studies of variability in HA activation in human populations and the factors that might influence this. Hence, a number of alternative approaches have been investigated and some of these are illustrated in Fig. 2. The concept of biomonitoring is now well established, and relies upon the use of a surrogate marker for the compound of interest, which is generally too short-lived to enable direct analysis. Hence, adducts with DNA or protein, in the target tissue or in other more accessible tissues or excreta have been widely used for this purpose. We and others have explored the use of such biomarkers to monitor HA activation *in vivo*, initially in animals and then in humans. Whilst some of these approaches may have just sufficient sensitivity to permit adducts to be detected in some individuals in target tissues, no method yet described has sufficient sensitivity and utility for widespread use. For example, adducts with serum proteins such as albumin and haemoglobin represent a very small fraction of the total dose ingested, which itself is very low, from dietary sources (Lynch et al. 1991, 1993). Further, the convenient sulphinamide release assay so effective for simpler aromatic amines such as 4-aminobiphenyl (Harris et al. 1987) is not appropriate for the cooked food HAs, since less than 10% of total protein-adducted material is present in this form (Lynch et al. 1991, 1993). Whilst the identity of the other adducts has not yet been established, total levels of adduction are such that it is unlikely that a simple chemical assay will provide the sensitivity required. Analysis of DNA adducts is even more problematical, and whilst P^{32} post-labelling is very sensitive, it is still not sufficiently sensitive to detect the levels of adducts that will be present in accessible tissues such as white blood cells. There have been reports that levels of DNA adducts in intestinal mucosa samples from some subjects are such that they are detectable by this method (Friesen et al. 1994) but again, this could not be used routinely as a general technique. Novel approaches are required to this problem, and the application of accelerator mass spectrometry (Turteltaub et al. 1995) to subjects ingesting very low levels of radiolabelled HA should provide at least some useful information on this question. The difficulty of devising a routine assay still exists and is likely to require considerable ingenuity to succeed.

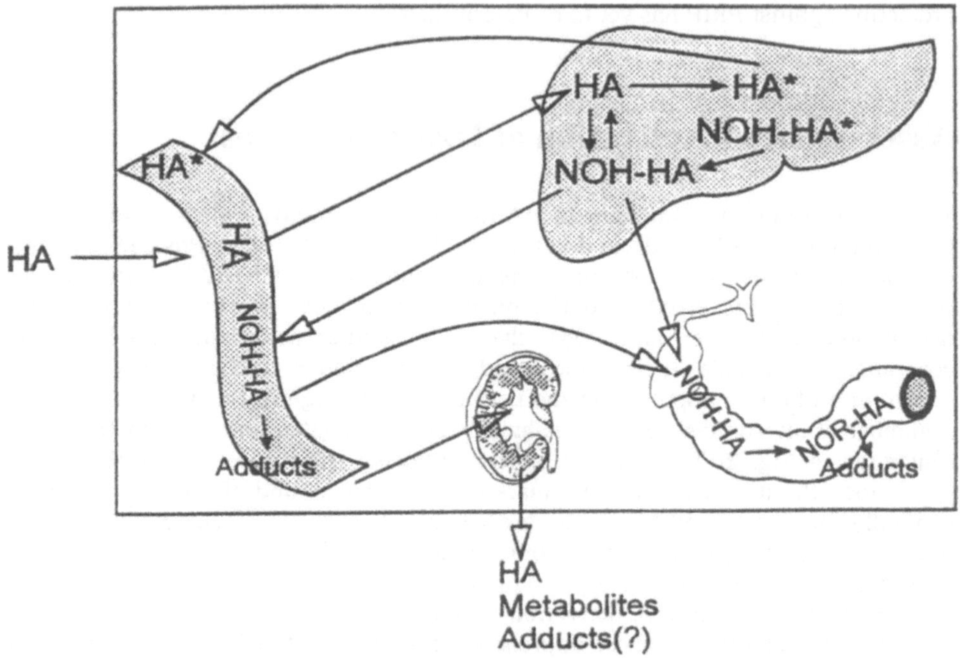

Fig. 2. Possible strategies for monitoring activation of HAs *in vivo* in humans.

Whilst not a direct measure of HA activation, it might be possible to obtain somesome information on their biological effects by identifying a mutagenic fingerprint characteristic of these compounds, as has been done for aflatoxin B_1, which produces a specific mutation in codon 249 of the P53 tumour suppressor gene (Aguilar et al. 1993). Studies in our own and other laboratories have been concerned to investigate this question utilising reporter genes such as HPRT and APRT initially. These studies have shown that the HA PhIP produces a characteristic transversion of GC to TA, which occurs with a high frequency (9/28 point mutations) at the second guanine residue in a GGA triplet, the mutation occurring on the non-transcribed strand, presumably due to selective repair of the transcribed strand. A similar finding has been reported by Morgenthaler and Holzhäuser (1995). In addition, PhIP induces a number of -GGG- to -GG- mutations, leading to truncated products. This has also been reported by Kakiudi et al. (1995) in the APC gene of PhIP-induced colon tumours. Such mutations may provide characteristic mutagenic fingerprints for PhIP *in vivo*, although their use to investigate variation in exposure to and activation of PhIP in human populations would require the refinement of current procedures for identifying low frequency mutations within a population of cells (Aguilar et al. 1993).

Conclusions

The common HAs, MeIQx and PhIP, differ markedly in their mutagenic potency, and in the tissue and species specificity of their carcinogenicity. Hence, these structurally-related compounds provide a very useful paradigm to investigate the causes of such differences. There is now good evidence that CYP1A2 plays a key role in determining the genotoxic effects of the HAs. The absence of CYP1A2 protects the cynomolgus monkey from both PhIP and MeIQx-induced cancer. The extent to which interindividual differences in CYP1A2 expression in humans contributes to differences in cancer susceptibility has yet to be determined. However, there is some evidence that increased CYP1A2 activity as a consequence of smoking may not be associated with the same risk as genetically-determined elevations of expression. This is because of parallel induction, by smoking, of protective pathways, particularly by glutathione conjugation in the case of PhIP. Attempts to assess the risk of HAs to humans from rodent cancer studies have generally failed to take full account of the marked enzymic differences that exist between the species. Hence, not only is human CYP1A2 intrinsically more active than the rat orthologue at activating HAs, humans metabolise almost all of a dose via this pathway, whilst rats detoxify a significant proportion by ring-hydroxylation. Whilst the latter will not affect HA activation *in vitro*, where substrate is not limiting, *in vivo* competition between activation and detoxication will serve to protect the rat compared to the human. The HAs MeIQx and PhIP as activated by only CYP1A2 in liver, and whilst CYP1A1 can activate the HAs to some extent, this enzyme is not normally expressed constitutively. Hence, HA activation appears to occurs primarily in the liver, via CYP1A2, target organ specificity then being determined by the intrinsic reactivity of the genotoxic products and by the distribution and specificity of the enzymes responsible for esterifying the N-hydroxylamines to their respective ultimate genotoxic form and for detoxifying these compounds, notably GST alpha. It is now apparent that none of the animal studies of HA carcinogenicity is such that it is adequate for estimating human risk from these compounds. Some means of assessing HA activation *in vivo* is necessary, to permit comparison with tumour incidence. Current analytical techniques lack the required sensitivity. Novel approaches are required, such as analysis, in non-tumour tissue, of the frequency of unique mutations that can be attributed to the HAs. Alternatively, human risk could be assessed from an intervention study in which HA absorption or activation is prevented. Whatever the approach adopted, there is sufficient evidence from the metabolism and genotoxicity of HAs to be concerned about their possible impact on human health.

References

Aguilar F, Hussain SP, Cerutti P (1993) Aflatoxin B1 induces the transversion of G→T in codon 249 of the p53 tumor suppressor gene in human hepatocytes. Proc Natl Acad Sci USA 90:8586-8590

Alexander J, Fossum BH, Holme JA (1994) Metabolism of the food mutagen 2-amino-1-methyl-6-phenylimidazo[4,5-b]pyridine (PhIP) in isolated liver cells from guinea pig, hamster, mouse and rat. Environ Health Perspect 102 (Suppl 6):109-114

Alexander J, Wallin H, Home JA, Becher G (1989) 4-(2-Amino-1-methylimidazo[4,5-b]pyrid-6-yl)phenyl sulfate - a major metabolite of the food mutagen 2-amino-1-methyl-6-phenylimidazo[4,5-b]pyridine (PhIP) in the rat. Carcinogenesis 10:1543-1547

Bogaards JJ, Verhagen H, Willems MI, van Poppel G, van Bladeren PJ (1994) Consumption of Brussels sprouts results in elevated alpha-class glutathione S-transferase levels in human blood plasma. Carcinogenesis 15:1073-1075

Boobis AR, Gooderham NJ, Rich KJ, Zhao K, Edwards RJ, Murray BP, Lynch AM, Murray S, Davies DS (1995) Enzymatic studies of the activation of heterocyclic food mutagens in man. In: Adamson RH, Gustafsson J-A, Ito N, Nagao M, Sugimura T, Wakabayashi K, Yamazoe Y (eds) Heterocyclic amines in cooked foods: possible human carcinogens. Princeton Scientific Publishing Co, Princeton, NJ, pp 134-144

Brooks RA, Gooderham NJ, Zhao K, Edwards RJ, Howard LA, Boobis AR, Winton DJ (1994) 2-Amino-1-methyl-6-phenylimidazo[4,5-b]pyridine is a potent mutagen in the mouse small intestine. Cancer Res 54:1665-1671

Buchthal J, Grund KE, Buchmann A, Schrenk D, Beaune P, Bock KW (1995) Induction of cytochrome P4501A by smoking or omeprazole in comparison with UDP-glucuronosyltransferase in biopsies of human duodenal mucosa. Eur J Clin Pharmacol 47:431-435

Buonarati M H, Felton J S (1990) Activation of 2-amino-1-methyl-6-phenylimidazo[4,5-b]pyridine (PhIP) to mutagenic metabolites. Carcinogenics 11:1133-1138

Buonarati MH, Roper M, Morris CJ, Happe JA, Knize MG, Felton JS (1992) Metabolism of 2-amino-1-methyl-6-phenylimidazo[4,5-b]pyridine (PhIP) in mice. Carcinogeneses 13:621-627

Buonarti MH, Turteltaub KW, Shen NH, Felton JS (1990) Role of sulfation and acetylation in the activation of 2-hydroxamino-1-methyl-6-phenylimidazo[4,5-b]pyridine to intermediates which bind to DNA. Carcinogenesis 11:1133-1138

Butler MA, Iwasaki M, Guengerich FP, Kadlubar FF (1989) Human cytochrome P450PA (P4501A2), the phenacetin O-deethylase, is primarily responsible for the hepatic 3-demethylation of caffeine and the N-oxidation of carcinogenic arylamines. Proc Natl Acad Sci USA 86:7696-7700

Clarke SE, Ayrton AD, Chenery RJ (1994) Characterization of the inhibition of P4501A2 by furafylline. Xenobiotica 24:517-526

Davis CD, Schut HA, Snyderwine EG (1993) Enzymatic phase II activation of the N-hydroxylamines of IQ, MeIQx and PhIP by various organs of monkeys and rats. Carcinogenesis 14: 2091-2096

Edwards RJ, Murray BP, Murray S, Schultz T, Neubert D, Gant TW, Thorgeirsson SS, Boobis AR, Davies DS (1994) Contribution of CYP1A1 and CYP1A2 to the activation of heterocyclic amines in monkeys and human. Carcinogenesis 15:829-836

Edwards RJ, Singleton AM, Murray BP, Davies DS, Boobis AR (1995) Short synthetic peptides exploited for reliable and specific targeting of antibodies to the C-termini of cytochrome P450 enzymes. Biochem Pharmacol 49:39-47

Felton JS, Knize MG (1991) Occurrence, identification and bacterial mutagenicity of heterocyclic amines in cooked food. Mutat Res 259:205-217

Felton JS, Knize MG, Shen NH, Andresen BD, Bjeldanes LF, Hatch FT (1986) Identification of mutagens in cooked beef. Environ Health Perspect 67:17-24

Friesen MD, Kaderlink K, Lin D, Garren L, Bartsch H, Lang NP, Kadlubar FF (1994) Analysis of DNA adducts of 2-amino-1-methyl-6-phenylimidazo[4,5-b]pyridine in rat and human tissues by alkaline hydrolysis and gas chromatography/electron capture mass spectrometry: validation by comparison with ^{32}P-postlabeling. Chem Res Toxicol 7:733-739

Fuhr U, Doehmer J, Battula N, Wölfel C, Kudla C, Keita Y, Staib AH (1992) Biotransformation of caffeine and theophylline in mammalian cell lines genetically engineered for expression of single cytochrome P450 isoforms. Biochem Pharmacol 43:225-235

Gerhardsson de Verdier M, Hagman U, Peters RK, Steineck G, Övervik E (1991) Meat, cooking methods and colorectal cancer: A case-referent study in Stockholm. Intl J Cancer Res 49:520-525

Giovannucci E, Rimm EB, Stampfer MJ, Colditz GA, Ascherio A, Willett WC (1994) Intake of fat, meat and fiber in relation to risk of colon cancer in men. Cancer Res 54:2390-2397

Hakkola J, Pasanen M, Hukkanen J, Boobis AR, Edwards RJ, Pelkonen O, Mäenpää J, Cresteil T, Raunio H (in press) Expression of xenobiotic-metabolizing cytochrome P450 forms in human full-term placenta. Biochem Pharmacol

Harris CC, Weston A, Willey JC, Trivers GE, Mann DL (1987) Biochemical and molecular epidemiology of human cancer; Indicators of carcinogen exposure, DNA damage, and genetic predisposition. Environ Health Perspect 75:109-119

Hayashi S, Watanabe J, Nakachi K, Eguchi H, Gotoh O, Kawajiri K (1994) Interindividual difference in expression of human Ah receptor and related P450 genes. Carcinogenesis 15:801-806

Ito N, Hasegawa R, Sano M, Tamano S, Esumi H, Takayama S, Sugimura T (1991) A new colon and mammary carcinogen in cooked food, 2-amino-1-methyl-6-phenylimidazo[4,5-b]pyridine (PhIP). Carcinogenesis 12:1503-1506

Kaderlik KR, Mulder GJ, Shaddock JG, Casciano DA, Teitel CH, Kadlubar FF (1994) Effect of glutathione depletion and inhibition of glucuronidation and sulfation on 2-amino-1-methyl-6-phenylimidazo[4,5-b]pyridine (PhIP) metabolism, PhIP-DNA adduct formation and unscheduled DNA synthesis in primary rat hepatocytes. Carcinogenesis 15: 1711-1716

Kakiuchi H, Watanabe M, Ushijima T, Toyota M, Imai K, Weisburger JH, Sugimura T, Nagao M (1995) Specific 5'-GGGA-3'→5'-GGA-3' mutation of the Apc gene in rat colon tumors induced by 2-amino-1-methyl-6-phenylimidazo[4,5,-b]pyridine. Proc Natl Acad Sci USA 92:910-914

Kato R, Yazmazoe Y (1987) Metabolic activation and covalent binding to nucleic acids of carcinogenic heterocylic amines from cooked foods and amino acid pyrolysates. Jpn J Cancer Res 78:297-311

Kerdar RS, Fasshauer I, Probst M, Blum M, Meyer UA, Wild D (1993) 32P-Postlabeling studies on the DNA adducts of the food mutagens/carcinogens IQ and PhIP-adduct formation in a chemical system, and by rat and human metabolism. IARC Sci Publ 124:173-179

Lang NP, Butler MA, Massengill J, Lawson M, Stotts RC, Hauer-Jensen M, Kadlubar FF (1994) Rapid metabolic phenotypes for acetyltransferase and cytochrome P4501A2 and putative exposure to food-borne heterocyclic amines increase the risk for colorectal cancer or polyps. Cancer Epidemiol Biomark Prevent 3:675-682

Lin D, Meyer DJ, Ketterer B, Lang NP, Kadlubar FF (1994) Effects of human and rat glutathione S-transferases on the covalent DNA binding of the N-acetoxy derivatives of heterocyclic amine carcinogens *in vitro*: a possible mechanism of organ specificity in their carcinogenesis. Cancer Res 54:4920-4926

Lodovici M, Dolara P, Caderni G, Carini M, Costantini A, Selli C, Barbagli G, Calzolai A. (1983) The effect of cigarette smoke on aryl-hydrcarbon hydroxylase (AHH) activity of the human kidney. Eur J Cancer Clin Oncol 19:1565-1568

Lynch AM, Murray S, Boobis AR, Davies DS, Gooderham NJ (1991) The measurement of MeIQx adducts with mouse haemoglobin *in vitro* and *in vivo*: implications for human dosimetry. Carcinogenesis 12:1067-1072

Lynch AM, Murray S, Zhao K, Gooderham NJ, Boobis AR, Davies DS (1993) Molecular dosimetry of the food-borne carcinogen MeIQx using adducts of serum albumin. Carcinogenesis 14:191-194

McKinnon RA, McManus ME (1995) Function and localization of cytochromes P450 involved in the metabolic activation of food-derived heterocyclic amines. In: Adamson RH, Gustafsson J-A, Ito N, Nagao M, Sugimura T, Wakabayashi K, Yamazoe Y (eds) Heterocyclic amines in cooked foods: possible human carcinogens. Princeton Scientific Publishing Co, Princeton, NJ, pp 145-153

McKinnon RA, Burgess WM, Hall P de la M, Zaiminuddin A-A, McManus ME (1992) Metabolism of food-derived heterocyclic amines in human and rabbit tissues by P4503A proteins in the presence of flavonoids. Cancer Res 52(Suppl):2108s-2113s

McKinnon RA, Hall P de la M, Quattrochi LC, Tukey RH, McManus ME (1991) Localization of CYP1A1 and CYP1A2 messenger RNA in normal human liver and in hepatocellular carcinoma by in situ hybridization. Hepatology 14:848-856

Morel F, Fardel O, Meyer DJ, Langouet S, Gilmore KS, Meunier B, Tu CP, Kensler TW, Ketterer B, Guillouzo A (1993) Preferential increase of glutathione S-transferase class alpha transcripts in cultured human hepatocytes by phenobarbital, 3-methylcholanthrene, and dithiolethiones. Cancer Res 53:231-234

Morgenthaler P-ML, Holzhäuser D (1995) Analysis of mutations induced by 2-amino-1-methyl-6-phenylimidazo[4,5-b]pyridine (PhIP) in human lymphoblastoid cells. Carcinogenesis 16: 713-718

Murray BP, Edwards RJ, Murray S, Singleton AM, Davies DS, Boobis AR (1993) Human hepatic CYP1A1 and CYP1A2 content, determined with specific anti-peptide antibodies, correlates with the mutagenic activation of PhIP. Carcinogenesis 14: 585-592

Nagao M, Fujita Y, Wakabayashi K, Sugimura T (1983) Ultimate forms of mutagenic and carcinogenic heterocyclic amines produced by pyrolysis. Biochem Biophys Res Commun 114:626-631

Nakajima M, Yokoi T, Mizutani M, Shin S, Kadlubar FF, Kamataki T (1994) Phenotyping of CYP1A2 in Japanese population by analysis of caffeine urinary metabolites: absence of mutation prescribing the phenotype in the CYP1A2 Gene. Cancer Epidemiol Biomark Prevent 3:413-421

Nakajima T, Elovaara E, Anttila S, Hirvonen A, Camus AM, Hayes JD, Ketterer B, Vainio H (1995) Expression and polymorphism of glutathione S-transferase in human lungs: risk factors in smoking-related lung cancer. Carcinogenesis 16:707-711

Nijhoff WA, Bosboom MA, Smidt MH, Peters WH (1995) Enhancement of rat hepatic and gastrointestinal glutathione and glutathione S-transferases by alpha-angelicalactone and flavone. Carcinogenesis 16: 607-612

Ozawa S, Chou HC, Kadlubar FF, Nagata K, Yamazoe Y, Kato R (1994) Activation of 2-hydroxyamino-1-methyl-6-phenylimidazo[4,5-b]pyridine by cDNA-expressed human and rat arylsulfotransferases. Jpn J Cancer Res 85:1220-1228

Potter JD, McMichael AJ (1986) Diet and cancer of the colon and rectum: a case-control study. J Natl Cancer Inst 76:557-569

Rich KJ, Murray BP, Lewis I, Rendell NB, Davies DS, Gooderham NJ, Boobis AR (1992) N-Hydroxy-MeIQx is the major microsomal oxidation product of the dietary carcinogen MeIQx with human liver. Carcinogenesis 13:2221-1226

Schiffman MH, Felton JS (1990) Fried foods and the risk of colon cancer. Am J Epidemiol 131:376-378

Sesardic D, Boobis AR, Murray BP, Murray S, Segura J, de la Torre R, Davies DS (1990) Furafylline is a specific and potent inhibitor of cytochrome P450IA2 in man. Br J Clin Pharmacol 29:651-663

Sesardic D, Cole KJ, Edwards RJ, Davies DS, Thomas PE, Levin W, Boobis AR (1990) The inducibility and catalytic activity of cytochromes P450c and P450d in rat tissues. Biochem Pharmacol 39:499-506

Shimada T, Yamazaki H, Mimura M, Inui Y, Guengerich FP (1994) Interindividual variations in human liver cyctochrome P-450 enzymes involved in the oxidation of drugs, carcinogens and toxic chemicals: studies with liver microsomes of 30 Japanese and 30 Caucasians. J Pharmacol Exp Ther 270:414-423

Snyderwine EG, Turesky RJ, Buonarati MH, Turteltaub KW, Adamson RH (1995) Metabolic processing and disposition of 2-amino-3-methylimidazo[4,5-f]quinoline (IQ) 2-amino-3,8-dimethylimidazo[4,5-f]quinoxaline (MeIQx), and 2-amino-1-methyl-6-phenylimidazo[4,5-b]pyridine (PhIP) in nonhuman primates. In: Adamson RH, Gustafsson J-A, Ito N, Nagao M, Sugimura T, Wakabayashi K, Yamazoe Y (eds) Heterocyclic amines in cooked foods: possible human carcinogens. Princeton Scientific Publishing Co, Princeton, NJ, pp 69-77

Sugimura T (1985) Carcinogenicity of mutagenic heterocylic amines formed during the cooking process. Mutat Res 150:33-41

Sugimura T (1995) History, present and future, of heterocyclic amines, cooked food mutagens. In: Adamson RH, Gustafsson J-A, Ito N, Nagao M, Sugimura T, Wakabayashi K, Yamazoe Y (eds) Heterocyclic amines in cooked foods: possible human carcinogens. Princeton Scientific Publishing Co, Princeton, NJ, pp 214-231

Turesky RJ, Aeschbacher HU, Würzner HP, Skipper PL, Tannenbaum SR (1988) Major routes of metabolism of the food-borne carcinogen 2-amino-3,8-dimethylimidazo[4,5-f]quinoxaline in the rat. Carcinogenesis 9:1043-1048

Turesky RJ, Lang NP, Butler MA, Teitel CH, Kadlubar FF (1991) Metabolic activation of carcinogenic heterocyclic aromatic amines by human liver and colon. Carcinogenesis 12:1839-1845

Turesky RJ, Markovic J, Bracco-Hammer I, Fay LB (1991) The effect of dose and cytochrome P450 induction on the metabolism and disposition of the food-borne carcinogen 2-amino-3,8-dimethylimidazo[4,5-f]quinoxaline (MeIQx) in the rat. Carcinogenesis 12:1847-1855

Turteltaub KW, Vogel JS, Frantz C, Felton JS, McManus M (1995) Assessment of the DNA adduction and pharmacokinetics of PhIP and MeIQx in rodents at doses approximating human exposure using the technique of accelerator mass spectrometry (AMS) and 32P-postlabeling. In: Adamson RH, Gustafsson J-A, Ito N, Nagao M, Sugimura T, Wakabayashi K, Yamazoe Y (eds) Heterocyclic amines in cooked foods: possible human carcinogens. Princeton Scientific Publishing Co, Princeton, NJ, pp 93-102

Wakabayashi K, Nagao M, Esumi H, Sugimura T (1992) Food-derived mutagens and carcinogens. Cancer Res (Suppl) 52:2092s-2098s

Wallin H, Mikalsen A, Guengerich FP, Ingelman-Sundberg M, Solberg KE, Rossland OJ, Alexander J (1990) Differential rates of metabolic activation and detoxication of the food mutagen 2-amino-1-methyl-6-phenylimidazo[4,5-b]pyridine by different cytochrome P450 enzymes. Carcinogenesis 11:489-492

Winton DJ, Blount MA, Ponder BAJ (1988) A clonal marker induced by mutation in mouse intestinal epithelium. Nature 333:463-466

Winton DJ, Gooderham NJ, Boobis AR, Davies DS, Ponder BAJ (1990) Mutagenesis of mouse intestine *in vivo* using the *Dlb*-1 specific locus test: studies with 1,2-dimethylhydrazine, dimethylnitrosamine, and the dietary mutagen 2-amino-3,8-dimethylimidazo[4,5-f]quinoxaline. Cancer Res 50:7992-7996

Yoshikawa M, Arashidani K, Kawamoto T, Kodama Y (1994) Aryl hydrocarbon hydroxylase activity in human lung tissue: in relation to cigarette smoking and lung cancer. Environ Res 65:1-11

Zhao K, Murray S, Davies DS, Boobis AR, Gooderham NJ (1994) Metabolism of the food derived mutagen and carcinogen 2-amino-1-methyl-6-phenylimidazo[4,5-b]pyridine (PhIP) by human liver microsomes. Carcinogenesis 15(6):1285-1288

Extrapolation of Heterocyclic Amine Carcinogenesis Data from Rodents and Nonhuman Primates to Humans

Richard H. Adamson[1,2], Unnur P. Thorgeirsson[1] and Takashi Sugimura[3]
[1]National Cancer Institute, Bethesda, MD 20892
[3]National Cancer Center Research Institute, Tokyo, Japan

Studies investigating the role of diet in the etiology of human cancer have generally been concerned with calories, macro molecules (fat and fiber), food additives or pesticides. More recently investigators have studied the role that naturally occurring compounds such as hydrazines in mushrooms, acquilide A/ptaquiloside in bracken fern, or fungal contaminants such as aflatoxin B_1, sterigmatocystin or fusariums may contribute to the causation of human cancer (Adamson and Thorgeirsson 1995, Sugimura, 1995). Following the report by Sugimura and coworkers of mutagenic compounds in pyrolytic products of broiled foods (Sugimura et al., 1977) numerous investigators in Japan, Europe, Australia and the United States of America have contributed to the identification, isolation and characterization of the biological properties of another class of mutagenic and carcinogenic compounds now known as heterocyclic amines (HCAs). These studies have been summarized in several reviews (Adamson 1990, Eisenbrand and Tang 1993, Felton and Knize 1990, Sugimura and Wakabayashi 1990, and Sugimura, 1995). Recently the 23rd International Symposium of the Princess Takamatsu Cancer Research has published a number of important contributions concerning the chemical, biological and epidemiological studies of the HCAs (Adamson et.al. eds. 1995). This paper will present HCA data concerning *in vitro* mutagenicity, amount in cooked foods, activation by various species, carcinogenicity in animals, epidemiological studies and assessment of the carcinogenic risk of HCAs to humans.

Mutagenicity of Heterocyclic Amines

The high degree of mutagenicity of the HCAs in the Ames/Salmonella test has made possible the extraction, isolation and identification of the HCAs. Thus far twenty different HCAs have been isolated and identified from cooked foods

---

[2]Present address: National Soft Drink Association, Washington, D.C. 20036, USA

primarily fish, chicken, and pork. Several additional HCAs have been reported but their structures have not been fully elucidated. The HCAs are formed primarily from cooking of muscle meats at high but normal cooking temperature and thus generally by broiling or grilling practices. The formation of the HCA's is primarily from the reaction of amino acids with creatine at high temperatures and although both time and temperature are important variables in HCA formation temperature appears to be the more important variable (Pariza et al., 1979, Bjeldanes et al., 1983). In some cases sugar was shown to enhance the formation of the HCAs (Skog et al., 1995). Recently however, it was reported that amino acids plus arginine or 1-methylguanidine, a fragment of arginine would also form mutagenic compounds and thus HCAs are formed, but to a less degree, in foods containing gluten (Knize et al., 1994).

The HCAs are generally divided into the imidazole type, as a result of having an amino group attached to the 2 position of an imidazole ring, or to the nonimidazoie type which has an amino group attached to a pyridine ring. These two types can be separated by treatment with nitrite under acid conditions since the imidazole type still demonstrate mutagenicity whereas the nonimidazole type is deaminated and does not demonstrate mutagenicity following treatment. Studies using cooked beef and sardines demonstrated a high degree of acid nitrite resistance suggesting that the imidazole type are the predominant type in foods (Tsuda et al., 1980, 1985). Since all the known HCAs have a planar structure and are intercalated into GC pair rich stretches of DNA the Salmonella typhimurium TA 98 strain is more susceptible to the mutagenic activities of the HCAs than the TA 100 strain. Table 1 summarizes the mutagenic activity of various HCAs as well as those of some known carcinogens in both the TA 98 and the TA 100 tester strain. There is a wide range of mutagenic activities with the various HCAs in both tester strains, with the imidazole type in general having higher activity than the nonimidazole type, the exceptions being high mutagenic activity with Trp-P-2 and low mutagenic activity with PhIP. Also the imidazole type HCAs have higher mutagenic activity than known carcinogens. The HCAs are also mutagenic in Chinese hamster lung cells but the differences seen in the Ames/tester strains between the mutagenicities of the HCAs and of known carcinogens were not observed in this system (Nakayasu et al., 1983). In contrast in Chinese hamster ovary (CHO) cells the relatively weaker HCAs in the Ames test, Trp-P-2 and PhIP, were the most potent (Trp-P-2 > PhIP > IQ > MeIQ > MeIQx (Thompson et al., 1987)

Amount of Heterocyclic Amines in Cooked Foods

The amounts of HCAs found in cooked foods, although in parts per billion, are important because of their high mutagenicity and carcinogenic potential. Although there is a lot of data in the literature on the amounts of various HCAs in beef, chicken, lamb, pork, fish and beef extract a comparison of the amounts of HCAs formed in different foods is difficult because of the variability in type of

Table 1 Mutagenicity of various heterocyclic amines compared to typical known carcinogens in Salmonella typhimurium TA98 and TA100[a]

Heterocyclic Amine	Revertants/mg TA98	Revertants/mg TA100
Nonimidazole type		
Trp-P-1	39,000	1,700
Trp-P-2	104,200	1,800
Glu-P-1	49,000	3,200
Glu-P-2	1,900	1,200
AaC	300	20
MeAaC	200	120
Imidazole type		
IQ	433,000	7,000
MeIQ	661,000	30,000
IQx	75,000	1,500
8-MeIQx	145,000	14,000
418-DiMeIQx	183,000	8,000
718-DiMeIQx	163,000	9,900
PhIP	1,800	120
Known Carcinogens		
Aflatoxin B_1	6,000	28,000
4-Nitroquinoline 1-oxide	970	9,900
Benzo(a) pyrene	320	660
N-Nitrosodiethylamine	0.02	0.15

[a]After Sugimura and Wakabayashi, 1990

cooking, temperature and time and generally not all the HCAs were measured in the same sample. One study did make a comparison of mutagenicity from patties made from beef, chicken and fish which were fried identically (Knize et al., 1988). This study reported that beef and chicken were the most mutagenic and fish meat was 5 to 8 times less. However, mutagenicity does not necessarily translate into carcinogenicity and high amounts of PhIP which has low mutagenicity, as compared to the other HCAs (see Table 1), have been reported in fish and PhIP is a rodent carcinogen. A data base on the concentrations of many of the HCAs in cooked foods is being maintained and has been published (Layton

Table 2 Amount of some heterocyclic amines in cooked foods[a]

HCA	Food type	Amount (ng/g)
Nonimidazole type		
Trp-P-1	Broiled beef	0.21-53
	Broiled sardine	13.3
Trp-P-2	Broiled/fried beef	0.25
	Beef extract	3.2
	Broiled sardine	13.1
Glu-P-2	Broiled cuttlefish	280
AaC	Broiled/fried beef steak	1.2-8.9
	Grilled beef	651
	Grilled chicken	180
	Broiled/barbecued salmon	2.8-109
MeAaC	Grilled beef	63
	Grilled chicken	15
Imidazole type		
IQ	Fried ground beef	0.02-0.6
	Broiled sardine	4.9-20
	Broiled salmon	0.3-1.8
8-MeIQx	Fried ground beef	0.1-10.8
	Broiled/fried beef steak	0.5-8.3
	Grilled bonito	5.2
	Fried bacon	10.7
	Bacon well done	26.7
418-DiMeIQx	Fried ground beef	0.06-9.35
	Broiled/fried beef steak	0.1-2
	Grilled bonito	5.4
	Broiled chicken	0.81
	Fried bacon	0.3-2
	Bacon well done	9.3
PhIP	Fried ground beef	0.15-21.5
	Broiled/fried beef steak	0.6-48.5
	Broiled chicken	38.1
	Fried pollock	69.2
	Fried bacon	1.5-2.7
	Bacon well done	36
	Baked salmon	5.9-18
	Broiled/barbecued salmon	2-73

[a]Summarized from Felton and Knize 1990 and Layton et.al. 1995

et al., 1995). The data in Table 2 gives the amounts of some of the HCAs foundin cooked foods. In the nonimidazole type of HCAs the highest amounts found are with AaC whereas in the imidazole type PhIP was highest. Both types of mutagens are also found in food grade beef extract.

Activation of Heterocyclic Amines by Various Species

The HCAs like most chemical carcinogens require metabolic activation before becoming mutagenic/carcinogenic. The HCAs are activated by the cytochrome P450s with conversion of an amino group to an hydroxylamino group. Although for most of the HCAs the highest rate of N-hydroxylation is observed with cytochrome P450 1A2 some activation is also catalyzed with cytochrome P450 1A1 and others (Kato and Yamazoe 1987, Kitada et al., 1991, Turesky et al., 1991, McKinnon and McManus, 1995). The mutagenic activation of the HCAs varies among the species but rodents, nonhuman primates and humans have the capacity to activate these compounds (Alldrick and Rowland 1985, Felton and Healy 1984, McManus et al., 1988, Davis et al., 1993). In a recent study of the comparative capacity of male rat, cynomolgus monkey and human microsomes to activate 6 nonimidazole HCAs and 4 HCAs of the imidazole type human hepatic microsomes showed the greatest capacity to activate all 4 of the imidazole type HCAs and 4 of the 6 nonimidazole HCAs. No difference was found among the three species in activating the other 2 nonimidazole HCAs (Davis et al., 1993). Also, cynomolgus monkeys had almost no capacity to activate the quinoxaline HCAs to mutagens as compared to rats and humans but were able to activate the quinoline, pyridoindole and pyridoimidazole compounds (Davis et al., 1993). HCAs have also been detected in the urine of healthy volunteers following ordinary diets or following restriction from meat or meat products for 24 hours prior to a test meal of cooked meat. Thus, an average of 2.1% of the dose of MeIQx ingested and 1.1% of the dose of PhIP ingested was excreted unchanged in the 8 hour urine following a fried beef meal (Davies et al., 1995). Comparable levels of these two HCAs and also measurable levels of Trp-P-1 and Trp-P-2 were found in the 24 hour urine of healthy volunteers but not from the urine samples of inpatients receiving only parental alimentation (Ushiyama et al., 1991). Therefore, some HCAs are well absorbed and extensively metabolized following ingestion by humans. Furthermore, following administration of furafylline, a potent inhibiter of human cytochrom P4501A2 there was a dramatic increase in urinary excretion of MeIQx (14.3 fold) and to a lesser extent in PhIP excretion (4.1 fold). Thus, *in vivo* metabolism of these two HCAs has also been demonstrated in humans (Davies et al., 1995).

Carcinogenicity of Heterocyclic Amines in Rodents and Nonhuman Primates

Of the twenty known HCAs ten have been synthesized in sufficient quantity for carcinogenesis studies in rodents and three are undergoing evaluation for carcinogenesis in nonhuman primates. Six of the ten are HCAs of the nonimidazole type and four are of the imizadole type. All ten were shown to be carcinogenic in rodents and this data is summarized in Table 3. In general, the target organ for the nonimidazole type HCA was the liver or blood vessels but two of the compounds, Glu-P-1 and Glu-P-2, also induced cancer of the small and large intestine in the rat. In contrast, the imidazole type HCA seem to have a wider spectrum of target organs inducing cancer of liver, skin, forestomach, oral cavity, Zymbal gland, mammary gland and small and large intestine.

In a collaborative study involving scientists at the National Cancer Institute in Bethesda, MD and the National Cancer Center Research Institute in Tokyo, Japan three HCAs were selected for evaluation of their carcinogenic potential in nonhuman primates (Adamson et al., 1990, Adamson et al., 1991). The HCAs were selected based on several factors including structure, mutagenic activity *in vitro*, concentration in cooked food, carcinogenic activity in rodents, and availability. Therefore, IQ, 8-MeIQx and PhIP were selected for studies in nonhuman primates, primarily cynomolgus monkeys. Compounds were administered by oral gavage 5 times a week at a dose of 10 or 20 mg/kg. Thus far, IQ has been shown to be carcinogenic in cynomolgus and rhesus monkeys inducing hepatocellular carcinoma in a high percentage of the animals. As shown in Table 4 at a dose of 20 mg/kg IQ induced tumors in 95% of the nonhuman primates with an average latent period of about 44 months. This is about one seventh of the life span of these animals. The tumors were well to moderately well-differentiated hepatocellular carcinoma and a trabecular pattern was generally seen. Pulmonary metastases were also found in several of the animals (Adamson et al., 1994, Thorgeirsson et al., 1995).

Epidemiological Studies

Several epidemiological studies have been performed that are relevant to the hypothesis that HCAs may play a role in the etiology of human cancer. None of the studies have directly measured the intake of total HCAs or any specific HCA but in two of the studies questions were asked about the cooking of the meat. Thus, as summarized in Table 5, the risk for colorectal or colon and rectal cancer was higher in three studies when the cooked meat was well done as compared to less well done (Schiffmann and Felton 1990, Gerhardsson de Verdier et al., 1991, Lang et al., 1994). Also, the risk of colon cancer in women who ate beef, pork or lamb as a main dish every day as compared to those who ate these meats less than once a month was 2.49, but the degree of doneness of the meat was not

Table 3 Carcinogenicity of nonimidazole and imidazole heterocyclic amines in rodents[a]

Heterocyclic amine	Species	% HCA in diet	Site(s) tumor occurred
Nonimidazole type			
Trp-P-1	Mouse	0.02	Liver
	Rat	0.015, 0.02	Liver
Trp-P-2	Mouse	0.02	Liver
	Rat	0.02	Liver
Glu-P-1	Mouse	0.05	Liver, blood vessels
	Rat	0.05	Liver, small and large intestine, Zymbal gland, clitoral gland
Glu-P-2	Mouse	0.05	Liver, blood vessels
	Rat	0.05	Liver, small and large intestine, Zymbal gland
AaC	Mouse	0.08	Liver, blood vessels
MeaC	Mouse	0.08	Liver, blood vessels
Imidazole type			
IQ	Mouse	0.03	Liver, lung, forestomach
	Rat	0.03	Liver, skin, small and large intestine, Zymbal gland, clitoral gland
MeIQ	Mouse	0.01, 0.04	Liver, forestomach
	Rat	0.03	Large intestine, oral cavity, mammary gland, Zymbal gland, skin
8-MeIQx	Mouse	0.06	Liver,lung, hematopoietic system
	Rat	0.04	Liver, skin, clitoral gland, Zymbal gland
PhIP	Mouse	0.04	Lymphoid tissue
	Rat	0.04	Colon, breast

[a]After Ohgaki et.al. 1991

documented (Willett et al., 1990). Several other case control studies in the literature also suggest that a diet high in fat and meat play a role in the etiology of colon cancer (Schottenfeld and Winawer 1982, Adamson 1990, Steineck et al., 1993). Furthermore, studies have also shown that humans who are phenotypically rapid metabolizers for both cytochrome P450 1A2 and the polymorphic acetyltranferase (NAT2) are at a significantly higher risk for the development of colon cancer and polyps than are controls. Also, PhIP-DNA adducts can be detected in human colon samples (Kadlubar et al., 1995). In 66 healthy nonsmokers it was demonstrated that cytochrome P450 1A2 activity was increased following consumption of pan-fried meat cooked at a high temperature (250o C) for seven days but not when the meat was cooked at a low temperature (100o C). The high temperature fried meat was found to contain MeIQx, DiMeIQx and PhIP. Thus, the HCAs are likely to induce their own activation in humans (Sinah et al., 1994). An investigation of cancer mortality was carried out in Japan involving more than 7500 adult subjects whose dietary habits included the intake of broiled fish. The results showed that the relative risks of cancer mortality associated with consumption of broiled fish twice or more weekly, compared to less frequent consumption were 1.3 for cancer at all sites and 1.7 for gastric cancer (Ikeda et al., 1983). A recent study demonstrated that the risk for breast cancer increased with a high intake of fried meat and that the relative risk was 1.8 for the highest tertile compared to the lowest (Knekt et al., 1994).

Table 4 Malignant liver tumors induced in nonhuman primates by Iq[a]

Dose of IQ	No. treated	No.with tumors (%)	Average total dose	Average latent period (months)
10mg/kg	20	11 (55%)	39.7 g	59.6
20mg/kg	20	19 (95%)	45.1 g	43.4

[a]After Adamson et.al. 1994

In addition to intake of HCAs in foods, humans may also be exposed to HCAs from fumes of cooking meat. Therefore, studies of cooks and food preparers occupationally exposed to HCAs in fumes warrants further study (Thieband et al., 1994, Rappaport et al., 1979, Berg et al., 1988, Lund and Borgan 1987). Future epidemiological studies need to be conducted to validate the hypothesis that HCAs play a role in the etiology of human cancer. Also, studies need to include measures of intake of HCAs, urinary excretion of HCAs, and ask questions on well doneness, cooking methods of the meat, amount of gravy consumed, ascertain fat intake, and also intake of greens whose chlorophyll content may modify the carcinogenic potential of the HCAs.

Table 5 Epidemiological studies of colorectal cancer and cooked meat

Cancer Site	Degree of Cooking	Odds ratio	Authors
Colorectal	rare to medium rare	1.0	Schiffman and Felton, 1990
	medium to medium well	0.9	
	well done	3.5	
Colorectal	well done	2.1	Lang et al., 1994
Colon	heavily browned	3.5	Gerhardsson de Verdier et al., 1991
	medium/lightly browned	1.0	
Rectum	heavily browned	7.0	Gerhardsson de Verdier et al., 1991
	medium/lightly browned	1.2	

Assessment of the Carcinogenic Risk of Heterocyclic Amines to Humans

Several authors have used published information on carcinogenic data in rodents or cynomolgus monkeys along with estimated human intake of the HCAs to calculate the risk of HCAs to humans. Generally the authors have also given the rationale and listed the assumptions underlying the calculated risk (Fennema and Hall 1990, Gaylor and Kadlubar 1991, Bogen 1994, Layton et al., 1995, Sugimara 1995, Adamson 1990). The risk assessments using a safety factor approach or an upper bound risk are summarized in Table 6. Listing seven assumptions, utilizing a safety factor approach, and deriving a TD_{50} value in rats for the imidazole type HCAs an apparent safety factor of 950 for human males and 1050 for females was estimated (Fennema and Hall, 1990) a risk of about 10^{-3}. Using published literature on tumor incidence data in rodents the maximum q1* of all tumor sites in both male and female mice and rats was estimated for 10 HCAs. Using human exposure levels from values of HCAs in various foods and assuming that 10% of the total human diet was cooked meat, upper limits of life time tumor risk from several HCAs were found to range from 1.0×10^{-4} to 2.2×10^{-8}. If several HCAs were present in the diet at 1ppb, or if one HCA was present at 10 ppb of the average total diet, the upper limit estimate was about 10^{-4} (Gaylor and Kadlubar, 1991). Another approach was used to reflect the fact that most of the HCAs induced tumors in multiple organs with each sex/species tested. Therefore, estimates of both single tumor type and estimates of aggregate potency were made using differing methods to scale the bioassay potency to humans. The potency estimates for an average lifetime cooked beef intake of 0.5 lb/day were an upper bound cancer risk of between 10^{-3} and 10^{-4} (Bogen 1994). To assess the magnitude of the cancer risk of HCAs to humans in the U.S. a recent and comprehensive study used the average intake of HCAs based on the concentration of HCAs in cooked foods from a dietary survey of the U.S. population and data from dose-response studies in animal bioassays The average consumption of HCAs was determined by analyzing 3 day dietary records of 3563 individuals. The dietary intakes of the five principal HCAs in descending order were PhIP > AaC > MeIQx > DiMeIQx > IQ whereas the carcinogenic potencies were IQ > DiMeIQx > MeIQx > PhIP > AaC. Based on body surface, an upper bound

Table 6 Risk assessment of heteocyclic amines in humans

Method	Estimated risk to humans or safety factor	Authors
Estimated intake in humans compared to potency in rat of imidazole HCAs	950 (M), 1050 (F)	Fennema and Hall, 1990
Linear multistage, assume several HCAs are present at 1ppb or one at 10ppb; use of rodent data	Upperbound risk 10-4	Gaylor and Kadlubar, 1991
Linear multistage on ten HCAs using rodent data and intake of approximately 0.5 lb/day cooked beef	Upperbound risk between 10^{-3} and 10^{-4}	Bogen, 1994
Use of consumption data on 3563 individuals for intake of five HCAs and cancer potencies from rodent data on body surface area basis	Risk 1.1×10^{-4}; upperbound risk 3.8×10^{-4}	Layton et al., 1995
Safety factor based on rat TD_{50} for PhIP and MeIQx and estimated human exposure	For PhIP 800, for MeIQx 4700	Sugimura, 1995
TD_{50} for IQ in cynomolgus, monkey, time to tumor one-tenth of life span	Between 10^{-3} to 10^{-4}	Adamson,1990; Adamson, personal communication

estimate of the cancer risk was 1.1 x 10^{-4} (Layton et al., 1995). The authors summarized factors which would increase the risk and factors that would decrease the risk. They estimated that the collective factors leading to increased risks are 3.5 while those decreasing the risk are 11. Thus, the conservative lower and upper bounds of risk were 1 x 10^{-5} to 3.8 x 10^{-4} (Layton et al., 1995). Using the calculated levels of human exposure to PhIP and MeIQx to be 0.1-13.8 and 0.2-2.6 mg/day/person, respectively, the safety factors for humans were calculated from the TD50 (mg/m2/day) value in rats. Therefore, the TD_{50} value of PhIP and MeIQx in the rats was 800 and 3,300 times the daily human exposure to these two HCAs (Sugimura 1995). However, if the risk is calculated on the basis of quantium toxicology the risk for PhIP would be 3.2 times and the risk for MeIQx 13.2 times the daily human exposure (Sugimura 1992, 1995). In assessing the carcinogenicity of IQ in cynomolgus monkeys and using the dose which produced hepatocellular carcinomas in one tenth of the lifetime of these animals, a risk of 10^{-3} to 10^{-4} for HCAs to humans was estimated (Adamson 1990, Adamson personal communication).

There are numerous factors which increase or decrease the above risk estimates. Among the modifying factors that would increase the risk are the following: additive or synergistic effects of more than one HCA; unidentified HCAs; HCAs for which consumption data is not available in the diet; greater activation of HCAs by humans than in animals; increased sensitivity of HCAs to neonates and children; under-predicting by the model used; under-reporting of food intake; and interaction with other factors (viruses, tumor promoters, agents causing cell proliferation, etc.). Indeed, it has been found that administration of five carcinogenic HCAs at one-fifth of the single compound carcinogenic doses resulted in a synergistic, rather than additive, effect on tumor incidence (Takayama et al., 1987). Also, both additive and synergistic effects were noted on the development of glutathione S-transferase placental form following the administration of five HCAs administered either in combination or as individual compounds at varying dose levels (Ito et al., 1995). Factors decreasing the risk estimates of the HCAs include: overpredicting risk by the model; greater detoxication of the proximate carcinogen by humans; inaccurate data on HCA consumption; use of the wrong scaling factor; and interaction with chlorophyll or other modifying compounds in the diet.

Methods to Minimize the Formation and/or Consumption of the Heterocyclic Amines

Although foods prepared by normal cooking practices contain mutagenic and carcinogenic HCAs, there are methods that can be taken to minimize the intake of HCAs (Adamson 1990, Felton 1990, Felton et al., 1994, Sugimura 1995). Among these are:

(1) vary the method of cooking
(2) eat beef medium or medium well instead of well done
(3) broil or pan fry meat less often
(4) microwave fish and poultry more often
(5) remove the blackened parts of charred food
(6) if barbecuing, microwave first, then discard the juice in the microwave dish before proceeding to barbecue
(7) wrap meat or fish in aluminum foil to prevent contact with an open flame
(8) poach fish more often, stew beef more often
(9) do not allow drippings from meat to become dry before making gravy
(10) use bacon grease sparingly for cooking

These ten practical and simple methods will minimize the formation of and/or decrease the intake of the mutagenic and carcinogenic HCAs.

Summary

Twenty different heterocyclic amines have been isolated and identified from cooked foods especially beef, fish, pork and fowl. Other HCAs have also been isolated but their structure remains to be elucidated and new HCAs are likely to be identified in the future. The HCAs are highly mutagenic and all ten HCAs that have been tested for carcinogenic activity, produce tumors in mice and rats. For humans the average daily intake of HCAs is in quantities of 10-20 mg/person/day. The HCAs are procarcinogens and are activated by the cytochrome P450 system especially CYP 1A2. Rodents, monkeys and humans have the capacity to activate HCAs. Studies using hepatic microsomes demonstrated that humans have a greater capacity to activate the majority of HCAs tested than rodents or cynomolgus monkeys. Three HCAs are currently under evaluation in nonhuman primates for carcinogenic activity and one, IQ, is highly carcinogenic inducing primary hepatocellular carcinomas in the majority of cynomolgus monkeys treated. Epidemiological studies, although not definitive, are supportive of an association of HCAs intake to the etiology of human cancer. Risk assessments from animal data show a risk of HCAs to humans in the range of 10^{-3} to 10^{-4} which is an order of magnitude greater than compounds currently regulated by the U.S. Food and Drug Administration or the Environmental Protection Agency. Taken together evidence from mutagenicity data, activation by various species including humans, carcinogenicity in animals, human consumption data, epidemiological studies and risk assessment, supports the conclusion that HCAs are probable human carcinogens.

References

Adamson RH (1990) Mutagens and carcinogens formed during cooking of foods and methods to minimize their formation. Cancer Prevention, November, 1-7.

Adamson RH (1995) Personal communication.

Adamson RH, Gustafsson J-A, Ito N, Nagao M, Sugimura T, Wakabayashi K and Yamazoe Y (1995) (eds.) Heterocyclic Amines in Cooked Foods: Possible Human Carcinogens. Princeton Scientific Publishing Co., Princeton, New Jersey, 1-306.

Adamson RH, Takayama S, Sugimura T and Thorgeirsson UP (1994) Induction of hepatocellular carcinoma in nonhuman primates by the food mutagen 2-amino-3-methylimidazo [4,5-f] quinoline. Environ. Health Perspect. 102, 190-193.

Adamson RH, Snyderwine EG, Thorgeirsson UP, Schut HAJ, Turesky RJ, Thorgeirsson SS, Takayama S and Sugimura T (1991) Metabolic processing and carcinogenicity of heterocyclic amines in nonhuman primates. In: Xenobiotics and Cancer. Ernster, L et al. (eds.) Japan Scientific Societies Press, Tokyo 289-301.

Adamson RH and Thorgeirsson UP (1995) Carcinogens in foods: heterocyclic amines and cancer and heart disease. In: Nutrition and Biotechnology in Heart Disease and Cancer. Longnecker JB et al. (eds.) Plenum Press, New York, 211-220.

Adamson RH, Thorgeirsson UP, Snyderwine EG, Thorgeirsson SS, Reeves J, Dalgard DW, Takayama S and Sugimura T (1990) Carcinogenicity of 2-amino-3-methylimidazo[4,5-f] quinoline in nonhuman primates: induction of tumors in three macaques. Jpn. J. Cancer Res. 81, 10-14,

Alldrick AJ and Rowland IR (1985) Activation of the food mutagens IQ and MeIQ by hepatic S9 fractions derived from various species. Mutat. Res. 144, 59-62.

Berg I, Overvik E, Nord C-E and Gustafsson (1988) Mutagenic activity in smoke formed during broiling of lean pork at 200, 250 and 300o C. Mutat. Res. 207, 199-204.

Bjeldanes LF, Morris MM, Timourian H and Hatch FT (1983) Effects of meat composition and cooking conditions on mutagenicity of fried ground beef. J. Agric. Food Chem. 31, 18-21.

Bogen KT (1994) Cancer potencies of heterocyclic amines found in cooked foods. Food Chem. Toxicol. 32, 505-515.

Davies DS, Gooderham NJ, Murray S, Lynch A, De La Torre R, Segura J and Boobis AR (1995) Systemic exposure to dietary heterocyclic amines in man. In: Heterocyclic Amines in Cooked Foods: Possible Human Carcinogens. Adamson, RH et al. (eds.) Princeton Scientific Publishing Co., Princeton, New Jersey, 190-196.

Davis CD, Adamson RH and Snyderwine EG (1993) Studies on the mutagenic activation of heterocyclic amines by cynomolgus monkeys, rats and human microsomes show that cynomolgus monkeys have a low capacity to N-oxidize the quinoxaline-type heterocyclic amines. Cancer Letters 73, 95-104.

Eisenbrand G and Tang W (1993) Food-borne heterocycle amines. Chemistry, formation, occurrance and biological activities. A literature review. Toxicology 84, 1-82.

Felton JS (1990) Heterocyclic amines in food; methods of analysis. Toxicology Forum, 1990 Annual Summer Meeting, 2-17.

Felton JS, Fultz E, Dolbeare FA, and Knize MG (1994) Effect of microwave pretreatment on heterocyclic aromatic-amine mutagens/carcinogens in fried beef patties. Food Chem. Tox. 32, No. 10, 897-903.

Felton JS and Healy SK (1984) Activation of mutagens in cooked ground beef by human-liver microsomes. Mutat. Res. 140, 61-65.

Felton JS and Knize MG (1990) Heterocyclic-amine mutagens/carcinogens in foods. In: Handbook of Experimental Pharmacology. Cooper, CS et al. (eds.) Springer-Verlag, Berlin, Vol 94/I, 471-502.

Fennema O and Hall R (1990) Public exposure to heterocyclic amines in food. Toxicology Forum, 1990 Annual Summer Mtg, 63-72.

Gaylor DW and Kadlubar FF (1990) Quantitative cancer risk assessment of heterocyclic amines in cooked foods. In: Mutagens in Food: Detection and Prevention. Hayatsu H (ed) CRC Press, BBoca Raton, Florida, 229-236.

Gerhardsson de Verdier M, Hagman U, Peters RK, Steineck G and Overvik E (1991) Meat cooking methods and colorectal cancer: A case- referent study in Stockholm. Int. J. Cancer 49, 520-525.

Ikeda M, Yoshimoto K, Yoshimura T, Kono S, Kato H and Kuratsune M (1983) A cohort study on the possible association between broiled fish intake and cancer. Gann 74, 640-648.

Ito N, Hasegawa R, Shirai T, Fukushima S, Hakoi K, Takaba K, Iwasaki S, Wakabayashi K, Nagao M, and Sugimura T (1991) Enhancement of GST-P positive liver cell foci development by combined treatment of rats with five heterocyclic amines at low doses. Carcinogenesis 12, No. 5, 767-772.

Kadlubar FF, Kaderlik RK, Mulder GJ, Lin D, Butler MA, Teitel CH, Minchin RF, Ilett KF, Friesen MD, Bartsh H, Nagao M, Esumi H, Sugimura T, and Lang NP (1995) Metabolic activation and DNA adduct detection of PhIP in dogs, rats, and humans in relation to urinary bladder and colon carcinogenesis. In: Heterocyclic Amines in Cooked Foods: Possible Human Carcinogens. Adamson, RH et al. (eds.) Princeton Scientific Publishing Co. Princeton NJ, 207-213.

Kato R and Yamazoe Y (1987) Metabolic activation and covalent binding to nucleic acids of carcinogenic heterocyclic amines from cooked foods and amino acid pyrolysates. Jpn. J. Cancer Res. 78, 297-311.

Kitada M, Taneda M, Itahashi T and Kamataki T (1991) Four forms of cytochrome P-450 in human fetal livers: purification and their capacity to activate promutagens. Jpn. J. Cancer Res. 82, 426-432.

Knekt P, Steineck G, Jarviven R, Hakulinen T and Aromaa A (1994) Intake of fried meat and risk of cancer: A follow-up study in Finland. Int. J. Cancer 59, 756-760.

Knize MG, Cunningham PL, Avila JR, Jones AL, Griffin EA Jr. and Felton JS (1994) Formation of mutagenic activity from amino acids heated at cooking temperatures. Food Chem. Toxic. 32, 55-60.

Knize MG, Shen NH and Felton JS (1988) The production of mutagens in foods. Proc. Air Pollution Control Ass. 88-130.3: 1-8.

Lang NP, Butler MA, Massengill J, Lawson M, Stotts RC, Hauer-Jensen M and Kadlubar FF (1994) Rapid metabolic phenotypes for acetyltransferase and cytochrome P4501A2 and putative exposure to food-borne heterocyclic amines increase the risk for colorectal cancer or polyps. Cancer Epid. Biomarkers and Prevention 3, 675-682.

Layton DW, Bogen KT, Knize MG, Hatch FT, Johnson VM and Felton JS (1995) Cancer risk of heterocyclic amines in cooked foods: an analysis and implications for research. Carcinogenesis 16, No. 1, 39-52.

Lund E and Borgan J-K (1987) Increased lung cancer mortality among Norwegian cooks. Scand. J. Work Environ. Health 13, 156.

McKinnon RA and McManus ME (1995) Function and localization of cytochromes P450 involved in the metabolic activation of food-derived heterocyclic amines. In: Heterocyclic Amines in Cooked Foods: Possible Human Carcinogens. Adamson, RH et al. (eds.) Princeton Scientific Publishing Co., Princeton, New Jersey, 145-153.

McManus ME, Burgess W, Stupans I, Trainor KJ, Renech M, Robson RA, Morley MA and Snyderwine EG (1988) Activation of the food-derived mutagen 2-amino-3-methylimidazo [4,5-f] quinoline by human liver microsomes. Mutat. Res. 204, 185-193.

Nakayasu M, Makasato F, Sakamato H, Terada M and Sugimura (1983) Mutagenic activity of heterocyclic amines in Chinese hamster lung cells with diphtheria toxin resistance as a marker. Mutat. Res. 118, 1147-1149.

Ohgaki H, Takayama S and Sugimura T (1991) Carcinogenicities of heterocyclic amines in cooked foods. Mutat. Res. 259, 399-410.

Pariza MW, Ashoor SH, Chu FS and Lund DB (1979) Effects of temperature and time on mutagen formation in pan-fried hamburger. Cancer Lett. 7, 63-69.

Rappaport SM, McCartney MC and Wei ET (1979) Volatilization of mutagens from beef during cooking. Cancer Letters 8, 139-145.

Schiffman MH and Felton JS (1990) Fried foods and the risk of colon cancer. Am. J. Epidemiol. 131, 376-378.

Schottenfeld D and Winawer SJ (1982) Large intestine In: Cancer Epidemiology and Prevention. Schottenfeld D et al. (eds.) WB Saunders, Philadelphia, 703-727.

Sinha R, Rothman N, Brown ED, Mark SD, Hoover RN, Caporaso NE, Levander OA, Knize MG, Lang NP and Kadlubar FF (1994) Pan-fried meat containing high levels of heterocyclic aromatic amines but low levels of polycyclic aromatic hydrocarbons induces cytochrome P4501A2 activity in humans. Cancer Res. 54, 6154-6159.

Skog K, Johansson M and Jägerstad M (1995). In: Heterocyclic Amines in Cooked Foods: Possible Human Carcinogens. Adamson RH et al. (eds.) Princeton Scientific Publishing Co., Princeton, New Jersey, 214-231.

Steineck G, Gerhardsson de Verdier M and Overvik E (1993) The epidemiological evidence concerning intake of mutagenic activity from the fried surface and the risk of cancer cannot justify preventive measures. Europe. J. Can. Prev. 2, 293-300.

Sugimura T (1992) Multistep carcinogenesis: a 1992 perspective. Science 258, 603-607.

Sugimura T (1995) History, present and future of heterocyclic amines, cooked food mutagens. In: Heterocyclic Amines in Cooked Foods: Possible Human Carcinogens. Adamson RH et al. (eds.) Princeton Scientific Publishing Co., Princeton, New Jersey, 214-231.

Sugimura T, Nagao M, Kawachi T, Honda M, Yahagi T, Seino Y, Sato S and Matsukura (1977) Mutagen-carcinogens in food, with special reference to highly mutagenic pyrolytic products in broiled foods. In: Origins of Human Cancer. Hiatt HH et al. (eds.) Cold Spring Harbor, New York, 1561-1577.

Sugimura T and Wakabayashi K (1990) Mutagens and carcinogens in food. In: Mutagens and Carcinogens in the Diet. Pariza MW et al. (eds.) Wiley-Liss, Inc., New York, 1-18.

Takayama S, Nakatsuru Y, and Sato S (1987) Carcinogenic effect of the simultaneous administration of five heterocyclic amines to F344 rats. Jpn. J. Cancer Res. 78, 1068-1072.

Thiebaud HP, Knize MG, Kuzmicky PA, Felton JS and Hsieh DP (1994) Mutagenicity and chemical analysis of fumes from cooking meat. J. Agric. Food Chem. 42, 1502-1510.

Thompson LH, Tucker JD, Stewart SA, Christensen ML, Salazar EP, Carrano AV and Felton JS (1987) Genotoxicity of compounds from cooked beef in repair-deficient CHO cells versus Salmonella mutagenicity. Mutagenesis 2, 483-487.

Thorgeirsson UP, Gomez DE, Nagy P, Sinha CC and Adamson RH (1995) Histopathology of IQ-induced hepatocarcinogenesis in nonhuman primates. In: Heterocyclic Amines in Cooked Foods: Possible Human Carcinogens. Adamson RH et al. (eds.) Princeton Scientific Publishing Co., Princeton, New Jersey, 274-280.

Tsuda M, Negishi C, Makino R, Sato S, Yamaizumi Z, Hirayama T and Sugimura T (1985) Uses of nitrite and hypochlorite treatments in determination of the contributions of IQ-type and non-IQ type heterocyclic amines to the mutagenicities in crude pyrolized materials. Mutat. Res. 147: 335-341.

Tsuda M, Takahashi Y, Nagao M, Hirayama T and Sugimura T (1980) Inactivation of mutagens from pyrolysates of tryptophan and glutamic acid by nitrite in acidic solution. Mutat. Res. 78, 331-339.

Turesky RJ, Lang NP, Butler MA, Teitel CH and Kadlubar FF (1991) Metabolic activation of carcinogenic heterocyclic aromatic amines by human liver and colon. Carcinogenesis 12, 1839-1845.

Ushiyama H, Wakabayashi K, Hirose M, Itoh H, Sugimura T and Nagao M (1991) Presence of carcinogenic heterocyclic amines in urine of healthy volunteers eating normal diet, but not of inpatients receiving parenteral alimentation. Carcinogenesis 12, 1417-1422.

Willett WC, Stampfer MJ, Colditz GA, Rosner BA and Speizer FE (1990) Relation of meat, fat and fiber intake to the risk of colon cancer in a prospective study among women. N. Engl. J. Med. 323, 1664-1672.

Clinical Toxicology of Peptides and Proteins

(Chairs: C. Bismuth, France, and D. Pelcová, Czech Republic)

Antibodies Proposed as Therapeutic Agents

Ch. Bismuth[1], F.J. Baud[1], S. W. Borron[1] and J.M. Scherrmann[1,2]
[1]Hôpital Fernand Widal - 200 rue du Faubourg Saint-Denis 75475 Paris Cedex 10 - France Université Paris 7
[2] INSERM U26 Hôpital Fernand Widal - 200 rue du Faubourg Saint-Denis 75475 Paris Cedex 10 - France Université Paris 7

History

A first approach : the Antivenoms (Dart 1994)

The history of antibody development is one of a breakthrough discovery, followed by repeated refinements. The discovery occurred in 1887 when Sewall reported that pigeons could be immunized with preventive inoculation of rattlesnake venom. It took another 100 years, however, for Sewall's discovery to evolve into a safe and effective treatment for humans. Curiously, Sewall did not extend his work into the realm of therapeutics. Fortunately, Calmette reported in 1894 that an animal could be protected against envenomation with the serum of another animal which had been immunized to the venom. The concept of serotherapy was thus born and the rush to produce new serotherapies was on. These pioneering investigators did not realize that their therapy involved a blood component we now term the antibody, nor that the foundation of their treatment was based on a highly specific interaction called antigen-antibody binding. Indeed, Calmette initially claimed that his cobra antiserum would cure the bite of all snakes ! But every new discovery produces misconceptions, which are corrected as further research emerges. In the case of serotherapy, Brazil (1901) soon demonstrated that Calmette's antiserum did not protect against Crotalus durissus terrificus. In 1900, Erlich proposed the concept of antibodies. Although his depiction was somewhat inaccurate, he made it possible to conceive how an antiserum could protect the recipient. His theoretical construct was needed as a basis for further research.

But until the 1960's, immunoglobulins and purified antibodies were essentially used for research and diagnostic purposes.

A specific story : anti-digitalis immunotherapy

The first documented therapeutic human application occurred in the 1970's and concerned digitalis intoxication. The scientific approach, step by step, has been indeed exemplary :

- **1967:** Coupling of the digoxin hapten to serum albumin resulting in an immunogenic conjugate and allowing the production of anti-digitalis specific antibodies (rabbit).
Two perspectives: analytical and therapeutic. (Butler and Chen)
- **1969:** Radioimmunologic quantification of therapeutic and toxic concentrations of digoxin. (Smith et al.)
- **1970:** 1) Radioimmunologic quantification of therapeutic and toxic concentrations of digitoxin. (Smith), 2) Isolation and characterization of the antibodies and control of their specificity. (Smith et al.), Elimination of cross reactions with similar structures (steroids)
- **1971:** First therapeutic steps:

1) Prevention by active immunization in the rabbit. (Schmidt and Butler 1971a)
2) Reversibility of intoxication by administration of antibodies in the dog. (Schmidt and Butler 1971b) (Curd et al.)

- **1973:** Separation and purification of ovine specific antibody Fab fragments (Butler et al.) Smaller load of foreign proteins with their attendant immunologic risks: anaphylaxis and serum sickness. Rapid elimination of Fab-digoxin complex (MW of Fab = 50,000).
- **1976:** Grave human digoxin intoxication treated with Fab fragments. (Smith et al.)
- **1977:** Efficacy of Fab fragments for digitoxin demonstrated in the dog. (Ochs and Smith)
- **1979:** Efficacy of Fab fragments for lanatoside demonstrated. (Hess et al.)
- **1980:** Efficacy of Fab fragments for digitoxin demonstrated in man. (Aeberhard et al.)

Requisites for the Successful Development of Immunotoxicotherapy (ITT)

Immuno-neutralization of the effects of toxins is potentially one of the most powerful detoxification procedures currently available (Scherrmann 1994).

It ideally is able to simultaneously sequester, extract or redistribute and eliminate the toxin by using specific active binding sites (SABS) derived from different antibody molecular entities.(Table 1). Some requisites depend on the specific toxin. Others depend on the SABS selection.

Requisites depending on the toxin

ITT concerns toxins :

- With a high risk of mortality and short-term effects (digoxin, tricyclic antidepressants, colchicine, paraquat, snakes or scorpion venoms) or with long-term effects and risk of tissue accumulation (environmental contaminants such as

hexachlorobiphenyl compounds).
- Which are capable of producing an antibody response in the host. (They appear extremely variable in nature (chemicals, glycoproteins) and molecular weight (haptens to macromolecules).
- Which are available for the antibody to bind : The extent of toxin distribution can vary from several hundred liters for low molecular weight toxins (Digoxin, Colchicine, Paraquat) to the volume of extra-cellular water for high molecular weight toxins (venoms).
- Their effects must be reversible (functional intoxicant). There is no evidence that therapeutic antibodies can reverse lesional organ injury. (This is probably the main limiting factor explaining their inactivity towards Paraquat). On the other hand, antibodies can neutralize circulating and/or residual toxin before its action, even if a great part of the toxin has already damaged the organism, thus transforming a lethal intoxication into a disease compatible with survival.
- Some toxins are dangerous at the scale of milligrams (digoxin, colchicine). Their stoichiometric neutralization with antibodies has proven to be effective. Other severe intoxications have toxic ranges 10 to 100 times higher (such a tricyclic antidepressants or chloroquine). The administration of large doses of drug-specific antibodies or their fragments in such cases has no clinical or experimental basis and would evoke ethical and economic reservations. In fact, partial neutralization may prove useful, as we will see.

Table 1. Toxins and their antibodies

Toxin	Antibody	Molecular Weight
Gram-negative bacterial lipopolysaccharide	IgM	600 000
Tetanus	IgG	180 000
Viper or scorpion venoms	Fab2,Fab'2	100 000
Digoxin, colchicine	Fab, Fab'	50 000
Digoxin	Fv	22 000

Requisites depending on the antibody

1 - The choice of the size of the antibody : IgM - IgG - Fab 2 - Fab depends on the distribution of the toxins :
For toxins with vascular distribution = IgM - IgG
For toxins with extra-vascular distribution = fragmentation
2 - To form very stable toxin antibody complexes, we need to work with antibody binding sites of high affinity for the toxin. For ex : the success for Digoxin or

Colchicine is mainly related to an affinity more than 109 M-1. This value seems to be the minimal limit to develop efficient antibody for immunotherapy.
3 - The withdrawal of Fc fragments from the IgG increases the safety of antibodies, and limits the risks of hypersensitivity reactions and of serum sickness.

Clinical Applications

Fragmentation : progress in the safety of antivenoms

The present tendency is to prepare polyclonal antibodies of ovine origin (Karlson-Stiber et al. 1994b) (Egen et al. 1994) (Dart 1994) rather than equine origin (Karlson-Stiber et al. 1994a), as was initially the case. This appears to be a pure problem of industrial manufacturing capacity.

The fragmentation of antibodies which is so useful for haptens (< 1000 daltons), clearly introduces a degree of progress in the safety of antivenoms. However in light of the high molecular weight of most venom toxins, which can give a toxin-Fab complex of more than 60 000 daltons (not capable of renal glomerular filtration), the practical usefulness of this procedure for anti-venoms (equine) Fab2 and (ovine) Fab must be challenged. Furthermore, the risk of high fragmentation is the reduction of the stability of the toxin-antibody complex with the risk of the release of the toxin.

A standard therapy : Antidigitalis ITT

Production[1*]: Because the digoxin molecule (780 daltons) is too small to be immunogenic, sheep are first immunized with a digoxin-serum albumin conjugate. Second, antisera are collected from the animals having high titers of antibody with high affinity and specificity for digoxin. The digoxin antibodies (IgG) are separated and highly purified. After digestion by papain, one IgG is cleaved into two Fab and one Fc fragments. Fab fragments with specificity for digoxin are then separated by affinity chromatography. Fab fragments are threefold more effective than whole IgG (150 000 daltons). Their small molecular mass (5000 daltons), provides a more rapid onset of action than intact IgG due to enhanced volume of distribution and allows rapid renal excretion of the Fab-digitalis complex. Because the antigenic determinants and complement binding sites are eliminated, the immunogenicity of Fab fragments is decreased, compared to intact IgG, diminishing the risk of immunogenic reaction.

[1*] A vial of Digidot (Boehringer Mannheim) contains 80 mg of Fab, neutralizes 1 mg of either digitoxin or digoxin and costs approximately $ 1000 US. Digidot is more expensive than one vial of Digibind (Wellcome) which contains 40 mg of Fab and costs $ 277 US.

Mechanism of action : Fab fragments have a dual action, toxicodynamic and toxicokinetic (Taboulet et al. 1993). When Fab fragments are administered intravenously, the active site of the antibody binds to a circulating digitalis molecule, either digoxin or digitoxin, and generates a Fab-digitalis complex, unable to bind to receptors. Although total digoxin levels may increase 5 to 20 times, free serum digitalis drops to zero within a few minutes. By mass action effect, the free intracellular digitalis and the digitalis bound to receptors are displaced and the membrane ATPases are reactivated. Intravascular concentrations of Fab-bound digitalis rise substantially but these complexes are inactive.

In current practice, immunotherapy reverses the clinical effects of both digoxin and digitoxin toxicity within 60 min. This toxicodynamic action of Fab framents is supplemented by a toxicokinetic action. With normal renal function the Fab-digitalis complex excretion half-life is 10 to 20 h compared to spontaneous half life of 160 h for digitoxin and 39 h for digoxin.

Adverse effects with Fab fragments are rare. To date, in the United States more than 1000 patients with potentially life-threatening digitalis toxicity have been treated with Fab fragments (Antman et al. 1990, Hickey et al. 1991, Smith 1988).

1 - Hypersensitivity reactions (adverse allergic responses such as pruritic rash) were reported in 0.8% of 717 adults during a surveillance study.

2 - The possibility exists that Fab fragments could lead to exacerbation of previous heart failure, however such an event is difficult to prove. In case of cardiac failure, after full neutralization of the body load of digitalis, an inotropic drug may be helpful.

3 - Recurrence of toxic signs may occur some time after immunotherapy, corresponding to the free digitalis rebound (spontaneous redigitalization). The recrudescent toxicity is often heralded by nausea and vomiting and may lead, once more, to severe arrhythmia. This risk exists in persons receiving < 50 % of the estimated adequate dose.

Because full neutralization with Fab fragments is expensive and sometimes not available, Fab administration is sometimes delayed until serious arrhythmia occurs. Under these emergency conditions, ventricular fibrillation and asystole too often lead to a fatal outcome and the benefit of immunotherapy is greatly reduced. Thus a "prophylactic" attitude should be adopted and the risk of arrhythmia countered by early administration of a half equimolar quantity of Fab fragments (Taboulet et al. 1993)

1 - The immediately life-threatening disturbances : ventricular arrhythmias such as ventricular tachycardia or ventricular fibrillation and bradyarrhythmias including sinus bradycardia under 40 bpm refractory to 1 mg of atropine, second or third-degree atrioventricular blocks with slow ventricular rate and ventricular asystole ; hyperkalemia over 5 mEq/l, mesenteric infarct or cardiogenic shock. In these patients, a curative dose of Fab (equimolar neutralization) must be rapidly administered.

2 - The potentially lethal intoxications : mild bradycardia under 60 bpm, regardless of the conduction disturbances, especially in patients with associated factors of poor prognosis. In these patients, when Fab fragment administration is delayed, fatal arrhythmias often occur suddenly, sometimes despite pacing. In this

population, when atropine fails to accelerate the cardiac rhythm to over 60 bpm, in adults, we rapidly administer a half equimolar neutralizing dose of Fab ('prophylactic'dose). If heart rate remains under 60 bpm, we administer the other half of the dose. In our experience, a full dose of Fab is not always necessary. In children, because of their low weight, equimolar neutralization is always chosen. It must be stressed that the clinical value of a half dose of Fab in this situation has not yet been assessed. If the clinician decides to treat atropine-resistant, isolated bradycardia with half the recommended dose of Fab, we emphasize the need for close and prolonged monitoring of patients (at least 72 h).

Skin tests are not needed to detect hypersensitivity reactions.

Fab fragment dosages: These are calculated from the body load of glycoside which is estimated either from the ingested amount or from serum concentration. Since the digoxin bioavailability factor (absorption of digoxin ranges from 20 to 80 %) and the delays between overdose and admission are variable, calculation of the body load of digoxin from the ingested dose can be overestimated. We suggest, if there is no emergency, using the digitalis blood level for calculation of Fab dosage (Table 2).

Table 2. Calculation of the Body Load of Glycoside (From Taboulet et al 1993)

From the ingested amount	
	= ingested amount (mg) x digoxin bioavailability (60 %) or digitoxin bioavailability (100 %)
From the serum glycoside concentration	
	= serum glycoside concentration (ng/mL) x volume of distribution (VD) x weight (kg) with VD = 5.6 L/kg for digoxin and 0.56 L/kg for digitoxin

For equimolar neutralization, the full dose of Fab is administered intravenously over 30 min or by direct intravenous injection if cardiac arrest is imminent. For "prophylactic" neutralization, a slower rate of administration may be appropriate, but this remains to be determined. In cases of severe adult overdosage, when no data are available for the patient the manufacturer recommends a dose of 800 mg of Digibind (which neutralizes 10 mg of digitalis). In our experience and that of others, 800 mg of Fab does not provide full neutralization of all intentional digitalis intoxications, since the ingested amount may be extreme. Thus, if initial administration of 800 mg of Fab fails to reduce digitoxin-induced arrhythmias, another dose of 800 mg should be tried. In contrast, for chronic overdosage a mean dose of 80-160 mg of Fab is generally adequate.

A disappointed hope : The anti-endotoxin monoclonal antibodies.

The desire to use antibodies in the treatment of sepsis occurred because of the poor outcome of patients presenting with gram-negative sepsis treated with current antimicrobial agents. Morbidity and mortality may be diminished, in theory, by interrupting the inflammation cascade initiated in sepsis. When a host becomes infected with a microorganism, a complex set of responses is initiated in an attempt to curtail and eradicate the pathogen. Once initiated, this inflammatory cascade has the potential to result in host tissue injury and even death. These toxic manifestations from gram-negative bacteria appear to be triggered by lipopolysaccharide (LPS), a component of the outer cell membrane of the bacteria. The lipopolysaccharide (the endotoxin) consists of three regions : the highly variable O polysaccharide chain, which is connected to the R core region, which in turn is linked to lipid A. Lipid A, the toxic part of the endotoxin, is conserved through many bacterial species. Because of this, monoclonal antibodies directed against lipid A theoretically interrupt the inflammatory cascade, in gram-negative infections.

Two monoclonal antibodies directed against bacterial endotoxin have been developed for adjunctive treatment of gram-negative sepsis (Warren et al. 1992; Fink 1993). E5 is a murine immunoglobulin M (IgM) monoclonal antibody, raised in mice immunized with Escherichia coli J5, that binds to an epitope on lipid A. HA-1A is a human IgM monoclonal antibody that binds to lipid A. It was derived from a heterohybridoma created from the spleen cells of patients who had been vaccinated with Escherichia coli J5 before splenotomy. Both E5 and HA-1A have been tested in placebo-controlled clinical trials, but neither has consistently prevented death in patients with sepsis, and therefore, have not been approved by the Food and Drug Administration (FDA). It is hoped that within the next decade a monoclonal antibody therapy will be available to treat patients presenting with gram-negative sepsis.

A new advance : Anticolchicine antibodies

Animal experiments : Until recently there were no successful specific measures in colchicine intoxication. Experimental work suggested in the last few years that specific therapy might be possible, although intracellular binding of colchicine was expected to limit the efficacy of immunotoxicotherapy for colchicine poisoning, a reversible effect was found *in vitro* (Wolf et al. 1980). A preliminary work in mice, found by one of us to be a suitable animal model for the study of colchicine poisoning, showed that administration of specific anticolchicine immunoglobulin G (IgG), even after the distribution phase, significantly decreased the mortality rate (Terrien et al. 1988, Terrien et al. 1990). In further studies by the same group (Terrien et al. 1989) IgG colchicine antibodies were produced in goats and given to mice who received a LD 100 of colchicine intraperitoneally. Despite a relatively low dose of antibody, a beneficial effect was shown and colchicine pharmacokinetic parameters were altered favorably. The steady-state volume of distribution decreased significantly in the treated mice. There was sequestration of free colchicine in the intravascular space (low free plasma toxin) and a decrease in colchicine concentration in most tissues of antidote-treated mice, indicating a tissue extraction process.

The practicality of immunotoxicotherapy for colchicine poisoning also was examined using an active immunization approach (Scherrmann et al. 1989b). A dose of 3 mg/kg of colchicine (LD 100) was given to rabbits previously immunized. The protective effects of immunization depended on the anticolchicine antibody titer.

Promising preliminary results using immunotoxicotherapy were also reported by Edmond-Rouan et al (1990). They prepared a specific high-affinity colchicine-binding monoclonal antibody and investigated its ability to reverse *in vitro* the effects of colchicine on ovary cells of the Chinese hamster. Colchicine-induced changes of polyploidy and chromosomal aberrations were reversible even when the antibody was administered up to 6 hours after colchicine exposure.

Clinical experience : The human use of colchicine is limited by its toxicity and colchicine overdose is associated with a high mortality rate (90 % above 0.9 mg/k). Patients with early hemodynamic collapse due to colchicine overdose have particularly poor prognosis, and there has been no effective treatment for this complication of severe colchicine intoxication. We report here the first clinical use of ITT in colchicine poisoning (Baud et al 1995).

Toxicodynamics: A 25 y. old female was admitted to Bordeaux Hospital 24 h after the ingestion of colchicine 1 mg/kg. She presented with vomiting, diarrhea, hypotension and a severe DIC. She was rehydrated and transported under Dobutamine to our hospital at H 36. She was conscious, in cardiac collapse with oliguria and ARDS. The Factor V level was 5 % with important bleeding at the vascular punctures. Hemodynamic monitoring showed a cardiogenic shock refractory to dobutamine. The prognostic features implied a 99 % chance of death in less than 24 H. After her informed consent, she received a 1/8 molar neutralization of the ingested dose with goat colchicine specific antibodies at H 40, representing about a 0.4 molar ratio of the body load of the colchicine present at this time. During the perfusion, the shock became reactive to inotropic drugs. The withdrawal of cardiotropic agents was possible 36 h after admission. ARDS took longer to subside.

On D3, a bone marrow aplasia was observed, concomittant with a bacteremia, responsive to antibiotics. The neutropenia lasted for 2 days. Eventually, there was a total hair loss. Transient sensitive polyneuritis of upper members was observed. The local tolerance to the immunotherapy was good. No serum sickness was observed. Finally, the patient survived, without avoiding the well-known sequential phases of this poisoning, but with a shortening of their duration.

Toxicokinetics: 40 hrs after ingestion of 60 mg colchicine, the patient had received 6.4 g of high affinity polyclonal Fab fragments specific to colchicine via the right ventricular lumen of a Swan Ganz catheter over a 7 h period. Fab fragment, free and protein-bound colchicine concentrations were assayed in plasma and urine from admission of the patient until day 13 by specific radioimmunoassays. On the patient's admission 26.5 hrs after colchicine ingestion, plasma concentration was 24 ng/ml. Fab fragment administration resulted in a 6-fold increase of total colchicine concentrations 10 min after starting Fab infusion. Free colchicine concentration fell to undetectable levels over a 7 h period. A 4.4-fold decrease of colchicine distribution volume (2.9 l/kg) under Fab treatment indicated that substantial amounts of colchicine were being removed from peripheral sites and redistributed into the extracellular space corresponding to the distribution volume of Fab fragments. Fab infusion resulted in a 6-fold increase of urinary excretion rate (24 and 134 μg/h respectively before and after Fab administration). Colchicine was excreted by the renal route bound at 98 % to Fab fragments over a 16 h period. The amount of Fab-bound colchicine renally excreted was equal to the binding capacity of the Fab fragment amount recovered in urine. So, the neutralization yield of Fab fragments was near 100 %. The amount of neutralized colchicine was estimated at 3.7 mg among 9 mg of colchicine present in the body prior to Fab dosing.

Anti-tricyclic antidepressants immunotherapy

We have presented evidence that drug-specific antibody fragments are useful clinically to treat snakebites, digoxin or colchicine overdose. Because tricyclic antidepressants are the most common cause of death from intentional drug overdose in the United States, (Litovitz et al. 1990) ITT has been proposed. The lethal dose of tricyclic antidepressants (TCAs) is 10 to 100 times greater than the drugs studied before, however, and will presumably require correspondingly higher doses of antibodies (up to several g/kg) to reverse toxicity. Preliminary studies with TCA-specific antibodies suggest that the administration of such doses is feasible.

High affinity TCA-specific monoclonal Fab'or sheep polyclonal Fab fragments rapidly reverse the cardiovascular toxicity of the TCA desipramine (DMI) in rats, and prolong survival. Therapeutic effect occurs within minutes and is evident with relatively low Fab or Fab' doses (10 to 30 % of the molar DMI dose), suggesting that these antibody fragments preferentially redistribute DMI out of a rapidly equilibrating compartment (Pentel et al. 1994)

TCA-specific Fab' or Fab is generally well tolerated in rats, but doses several times higher than those anticipated for human use may lead to cardiovascular deterioration or death. 1/8 of the rats receiving the larger doses of TCA specific Fab died (2 gr/Kg representing a molar ratio of TFab to DMI of 0,22). This death may relate to the rapid rate of infusion of relatively large amounts of Fab. Strategies for further minimizing the required dose are desirable. Combining TCA-specific Fab with sodium bicarbonate, the standard clinical treatment for TCA overdose, is more effective than either treatment alone and thus represents one such strategy (Brunn et al. 1992). Also possible is the development of smaller antibody fragments that may have improved therapeutic properties. A recombinant single chain Fv fragment (sFv), one half the size of Fab, has been cloned which retains a high affinity for DMI and is able to alter DMI distribution *in vivo*. Because sFv has a shorter elimination half-life and more extensive renal excretion than Fab, it may have therapeutic advantages (Pentel 1995). Moreover, this fragment can potentially be engineered to improve its therapeutic properties.

These data suggest that the use of drug-specific antibody fragments represents a viable general strategy for the treatment of human tricyclic antidepressant overdose.

Future trends

Future trends will probably depend on immunologic progress : monoclonal, human-human antibodies, and semi-synthetic antibodies, as well as further antibody fragmentation (Scherrmann et al. 1989 a).

Conclusion

Immunotoxicotherapy (ITT) has proven its efficacy in man :
- in snake envenomation for a century
- in Digitalis and Colchicine poisoning in the last 20 years
- a preclinical study for tricyclic antidepressant poisoning is under way.

ITT efficacy does not suppress the necessity to apply aggressive symptomatic treatment in life-threatening intoxications.

ITT toxicity appears rare and minor : it remains due to active binding site heterogeneity and immunogenicity (leading to hypersensitivity and serum sickness). The fragmentation of antibodies has made them safer.

The limitations of ITT are its high cost and its restriction, for the moment, to drugs having low toxic doses (in the range of a few mg).

References

Aeberhard P, Butler VP, Smith TW (1980) Traitement d'une intoxication digitalique massive (20 mg de digitoxine) par les anticorps antidigoxine fractionnés (Fab). Arch Mal Coeur 12:1471-1478

Antman EM, Wenger TL, Butler VP, Haber E, Smith TW (1990) Treatment of 150 cases of life-threatening digitalis intoxication with digoxin-specific Fab antibody fragments. Ciruclation 81:1744-1752

Baud FJ, Sabouraud A, Vicaut E, Taboulet P, Lang J, Bismuth Ch, Rouzioux JM, Scherrmann JM (1995) Treatment of severe colchicine overdose with colchicine-specific Fab fragments. N Engl J Med 332 10:642-645

Brazil V (1901) Contribucao ao estudo de venemo phidico. III. Tratamentos das mordeduras das cobras. Rev Med Sao Paulo 4:375-380

Brunn GJ, Keyler DE, Pond SM, Pentel PR (1992) Reversal of desipramine toxicity in rats using drug-specific antibody Fab'Fragment : effects on hypotension and interaction with sodium bicarbonate. J Pharmacol Exp Ther 260 3:1392-1399
Butler VP Jr, Chen JP (1967) Digoxin specific antibodies. Proc Nat Acad Sci USA 57:71-78
Butler VP Jr, Schmidt DH, Smith TW (1973) Effect of Fab fragments of sheep antidigoxin antibody (Dig Ab) on serum concentrations and urinary excretion of digoxin (dig) Circulation:Suppl 4:139
Calmette A (1908) Venoms, venomous animals and serotherapy. Paris, Masson
Curd J, Smith TW, Jaton JC et al (1971) The isolation of digoxin specific antibodies and its use in reversing the effects of digoxin. Proc Nat Acad Sci, USA 68:2401-2406
Dart RC (1994) Antibodies as therapeutic agents : the antivenoms. The 5th World Congress of the World Federation of Associations of Clinical Toxicology Centers & Poison Control Centers. November 8-11 Taipei, Taiwan
Edmond-Rouan SK, Otterness IG, Cunningham AC et al (1990) Reversal of colchicine-induced mitotic arrest in Chinese hamster cells with a colchicine-specific monoclonal antibody. Am J Pathol 137:779-787
Egen N, Russell FE, Consroe P, Gerrish K (1994) A new ovine antivenom for north - american venomous snakes. Vet Human Toxicol 36:362
Erlich P (1900) On immunity with special reference to cell life. Proc Royal Soc London 66:424-448
Fink MP (1993) Adoptive immunotherapy of gram-negative sepsis : use of monoclonal antibodies to lipopolysaccharide. Crit Care Med 21(suppl 1):32-39
Hess T, Stucki P, Baraudun J, Sholttsik G, Roesen W (1979) Treatment of a case of lanatoside intoxication with digoxin specific Fab antibody fragments. Am Heart J 98:767-771
Hickey AR, Wenger TL, Carpenter VP et al. (1991) Digoxin immune Fab therapy in the management of digitalis intoxiation : Safety and efficacy results of an observational surveillance study. J Am Coll Cardio 17:590-598
Karlson-Stiber C, H. Persson (1994a) Antivenom treatment in Vipera berus envenoming. Report of 30 cases. Journal of Internal Medicine 235:57-61
Karlson-Stiber C, Persson H, Heath A, Smith D, Al Abdullah I (1994 b) Specific sheep Fab fragments in the treatment of vipera berus envenoming. Report of a multicentre study. The 5th World Congress of the World Federation of Associations of Clinical Toxicology Centers & Poison Control Centers. November 8-11 Taipei, Taiwan
Litovitz T, Schmitz BF, Bailey KM (1990) 1989 annual report of the American Association of Poison Control Centers National Data Collection System. Am J Emerg Med 8:394-442
Ochs HR, Smith TW (1997) Reversal of advanced digitoxin toxicity and modification of pharmacokinetics by specific antibodies and Fab fragments. J Clin Invest 60:1303-1313
Pentel PR (1995) Immunotherapy for tricyclic antidepressant overdose : human implications. Congress International Union of Toxicology - Horizion in Toxicology : preparing for the 21st Century - Seattle 2 - 6 July 1995
Pentel PR, Ross CA, Landon J, Sidki A, Shelver WL, Keyler DE (1993) Reversal of desipramine toxicity in rats with polyclonal drug-specific antibody Fab fragments. J Lab Clin Med 123 3:387-393
Scherrmann JM (1994) Antibody treatment of toxin poisoning. Recent advances. Clinical Toxicology 32(4):363-375
Scherrmann JM, Terrien N, Urtizberea M, Pierson P, Denis H, Bourre JM (1989a) Immunotoxicotherapy : present status and future trends. Clinical Toxicology 27(1&2): 1-35

Scherrmann JM, Urtizberea M, Pierson P et al (1989b) The effect of colchicine-specific active immunization on colchicine toxicity and disposition in the rabbit. Toxicology 59:213-222

Schmidt DH, Butler VP Jr (1971a) Immunological protection against digoxin toxicity. J Clin Invest 50:866-871

Schmidt DH, Butler VP Jr (1971b) Reversal of digoxin toxicity with specific antibodies. J Clin Invest 50:1738-1744

Sewall H (1887) Experiments on the preventive inoculation of rattlesnake venom. J Physiol Lond 1887 8:203-210

Smith TW (1970) Radio immuno assay for serum digitoxin concentrations. Methodology and clinical experience. J Pharmacol Exp Ther 175:352-360

Smith TW (1988) Digitalis. Mechanisms of action and clinical use. N Engl J Med 6:358-365

Smith TW, Butler VP Jr, Haber E (1969) Determination of therapeutic and toxic serum digoxin concentration by radio immuno assay. N Engl J Med 281:1212-1216

Smith TW, Butler VP Jr, Haber E (1970) Characterization of antibodies of high affinity and specificity for the digitalis glycoside digoxin. Biochemistry 9:331-337

Smith TW, Haber E, Yeatmen L, Butler VP Jr (1976) Reversal of advanced digoxin intoxication with fab fragments of digoxin specific antibodies. N Engl J Med 294:797-800

Taboulet P, Baud FJ, Bismuth Ch (1993) Clinical features and management of digitalis poisoning. Rationale for immunotherapy. Clinical Toxicology 31(2):247-260

Terrien N, Urtizberea M, Scherrmann JM (1988) Immuno-toxicotherapy for colchicine poisoning : Toxicological and pharmacological features of an experimental model simulating human poisoning. Arch Toxicol Suppl 12:322-325

Terrien N, Urtizberea M, Scherrmann JM (1989) Influence of goat colchicine-specific antibodies on murine colchicine disposition. Toxicology 59:11-22

Terrien N, Urtizberea M, Scherrmann JM (1990) Reversal of advanced colchicine toxicity in mice with goat colchicine-specific antibodies. Toxicol Appl Pharmacol 104:504-510

Warren HS, Danner RL, Munford RS. (1992) Anti-endotoxin monoclonal antibodies. N Engl J Med 326:1,153-1,157

Wolf J, Capraro HG, Brossi A et al (1980) Colchicine binding to antibodies. J Biol Chem 255:7144-7149

Unanticipated Human Toxicology of Recombinant Proteins

Bernhard Ryffel
Institute of Toxicology, University Zurich Fax 0041 61 302 13 29

Abstract

Recombinant human proteins play already an important role in therapy, e.g. erythropoietin and colony stimulating factors, while several promising candidates such IL-6, IL-12, thrombopoietin and others are in clinical development. Since the recombinant proteins are copies of endogenous proteins, it was assumed that they would be well tolerated. While this assumption is correct for some, other proteins proved to be a highly toxic. Therefore, preclinical safety assessment of these proteins is necessary. Based on the experience with several proteins, some guidance for the safety assessment can be given. Furthermore, data are presented demonstrating that preclinical toxicity studies have a predictive value for man. Limitations of the classical approach of safety tests and new concepts are discussed.

Introduction

With the recent discovery and large scale production of potent regulatory proteins/cytokines, novel therapeutic strategies emerged for certain diseases. Cytokines with regulatory functions in haematopoiesis, immune response, inflammation and tissue repair (Aggarwal, 1992, 1995; Dexter, 1990; Arai, 1990; Moore, 1991; Gianella-Borradori, 1992) are presently in the pipeline of the pharmaceutical development and a few have already reached the market (erythropoietin, G-CSF). The proteins obtained by recombinant technologies are quite faithful copies of the endogenous protein. Therefore, it was assumed that the administration of the recombinant proteins should be well tolerated by the patients. While this assumption was correct e.g. for erythropoietin, it was clearly erroneous for others e.g. interferons. Reasons for this unanticipated toxicity were systemic administration of mostly local tissue factors delivered, at high doses, species-specificity and the poorly understood physiological roles of several cytokines (Sundwall, 1990; Zbinden, 1990; Ryffel ,1993 a,b).

The success and acceptance of a novel therapy such as recombinant proteins depends largely on good knowledge of its physiological roles, which allows deductions of potential side effects. It has been argued, whether animal and in vitro tests with recombinant human proteins have a predictive value for the safety evaluation for man. I present data that demonstrate that most side effects in man of a therapy with recombinant proteins may be predicted by data from experimental toxicological studies. There are limitations in the experimental approach and failures occurred, which will be discussed. The reasons for the discrepancies include insufficient understanding of the physiology of the protein, quantitative differences, different biological response in man or misinterpretation of experimental toxicological data. Therefore, a dialogue between the clinician and toxicologist is important. The feedback from clinical studies must guide future experimental toxicity studies.

Interleukin-6 (RhIL-6)

IL-6 is a rather pleiotropic cytokine with stimulatory effects in vitro on the lympho-haematopoietic systems, on the hepatic acute phase proteins, the renal mesangium and central nervous system. A beneficial effect on thrombopoiesis is one important property of IL-6, which attracted the interest for a clinical development. The essential question raised from the safety side were, whether a dose exists which results in increased platelet counts in the absence of other systemic effects known to occur at higher doses. Possible side effects limiting further development would have been other biological properties reported for IL-6, e.g. glomerulonephritis, severe acute phase reaction with fever and immunostimulation resulting in autoimmune disease (reviewed in Hirano et al., 1991).

RhIL-6 was given to primates, rats and mice, and the main findings have been recently reviewed (Ryffel et al., 1992). RhIL-6 induced a sustained increase in blood platelets, stimulation of the immune system and the synthesis of acute phase proteins. These alterations were not complicated by any secondary disease at the highest dose tested, which was 1000mg/kg/day given for 4 or 8 weeks. Despite the pleiotropic nature in vitro, rhIL-6 had a quite selective effect in primates and was extremely well tolerated (Ryffel et al. 1994). Despite a significant increase in acute phase proteins, there was no significant rise in body temperature in cynomolgus monkeys (Klug et al., 1995). Since rhIL-6 has also biological effects in rodents, we tested the general tolerability in mice and rats and obtained similar results. In all expèrimental species antibody against rhIL-6 developed, which neutralised the biological effects of rhIL-6. It is important to note, that these studies did not reveal any renal or hepatotoxic potential.

Data from clinical phase 1 and 2 trials confirm that rhIL-6 is generally well tolerated except for fever in a substantial fraction of patients, which was not predicted from the experimental studies. Rodent and primate studies predicted an acute phase response as evidenced by increased acute phase protein synthesis, but

Table 1. Comparison of IL-6 induced adverse effects across the species

Finding	Mouse	Rat	Marmoset	Rhesus	Man
Increased platelets	+	+	+	+	+
Acute phase proteins	+	+	+	+	+
Immunostimulation	+	+	+	+	+
Glomerulonephrits	-	-	-	-	-
Fever	-	-	-	-	+

there was no evidence of fever or general malaise in primates as observed in most patients receiving rhIL-6. Therefore, the severity of the acute phase response with fever in man was unexpected.

Granulocyte-Macrophage Colony Stimulating Factor (GM-CSF)

GM-CSF is a member of the family of CSF with differing specificity on hematopoietic precursor cells in the bone marrow. In contrast to IL-6, rhGM-CSF is inactive in rodents, but has a strong myelopoietic effect in non-human primates such as Rhesus, cynomolgus and marmoset. GM-CSF causes not only the proliferation and differentiation of myeloid precursor cells resulting in a 20 fold increase of mature granulocyte counts in the blood, but also activates granulocytes and monocytes. The proinflammatory cytokines released by these cells cause a wide-spread inflammation in Rhesus monkeys at high GM-CSF doses (Robison and Myers, 1993). The inflammation was especially apparent on the serosal surfaces (pleuritis, pericarditis and peritonitis), the liver and the local injection site in the skin. The dramatic increase of granulocyte counts resulted in circulatory complications such as thrombosis, a finding which was also reported with the more lineage specific G-CSF at high doses.

The data from Rhesus monkeys predicted the clinical adverse effects. Based on the present experimental and clinical findings it is obvious that a therapy with GM-CSF is difficult to manage and requires extremely careful monitoring. GM-CSF has likely advantages for the treatment of certain diseases over G-CSF, since it stimulates monocytes and dendritic cell functions.

It is important to note, if the safety studies with rhGM-CSF were performed in rodents, which only respond to the murine homologue of GM-CSF, no toxicity could have been predicted.

Table 2. Comparison of the GM-CSF adverse effects across the species

Finding	Mouse	Rhesus	Man
Increased granulocytes	-	+	+
Serositis	-	+	+
Thrombosis	-	+	+
Inflammation (injection)	-	+	+

Interferons (IFN-αβ−γ)

Interferons are a large family of related proteins which can be devided into type I (antiviral and antiproliferative properties, α and β interferons) and type II (immunostimulation, γ-interferon) interferons. The safety of recombinant human interferons has been explored in several animal species including primates. Most of these studies suggested that interferons are well tolerated; however it was realized later that most animal species with the exception of baboons do not recognize human interferons (Terrell and Green, 1993). Interferons are an example of extremely species-specific proteins, which raised the fundamental question whether experimental toxicological studies are of any value for recombinant human proteins. As we learned later, this high selectivity is rather the exception. Several human cytokines e.g. GM-CSF, IL-3, IL-4 and IL-12 have no biological effects in rodents,but are recognized by primates (Table 3). The case

Table 3. Species-specificity of recombinant human cytokines

Rh protein	response in rodents	or in primates
IL-2	+	+
IL-3	-	+
IL-4	-	+
IL-6	+	+
IL-12	-	+
Interferons	-	-
G-CSF	+	+
GM-CSF	-	+

The main adverse effects of interferons are hepato- and splenomegaly due to macrophage, NK- and T-cell infiltration, anaemia and myelodepression (Terrell and Green, 1993). As an alternative homologous proteins may be used. Mouse interferon-g, when given to mice, caused a similar pattern of toxicity.

Interleukin-12 (IL-12)

IL-12 is a macrophage derived heterodimeric protein, has effects in vitro and in vivo on the hematopoietic and immune systems, resulting in NK and CTL activation with the release of interferon-g (Brunda, 1994; Gately, 1993; Trinchieri and Scott, 1994). The source and site of action of IL-12 are given in the figure. Experimental data suggest that IL-12 together with interferon-g, which is induced by IL-12, results in a strong cellular immune response (TH1 response). This immune response may be exploited clinically for the defence against cancer and viral disease. Preliminary toxicity studies in primates (Sarmiento et al., 1994) and mice (in preparation) demonstrate a similar adverse effect profile as for interferon-g. We provided evidence which suggest that interferon-g is mediating toxicity of an IL-12 treatment in mice, since no toxicity is observed in IFNg-R deficient mice (Eng et al., 1995, Ryffel, 1995).

Tumor necrosis factor (TNF): A proinflammatory cytokine

TNF is an interesting and highly cytoxtoxic cytokine (Vassalli, 1992), which is presently used for the local therapy of sarcoma in patients. The toxicological data confirm the high and broad biological effects, which precludes a systemic administration of this cytokine (Beutler, 1992; Aggarwal, 1992; Grau et al., 1993).

Use of Knockout mice to investigate the role of TNF in toxicopathogenesis

Since high levels of endogenous TNF are induced by many toxic agents (Grau et al., 1993), we are presently investigating the role of TNF as mediator of toxicity using mice with a disruption of the TNF and related lymphotoxin (LT a/b) gene locus.

We have inactivated TNF and LTa (Eugster et al., unpublished) and demonstrate that these mice are resistant to several pathologies thought to be mediated by TNF, e.g. endotoxic shock, bleomycin or oxygen induced pulmonary toxicity, cerebral malaria and many other toxic syndromes.

These data demonstrate the possible use of mice with inactivation of specific genes thought be involved in toxic processes e.g. TNF, interferons etc. (Ryffel, 1995). Such studies in mice play only a role to understand the pathogenesis of a toxic syndrome.

Recommendations for safety studies with rh proteins

Based in the available experience, it is of prime importance to select a species responding to the test protein. While the non-human primate is often the only choice, data from rodents may provide useful safety data as exemplified for rhIL-6,IL-2, IL-12 and TNF. Since human proteins have an immunogenic potential in animals including primates, the antibody response has to be monitored. Our experience tells us that this immune response neutralises the biological effect, and thus, the information gained from studies lasting beyond 4 to 8 weeks is of questionable value.

Table 4 Consideration for toxicity studies with rh proteins

Species selection:	Responsiveness
Dose:	10-30 fold clinical dose
Duration:	4-8 weeks
Toxicokinetics:	required
Antibody formation:	test after 2-4 weeks for neutralising Ab
Carinogenicity:	No
Reproduction:	(No)

Future: From protein to gene therapy

The recent discovery of the molecular-genetic basis of some diseases (cystic fibrosis, neurological disorders, cancer, etc) encouraged an alternative approach to protein therapy by delivering the defective gene (DNA) to the host cell (somatic gene therapy). Several viral and non-viral delivery systems have been developed. Preliminary data from clinical studies are promising, but clearly show that the gene transfer is only partial and transient. Possible risks include viral infection (adenoviral vectors), integration into the genome (retroviral vectors) and tumor development (Zabner, 1994, Davies and Rosenberg, 1994; Huber, 1994).

Efficacy studies to prove the therapeutic concepts are performed in experimental disease models, and safety studies are requested by the authorities. The design of toxicological studies for gene therapy is a field of intense discussions,

Extrapolation from the laboratory to man

The question raised at the outset, whether animal and in vitro toxicity tests have a predictive value in assessing the risk of a novel gene or protein therapy in man, can be answered with Yes.

The presented cases and many others from our and other laboratories suggest that experimental data can be extrapolated to man, although not every single adverse effect may be correctly predicted.

For gene therapy it is clear that essential information can be gained by experimental animal studies including toxicity studies; However in this area much has to be learned. It is important to realise that safety studies are not a routine business, but should be conducted in a case-by-case fashion in accordance with the present knowledge of the drug action. This scientific knowledge will be a very useful guide to obtain predictive data for the human protein and gene therapy.

Acknowledgement

This work was supported by grants from 3R Foundation, Swiss National Science Foundation and Sander Foundation.

References

Aggarwal BB, Pocisik E (1992) Cytokines: From clone to clinic. Arch Biochem Biophys 292:335-359

Aggarwal BB, Puri RK (1995) Human cytokines: Their role in disease and therapy. Blackwell Science Publ. Cambridge

Arai K, Lee F, Miyajima A, Miyatake S, Arai N, Yokota T (1990) Cytokines of immune and inflammatory responses. Annu Rev Biochem 59:783-836

Beutler B (1992). Tumor necrosis factors: The molecules and their emerging role in medicine. Raven Press, NY.

Brunda MJ (1994) Interleukin-12. Review, J. Leukocyte Biol 55:280-296

Car BD, Eng VM, Schnyder B, Ozmen L, Huang S, Gallay Ph, Heumann D, Aguet M, Ryffel B (1994) INF-g receptor deficient mice are resistant to endotoxic shock. J Exp Med 179:1437-1444

Car BD, Eng VM, Schnyder B, LeHir M, Shakhov A, Ryffel B (submitted) Role of interferon-g in interleukin 12 induced pathology in mice.

Davies J, Rosenberg M (1994) Mammalian gene studies. Current Opinion Biotechology 5:604-636
Dexter TM, Garland JM, Test NG (1990) Colony stimulating factors. Marcel Dekker, New York
Eng VM, Car BD, Schnyder B, Lorenz M, Lugli S, Aguet M, Anderson TD, Ryffel B, Quesniaux V (1995) The stimulatory effect of IL-12 on hematopoiesis are inhibited by IL-12 IFNg in vivo. J Exp Med 181:1893-1898
Gately MK, Warrier RR, Honasoge S, Carvajal DM, Faherty DA, Connaughton SE, Anderson TD, Sarmiento U, Hubbard BR, and Murphy M (1993) Administration of recombinant IL-12 to normal mice enhances cytolytic lymphocyte activity and induces production of IFN-g in vivo. Int Immunol.6:157-167
Gianella-Borradori A (1992) Impact of hemopoietic growth factors in clinical oncology: Present and future. Triangle (Sandoz) 31:35-43
Grau GE, Lambert PH, Vassalli P, Piguet PF (1993) Tumor necrosis factor/Cachectin as an effector of T cell-dependent immunopathology. In: Experimental Pathology, Vol 34, Cytokine induced Pathology B, Academic Press, Inc., pp. 159-171
Hirano T, Akiro S, Taga T, Kishimoto T (1990) Biological and clinical aspects of interleukin 6. Immunology Today 11:443-449
Huber B (1994) Gene therapy. Annu NY Acad Sci 716:1-366
Klug St, Neubert R, Stahlmann R, Thiel R, Ryffel B, Car BD and Neubert D (1994) Effects of recombinant human interleukin 6 (rhIL-6) in marmosets (Callithrix jacchus). Arch Toxicol 68:619-631
Moore MAS (1991) Hemopoetic growth factors. Annu Rev Immunol 9:159-191
Robison RL, Myers LA (1993) Preclinical safety assessment of recombinant human GM-CSF in Rhesus Monkeys. In: Experimental Pathology, Vol 34, A, Cytokine-induced Pathology, Academic Press, Inc. pp. 149-172
Ryffel B, Kammüller M, Robison R, Myers L (1992) Pathology induced by interleukin-6 Toxicol Letters 64/65:311-319
Ryffel B (1993) Zytokine, Therapeutika, Toxine und Mediatoren der Toxizität. SWISSMED 15:16-20
Ryffel B (1994) Pathology induced by cytokines. Int Rev Exp Pathology 34 (A and B): A concise review on safety studies on a variety of cytokines, which are contributed by the experts in the specific area
Ryffel B, Car BD, Gunn H, Roman D, Hiestand P, Mihatsch MJ (1994) Interleukin-6 exacerbates glomerulonephritis in (NZBxNZW F1) mice. Am J Pathol 4, 144:927-937
Ryffel B, Car BD, Woerly G, Weber M, DiPadova F, Kammüller M, Klug S, Neubert R, Neubert D (1994) Long-term interleukin-6 administration stimulates sustained thrombopoiesis and acute phase protein synthesis in a small primate - the marmoset. Blood 83:2093-2102
Ryffel B (1995) Interleukin-12: Adjuvans in der Tumortherapy? Therapiewoche Schweiz 11(5) 241-244
Ryffel B (1995) Use of knockout mice in pharmacological-toxicological research. Toxicology (1995, in press)
Sarmiento UM, Riley JH, Knaack PA, Lipman JM, Becker JM, Gately K, Chizzonite R, Anderson TD (1994) Biologic effects of recombinant human interleukin-12 in Squirrel Monkeys (Sciureus saimiri). Laboratory Investigation, 71:862-873
Sundwall A, Ekman L, Johansson HE, Sjöberg B, Sjöholm I (1990) Preclinical evaluation of peptides and recombinant proteins. Skogs Grafiska AB Malmö, Workshop Stockholm Schweden May 7-8 1993
Terrell TG, Green JD (1993) Comparative pathology of recombinant murine interferon-g in mice and recombinant human interferon-g in cynomolgus monkeys. In: Experimental Pathology, Vol 34 B, Cytokine-induced Pathology, Academic Press, Inc. pp 73-101

Trinchieri G, Scott Ph (1994) The role of interleukin 12 in the immunoresponse, disease and therapy. Immunology Today, 15 (10):460-463

Vassalli P (1992) The pathophysiology of tumor necrosis factor. Annu Rev Immunol 10:411-452

Woerly G, Ryffel B (1993) The use of receptor assays for the rapid selection of an appropriate animal species for the preclinical investigations of human interleukins. In Vitro Toxicology 6:67-79

Zabner J, Pertersen DM, Welsh MJ (1994) Safety and efficacy of repetitive adenovirus-mediated transfer of CFTR cDNA to airway epthelia of primates and cotton rats. Nature Genetics 6:75-81

Zbinden G (1990) Safety evaluation of biotechnology products. Drug Safety 5(1): 58-64

The Mechanism of Action of Tetanus and Botulinum Neurotoxins

Cesare Montecucco, Giampietro Schiavo and Ornella Rossetto
Centro CNR Biomembrane and Dipartimento di Scienze Biomediche, Università di Padova, 35100 Padova, Italy

Introduction

Tetanus and botulinum neurotoxins were identified about a century ago as the cause of the neuroparalytic syndromes of tetanus and botulism (Faber, 1890; Tizzoni & Cattani, 1890; van Ermengem, 1897). Bacteria of the genus Clostridium produce one tetanus neurotoxin (TeNT) and seven different serotypes of botulinum neurotoxins (BoNT/A, /B, /C, /D, /E, /F, and /G). These bacteria show a widespread distribution in the environment, mainly as spores, which can germinate under appropriate anaerobic conditions. Tetanus follows the contamination of necrotic wounds with spores of Clostridium tetani, limited bacterial proliferation and toxin production. Botulism derives from ingestion of uncooked anaerobic food, where spores of Clostridium botulinum, or related species, have germinated with the production of the toxin (foodborne botulism). More rarely, spores germinate in anaerobic areas of the newborn intestine giving rise to infant botulism (Simpson, 1989; Montecucco, 1995). Since tetanus is caused solely by intoxication with TeNT, it was possible to prepare a anti-tetanus vaccine simply by treating TeNT with paraformaldheyde, a chemical modification that inactivates the toxin, but preserves its immunogenic properties (Ramon & Descombey, 1925).

These neurotoxins bind specifically neuronal cells, enter the cytosol and block neurotransmitter release (Montecucco & Schiavo, 1995). As depicted in fig.1, TeNT acts at the inhibitory interneurons of the spinal cord and impairs the neuronal circuit that ensures balanced voluntary muscle contraction, thus causing a spastic paralysis. BoNTs act at the neuromuscular junction where they block acetylcholine release and hence cause a flaccid paralysis (Burgen *etal.*, 1949). The opposite clinical symptoms of tetanus and botulism, spastic versus flaccid muscle paralysis, results from the different sites of toxin action and not from a different mechanism of cell intoxication. Given the central importance of the nervous system in vertebrate physiology and behaviour, this intoxication results frequently in the death of the intoxicated animal. The neurospecificity of these toxins as well as their catalytic type of activity explains why they are the most poisonous substances known. The mouse lethal dose of these neurotoxins is comprised between 0.1 and 1ng/Kg of body weight and these figures are probably much lower in the wilderness, where even very small deficits in motility may be sufficient to prevent survival.

The notion of the neuromuscular junction blockade of BoNT has lead to the therapeutic use of BoNTs in a variety of dystonias. Minute amounts of BoNT are injected into the muscles to be paralyzed and the beneficial effect of such a treatment last for few months. In fact, the synapse intoxicated by BoNT degenerates and a new functional synapse reforms nearby in a period of few months and BoNT injection has then to be repeated (Jankovic & Hallett, 1994). Hence, also BoNTs are very relevant to human health care.

Structure of tetanus and botulinum neurotoxins

The similar function of TeNT and BoNT is paralleled by a closely related structural organization. As depicted in fig. 1, clostridial neurotoxins consist of three 50 kDa domains (Montecucco & Schiavo, 1993, 1995), each of which plays a different role in cell intoxication: the L chain is responsible for the intracellular toxin activity, the amino-terminal 50 kDa H chain domain, termed HN, is implicated in membrane translocation, whereas the carboxy-terminal part of the H chain, termed HC, is mainly responsible for binding to neuronal cells. The activity of L is freed only upon reduction of the single interchain disulfide bond which takes place inside neurons (Schiavo *etal.*, 1990; de Paiva *etal.*, 1993).

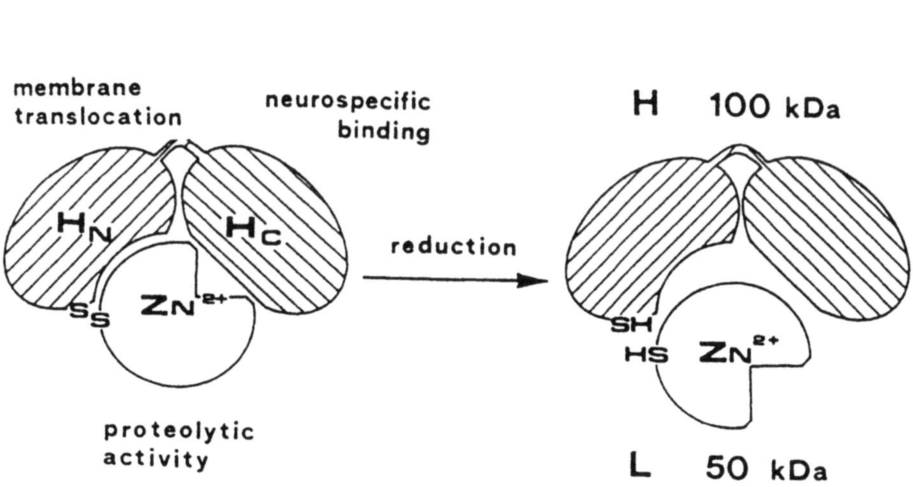

Figure 1. Structure and activation of clostridial neurotoxins. They consist of two disulfide-linked polypeptide chains: H and L. H (100 kDa) has two domains involved in cell binding and penetration, L is a zinc-endopeptidase whose activity is freed only upon reduction.

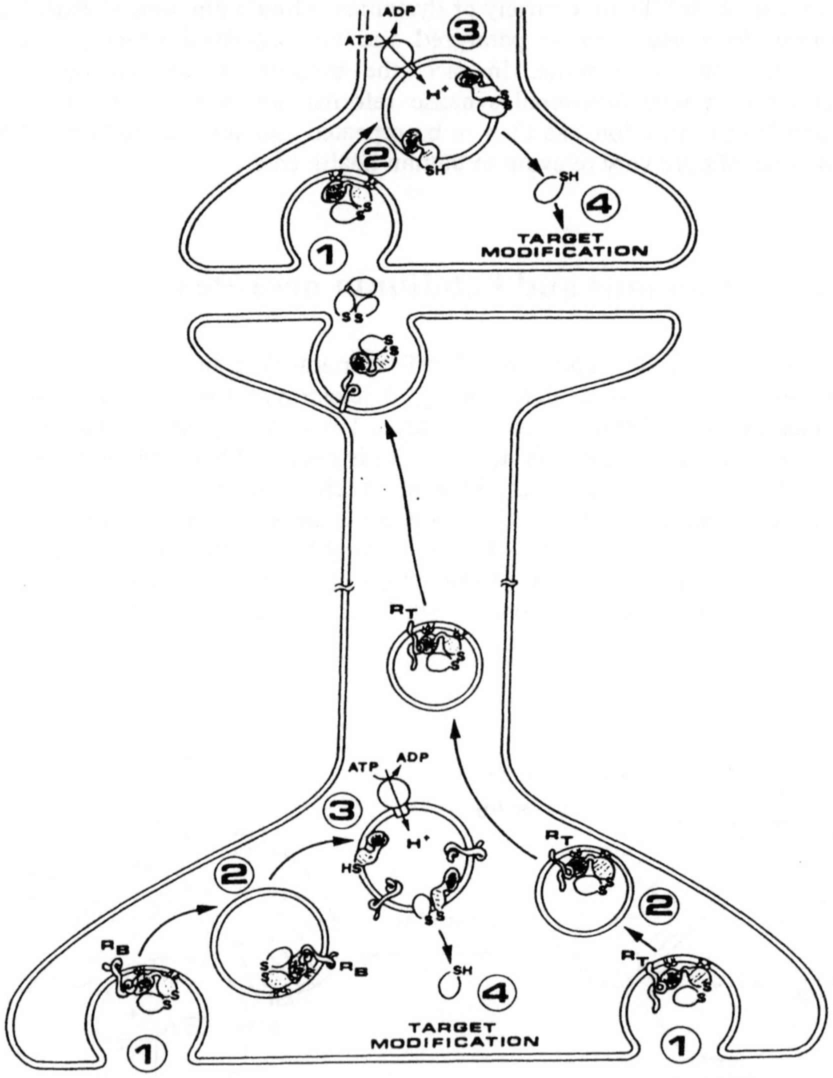

Figure 2. Tetanus and botulinum neurotoxins intoxicate central and peripheral neurons respectively, via a four-step mechanism. TeNT and BoNTs bind to the neuromuscular junction. The protein receptor RB is a synaptic vesicle protein which exposes its toxin binding domain upon fusion of the vesicle with the presynaptic membrane. BoNTs bind and are internalized following vesicle endocytosis (step 2). The lumen of the synaptic vesicle is acidified by a vacuolar ATPase proton pump. This induces a structural transition of BoNT, which adopts a "acid structure" able to insert in the vesicle membrane and to translocate the L chain in the cytosol (step 3). Inside the neuron cytosol, L blocks neuroexocytosis via selective zinc-dependent proteolysis of a protein component of the neuroexocytosis apparatus (step 4). TeNT binds to RT, a protein receptor that mediates its internalization inside cell vesicles that are transported retroaxonally up to the cell body and discharged in the intersynaptic space between the motoneuron and the spinal cord inhibitory interneuron. TeNT is internalized inside the inhibitory interneuron, enters the cytosol, as BoNTs do at the neuromuscular junction, and inhibits the release of GABA and glycine.

Cell binding and penetration

After spreading from the site of production or adsorption into body fluids, BoNTs and TeNT bind to the presynaptic membrane of the neuromuscular junction. The recptors specific for such a specific and high affinity binding have not yet been characterized. Available data fit in a double receptor mode of binding of clostridial neurotoxins to nerve cells (Montecucco, 1986). The toxin is proposed to adsorb to the negatively charged surface of the presynaptic membrane via a low affinity high capacity interaction(s). The toxin then moves laterally to bind a protein receptor that is responsible for neurospecificity. A high affinity binding is thus obtained because the cell association constant of the toxin to the nerve cell is the product of the association constants of binding to the protein receptor and of binding to acidic lipids. The BoNTs enter the neuronal cytoplasm at the neuromuscular junction and TeNT is driven by its receptor inside intracellular vesicles that move retrogradely, along the motorneuron axon, to the soma. Eventually, TeNT is discharged in the intersynaptic space between the motorneuron and the inhibitory interneurons of the spinal cord (Schwab *etal.*, 1979; Wellhoner, 1992). Here it binds to unidentified receptors and is again internalized inside vesicles, wherefrom the L chain escapes in the cytosol to block GABA and glycine release.

To reach the cytosol, both TeNT, inside vesicles of the inhibitory interneurons, and BoNTs, inside vesicles of the motorneuron synapse, must necessarily cross the hydrophobic barrier of the vesicle membrane (see fig.2).

The opposite membrane trafficking of TeNT and BoNT at the NMJ clearly indicates that internalization is not necessarely followed by membrane translocation in the cytosol (fig. 2). Step 3 has been studied mainly with model membrane systems and scarce data were obtained in vivo. Available evidence indicate that at low pH TeNT and BoNTs undergo a conformational change from a water soluble "neutral" form to a "acid" form, characterized by the exposure of hydrophobic segments that enables it to insert with both the H and L chains in the lipid bilayer of the membrane (Boquet & Duflot, 1982; Roa & Boquet, 1985; Montecucco *etal.*, 1986, 1988, 1989; Menestrina *etal.*, 1989; Schiavo *etal.*, 1991a). There is indirect evidence that these toxins escape from the vesicle at the neuromuscular junction via a low pH-dependent step (Simpson *etal.*, 1994; Williamson & Neale, 1994).

A common property of TeNT and BoNT is their ability to form at low pH ion channels in planar lipid bilayers and this property is associated to the HN domain (Boquet & Duflot, 1982; Hoch *etal.*, 1985; Donovan & Middlebrook, 1986; Shone *etal.*, 1987; Gambale & Montal, 1988; Blaustein *etal.*, 1989; Beise *etal.*, 1994). There is a general consensus that this toxin channel is related to the physical entity that mediates the translocation of the L chain across the vesicle membrane into the nerve cytosol. However, there are three alternative views on how this process may take place (discussed in Montecucco & Schiavo, 1995) and further experiments are needed to clarify this third passage of the mechanism of cell intoxication of clostridial neurotoxins.

The metallo-proteinase activity of tetanus and botulinum neurotoxins

The most conserved segment of the L chains is located in the middle of the polypeptide chain (Minton, 1995). It contains the His-Glu-Xaa-Xaa-His zinc-binding motif of zinc-endopeptidases (Schiavo *etal.*, 1992a, c; Wright *etal.*, 1992). Atomic absorption measurements of the zinc content of clostridial neurotoxins showed that they contain one atom of zinc per molecule of toxin, bound to the L chain, except for BoNT/C which binds two zinc atoms (Schiavo *etal.*, 1992a, c; 1993a; 1995).

Zinc-endopeptidases can be divided into two groups depending on the involvement of two or three histidines in the coordination of the zinc atom. Thermolysin-like enzymes bind zinc via two histidines and one carboxylate

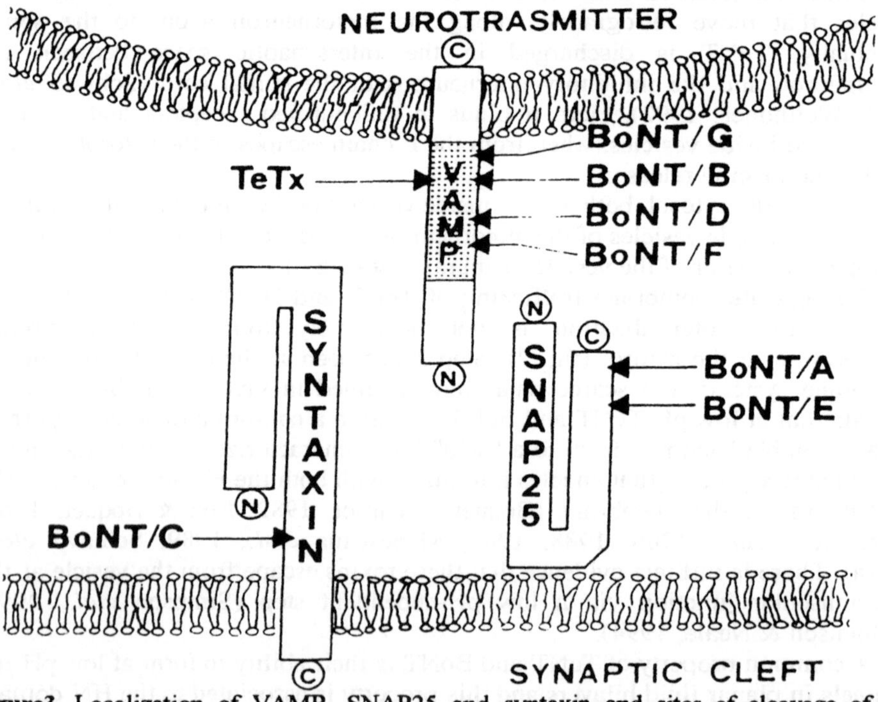

Figure3. Localization of VAMP, SNAP25 and syntaxin and sites of cleavage of the clostridial neurotoxins. VAMP is bound to the synaptic vesicle membrane, whereas SNAP-25 and syntaxin are bound to the target membrane. TeNT and BoNT/B, /D, /F or /G act on the conserved central portion of VAMP (dotted area) and release the amino-terminal part of VAMP in the cytosol. BoNT/A and/E cleave SNAP25 within the carboxyl terminus. Also BoNT/C treatment of neurons in culture leads to SNAP25 cleavage at the CCOH-terminus (our unpublished data). BoNT/C cleaves syntaxin at a single site near the cytosolic membrane surface. The action of TeNT and BoNT/B, /C, /D, /F and /G causes the release of a large portion of the cytosolic domain of VAMP and syntaxin. At variance, only a small segment of SNAP-25 is released by the selective proteolysis of BoNT/A, /C and /E.

(Matthews *et al.*, 1972). At variance, astacin-like proteinases coordinate zinc via three histidines and one tyrosine, while adamalysin and collagenases via three histidines (Stocker *etal.*, 1995). In all cases one molecule of water is an additional zinc ligand and is involved in the hydrolysis of the peptide bond. Available evidence indicate that clostridial neurotoxins coordinate zinc in a novel mode via two histidines, a Glu-bound water molecule and aromatic residue(s) tentatively identified as a conserved tyrosine (Montecucco & Schiavo, 1995).

TeNT and BoNTs were recently shown to be zinc-endopeptidases specific for protein components of the neuroexocytosis apparatus (Montecucco & Schiavo, 1993, 1994), as detailed in fig. 3.

TeNT and BoNT/B, /D, /F and /G act on the same intraneuronal target: they cleave VAMP, also called synaptobrevin, a membrane protein of synaptic vesicles (Schiavo *etal.*, 1992a, b; Schiavo *etal.*, 1993a, b; Schiavo *etal.*, 1994). At variance, BoNT/A, /C and /E act on proteins of the presynaptic membrane: BoNT/A and /E cleave SNAP-25, while serotype C cleaves syntaxin (Blasi *etal.*, 1993a, b; Schiavo *et l.*, 1993b, c; Binz *etal.*, 1994; Schiavo *etal.*, 1995). Except for TeNT and BoNT/B, each one of the different clostridial neurotoxins catalyses the hydrolysis of different peptide bonds, as summarized in the following table.

Table1. Protein substrates and peptide bond specificities of tetanus and botulinum neurotoxins

Toxin Type	Target	Peptide Bond Cleaved P3 - P2 - P1 -| - P1' - P2' - P3'
TeNT	VAMP	Ala-Ser-Gln- | -Phe-Glu-Thr
BoNT/A	SNAP-25	Ala-Asn-Gln- | -Arg-Ala-Thr
BoNT/B	VAMP	Ala-Ser-Gln- | -Phe-Glu-Thr
BoNT/C	Syntaxin	Thr-Lys-Lys-| -Ala-Val-Lys
BoNT/D	VAMP	Ile-Asp-Arg- | -Ile-Met-Glu
BoNT/E	SNAP-25	Asp-Gln-Lys- | -Leu-Ser-Glu
BoNT/F	VAMP	Arg-Asp-Gln- | -Lys-Leu-Ser
BoNT/G	VAMP	Thr-Ser-Ala- | -Ala-Lys-Leu
BoNT/C	SNAP-25	*

*in BoNT/C treated neurons SNAP-25 is cleaved, in addition to syntaxin, at a yet undetermined peptide bond near the carboxyl-terminus (Osen-Sand *etal.*, submitted).

No common features of the amino acid sequences around the cleavage sites of the various neurotoxins in terms of amino acid lateral chain charge, hydrophilicity/hydrophobicity and volume can be identified. Hence, the geometry of the active sites of these proteinases must be different in order to accomodate and hydrolyse such different peptide structures. Recently, it has been found that SNAP-25, in addition to syntaxin, is cleaved in neurons exposed for many hours to BoNT/C (Osen-Sand *etal.*, submitted; Williamson *etal.*, submitted). This result may be taken as an indication that BoNT/C cleaves both syntaxin and SNAP-25.

These neurotoxin metallo-proteinases have the peculiar property of not cleaving short peptides of sequences correspondent to the region cleaved in the intact substrate molecule (Schiavo *etal.*, 1992a; Shone *etal.*, 1993; Shone & Roberts, 1994). On the other hand, peptides encompassing the cleavage sites do bind to the respective toxin and inhibited it. BoNT/B proteolysis on VAMP peptides was appreciable only with fourthy-residue long peptides and that maximal clevage rates were only observed with the 33-96 VAMP segment (Shone *etal.*, 1993; Shone & Roberts, 1994). Yamasaki *etal.* (1994) found that BoNT/F and /D appear to require peptides more extended at their amino-terminal that those hydrolysed by TeNT and BoNT/B. A requirement for long peptides is also shown by BoNT/A and /E, which hydrolyse the SNAP-25 fragment 136-206 (Binz *etal.* 1994). Another peculiarity of these metalloproteinases is that they hydrolyse only one out of other identical peptide bonds present in the target protein sequence. These features indicate that TeNT and BoNTs are proteinases that recognize the tertiary structure of their targets and that the cleaved peptide bond with flanking residues is not the sole determinant of the very specific recognition of VAMP, SNAP-25 and syntaxin (Schiavo *et al.* 1993a; Montecucco & Schiavo, 1993, 1995).

The strong similarity of the eight known L chains indicates that they derive from a common ancestral metalloprotease. As a consequence, their three substrate proteins could share a region involved in toxin binding. In fact, VAMP, SNAP-25 and syntaxin, show the common presence of a short motif, which is comprised within regions predicted to adopt an alpha-helical conformation (Rossetto *etal.*, 1994). The motif has a face with three negative charges contiguous to a face formed by hydrophobic residues. There are two copies of the motif in VAMP, four copies in SNAP-25 and two copies in syntaxin. Several experimental evidence indicate that the motif is exposed on the protein surface and is involved in neurotoxin bonding (Rossetto *etal.*, 1994; Pellizzari *etal*, manuscript in preparation). Thus, we have proposed that clostridial neurotoxins recognize their protein substrates via two regions that interact with: a) a segment that includes the peptide bond to be cleaved and b) another segment closely similar in VAMP, SNAP-25 and syntaxin, which accounts for antibody cross-reactivity and cross-inhibition of the different neurotoxin types. The relative contribution of segment a) and b) to the specificity and strenght of neurotoxin binding remains to be determined. It can be anticipated that hydrolysis of region a) leaves the toxin bound to its substrate only via its interaction with b) with a large decrease in binding affinity, which is expected to lead to a rapid release of the hydrolysed substrate. Such a model also explains the available data on the minimal size of the target still cleaved by the neurotoxins (Shone *etal.*, 1993; Binz *etal.*, 1994; Yamasaki *etal.*, 1994; Shone & Roberts, 1994), because, in order to be cleaved, a peptide must include both a) and b). Moreover, this model explains the recent report that the n-syb isoform of Drosophila VAMP with two non-conservative mutations in the b) region and identical site of cleavage, is not proteolysed by TeNT (Sweeney *etal.* 1995). Also the syntaxin isoform specifity of BoNT/C is accounted for by this model (Schiavo *et al.*, 1995). Preliminary results of side directed mutagenesis experiments indicate that the negative charges of the carboxylate residues are essential for substrate hydrolysis and antibody recognition (Pellizzari *etal.* manuscript in preparation).

The regions of TeNT and BoNTs involved in substrate binding are unknown. It is possible that the strongly conserved one hundred-residue long amino-terminal region is implicated. Removal of more than eight residues from the amino-terminus leads to complete loss of activity (Kurazono *etal.* 1992).

Tetanus and botulinum neurotoxins and the inhibition of neuroexocytosis

VAMP was first identified in the Torpedo electric organ and then in rat, bovine, human, Drosophila and yeast (Bennett & Scheller, 1994; Südhof, 1995). Rat brain contains two isoforms, termed VAMP-1 and VAMP-2, of 118 and 116 amino acids respectively, which differ almost exclusively in the 33 residue-long amino-terminal part. VAMP1 and 2 are present in all tissues, though in different amounts and ratios (Rossetto *etal.*, submitted). Cellubrevin, another TeNTcleavable isoform, has been identified in non neuronal cells (McMahon *etal.*, 1993). These findings indicate that clostridial neurotoxins restrict their activity to the nerve tissue due to the presence of toxin receptors and not because of lack of cleavable targets.

The central part (residues 33-96) of VAMP is highly conserved among species and isoforms. The protein is anchored to vesicles via a carboxy-terminal hydrophobic tail and exposes to the cytosol the largest part of its mass, as depicted in fig. 3. Clostridial neurotoxins cleave VAMP, provided that the isoforms possess an appropriate cleavage site. This is the case for all the variants known so far, except for the brain isoform I of chickens and rats, which carries a Val residue in place of the Gln residue present in human and mouse VAMP-1 (fig. 3). This single amino acid replacement at the cleavage site of TeNT and BoNT/B has been proposed to be associated with the resistance of rats and chickens, and the sensitivity of humans and mice to tetanus and to type B botulism (Payling-Wright, 1955; Patarnello *etal.*, 1993). BoNT/D, /F and /G cleave both VAMP isoforms with similar rates.

SNAP-25 is a 206 residue-long protein anchored to the cytosolic surface of the nerve plasmalemma via palmitoylation of four cysteines, located in the center of the molecule (Oyler *etal.*, 1989) (fig. 3). SNAP-25 is required for axonal growth during development and possibly nerve terminal plasticity in the mature nervous system (Osen-Sand *etal.*, 1993). The protein is highly conserved from Drosophila to mammals and there are two human isoforms differentially expressed during development (Bennett & Scheller, 1994).

Also syntaxin is located on the cytosolic surface of the nerve plasmalemma (Bennett *etal.*, 1992; Inoue *etal.*, 1992). Similarly to VAMP, it is membrane-anchored via a carboxy-terminal segment and most of its molecular mass is exposed to the cytosol (fig. 3). It is present at the active zones of the presynaptic membrane, where neurotransmitter release takes place. Several isoforms of different lenght have been identified (Bennett and Scheller, 1994; Südhof, 1995).

Söllner *et al.* (1993) have shown that VAMP, SNAP-25 and syntaxin together with a group of cytosolic proteins form a 20S multi-protein complex, proposed to mediate vesicle docking and fusion. The fact that cleavage of these three proteins leads to a persistent blockade of neuroexocytosis is a strong evidence in favour of the involvement of the 20S complex in neuroexocytosis. The cleavage of VAMP and of syntaxin by TeNT, BoNT/B, /D, /F, /G and /C leads to the release in the cytosol of a large part of the cytosolic target molecule and this prevents assembly of a VAMP, SNAP-25 and syntaxin heterotrimer (Hayashi *et al.*, 1994). At variance, BoNT/A removes only nine residue from the SNAP-25 carboxy-terminus (Schiavo *et al.*, 1993c). Nonetheless, this is sufficient to impair neuroexocytosis, thus indicating that this part of the molecule is essential for SNAP-25 folding. Alternatively, the carboxyl end of SNAP-25 plays an important role in the function of the neuroexocytosis complex.

Evidence that TeNT and BoNTs block neuroexocytosis via their specific cleavage of VAMP, SNAP-25 and syntaxin are rapidly accumulating:

a) the intracellular activity of the toxins is inhibited by specific inhibitors of zinc-endopeptidases such as phosphoramidon and captopril (Schiavo *et al.*, 1992a; Dayanithi *et al.*, 1994; Poulain *et al.* unpublished results);

b) VAMP or SNAP-25 or syntaxin are cleaved in synaptosomes and neurons in culture, whose neurotransmitter release was blocked by TeNT or BoNTs (Blasi *et al.*, 1993a, b; Schiavo *et al.*, 1993b; Osen-Sand *et al*, submitted);

c) peptides spanning the cleavage site of VAMP inhibit TeTx and BoNT/B in Aplysia neurons or chromaffin cells (Schiavo *et al.*, 1992b; Dayanithi *et al.*, 1994)

d) an antibody highly specific for VAMP prevents the blockade of neurotrasmitter release in Aplysia neurons induced by TeNT and BoNT/B, but not that caused by BoNT/A (Poulain *et al.*, 1993). These results do not exclude that these toxins may act on other protein substrates inside neurons, but clearly indicate that VAMP, SNAP-25 and syntaxin are major targets of the intraneuronal activity of these neurotoxins.

Conclusions

The neurotoxins produced by Clostridia responsible of tetanus and botulism form a new group of zinc-endopeptidases endowed with peculiar properties (Montecucco & Schiavo, 1993, 1995). They are produced as inactive precursors which are activated by specific proteolysis followed by reduction of a single disulfide bond. Their amino acid sequence around the zinc binding motif does not fit in any of the various groups of zinc-endopeptidases. They act in the cell cytosol and are very specific in terms of both protein target and peptide bond cleaved. A future challenge is that of defining the structural basis of the recognition and cleavage of their substrates. This problem is related to that of finding specific inhibitors of their zinc-endopeptidase activity that are able at the same time to cross the nerve plasmalemma. Such inhibitors would be potential therapeutic agents in the treatment of tetanus and botulism.

Another important line of development is that aimed at determining the three-dimensional structure of these neurotoxins. Highly diffracting crystals of BoNT/A are already under analysis. Another important development will be the determination of their receptors, which is related to the possibility of using the H chain of these toxins as carriers of biologics inside nerve cells.

Acknowledgements

The authors are supported by CNR-Biotecnologie, Telethon-Italia grant 473 and MURST.

References

Bennett, M.K., Calakos, N. & Scheller, R.H. (1992) Syntaxin: a synaptic protein implicated in docking of synaptic vesicles at presynaptic active zones. Science, 257, 255-259.

Bennett, M.K. & Scheller, R.H. (1994) A molecular description of synaptic vesicle membrane trafficking. Annu. Rev. Biochem. 63, 63-100.

Beise, J., Hahnen, J., Andersen-Beckh, B. & Dreyer, F. (1994) Pore formation by tetanus toxin, its chain and fragments in neuronal membranes and evaluation of the underlying motifs in the structure of the toxin molecule. Naunyn-Schmiedeberg's Arch. Pharmacol., 349, 66-73.

Binz, T., Blasi, J., Yamasaki, S., Baumeister, A., Link, E., Südhof, T.C., Jahn, R. & Niemann, H. (1994) Proteolysis of SNAP-25 by types E and A botulinal neurotoxins. J. Biol. Chem., 269, 1617-1620.

Blasi, J., Chapman, E.R., Link, E., Binz, T., Yamasaki, S., De Camilli, P., Südhof, T., Niemann, H. & Jahn, R. (1993) Botulinum neurotoxin A selectively cleaves the synaptic protein SNAP-25. Nature, 365, 160-163.

Blasi, J., Chapman, E.R., Yamasaki, S., Binz, T., Niemann, H. & Jahn, R. (1993) Botulinum neurotoxin C blocks neurotransmitter release by means of cleaving HPC-1/syntaxin. EMBO J., 12, 4821-4828.

Blaustein, R.O., Germann, W.J., Finkelstein, A. & DasGupta, B.R. (1989) The N-terminal half of the heavy chain of botulinum type A neurotoxin forms channels in planar phospholipid bilayers. FEBS Lett., 226, 115-120.

Boquet, P. & Duflot, E. (1982) Tetanus toxin fragment forms channels in lipid vesicles at low pH. Proc. Natl. Acad. Sci. USA, 79, 7614-7618.

Burgen, A.S.V., Dickens, F. & Zatman, L.Y. (1949) The action of botulinum toxin on the neuromuscular junction. J. Physiol. (London), 109, 10-24.

de Paiva, A., Poulain, B., Lawrence, G.W., Shone, C.C., Tauc, L. & Dolly, J.O. (1993) A role for the interchain disulfide or its participating thiols in the internalization of botulinum neurotoxin A revealed by a toxin derivative that bindl to ecto-acceptors and inhibits transmitter release intracellularly. J. Biol. Chem., 268, 20838-20844.

Dayanithi, G., Stecher, B., Hohne-Zell, B., Yamasaki, S., Binz, T., Weller, U., Niemann, H. & Gratzl, M. (1994) Exploring the functional domain and the target of the tetanus toxin light chain in neurohypophysial terminals. Neuroscience, 58, 423-431.

Donovan, J.J. & Middlebrook, J.L. (1986) Ion-conducting channels produced by botulinum toxin in planar lipid membranes. Biochemistry, 25, 2872-2876.
Faber, K. (1890) Die Pathogenie des Tetanus. Berl. klin. Wochenschr., 27, 717-720.
Gambale, F. & Montal, M. (1988) Characterization of the channel properties of tetanus toxin in planar lipid bilayers. Biophys. J., 53, 771-783.
Hayashi, T., McMahon, H., Yamasaki, S., Binz, T., Hata, Y., Südhof, T.C. & Niemann, H. (1994) Synaptic vesicles membrane fusion complex: action of clostridial neurotoxins on assembly. EMBO J. 13, 5051-5061.
Hoch, D.H., Romero-Mira, M., Ehrlich, B.E., Finkelstein, A., DasGupta, B.R. & Simpson, L.L. (1985) Channels formed by botulinum, tetanus, and diphtheria toxins in planar lipid bilayers: relevance to translocation of proteins across membranes. Proc. Natl. Acad. Sci. USA, 82, 1692-1696.
Hohne-Zell, B., Stecher, B. & Gratzl, M. (1993) Functional characterization of the catalytic site of the tetanus toxin light chain using permeabilized adrenal chromaffin cells. FEBS Lett., 336, 175-180.
Kurazono, H., Mochida, S., Binz, T., Eisel, U., Quanz, M., Grebenstein, O., Wernars, K., Poulain, B., Tauc, L. & Niemann, H. (1992) Minimal essential domains specifying toxicity of the light chains of tetanus toxin and botulinum nurotoxin type A. J. Biol. Chem. 267, 14721-14729.
Inoue, A., Obata, K. & Akagawa, K. (1992) Cloning and sequence analysis of cDNA for a neuronal cell membrane antigen, HPC-1. J. Biol. Chem., 267, 10613-10619.
Jankovic, J. & Hallett, M. (eds.) (1994) Therapy with botulinum toxin. Marcel Dekker, New York.
Matteoli, M., Takei, K., Perin, M.S., Südhof, T.C. & De Camilli, P. (1992) Exo-endocytotic recycling of synaptic vesicles in developing processes of cultured hippocampal neurons. J. Cell. Biol., 117, 849-861.
Matthews, B.W., Jansonius, J.N. & Colman, P.M. (1972) Three-dimensional structure of thermolysin. Nature New. Biol., 238, 37-41.
McMahon, H.T., Ushkaryov, Y.A., Edelmann, L., Link, E., Binz, T., Niemann, H., Jahn, R. & Südhof, T.C. (1993) Cellubrevin is a ubiquitous tetanus-toxin substrate homologous to a putative synaptic vesicle fusion protein. Nature, 364, 346-349.
Menestrina, G., Forti, S. & Gambale, F. (1989) Interaction of tetanus toxin with lipid vesicles: effects of pH, surface charge, and transmembrane potential on the kinetics of channel formation. Biophys. J., 55, 393-405.
Minton, N. (1995) Molecular genetics of clostridial neurotoxins. In Clostridial Neurotoxins (ed. C. Montecucco), Curr. Top. Microbiol. Immunol. vol. 195.
Montecucco, C. (1986) How do tetanus and botulinum toxins bind to neuronal membranes? Trends Biochem. Sci., 11, 314-317.
Montecucco, C., Editor (1995) Clostridial Neurotoxins, Curr. Top. Microbiol. Immunol. 195, in press
Montecucco, C., Schiavo, G., Brunner, J., Duflot, E., Boquet, P. & Roa, M. (1986) Tetanus toxin is labeled with photoactivatable phospholipids at low pH. Biochemistry 25, 919-924.
Montecucco, C., Schiavo, G., Gao, Z., Bauerlein, E., Boquet, P. & DasGupta, B.R. (1988). Interaction of botulinum and tetanus toxins with the lipid bilayer surface. Biochem. J., 259, 47-53.
Montecucco, C., Schiavo, G. & DasGupta, B.R. (1989). Effect of pH on the interaction of botulinum neurotoxins A, B and E with liposomes. Biochem. J., 259, 47-53.
Montecucco, C. & Schiavo, G. (1993) Tetanus and botulism neurotoxins: a new group of zinc proteases. Trends Biochem. Sci. 18, 324-327.
Montecucco, C. & Schiavo, G. (1995) Structure and function of tetanus and botulism neurotoxins Quart. Rev. Biophys., 28, in press.

Montecucco, C., Papini, E. & Schiavo, G. (1994) Bacterial protein toxins penetrate cells via a four-step mechanism. FEBS Lett. 346, 92-98.

Osen-Sand, A., Catsicas, M., Staple, J.K., Jones, K.A., Ayala, G., Knowles, J., Grenningloh, G. & Catsicas, S. (1993) Inhibition of axonal growth by SNAP-25 antisense oligonucleotides *in vitro* and *in vivo*. Nature, 364, 445-448.

Oyler, G.A., Higgins, G.A., Hart, R.A., Battenberg, E., Billingsley, M., Bloom, F.E. & Wilson, M.C. (1989) The identification of a novel synaptosomal-associated protein, SNAP-25, differently expressed by neuronal subpopulations. J. Cell. Biol. 109, 3039-3052.

Patarnello, T., Bargelloni, L., Rossetto, O., Schiavo, G. & Montecucco, C. (1993) Neurotransmission and secretion. Nature, 364, 581-582.

Payling-Wright, G. (1955) The neurotoxins of Clostridium botulinum and Clostridum tetani. Pharmacol. Rev., 7, 413-465.

Ramon, G. & Descombey, P.A. (1925) Sur l'immunization antitetanique et sur la production de l'antitoxine tetanique. Compt. Rend. Soc. Biol., 93, 508-598.

Roa, M. & Boquet, P. (1985) Interaction of tetanus toxin with lipid vescicles at low pH. J. Biol. Chem., 260, 6827-6835.

Rossetto, O., Schiavo, G. Montecucco, C., Poulain, B., Deloye, F., Lozzi, L. & Shone, C.C. (1994) SNARE motif and neurotoxin recognition. Nature 372, 415-416.

Schiavo, G., Papini, E., Genna, G. & Montecucco, C. (1990) An intact interchain disulfide bond is required for the neurotoxicity of tetanus toxin. Infect. Immun., 58, 4136-4141.

Schiavo, G., Demel, R. & Montecucco, C. (1991a) On the role of polysialoglycosphingolipids as tetanus toxin receptors: a study with lipid monolayers. Eur. J. Biochem. 199, 705-711.

Schiavo, G., Rossetto, O., Ferrari, G. & Montecucco, C. (1991b) Tetanus toxin receptor. Specific cross-linking of tetanus toxin to a protein of NGF-differentiated PC12 cells. FEBS Lett., 290, 227-230.

Schiavo, G., Poulain, B., Rossetto, O., Benfenati, F., Tauc, L. & Montecucco, C. (1992a) Tetanus toxin is a zinc protein and its inhibition of neurotrasmitter release and protease activity depend on zinc. EMBO J., 11, 3577-3583.

Schiavo, G., Benfenati, F., Poulain, B., Rossetto, O., Polverino de Laureto, P., DasGupta, B.R. & Montecucco, C. (1992b) Tetanus and botulinum-B neurotoxins block neurotransmitter release by a proteolytic cleavage of synaptobrevin. Nature, 359, 832-835.

Schiavo, G., Rossetto, O., Santucci, A., DasGupta, B.R. & Montecucco, C. (1992c) Botulinum neurotoxins are zinc proteins. J. Biol. Chem., 267, 23479-23483.

Schiavo, G., Shone, C.C., Rossetto, O., Alexandre, F.C.G. & Montecucco, C. (1993a) Botulinum neurotoxin serotype F is a zinc endopeptidase specific for VAMP/synaptobrevin. J. Biol. Chem., 268, 11516-11519.

Schiavo, G., Rossetto, O., Catsicas, S., Polverino de Laureto, P., DasGupta, B.R., Benfenati, F. & Montecucco, C. (1993b) Identification of the nerve-terminal targets of botulinum neurotoxins serotypes A, D and E. J. Biol. Chem., 268, 23784-23787.

Schiavo, G., Santucci, A., DasGupta, B.R., Metha, P.P., Jontes, J., Benfenati, F., Wilson, M.C. & Montecucco, C. (1993c) Botulinum neurotoxins serotypes A and E cleave SNAP-25 at distinct COOH-terminal peptide bonds. FEBS Lett., 335, 99-103.

Schiavo, G., Malizio, C., Trimble, W.S., Polverino de Laureto, P., Milan, G., Sugiyama, H., Johnson, E.A. & Montecucco, C. (1994). Botulinum G neurotoxin cleaves VAMP/synaptobrevin at a single Ala/Ala peptide bond. J. Biol. Chem. 269, 20213-20216.

Schiavo, G., Shone, C.C., Bennett, M.K., Scheller, R.H. & Montecucco, C. (1995) Botulinum neurotoxin type C cleaves a single Lys-Ala bond within the carboxyl-terminal region of syntaxins. J. Biol. Chem., 270, 10566-10570.

Schwab, M.E., Suda, K. & Thoenen, H. (1979) Selective retrograde trans-synaptic transfer of a protein, tetanus toxin, subsequent to its retrograde axonal transport. J. Cell. Biol., 82, 798-810.

Simpson, L.L. (1989) (ed) Botulinum neurotoxins and tetanus toxin. Academic Press, New York.

Simpson, L.L., Coffield, J.A. & Bakry, N. (1994) Inhibiton of vacuolar adenosine triphosphatase antagonizes the effects of clostridial neurotoxins but not phospholipase A2 neurotoxins. J. Pharmacol. Exp. Ther. 269, 256-262.

Shone, C.C., Hambleton, P. & Melling, J. (1987) A 50 kDa fragment from the N-terminus of the heavy subunit of Clostridium botulinum type A neurotoxin forms channels in lipid vesicles. Eur. J. Biochem., 167, 175-180.

Shone, C.C., Quinn, C.P., Wait, R., Hallis, B., Fooks, S.G. & Hambleton, P. (1993) Proteolytic cleavage of synthetic fragments of vesicle-associated membrane protein, isoform-2 by botulinum type B neurotoxin. Eur. J. Biochem. 217, 965-971.

Shone, C.C. & Roberts, A.K. (1994) Peptide substrate specificity and properties of the zinc-endopeptidase activity of botulinum type B neurotoxin. Eur. J. Biochem. 225, 263-270.

Söllner, T., Whiteheart, S.W., Brunner, M., Erdjument-Bromage, H., Geromanos, S., Tempst, P. & Rothman, J.E. (1993) SNAP receptors implicated in vesicle targeting and fusion. Nature 362, 318-324.

Stocker, W., Grams, F., Baumann, U., Reinemer, P., Gomis-Ruth, F.X., McKay, D.B. & Bode, W. (1995) The metzincins-Topological and sequential relations between the astacins, adamalysins, serralysins and matrixins (collagenases) define a superfamily of zinc-peptidases. Protein Science 4, 823-840.

Südhof, T.C. (1995) The synaptic vesicle cycle: a cascade of protein protein interactions. Nature, 375, 645-653.

Sweeney, S.T., Broadie, K., Keane, J., Niemann, H. & O'Kane, J. (1995) Targeted expression of tetanus toxin light chain in Drosophila specifically eliminates synaptic transmission and causes behavioural defects. Neuron 14, 341-351.

Tizzoni, G. & Cattani, G. (1890) Untersuchungen über das Tetanusgift. Arch. exp. Pathol. Pharmakol. 27, 432-450.

van Ermengem, E. (1897) šber einem neuen anaeroben Bacillus und seine Beziehungen zum Botulismus. Z. Hyg. InfektKr. 26, 1-56.

Yamasaki, S., Baumeister, A., Binz, T., Blasi, J., Link, E., Cornille, F., Roques, B., Fykse, E.M., Südhof, T.C., Jahn, R. & Niemann, H. (1994b) Cleavage of members of the synaptobrevin/VAMP family by types D and F botulinal neurotoxins and tetanus toxin. J. Biol. Chem. 269, 12764-12772.

Wellhoner, H.H. (1992) Tetanus and botulinum neurotoxins. In Handbook of Experimental Pharmacology, vol. 102 (eds H. Herken & F. Hucho), Springer-Verlag, Berlin, pp. 357-417. Williamson, L.C. & Neale, E.A. (1994) Bafilomycin A1 inhibits the action of tetanus toxin in spinal cord neurons in cell culture. J. Nerochem. 63, 2342-2345.

Wright, J.F., Pernollet, M., Reboul, A., Aude, C. & Colomb, M. (1992) Identification and partial characterization of a low affinity metal-binding site in the light chain of tetanus toxin. J. Biol. Chem., 267, 9053-9058.

Individual Susceptibility to Occupational Toxicants

(Chairs: H. Bolt, Germany, and M. Cikrt, Czech Republic)

Specific Factors Leading to Interindividual Variability in Response to Occupational Toxicants

H.W. Goergens
Zentrum "Arbeit und Gesundheit", Ardeystr. 67, D-44139 Dortmund

Among the many enzymic polymorphisms discovered to be important for toxicity of chemicals in humans (Watanabe and Ikawa, 1995) especially the polymorphic glutathione-S-transferases are of utmost interest in occupational toxicology.

Class μ polymorphism (GSTM1): The GST class μ isozyme is polymorphic in man, as first observed on incubations of human leukocytes with the classic substrate trans-stilbene oxide (Seidergård et al., 1985). About half the subjects in populations studied so far express the μ isozyme in leukocytes. The polymorphism is due to deletion of the entire GSTM1 gene, resulting in a null allele, (Seidergård et al., 1988).

Due to this polymorphism, GST activities in human lymphocytes towards *trans*-stilbene oxide were found to vary between practically zero and nearly 400pmol/min/10^6 cells; activities were low or absent in homozygously deleted subjects, followed by heterozygotes and then by homozygotes for the GSTM1 gene (Bell et al., 1992).

One can distinguish between the allelic variants ψ (GSTM1a) and μ (GSTM1b); these differ by one G-C_{534} exchange, leading to a change of lysine into asparagine. With test substrates, the resulting differences in enzyme activities between these variants are minimal (Brockmöller et al., 1993; Seidergård et al., 1988). There is a discussion as to whether non-expression of the GSTM1 gene is accompanied by a higher risk of lung cancer (Idle et al., 1992; Brockmöller et al., 1993; Kihara et al., 1994).

The hypothesis has been put forward that genetic polymorphism of GSTM1 might be linked with the occurence of human bladder cancer (Bell et al., 1993; Lin et al., 1994). A study of 229 patients with transitional cell carcinoma of the urinary epithelium and 211 control subjects in the USA led to the result that the GSTM1 O/O genotype conferred a 70% increased risk of bladder cancer, probably by a slower detoxification of tobacco-smoke related carcinogens. It was hypothesized that 25% of all bladder cancer (in the USA) might be attributable to this at-risk genotype (Bell et al., 1993).

From the standpoint of occupational toxicology, a polymorphism in the metabolism of a very special laboratory chemical like *trans*-stilbene oxide would be of limited interest only. However, it has been established that GST class μ catalyses also the conjugation of other epoxides, especially of styrene oxide , the intermediate epoxide metabolite of styrene. Warholm et al. (1983) have compared the turnover of styrene oxide *in vitro* by GST, α, μ and π; the class μ enzyme displayed a more than 20-fold higher reaction velocity than class π, and almost no enzyme reaction was seen with class α.

Recent work on the excretion of mercapturic acid conjugates of styrene metabolites in styrene-exposed workers has revealed major interindividual variations. This points to large individual differences in the biological inactivation of styrene oxide and is consistent with the long known variations in excretion of mandelic acid, which is the alternative end product of styrene oxide metabolism (Hallier et al., 1995).

Class θ polymorphism (GSTT1): A hitherto unknown glutathione transferase activity for C_1/C_2 compounds in human erythrocytes has been characterized which was later linked with the θ class of GST (Schröder et al., 1992). This enzyme activity also displays polymorphism. It is found in approximately three-quarters of the population, called "conjugators", whereas "non-conjugators" lack this specific enzyme activity. No animal model has been found so far for this unique human phenomenon in red blood cells. Animals which show class θ related activity in the liver do not express this enzyme in erythrocytes.

Previous investigations in our laboratory have identified a number of substrates for this polymorphic enzyme activity, namely methyl chloride (Peter et al., 1989), methyl bromide, methyl iodide (Hallier et al.,1990), dichloromethane (Thier et al. 1989) and ethylene oxide (Föst et al., 1991). Among these substrates, methyl bromide has the highest affinity for the new glutathione transferase isozyme in human erythrocytes and can therefore be used as a standard test substrate for the phenotyping of "conjugators" and "non-conjugators" from human blood specimens.

Support for the existence of great interindividual differences in the metabolism of methyl bromide and methyl chloride may also be inferred from published reports on subjects occupationally exposed to these compounds (Bolt and Gansewendt 1993). A field study on methyl bromide workers was performed in Japan by using the hemoglobin adduct S-methylcysteine as a biological monitor (Iwasaki et al., 1989). In a subgroup of seven workers with the highest exposure levels (filling of spray cans and gas cylinders), three individuals had very high adduct levels whereas the other four had only low levels, close to the background of persons not occupationally exposed to methyl bromide. A field study of six workers occupationally exposed to methyl chloride in the Netherlands (van Doorn et al., 1980) showed that four of the subjects excreted higher amounts of the metabolite S-methylcysteine in urine whereas two others "hardly excreted any S-methylcysteine after exposure to methyl chloride".

Similar interindividual differences have also been reported for ethylene oxide *in vivo*. In a clinical study, hemodialysis patients showing immediate-type allergic reactions during treatment, called "reactors", had significantly higher levels of ethylene oxide adducts with serum albumin than those without allergic reactions ("non-reactors") (Grammer et al., 1984). In investigations of DNA damage using the "alkaline elution" method in subjects occupationally exposed to ethylene oxide, indications were found for the existence of two subgroups with different susceptibility to ethylene oxide within the group of non-smokers (Fuchs et al., 1994).

Incubation of whole blood samples with any of the three chemicals methyl bromide, ethylene oxide and dichloromethane led to a marked induction of sister chromatid exchanges in the lymphocytes of "non-conjugators". An effect of the chemicals on the fequency of SCE was not or only hardly observable in the lymphocytes of the "conjugators" (Hallier et al., 1993). This was thought to be due to the reduction in the blood samples of the conjugators throught metabolic elimination by the GST in erythrocytes. The chemically highly reactive substance methyl bromide led to practically no cytogenetic effect on the lymphocytes of conjugators due to its high affinity for the metabolizing enzyme in erythrocytes. From previous investigations *in vitro* it was known that the distribution of radiolabelled ethylene oxide (Föst et al., 1991) and dichloromethane (Thier et al., 1989) in the cellular and plasma compartments of human blood, upon incubation, differed strikingly between blood from "conjugators" and from non-conjugators". The polymorphic GSTT1-1 in human erythrocytes converts dichloromethane to formaldehyde (Hallier et al., 1994).

In this context, it is of great interest that similar interindividual differences in cytosolic GST in human livers were also found, using the substrate dichloromethane (Bogaards et al., 1993).

The human polymorphic GST catalysing conjugation of dichloromethane could be characterized as a θ enzyme (GSTT1) by means of molecular biology. A cDNA was isolated which enclosed the human class θ GSTT1 and which had 82% identitiy to rat GSTsu6. Using PCR and Southern blot analyses it was shown that the GSTT1 gene is absent from approximately one-third of the population. It was also shown that the "conjugator" and "non-conjugator" phenotypes (v.s.) coincide with the presence and absence of the GSTT1 gene, respectively (Pemble et al., 1994).

The introduction of a plasmid encoding for the rat class θ GST-5-5 into S. typhimurium TA 1535 made it possible to demonstrate the effectiveness of GST class θ to transform dichloromethane to mutagenic metabolites (Thier et al., 1993).

The influence of GSTT1 enzyme polymorphism on the cytogentic toxicity of methyl bromide, ethylene oxide and dichloromethane bromide *in vitro* suggests that this may be an important factor in individual susceptibility to the toxic effects of these substances. It is now clear that 11-50% of various populations studied do not possess the GSTT1 gene (Hallier et al., 1993; Warholm et al., 1994; Bell, 1994). The gene product in indiividuals possessing this gene is expressed at least in liver and in erythrocytes. This means that human phenotyping is easily possible with the GST activity towards methyl bromide or ethylene oxide in erythrocytes, and that this is representative of the metabolic GSTT1 competence of the entire organism.

According to very recent data by Guengerich et al. (1995) glutathione-S-tranferase class theta is also involved in metabolism of the important 1,3-butadiene metabolite diepoxybutane to highly genotoxic intermediates. This stresses the potential importance of this particular polymorphic isoenzyme for human occupational toxicology.

References

Bell DA, Thompson CL, Taylor J, Miller CR, Peiera F, Hsieh LL, Lucier GW (1992) Genetic monitoring of human polymorphic cancer susceptibility genes by polymerase chain reaction: Application to Glutathione Transferase μ. Environ Hlth Perspect 98:113-117.

Bell DA, Taylor JA, Paulson DF et al. (1993) Genetic risk and carcinogen exposure: a common inherited defect of the carcinogen-metabolism gene glutathione-S-transferase M1 (GSTM1) that increases susceptibility to bladder cancer. J Natl Cancer Inst 85:1159-1164.

Bogaards JJP, van Ommen B, van Bladeren PJ (1993) Interindividual differences in the *in vitro* conjugation of methylene chloride with glutathione S-transferase in 22 human liver samples. Biochem Pharmacol 45:2166-2169

Bolt HM, Gansewendt, B (1994) Mechanism of carcinogenicity of methyl halides. CRC Crit Rev Toxicol 23:237-253

Brockmöller J, Kerb R, Drakoulis N et al. (1993) Genotype and phenotype of glutathione S-transferase class μ isoenzymes μ and ψ in lung cancer patients and controls. Cancer Res 53:1004 -1011

Doorn R van, Borm PJA, Leijdekkers CM, Henderson PT, Reuvers J, van Bergen TJ (1980) Detection and identification of S-methylcysteine in urine of workers exposed to methyl chloride. Int Arch Occup Environ Health 46:99-109

Föst U, Hallier E, Ottenwälder H et al. (1991) Distribution of ethylene oxide in human blood and its implications for biomonitoring. Hum Exp Toxicol 10:25-31

Fuchs J, Wullenweber U, Hengstler JG, Bienfait HG, Hiltl G, Oesch F (1994) Genotoxic risk for humans due to workplace exposure to ethylene oxide: remarkable individual differences in susceptibility. Arch Toxicol 68:343348

Grammer LC, Roberts M, Nicholls AJ et al. (1984) IgE against ethylene oxide-altered human serum albumin in patients who have had acute dialysis reactions. J Allergy Clin Immunol 74:544-549

Guengerich FP, Thier R, Persmark M, Taylor JB, Pemble SE, Ketterer B (1995) Conjugation of carcinogens by theta class glutathione-S-transferase: mechanisms and relevance to variations in human risk. Pharmacogenetics 5, Suppl.:S103-S107

Hallier E, Deutschmann S, Reichel C et al. (1990) A comparative investigation of the metabolism of methyl bromide and methyl iodide in human erythrocytes. Int Arch Occup Environ Health 62:221-225

Hallier E, Langhof T, Dannappel D et al. (1993) Polymorphism of glutathione conjugation of methyl bromide, ethylene oxide and dichloromethane in human blood: influence on the induction of sister chromatid exchanges (SCE) in lymphocytes. Arch Toxicol 67:173-178

Hallier E, Schröder KR, Asmuth K, Dommermuth A, Aust B, Goergens HW (1994) Metabolism of dichloromethane (methylene chloride) to formaldehyde in human erythrocytes: influence of polymorphism of glutathione transferase Theta (GSTT1- 1). Arch Toxicol 68:423-427

Hallier E, Goergens HW, Karels H, Golka K (1995) A note on individual differences in the urinary excretion of optical enantiomers of styrene metabolites and of styrene-derived mercapturic acids in humans. Arch Toxicol 69:300-305

Idle JR, Armstrong M, Boddy AV et al. (1992) The pharmacogenetics of chemical carcinogenesis. Pharmacogenetics 2:246-258

Iwasaki K, Ito I, Kagawa J (1989) Biological exposure monitoring of methyl bromide workers by determination of hemoglobin adducts. Ind Health 27:181-183

Kihara M, Noda K (1994) Lung cancer risk of GSTM1 null genotype is dependent on the extent of tobacco smoke exposure. Carcinogenesis 15:415-418

Lin HJ, Han CY, Bernstein DA, Hsiao W, Liu BK, Hardy S (1994) Ethnic distribution of the glutathione transferase Mu1-1 (GSTM1) null genotype in 1473 individuals and application to bladder cancer susceptibility. Carcinogenesis 15:1077-1081

Pemble S, Schroeder KR, Spencer SR, Meyer DJ, Hallier E, Bolt HM, Ketterer B, Taylor J (1994) Human glutathione S-transferase theta (GSTT1): c-DNA cloning and the characterization of a genetic polymorphism. Biochem J 300:271-276

Peter H, Deutschmann S, Reichel C, Hallier E (1989) Metabolism of methyl chloride in human erythrocytes. Arch Toxicol 63:351-355

Schröder K, Hallier E, Peter H, Bolt HM (1992) Dissociation of a new glutathione S-transferase activity in human erythrocytes. Biochem Pharmacol 43:1671-1674

Seidergård J, De Pierre JW, Pero RW (1985) Hereditary interindividual differences in the glutathione transferase activity towards trans-stilbene oxide in resting human mononuclear leucocytes are due to a particular isozyme(s). Carcinogenesis 6:1211-1216

Seidergård J, Vorachek WR, Pero RW, Pearson WR (1988) Hereditary differences in the expression of the human glutathione S-transferase activity on trans-stilbene oxide are due to a gene deletion. Proc Natl Acad Sci USA 85:7293-7297

Thier R, Föst U, Deutschmann S et al. (1989) Distribution of methylene chloride in humane blood. Arch Toxicol, Suppl. 14:254-258

Thier R, Taylor JB, Pemble SE et al. (1993) Expression of the rat theta class glutathione S-transferase 5-5 in S. typhimurium TA 1535 leads to base-pair mutation upon exposure to dihalomethanes. Biol Chem Hoppe-Seyler 374:796 (abstract F 105)

Warholm M, Guthenberg C, Mannervik B (1983) Molecular and catalytic properties of glutathione transferase μ from human liver, an enzyme conjugating epoxides. Biochemistry 22:3610-3617

Warholm M, Alexandrie AK, Högberg J et al. (1994) Polymorphic distribution of glutathione transferase activity with methyl chloride in human blood. Pharmacogenetics 4:307-311

Watanabe M, Ikawa S (1995) 14th Sapporo Cancer Seminar on Genetic Polymorphism and Cancer Susceptibility. Pharmacogenetics 5:i

Genetic Predisposition in Occupational Toxicology

Hermann M. Bolt
Institut für Arbeitsphysiologie an der Universität Dortmund
Ardeystr. 67, D-44139 Dortmund, Germany

Introduction

We are now witnessing the rise of a new era of research in toxicology which Zbinden has called the "era of individual expression" (Zbinden, 1992). It has been recognized that inherited genetic differences determine the susceptibility of an individual to a toxic chemical. Most of the specific factors described so far refer to the metabolism of xenobitocs which is critical for the balance of toxification and detoxification.

Ideas in this field were first developed in clinical pharmacology. For instance, the classical twin studies by Elliot S. Vesell (1975) have clearly demonstrated that the individual disposition of drugs may be largely determined by inherited factors, and population studies of pharmacokinetics of contraceptive estrogens by Goldzieher et al. (1980) have shown interesting "ethnic" differences in the metabolism of ethinylestradiol which now may be linked with the cytochrome P-450 isozyme CYP3A4 (Brian et al. 1990, Kerlau et al. 1992). Another classical field is the molecular genetics of alcohol metabolism and alcoholism (Agarwal et al., 1992; Yoshida, 1992). In clinical pharmacology, attention is being paid to the genetically variable metabolism of a variety of drugs (Dahl et al., 1993).

In toxicology, special emphasis has been put on the pharmacogenetics (toxicogenetics) of chemical carcinogens, and specifically on genetic polymorphisms related to cancer susceptibility (Carporoso et al. 1991; Idle et al. 1992).

It is obvious that these developments have a general drawback to occupational toxicology. Specifically, interference of the genetic variabilities of xenobiotic metabolism with biological monitoring and with the toxic actions of common workplace chemicals are of paramount interest.

Cholinesterases, a classic field: Pseudocholinesterase and "paraoxonase" deficiencies have been widely studied (Idle et al. 1992); different genotypes and phenotypes have been characterized very early (Kalow et al. 1957). In anesthesiology, atypical prolongations of the clinical effect of succinylcholine, due to polymorphism of pseudocholinesterase with reduced enzyme activity, are of have revealed systematic differences in excretion of the metabolite 2,5-hexandione between Japanese and European populations exposed to n-hexane (Deutsche Forschungsgemeinschaft, 1993). Such unexplained geographic differences may well be related to genetic factors.

N-Acetylation of aromatic amines: The molecular genetics of N-acetyltransferases (NAT-1, NAT-2) is a field of great current interest (Grant, 1993). The product of the NAT-1 gene, showing selectivity for compounds like p-amino-salicylic acid, is unrelated to the classically defined acetylation polymorphism governed by the NAT-2 genes.

Analysis of the ratio between two major caffeine urinary metabolites (5-acetylamino-6-formyl-amino-3-methyluracil and 1-methylxanthine) as a means of determining the NAT-2 phenotype has significantly facilitated the conduction of field studies in occupational toxicology (Butler et al., 1992).

In the pathogenesis of human bladder cancer due to occupational exposure to "classical" aromatic amines (e.g., benzidine, 4-aminodiphenyl, 1-naphthylamine) acetylation is mostly regarded as a detoxication step. Accordingly, studies on industrial workers exposed to such compounds in Poland (Hanke et al., 1990) and the United States (Weber et al., 1983), showed much higher risks of bladder cancer, associated with the slow acetylator phenotype. A study on benzidine-exposed workers in China did not confirm these findings (Hayes et al., 1993). However, it is important to note that in European populations the groups of slow *vs.* intermediate plus fast acetylators are about evenly distributed whereas in the Far East (Japan, China) slow acetylators represent only about 10% of the population (Deguchi et al., 1990); therefore, the number of slow acetylators in the cohorts studied is especially important.

A German study on 331 workers formerly employed (years 1951-1967) in the production of benzidine at BAYER, Leverkusen, is of particular interest for occupational toxicology. The distribution of slow (160 persons) and rapid (171 persons) acetylators in the entire cohort of workers corresponded well with the figures in the general popolation. Within the entire cohort, 92 persons suffered lastly from urinary bladder cancer. Among these cancer patients, however, were 75 slow, but only 17 fast acetylators (Lewalter et al., 1991).

However, it must be noted that the exposure situation has changed considerably since that time, as indicated by a recent hospital-based study performed at Leverkusen (Schöps et al., 1995) which showed an only very slight trend in local chemical workers suffering from urothelial cancer towards the slow-acetylator status.

A topic which has recently been very much debated is the carcinogenicity of benzidine-based azo dyes, due to metabolic liberation of benzidine (Bolt, 1995). Specifically, the former use of such materials by German painters has been debated. In this situation, it is of interest that in painters suffering from urothelial cancer a shift towards the slow-acetylator status can be observed (Kempkes et al., 1995) which confirms a connection with a former "internal exposure" to benzidine, due to working with carcinogenic colorants, and the disease.

have revealed systematic differences in excretion of the metabolite 2,5-hexandione between Japanese and European populations exposed to n-hexane (Deutsche Forschungsgemeinschaft, 1993). Such unexplained geographic differences may well be related to genetic factors.

N-Acetylation of aromatic amines: The molecular genetics of N-acetyltransferases (NAT-1, NAT-2) is a field of great current interest (Grant, 1993). The product of the NAT-1 gene, showing selectivity for compounds like p-amino-salicylic acid, is unrelated to the classically defined acetylation polymorphism governed by the NAT-2 genes.

Analysis of the ratio between two major caffeine urinary metabolites (5-acetylamino-6-formyl-amino-3-methyluracil and 1-methylxanthine) as a means of determining the NAT-2 phenotype has significantly facilitated the conduction of field studies in occupational toxicology (Butler et al., 1992).

In the pathogenesis of human bladder cancer due to occupational exposure to "classical" aromatic amines (e.g., benzidine, 4-aminodiphenyl, 1-naphthylamine) acetylation is mostly regarded as a detoxication step. Accordingly, studies on industrial workers exposed to such compounds in Poland (Hanke et al., 1990) and the United States (Weber et al., 1983), showed much higher risks of bladder cancer, associated with the slow acetylator phenotype. A study on benzidine-exposed workers in China did not confirm these findings (Hayes et al., 1993). However, it is important to note that in European populations the groups of slow *vs.* intermediate plus fast acetylators are about evenly distributed whereas in the Far East (Japan, China) slow acetylators represent only about 10% of the population (Deguchi et al., 1990); therefore, the number of slow acetylators in the cohorts studied is especially important.

A German study on 331 workers formerly employed (years 1951-1967) in the production of benzidine at BAYER, Leverkusen, is of particular interest for occupational toxicology. The distribution of slow (160 persons) and rapid (171 persons) acetylators in the entire cohort of workers corresponded well with the figures in the general popolation. Within the entire cohort, 92 persons suffered lastly from urinary bladder cancer. Among these cancer patients, however, were 75 slow, but only 17 fast acetylators (Lewalter et al., 1991).

However, it must be noted that the exposure situation has changed considerably since that time, as indicated by a recent hospital-based study performed at Leverkusen (Schöps et al., 1995) which showed an only very slight trend in local chemical workers suffering from urothelial cancer towards the slow-acetylator status.

A topic which has recently been very much debated is the carcinogenicity of benzidine-based azo dyes, due to metabolic liberation of benzidine (Bolt, 1995). Specifically, the former use of such materials by German painters has been debated. In this situation, it is of interest that in painters suffering from urothelial cancer a shift towards the slow-acetylator status can be observed (Kempkes et al., 1995) which confirms a connection with a former "internal exposure" to benzidine, due to working with carcinogenic colorants, and the disease.

These examples show that assessing the genetic status with respect to susceptibility towards toxicants is a tool of increasing value in occupational toxicology.

References

Agarwal DP, Goedde HW (1992) Pharmacogenetics of alcohol metabolism and alcoholism. Pharmacogenetics 2:48-62

Bell DA (1994) Ethnic variation in xenobiotic metabolism and increased risk for carcinogen-induced bladder cancer. ISSX Proceedings 6:51

Bogaards JJP, van Immen B, van Bladeren PJ (1993) Interindividual differences in the *in vitro* conjugation of methylene chloride with glutathione S-transferase in 22 human liver samples. Biochem Pharmacol 45:2166-2169

Bolt HM (1995) Special points in the toxicity assessment of colorants (dyes and pigments). In: Thomas H, et al., Toxicology of industrial compounds. Taylor & Francis, London, in press

Bolt HM, Gansewendt, B (1994) Mechanism of carcinogenicity of methyl halides. CRC Crit Rev Toxicol 23:237-253

Brian WR, Sari MA, Iwasaki M (1990) Catalytic activities of human liver cytochrome P-450 IIIA4 expressed in Saccaromyces cerevisiae. Biochemistry 29:11280-11292

Butler MA, Lang NP, Young JF et al. (1992) Determination of CYP1A2 and N-acetyltransferase 2 phenotypes in human population by analysis of caffeine urinary metabolites. Pharmacogenetics 2:116-127

Caporaso N, Landi MT, Vineis P (1991) Relevance of metabolic polymorphisms to human carcinogenesis; evaluation of epidemiologic evidence. Pharmacogenetics 1:4-19

Caporaso N, De Baum MR, Rothman N (1995) Lung cancer and CYP 2D6 (the debrisoquine polymorphism): sources of heterogenicity in the proposed association. Pharmacogenetics 5, Suppl.:S129-S134

Dahl ML, Bertilsson L (1993) Genetically variable metabolism of antidepressants and neuroleptic drugs in man. Pharmacogenetics 3:61-70

Deguchi T, Mashimo M, Suzuki T (1990) Correlation between acetylator phenotypes and genotypes of polymorphic arylamine N-acetyltransferase in human liver. J Biol Chem 265:12757-12760

Deutsche Forschungsgemeinschaft (1994) Maximum concentrations at the workplace and biological tolerance values for working materials. VCH Publ., Weinheim

Deutsche Forschungsgemeinschaft (1993) Biologische Arbeitsstoff-Toleranzwerte (BAT-Werte). Arbeitsmedizinisch-Toxikologische Begründungen. 7. Lfg., Henschler D, Lehnert G (eds). VCH Verl.Ges., Weinheim

Grant DM (1993) Molecular genetics of aromatic amines. Pharmacogenetics 3:45-50

Goldzieher JW, Dozier TS, de la Pena A (1980) Plasma levels and pharmacokinetics of ethynyl estrogens in various populations. Contraception 21:1-15

Hanke J, Kajewska A (1990) Acetylation phenotype and bladder cancer. J Occup Med 32:917-918

Hayes RB, Rothman N, Broly F et al. (1993) N-Acetylation phenotype and risk of bladder cancer in benzidine-exposed workers. Carcinogenesis 14:675-678

Idle JR, Armstrong M, Boddy AV et al. (1992) The pharmacogenetics of chemical carcinogenesis. Pharmacogenetics 2:246-258

Kalow W, Staron N (1957) On the distribution and inheritance of a typical forms of human serum cholinesterase, as indicated by dibucaine numbers. Can J Biochem 35:1305

Kempkes M, Golka K, Bolt HM (1995) Polymorphismus der N-Acetyltransferase-2 in der urothelialen Tumorgenese bei Malern. Verh Dt Ges Arbeitsmed Umweltmed 35:in press

Kerlau V, Dreano Y, Bercovici JP (1992) Nature of cytochromes P-450 involved in the 2/4-hydroxylations of estradiol in human liver microsomes. Biochem Pharmacol 44:1745-1756

Laib RJ, Tucholski M, Filser JG, Csanady GA (1992) Pharmacokinetic interaction between 1,3-butadiene and styrene in Sprague-Dawley rats. Arch Toxicol 66:310-314

Lewalter J, Miksche LW (1991) Empfehlungen zur arbeitsmedizinischen Prävention expositions- und dispositionsbedingter Arbeitsstoff-Beanspruchungen. Verh Dt Ges Arbeitsmed 31:135-139

Löhr GW (1970) Über die Bedeutung genetisch bedingter Enzymdefekte für die Pharmakotherapie. In: Löhr GW, Blum KU, Wiedemann OJ (eds) Enzyme und Pharmaka. Editiones "Roche", Grenzach, pp 79-101

Nebert DW, Adesnik M, Coon MJ, et al. (1987) The P-450 gene superfamily: recommended nomenclature. DNA 6:1-11

Rancy JL, Kraner JC, Lasker JM (1993). Bioactivation of halogenated hydrocarbons by cytochrome P-450E1. CRC Crit Rev Toxicol 23:1-20

Schöps W, Prior V, Golka K, Blaszkewicz M, Cascorbi J, Roots J, Bolt HM, Kierfeld G (1995) Untersuchung zur klinischen Relevanz der Acetyliererphänotypisierung bei 196 Urotheltumorpatienten. Der Urologe, in press

Sugimura H, Hamada GS, Suzuki J, Iwase T, Kiyokawa E, Kini J, Tsugane S (1995) CYP1A1 and CYP2E1 polymorphism and lung cancer, case-control study in Rio de Janeiro, Brasil. Pharmacogenetics 5, Suppl.:S145-S148

Vesell ES (1975) Pharmacogenetics. Biochem Pharmacol 24:445-450

Weber WW, Hein DW, Litwin A, Lower GM (1983) Relationship of acetylator status to isonioazid toxicity, lupus erythematosus and bladder cancer. FASEB J 42:3086-3097

Yoshida A (1992) Molecular genetics of human aldehyde dehydrogenase. Pharmacogenetics 2:139-147

Zbinden G (1992) The three eras of research in experimental toxicology. TIPS 13:221-223

Individual Susceptibility to Occupational Toxicants: Practical Consequences for Risk Management

T. Norseth
National Institute of Occupational Health, Box 8149 Dep 0033 Oslo, Norway

Introduction

Individual susceptibility or fitness to work have always been of major importance for general participation in the working life. Susceptibility may be defined as the sum of inherited and acquired human qualities and characteristics, and as such no human being is alike. This paper deals with the practical consequences and some ethical problems of individual susceptibility to occupational toxicants for risk management. Having accepted that no human being is alike, the concern is that this different susceptibility should neither lead to exploitation nor to discrimination of workers.

The remarkable advances in genetics during the past few years have led to dramatic improvements not only in the understanding of individual susceptibility to human diseases, but also in the understanding of the genetic basis for human behaviour in general. Within the next three years or so we must face that the International Human Genome Project will have identified the structure of every human gene, totalling 100,000. It is also expected to have identified genes associated with 4000 known genetic conditions. Even if susceptibility based on the genotype in recent years has been the focus of interest, it should be remembered that most health conditions - good or bad - are caused by some host-environment interactions. By definition, all work-related diseases belong to this category, and testing for susceptibility has already been in use for many years (Lappè 1983; Uzych 1986). In some aspects, genetic data do not seem different from all other medical data or data on socio-economic conditions, but there are some differences which will be discussed later (Gevers, 1993). Even level of education and specific skills generally accepted as a basis for selection may depend on genetic predisposition (susceptibility).

A simultaneous assessment of individual susceptibility and occupational risk factors offers risk management based on individual risk profiles for the benefit of the workers, but also carries the inherent possibility for discrimination based on alleged loss of working capacity or economical loss in the future for the employer. In this paper this controversial will be discussed based on present knowledge about the importance of individual susceptibility for some exposures and for some selected biological markers and diseases.

Even if risk assessment usually is not included in the risk management concept, some aspects of individual susceptibility in the risk assessment process will also be discussed because of the close linkage between risk assessment and risk management. Risk assessment always is the first step in a risk management process.

Effects, Dose-Response and Standard Setting

So-called stochastic individual susceptibility is expressed by the ideal dose-response curve, the normal distribution curve. The dose-response curve thus represents the frequency of a given effect in a population at different dose levels. Individual susceptibility may both relate to effects recorded only in the susceptible population, to different effects at corresponding dose levels or to a higher probability of an effect at a lower dose level.

For statistical reasons, neither animal experiments nor epidemiological investigations are useful for accurate estimation of the lower part of the dose-response curve. Experimentally the problem of low hazard identification has been solved by animal experiments with high doses and thus increased risk. Different extrapolation models accounting for extrapolation from high to low doses, for species differences and route of exposure have been applied. The so-called uncertainty factor is influenced by the nature and severity of the principal effect, the population in question and the degree of control achievable (Fairhurst 1995). Uncertainty factors as multiplications or fractions of 10 have a strong tradition in toxicological risk management, but are most often based on political acceptability rather than on scientific evidence. Both "health based standards", "no adverse effect level (NOAEL)" or similar terms are thus scientifically meaningless. A scientific basis for the uncertainty factor can be obtained only by studies of mechanisms.

Biological Markers of Individual Susceptibility

Causal inference in epidemiological investigations depends on the presence and frequency of individual susceptibility in the population under investigation. Khoury et al. (1987) have illustrated how the presence of an unknown genetic susceptibility to the effect of an environmental agent can dilute both measures of environmental and familiar risks and may lead to erroneous etiologic inferences. For environmental exposure the measured relative risk is diluted by the proportion of non-susceptibles in the population. The relative risk reaches its maximum value (with a given exposure) when all individuals are equally susceptible.

The power of epidemiological studies to identify effects caused by human exposure to toxicants will be greatly increased as markers of disease susceptibility are established allowing for the separation of people at high risk from those of low or no risk. This implies that it is possible to establish discrete dose-response curves for hyper- or hypo-susceptible individuals. But even within the hyper- or hypo-susceptible individuals there is a stochastic individual susceptibility with the inherent statistical uncertainty at low exposure (dose) levels.

Markers as discussed by Khoury et al. (1987) relate to genetic susceptibility, but the power of epidemiological investigations to identify effects caused by environmental exposures may also be improved by the use of biological markers for exposure or effects, even with the use of biological markers for prognosis. Such "surrogate end points" may imply genetic susceptibility, but may also be used for establishing cohorts with defined exposure or pre-clinical disease (sub-critical effects). Single individuals with a marker or a surrogate end point are defined in the populations and new cohorts are formed for investigating other relevant end points. A stratification of susceptibility will have the same effect for epidemiological hypothesis testing as stratification of exposures, direct the risk ratio away from unity. The biological marker or surrogate end point may relate to genetic predisposition, to intercurrent diseases, to life style (smoking) or to effects caused by previous and/or mixed exposures.

Traditionally, biological markers have been used for risk management as markers of exposure or early effects. Theoretically, however, all markers of susceptibility, genetic traits or non-genetic varieties, could be used for susceptibility stratification. To be useful in epidemiology, the use of biomarkers should reduce misclassification of exposure and/or effect, enhance detection of exposure-effect associations, and thus increase the opportunities for risk management (Shields and Harris 1991; Schulte 1992).

The number of known genetic traits for a variety of illnesses increases rapidly, probably the number will increase at least 10-fold over the next decade. Very few (less than 3 %) of human diseases are caused by defects in a single gene. Environmental or occupational factors may thus influence an otherwise benefictional or disastrous outcome foretöld in the DNA. Genetic tests can therefore very seldom predict an individual's future with perfect clarity, but much should be learned by epidemiological susceptibility stratification.

Cancer as an End Point for Susceptibility

Genetic polymorphism has been used extensively to classify cancer patients by a posteriori susceptibility in the case-control model. Very few, if any papers, have been presented for improving the power of epidemiological investigations for risk management purposes (Sidransky 1995). Candidates for prediction would be proto-oncogenes (H-RAS-1 and L-MYE) and p53 polymorphisms (Weston et al. 1992; Ryberg et al. 1994), cytochrome P450 and the human glutathione S-transferase supergene family for lung cancer risk (Guengerich et al. 1992;

Ketterer et al. 1992), and N-acetyltransferase polymorphism for bladder cancer (Vineis and Ronco 1992). The debrisoquine metabolic phenotype offers an interesting possibility for susceptibility testing (Caporaso et al. 1992). A phenotyping approach is more sensitive, but less specific than the genotyping approach. Epidemiological investigations at present are most commonly based on the phenotyping approach.

Stratification for smoking is important for carcinogenic risk assessment and management, but such information is still often insufficient for accurate risk assessment. The traditional example is smoking and asbestos exposure (Hammond and Selikoff 1973). Definite information on cancer risk in non-smoking asbestos workers was not obtained until also the reference population was stratified into smokers and non-smokers with a sufficient observation period (Hammond et al. 1978).

A further discussion of individual susceptibility based on mixed exposure and interaction between chemicals is outside the scope of this paper. Genetic traits for behavioral patterns have been identified although at present the results are highly controversial, and the consequences are probably heavily modified by the phenotype. Such traits are of potential importance for risk management and may also relate to occupational exposure to toxicants, but will not be further discussed.

Other End Points for Susceptibility

Susceptibility to lung cancer has been discussed above, but with every breath we inhale unwanted particles and gases, most of which are innocuous, some few may cause cancer, but probably many more injure the lungs through irritative effects or effects on the immune system.

Studies of occupational asthma have revealed some significant risk factors. These are prior or pre-existing asthma, atopy and smoking (Hendrick 1994). Both atopy and a family or personal history of asthma have been used as markers for increased susceptibility to occupational diseases. For unspecific irritative effects in the respiratory organs, non-specific airway responsiveness using metacholine or histamine have been suggested, but limited predictive power has been reported (Søyseth et al. 1994).

In some industries with exposure to high molecular weight agents, atopy has been considered to be an important individual risk factor. It appears, however, that both the specific airway responsiveness and atopy are unimodally distributed, even the specific IgE response (Hendrick 1994). It is thus concluded that neither atopy nor airway responsiveness are useful as a discriminant in selection for employment with exposure to allergenic substances even with a known IgE mediated effect as for laboratory animal allergy (Slovak and Hill 1987). Smoking as a fact turns out to be a better predictor (Venables et al. 1989). A justification of excluding atopic individuals who smoke and have a past history suggesting active asthma has been suggested (Hendrick 1994). It has been reported that atopy is a predisposing factor for chronic bronchitis, the practical consequences of this report do not seem different from other atopic conditions (Terho et al. 1995). It should be remembered that no occupational respiratory disease is limited to atopics exclusively (Terr 1986).

The genetic mechanism of asthma is not clear, neither for other allergies (Mapp et al. 1994). Recessive inheritance of a single major gene for high levels of IgE has been demonstrated, but the results are inconclusive. Genetic factors have been demonstrated with low molecular weight substances as isocyanate - a very potent allergy inducer. Other genetic mechanisms have been described, but the practical consequences of these are uncertain (Chan-Yeung and Malo 1994). Testing for α_1-antitrypsin may be of pratical value, the importance of the cystic fibrosis gene remains to be seen. Recently a genetic marker of beryllium disease was reported (Richeldi et al. 1993).

The information on the influence of atopy on the development of occupational skin disease is conflicting. Studies to date fail to show clear-cut correlation between allergic contact sensitivity and contact dermatitis even if a higher prevalence of contact dermatitis has been demonstrated with hay fever and other examples of IgE mediated allergy. However, the dryness and hyper-irritability of the skin in atopic dermatitis render the skin more susceptible to unspecific irritation (Terr 1986). Testing for atopy with exposure to skin sensitizers or irritants may have practical consequences for risk assessment, but the sensitivity and specificity do not seem sufficient for risk management. A careful history of skin diseases seems, however, warranted (Shmunes 1986).

Immunological mechanisms of alleged genetic nature have been suggested and are currently actively sought for in unspecific and poorly understood conditions like amalgamism or multiple chemical sensitivity. Such mechanisms are highly speculative and only needs to be mentioned at the present stage of knowledge.

A review of all known pharmacokinetic polymorphisms is not within the scope of this paper (Shields 1994). Most polymorphisms have been related to cancer developement. For other effects, the genetic polymorphism in the delta-aminolevulinic acid dehydrogenase which has been reported may be of interest (Smith et al. 1995).

Susceptibility differences based on sex have been repeatedly discussed, and many cases have been reported (Uzych 1986). Susceptibility of the fetus during pregnancy and of the child during lactation are, with few exceptions (glucose-6-phoshate deficiency, sickle cell trait) highly controversial topics for risk management procedures. A thorough discussion of this problem is outside the scope of this paper, but the well known Johnsons Controls case is of interest for the discussion of employment discrimination (Annas 1991). The problem in the context of this paper is whether the susceptibility of the fetus warrants risk management procedures which may be perceived as sex discrimination.

Individual Susceptibility for Risk Assessment

A discussion of the scientific principles of risk assessment is outside the scope of this paper, but some comments are given to ethical problems as they relate closely to the problems of risk management.

Even if risk assessment by some is regarded as a scientific process, risk assessment includes a number of value loaded questions which imply ethical problems. If risk assessment is regarded as research, the Helsinki declaration on "International ethical guidelines for biomedical research involving human subjects" is applicable (CIOMS 1993), and also the ethical principles applied to epidemiology (CIOMS 1991). These include the general principles of respect for persons (autonomy and protection), an ethical obligation to maximize benefits and to minimize harms and wrongs, and justice which requires equitable distribution of both the burdens and the benefits of participation in research.

Practical problems in the molecular biology may include the right to know contrasted to the right to freedom from knowledge, or a moral feeling for the investigator of a duty to inform to prevent irreversible effects or re-assure safety, even after a mutual agreement of freedom from knowledge. Khoury et al. (1987) underline the importance of screening for susceptibles. They state that when the relative risk is <10 and the allele frequency <0.1, the observed relative risk will always be <3. Such observed relative risk might suggest to the investigator that the agent is not a risk factor for disease. A moral obligation to apply all possible tests for susceptibility may follow for maximizing the probability for demonstrating a real risk even if it only applies to few persons. On the other hand, the results obtained may be misused as will be discussed later. The genetic approach to individual susceptibility arises more controversy than other methods probably because of a stronger feeling of privacy intrusion, even if genetic susceptibility factors also can be revealed by other methods (Gevers 1993).

Individual Susceptibility for Risk Management

Risk management on the governmental agency level includes the development of regulatory options ideally followed by an evaluation of the public health, economic, social and political consequences of the regulatory actions. Stochastic individual susceptibility has traditionally been the basis for such regulatory action, and the individual susceptibility problem has been solved by uncertainty factors and extrapolation. Uncertainty factors are, however, of limited value when the individual susceptibility is characterized by discrete dose-response curves. Occupationally, such individuals or groups have traditionally been indentified by pre-employment screening, for the general public a larger uncertainty factor has been applied.

Risk management on the local level has two components; one related to the working environment, the other related to the worker. A discussion of the working environment component is outside the scope of this paper. The host-environment interaction has been mentioned earlier, and the general principle of adapting the work to the worker and not the opposite must be remembered. Periodical health examination has been mandatory for some group and is believed to be for the benefit of the worker even if not always so. The more questionable pre-employment examination has, however, also been extensively used. Such examination may not be different from genetic screening and should in most cases be questioned.

From the risk management point of view there are three causes for screening. These relate both to genetic and other methods. The first one concerns if there are specific requirements for the safe appointment of a given worker. The safety aspects in this case relates to the health of the worker and may include exposure to chemicals. The second one relates both to the worker and to the management. Can the worker perform the expected tasks within the framework given, or are special precautions needed. The third relates more to the management. Does this worker represent a present or future economical responsibility to the management? The last purpose will weight differently according to the national insurance scheme, to what extent the company, private insurance or national social insurance come into effect. Thus, even if the ethics and practical consequences of genetic screening usually are described separately for employment and insurance, there is a definite overlap. All the three causes for screening may include susceptibility to occupational toxicants and are thus relevant for the problem to be discussed.

A discussion of risk assessment and management in parenthood falls outside the scope of this paper, but both employment, parenthood and insurance at the end relate to the issue of public health.

The progress of molecular biology and the use of biomarkers have rapidly increased the possibility for identifying individual risk profiles or risk profiles for groups which deviate from the generally assumed stochastic susceptibility. But this problem is not new. The Office of Technology Assessment in the US in 1982 surveyed 500 major American companies and 366 responded (Uzych 1986). Six companies were currently using biochemical or cytogenetic testing, 17 had used testing during the last 12 years and 59 reported that they would use such tests in the future. Most tests were related to the sickle cell trait and there were many examples of restrictions placed on carriers without any documentation of increased risk. The predicted increase in screening has not taken place probably because of costly legal actions based on false positives or false negatives. Restrictive legislation in screening for ethnic minority-related disorders (sickle-cell trait) may also have been important (Hatchwell 1994).

There are differences between tests. Cytogenetic testing, even if inheritable genetic traits may be discovered by such testing, is in reality an early effect applied as an indicator of a potential serious health hazard. Such testing is even a requirement for exposure to some substances in the US and has been used extensively. On a group basis such testing seems to predict an increased risk, on an individual basis the results are still uncertain (Nordic Study Group 1990).

As mentioned above, some tests have already been used for many years in risk management (Ashford et al. 1990). The most common include screening for glucose-6-phosphate dehydrogenase deficiency, sickle-cell trait and α_1-antitrypsin deficiency, but several other have been suggested (Lappè 1983). Such tests with rather specific and defined preventive aims have been regarded as part of a general medical testing and have not been subject to specific regulations.

Discrimination based on such tests has, however, recently been recorded (Billings et al. 1992). The examples include both employment and insurance discrimination, but these overlap. The themes can be grouped in three categories: 1)"The asymptomatic ill," 2)"The problem of variability" and 3)"The at risk: to test or not to test dilemma." The important lesson to be learned is that many individuals with "abnormal" genotype will either be completely healthy or suffer a minor controllable disease and be stigmatized by the test results even if it bears no relation to the job in question.

Genetic testing for cancer risk applying some of the previously mentioned methods has been suggested, but no reports of practical use for cancer prevention have been found. The same applies to the use of atopy as a general risk factor for disease even if specific allergies should be elucidated in some occupations, may be even before hiring is offered. The medical history is, however, often a better tool for such information than any testing procedure.

Genetic screening at present does not seem to add more information in this area. No one would argue against screening for colour-blindness in certain occupations, active epilepsia in heavy vehicle drivers or certain requirement for air traffic pilots. There is of course an extensive misuse of poorly validated psychological and "personality" formulas in the hiring procedure, but genetic testing will probably (and hopefully) not add to this in the foreseen future.

The Ethics of Risk Management Based on Susceptibility Screening

Screening for susceptibility involves a number of methodological problems like all screening procedures. The problems of sensitivity and specificity deserve to be mentioned, but they will not be further discussed. Ethical problems relate to reflections of morality, and because moral concepts like justice and fairness are important for the practical use of new knowledge in molecular biology, genetics and data processing, a discussion of ethics seems reasonable. The concepts of morality and ethics imply that scientifically correct answers cannot be given even if there seem to be some moral principles which are generally accepted as "true". Ethical problems must be solved by oneself, not by others.

The genome project implies that risk management based on genetic susceptibility will constitute a major activity even if other more traditional health examination methods also will be used. Ethical issues of risk management based on genetic susceptibility screening include consent to being screened, confidentiality and privacy , storage, stigmatization, and the use of genetic information for legal purposes.

There seems to be no ethical problems related to screening for the small number of treatable diseases like phenylketonuria or congenital hypothyroidism with only a remote relationship to employment. On the other hand, screening for sickle cell anemia, Huntington's chorea, Charcot-Marie Tooth disease and Turners syndrome appears to have attracted some controversy. There are examples of totally irrelevant stigmatization and denial of job for the two latter syndromes (Billings et al. (1992). Stigmatization may also be caused by genetic traits which are ethnical or race determined as for the sickle cell trait. As a result some American states have outlawed such screening. Such general genetic discrimination has been discussed extensively by Natowicz et al. (1992). Genetic screening has the potential of creating a new group of disadvantaged people who will need the same protection now accorded those suffering from rase or sex discrimination (Gostin 1991). An example of an ethical problem and sex discrimination based on fetal susceptibility has been mentioned and relates to this problem of group discrimination (Annas 1991).

Except for the sex discrimination based on fetal hypersusceptibility to lead, these problems may not relate to the problem to be discussed - individual susceptibility to occupational toxicants. The examples given relate, however, to future possibilities for discrimination based on the previously mentioned genetic traits involving cancer or immunological diseases.

It seems reasonable that genetic screening for employment should be optional, as for general health screening before or during employment, in order to provide the individual with information relevant to the promotion and maintenance of health. This is also the recommendation of the so-called "Nuffield Council Report“ from UK published in 1993 (Hatchwell 1994). Genetic screening should not be used to exclude people from employment, even if this has been done based on other forms for screening, or to avoid the implementation of safer practices or working environments. The last statement is an example of the close relationship between the risk assessment and risk management applications of individual susceptibility. Should work practices be organized for the most susceptible individuals, or are the most susceptible free to damage their own health with company and/or social consequences. A comparison may be given to smoking, or to driving without a speed limit. Further, employees or prospective employees must have their right to refuse genetic screening without prejudice to their employment prospects.

The predictive power of genetic screening is at present overrated by most people. The low percentage of single gene diseases should be remembered. All diseases have an indefinite number of causes. A philosophical discussion of the cause-effect problem should be included in the discussions of use and misuse of genetic screening. There is a difficult distinction between genetic and non genetic medical conditions, as between the practical consequences of genetic and non genetic monitoring, screening and surveillance. People are different, they have different possibilities. This fact cannot be changed, our problem is to make sure that the knowledge of these differences is used in a socially and morally acceptable and just way wheather obtained by genetic or other methods.

Conclusive remarks

The rapid development of information technology signals a new concept of the working environment. The worker as part of a formal group, as well as the working environment as a geographical entity may disappear in the so-called post-industrial society. The trend indicates a breakdown of large companies into small groups, into small organizations which hire single workers or small groups for time limited tasks.

Within this organization individual susceptibility and fitness to the specific task will be an important factor for opportunities and thus highly demanded by the employers, and interesting for the worker to demonstrate. I feel that it is not possible to stop this organizational trend, the problem is how to use the increasing knowledge of individual susceptibility socially just. In other words, test for individual susceptibility should provide both benefits to the new work organizations and health benefits without compromising values considered to be equal or higher, for example autonomy, privacy, social just or human respect.

Practically the problem of low risk quantification and stochastic individual susceptibility is solved by extrapolation. This extrapolation may not, however, be valid for certain high risk individuals. Pre-employment investigation has traditionally been used for individual risk estimation in the practical occupational health setting and is still recommended.

Practically the problem of stochastic individual susceptibility and risk management is solved by uncertainty factors. Such factors are of limited value when the individual susceptibility is characterized by discrete dose- response curves. The progress of molecular biology rapidly increases such knowledge. The number of genetic probes increases very rapidly, but there are still only a few practical applications for genetic screening in the occupational setting. The obvious candidates for such screening are cancer and allergic diseases where the genetic aspects are accepted even if the predictive power of the available tests are still limited.

As a final conclusion, a better knowledge of individual susceptibility in the most extensive definition should be beneficial both to the worker and to management, and as such to the society at large. The possibility for misuse has increased with the introduction of genetic screening. Regulations are needed, but so-called scientific progress cannot be stopped. The moral and ethical standards of the involved professionals are more important than ever before.

References

Annas GJ (1991) Fetal Protection and Employment Discrimination - The Johnson Controls Case. N Engl J Med 325:740-743

Ashford NA, Hattis DB, Spadafor CJ, Caldart CC (1990) Monitoring the worker for exposure and disease. The Johns Hopkins University Press, Baltimore

Billings PR, Kohn MA, de Cuevas M, Beckwith J, Alper JS, Natowicz MR (1992) Discrimination as a Consequence of Genetic Testing. Am J Hum Genet 50:476-482
Caporaso NE, Shields PG, Landi MT, Shaw GL, Tucker MA, Hoover R, Sugimura H, Weston A, Harris CC (1992) The Debrisoquine Metabolic Phenotype and DNA-based Assays: Implications of Misclassification for the Association of Lung Cancer and the Debrisoquine Metabolic Phenotype. Environ Health Perspect 98:101-105
Chan-Yeung M, Malo JL (1994) Aetiological Agents in Occupational Asthma. Eur Respir J 7:346-371
CIOMS (1991) International Guidelines for Ethical Review of Epidemiological Studies. The Council for International Organizations of Medical Sciences (CIOMS), Geneva
CIOMS (1993) International Ethical Guidelines for Biomedical Research Involving Human Subjects. The Council for International Organizations of Medical Sciences (CIOMS), Geneva
Fairhurst S (1995) The Uncertainty Factor in the Setting of Occupational Exposure Standards. Ann Occup Hyg 39:375-385
Gevers S (1993) Use of Genetic Data, Employment and Insurance: An International Perspective. Bioethics 7:126-134
Gostin L (1991) Genetic Discrimination: The Use of Genetically Based Diagnostic and Prognostic Tests by Employers and Insurers. Am J Law Med 17:109-144
Guengerich FP, Shimada T, Raney KD, Yun C-H, Meyer DJ, Ketterer B, Harris TM, Groopman JD, Kadlubar FF (1992) Elucidation of Catalytic Specificities of Human Cytochrome P450 and Glutathione S-Transferase Enzymes and Relevance to Molecular Epidemiology. Environ Health Perspect 98:75-80
Hammond EC, Selikoff IJ, Seidman H (1979) Asbestos Exposure, Cigarette Smoking and Death Rates. Ann N Y Acad Sci 330:473-490
Hammond EC, Selikoff IJ (1973) Relation of cigarette smoking to risk of death of asbestos-associated disease among insulation workers in the United States. In: International Agency for Research on Cancer (IARC). Scientific Publications No. 8. Lyon pp 312-317
Hatchwell P (1994) The Ethics of Genetic Screening. Occupational Health Review March/April:10-12
Hendrick DJ (1994) Management of Occupational Asthma. Eur Respir J 7:961-968
Ketterer B, Harris JM Talaska G, Meyer DJ, Pemble SE, Taylor JB, Lang NP, and Kadlubar FF (1992) The Human Glutathione S-Transferase Supergene Family, Its Polymorphism, and Its Effects on Susceptibility to Lung Cancer. Environ Health Perspect 98:87-94
Khoury MJ, Stewart W, Beaty TH (1987) The Effect of Genetic Susceptibility on Causal Inference in Epidemiologic Studies. Am J of Epidemiol 126:561-567
Lappé M (1983) Ethical Issues in Testing for Differential Sensitivity to Occupational Hazards. J Occup Med 25:797-808
Mapp EC, Saetta M, Maestrelli P, Di Stefano A, Chitano P, Boschetto P, Ciaccia A, Fabbri LM (1994) Mechanisms and Pathology of Occupational Asthma. Eur Respir J 7:544:554
Natowicz MR, Alper JK, Alper JS (1992) Genetic Discrimination and the Law. Am J Hum Genet 50:465-475
Nordic Study Group on the Health Risk of Chromosome Damage (1990) A Nordic data base on somatic chromosome damage in humans. Mut Res 241:325-337
Richeldi L, Sorrentino R, Saltini C (1993) HLA-DPB1 Glutamate 69: A Genetic Marker of Beryllium Disease. Science 262:242-244

Ryberg D, Kure E, Lystad S, Skaug V, Stangeland L, Mercy I, Børresen A-L, Haugen AA (1994) p53 Mutations in Lung Tumors: Relationship to Putative Susceptibility Markers for Cancer. Cancer Res 54:1551-1555

Schulte PA (1992) Biomarkers in Epidemiology: Scientific Issues and Ethical Implications. Environ Health Perspect 98:143-147

Shields PG, Harris CC (1991) Molecular Epidemiology and the Genetics of Environmental Cancer. JAMA 266(5):681-687

Shields PG (1994) Pharmacogenetics: Detecting Sensitive Populations. Environ Health Perspect 102:81-87

Sidransky D (1995) Molecular Markers in Cancer: Can We Make Better Predictions?. Int J Cancer 64:1-2

Shmunes E (1986) The Role of Atopy in Occupational Skin Diseases. State of the Art Rev Occup Med 1(2):219-228

Smith CM, Wang X, Hu H, Kelsey KT (1995) A Polymorphism in the delta-aminolevulinic Acid Dehydratase Gene May Modify the Pharmacokinetics and Toxicity of Lead. Environ Health Perspect 103:248-253

Slovak AJM, Hill RN (1986) Does Atopy Have Any Predictive Value for Laboratory Animal Allergy? A Comparison of Different Concepts of Atopy. Br J Ind Med 44:129-132

Søyseth V, Kongerud J, Kjuus H, Boe J (1994) Bronchial Responsiveness and Decline in FEV, in Aluminium Potroom Workers. Eur Respir J 7:888-894

Terho EO, Koskenvuo M, Kaprio J (1995) Atopy: A Predisposing Factor for Chronic Bronchitis in Finland. Journal of Epidemiology and Community Health 49:296-298

Terr AI (1936) The Atopic Worker. Clin Rev Allergy 4:267-288

Uzych L (1986) Genetic Testing and Exclusionary Practices in the Workplace. Publ Health Policy 7:37-57

Venables KM, Dally MB, Nunn AJ, Stevens JF, Stephens R, Farrer N, Hunter JV, Stewart M, Hughes EG, Taylor AJN (1989) Smoking and Occupational Allergy in Workers in a Platinum Refinery. BMJ 299:939-942

Vineis P, Ronco G (1992) Interindividual Variation in Carcinogen Metabolism and Bladder Cancer Risk. Environ Health Perspect 98:95-99

Weston A, Caporaso NE, Perrin LS, Sugimura H, Tamai S, Krontiris TG, Trump BF, Hoover RN, Harris CC (1992) Relationship of H-ras-1, L-myc, and p53 Polymorphisms with Lung Cancer Risk and Prognosis. Environ Health Perspect 98:61-67

Ecotoxicology in Central Europe

(Chairs: M. Anke, Germany, and I. Holoubek, Czech Republic)

Persistent Organic Pollutants - the State of Contamination of Ambient Air in Central Europe. Possible Sources and Effects

I. Holoubek
RECETOX (Research Centre for Atmospheric and Environmental Chemistry and Ecotoxicology), Masaryk University, Kotlárská 2, 611 37 Brno, Czech Republic

Introduction

Environmental problems are often complex, with multiple causes and diverse ecological effects. Included for example can be persistent environmental traces of industrial chemicals, pesticides, and wastes which have been studied intensively, especially within the group of organochlorines. Dealing with such problems requires a flexible decision-making process that can accommodate this diversity while providing some measure of the uncertainty with which decisions are being made.

Persistent Organic Pollutants in the Environment

During the last decade several studies have been carried out aiming at the improvement of our knowledge on the behavior of persistent organic pollutants (POPs) in the environment (Pacyna et al., 1995). Evidence has been found that many of these compounds create serious problems due to their toxicity, persistence, and bioaccumulation in varrious environmental compartments (UN ECE 1994).

Many POPs are anthropogenic in origin, heavily used and released to the environment in large quantities. They have long environmental half-lives, resulting in continued increases in the global inventory of POPs in the environment. Being often semivolatile, POPs can be transported through ecosystems and travel distances on a local to global scale with the atmosphere as a primary transport medium. For many POPs there is a systematic migration to cooler latitudes. Finally, a number of bioaccumulative POPs produce a broad range of immunological, reproductive, developmental, behavioral, and carcinogenic effects (Han 1994).

Although work has been done in order to understand the fate, behavior, and effects of POPs, less information is available on environmental cycling (migration) of these compounds through various environmental media. The major reason for this dificiency in our knowledge is related to a lack of detailed information on emissions of POPs and the parameters affecting their migration through the environment under various meteorological, chemical and physical conditions.

Persistent organic pollutants (POPs) are organic compounds that are resistant to photolytic, biological and chemical degradation. They can remain in the environment for long periods (weeks to years) and can cause adverse environmental effects (UN ECE 1994).

Many POPs are characterised by low water solubility and high lipid solubility. leading to their bioaccumulation in fat tissues. POPs with these characteristics are typically semi-volatile, enabling them to move long distances and condense over colder regions of the earth.

Examples of POPs that are well known to the bioccumulative and semivolatile include: many persistent pesticides (e.g. dieldrin. DDT, toxaphene, chlordanes, hexachlorocyclohexanes), several industrial compounds (e.g. polychlorinated biphenyls (PCBs), polybrominated biphenyls (PBBs), chloroparaffins), and some degradation, industrial and combustion by-products (e.g. polycyclic aromatic hydrocarbons (PAHs), polychlorinated dibenzo-p-dioxins (PCDDs), polychlorinated dibenzofurans (PCDFs), hexachlorobenzene (HCB)).

POPs of concern are those which are produced in large quantities, have been widely used in the past or present and which have been dispersed into the environment. POPs are largely synthetic products that have no significant natural sources. Their are associated with the manufacture and use of certain chemicals, with leaks, spills and dumping of these materials, and with the combustion of both fuels and wastes. Many factors influence the scale of the releases.

The properties of these POPs (i.e. their persistence and ability to bioaccumulate) led to increased concerns for the toxic effects that they can exert on a range of biota, in particular top-of-the-food-chain-species (including humans) even at extremely low levels in air at ambient temperatures.

Halogenated hydrocarbons are a major group of POPs, of which the organochlorines have had by far the widest production, use and release, although brominated and iodinated hydrocarbons are also of some environmental concern. Polycyclic aromatic hydrocarbons (PAHs) are the other major group of POPs, and benzo[a]pyrene (BaP) is commonly used as an indicator of presence and carcinogenicity of PAHs in air.

Organochlorine compounds are a large class of chemicals, including industrial organics (e.g. PCBs and PCBs), agricultural organics (e.g. DDT, HCHs, polychlorinated camphenes, chlordanes, cyclodienes), by-products of anthropogenic activities (e.g. PCDDs/Fs), chemical transformation products (e.g. DDE and DDD from metabolism of DDT, dieldrin from oxidation of aldrin) and biotransformation products (e.g. methylsulfonyl derivatives).

Chlorine is used extensively in a wide range of processes, including municipal water supply and sewage treatment, manufacturing of plastics, bleached paper, cleaning solvents, and pharmaceutical products (almost 85 % of all pharmaceutical products are based on chlorine chemistry). It is estimated that over 50 % of all commercial chemicals involve chlorine. The potential products, by-products or wastes associated with these processes may often be persistent, toxic, organochlorine compounds. The persistent organochlorines form only a fraction of total chlorinated organic production, which includes products such as solvents and waste products from pulp mills.

PAHs, which are compounds consisting of multiple fused aromatic rings, are generated primarily from incomplete combustion of organic matter, particularly fossil fuels. Anthropogenic high-temperature processes are the major sources and involve a wide range of applications such as internal combustion engines (especially diesel), coal and wood combustion, coke ovens, aluminium melting, iron and steel production, and petroleum products such as creosote and asphalt.

The term "semi-volatility" is often used to describe the properties that permit POPs to occur either in the gas phase, or associated with small particles. Under normal environmental conditions, substantial fractions of the mass of a semi-volatile compound can exist in the vapour phase, on atmospheric particles and adsorbed or dissolved on the Earth's surface. Semi-volatility of most POPs favours their long-range transport. In the case of organochlorines, volatility declines with increasing chlorination.

The temperature dependence of physico-chemical properties such as vapour pressure results in the phenomena of "cold condensation". This process involves a cycle of contaminant evaporation, transport and condensation/deposition, leading to systematic transfer to and accumulation in colder and cooler latitudes (Wania and Mackay 1993). Some of the more volatile POPs, such as HCH and toxaphene, have been found in the Arctic at higher levels than at more southern latitudes (Muir et al. 1992).

POPs have relatively high octanol-water partitioning coefficients (a measure of a compound's ability to partition into lipid relative to water) and are therefore lipophilic, reflected in their strong ability to cross and/or interact with the phospholipid structure of biological membranes. This is an important factor, since membrane proteins tend to be sites for enzyme activity, hormone receptors, antigens and cellular uptake by active transport. In addition, this characteristic is responsible for the low levels of POPs in water-based excretion products, and therefore leads to a high potential for bioaccumulation and biomagnification. Accumulation of POPs in biological tissues tends to be in lipids. The properties of POPs also permit them to reach successively higher levels at higher trophic levels (biomagnification).

Organochlorines are generally the most persistent and biologically active of the organohalogens. A comparison of brominated vs. chlorinated organohalogens, has shown that in juvenile Atlantic salmon uptake of bromobiphenyls with four or fewer substitutions is similar to the uptake of the analogous chlorobiphenyls. However, bromobiphenyls with five or more substitutions accumulate to a lesser extent than analogous chlorobiphenyls. It has been shown that more highly chlorinated biphenyl compounds tend to accumulate to a greater extent than less chlorinated PCBs. In addition, metabolism and elimination is more rapid for the less chlorinated PCBs than it is for the more highly chlorinated biphenyls. Generally, the greater the number of chlorine substitutions/functional groups, the greater the persistence and resistance to biological and photolytic degradation. This is in part due to the fact that the carbon-chlorine bond is very stable. The toxicity of chlorinated compounds appears to be related to the position of the chlorine substituents on the biphenyl ring.

Data on bioaccumulation and biomagnification of PAHs is limited. Biomagnification is not as well known for PAHs as it is for organochlorines, although bioaccumulation has been shown in organisms which are chronically exposed. This appears to be highly variable according to species. Most vertebrates possess enzyme systems which can metabolize PAHs, however the metabolic intermediates are often cytotoxic or carcinogenic.

Highly persistent compounds among the POPs have been shown to bioaccumulate in wildlife. In Baltic wildlife, HCB, PCBs, DDT and chlordane residues showed the highest overall enrichment rates (Paasivirta 1992). In contrast, HCHs and toxaphene components were not biomagnified from fish to seals or to predatory birds. Coplanar PCBs represent in most cases higher toxic loads in biota than the toxic PCDDs/Fs. Comparison to Arctic species indicated that polychloronaphthalenes and PCBs are mostly local, but HCB and toxaphene the most global organochlorine pollutants.

POPs generally have a high potential for chronic toxicity. This is, in part, due to their long residence time in living organisms, which permits them to exert their effects over extended periods of time, and enables them to accumulate to levels which can cause adverse effects. Bioaccumulation from chronic low level exposure results in rising tissue concentrations in individual organisms which are correspondingly subject to the cumulative and additive effects of repeated doses. The fact that these compounds bioaccumulate and are very persistent also mean that body burdens can increase to levels which are toxic to the chronically exposed organisms, or to offspring exposed in utero by maternal transfer of contaminants. In the case of PAHs, it is thought that the effects may be additive, since reaction with DNA (formation of adducts) appears to be irreversible (UN ECE 1994).

Research on laboratory animals and corresponding observations on wildlife exposed to relatively high ambient levels of contamination demonstrate that a broad range of potentially serious effects can be associated with chronic low-level exposure to POPs. Organochlorines exert estrogenic effects, thereby causing endocrine disruption and a number of associated effects such as reproductive and immune system dysfunction, neurobehavioral and developmental disorders, and cancer.

Several PAHs are known to have cytotoxic and carcinogenic effects. It has been demonstrated that the developing embryo or fetus is highly sensitive to disruption of developmental processes caused by in utero exposure to chemical compounds. Although the direct evidence for chronic human effects is less extensive, enough similarities or parallel exist with wildlife and laboratory work to raise concern (UN ECE 1994).

Sources and releases of POPs

As it was noted earlier, POPs are associated with the manufacture and use of certain chamicals, with leaks, spills and dumping of these materials, and with the combustion of both fuels and wastes (Holoubek et al. 1995a).

In general, the main sources of persistent organic pollutants are:

(a) combustion processes;
(b) industrial processes;
(c) diffusing sources (mobile sources, open burning etc.);
(d) secondary sources (volatilisation from landfills etc.); and
(e) accidental sources (industrial accident, transport accident etc.).

PAHs, in particular, are released during a variety of high temperature processes. Manufacture and use of pesticides is a major source of organochlorine POPs. With the exception of PCBs and PCDDs/Fs, the organochlorine compounds that frequently generate environmental concern are herbicides and pesticides. These products are by design released to the environment and are not typically used in closed systems. There is much uncertainty in estimating the total use, production and emission of pesticides. In part, this is because in many countries records of this information are not kept, or are considered proprietary information. When information on releases is not available, information on production, use, import, export and sale, feed stock may serve as a substitute. Estimates may also utilize extrapolations from other parameters such as population or area of arable land.

Pesticides may enter the atmosphere from spray drift during application, usually as aqueous solutions or emulsions. They may also volatilize from plant and soil surfaces. For a large number of pesticides, the amount lost to volatilization approaches the total amount applied. Estimates of direct atmospheric emissions are not adequate for characterization of input to the atmopshere. Releases to soils, land fills, sewage lagoons, holding ponds, lakes and oceans may lead to volatilization or entrainment to the atmosphere. In addition, meteorological factors are known to play an important role in emission rates to the atmosphere. Water evaporation rate, soil sorption and micrometeorology are factors which contribute to loss of contaminants from soil .

Several of the most toxic pesticides have been banned or restricted in a number of countries - DDT, aldrin, dieldrin, chlordane and toxaphene. However, these chemicals remain as a source of concern, both because of continued release in those countries in which they are not controlled, and because in the environment significant quantities of these chemicals or their transformation products are remaining, which are subject to continued environmental transport and cycling.

The use of some other POPs which were produced commercially e.g. PCBs, hexachlorobenzene (HCB) has also been controlled in many countries. However environmental inputs of these compounds are still occurring. Thus, PCBs are being released from; fires, spills and leakages from the 65 % of world production still in use (UN ECE 1994), from evaporation and leakage from landfills or storage sites and from the incineration of PCBs. The release of HCB to the environment now stems primarily from the use of chemicals such as tetrachloromethane, tetrachloroethylen and chlorinated benzenes, in which it is an impurity.

Lindane (gamma-HCH) is a pesticide still in current use as an insecticide, rodenticide and fungicide. Technical HCH is a mixture of related compounds containing varying proportions of the inactive alpha and the highly toxic beta HCH-isomer. Technical HCH is still in widespread use in the former Soviet Union and Asia (Barrie et al. 1992. It enters the atmosphere by drift during tilling of soil (Scholz and Voldner 1993) and by evaporation from treated wood.

PCDDs/Fs, commonly referred to as dioxins and furans, are formed unintentionally in a wide range of processes in which carbon and chlorine are heated together e.g. during many types of combustion processes, including waste incineration and vehicle operation, primary and secondary iron production, secondary copper refining and paper bleaching.

PAHs emissions occur from a range of combustion sources, including incomplete combustion of fossil fuels, organic matter and waste. PAHs are a component of petroleum products such as creosote and asphalt, and are also associated with cigarette smoke and vehicle combustion. The extent of release depends on the choice of process, the efficient operation of the process and the efficiency of abatement equipment fitted.

Monitoring of POPs

POPs are ubiquitous in the global ecosystem, consistently detected in remote regions, such as the Arctic and the Antarctic, far removed from any industrial or agricultural activities. These compounds contribute to environmental problems at local, regional, hemispheric and global scales (Holoubek et al. 1995b).

In order to assess the effects of the steps taken to reduce the emissions of the POPs with environmental impacts, it is necessary to monitor the concentration levels of these compounds. In many cases, emission to the atmosphere is the way over which the greatest portion of the compound is released to the environment, and therefore it is of interest to monitor the atmospheric concentrations of these compounds. Such monitoring must take into account the differences in the physical properties between the compounds (i.e. by monitoring of the gas phase and of the particulate fraction).

At the moment, there are no instruments available which are able to continuosly measure the air concentrations of many of the organic compounds. Thus, monitoring of these compounds must be done by discrete sampling and subsequent analysis in the laboratory.

In order to study the importance of long range transport and deposition of POPs, such air measurements are currently being carried out in the Czech Republic. The aim of these measurements is to obtain knowledge concerning the concentrations present in the air and the quantities that are deposited. Comparison of the measured organic compounds in relation to air concentrations of other types of pollutants and in relation to meteorological parameters is also carried out.

Three types of parallel ambient air monitoring activities concerning POPs exist now in the Czech Republic (Holoubek et al. 1995). The first is a cooperation of the Ministry for the Environment with RECETOX (Regional Research Centre for Atmospheric Chemistry and Effects of Atmospheric Pollutants), Masaryk University, Brno. RECETOX has prepared a preliminary monitoring system of POPs in ambient air, the so-called SYMOS. During a pilot study of SYMOS (July to December 1994) PAHs, PCBs, chlorinated pesticides and PCDDs/Fs were monitored in the area of the observatory GEMS, Košetice, South Bohemia. This observatory is a background locality of the EMEP and GAW Programs, too. The Project SYMOS is based on the philosophy of the Project TOCOEN (Toxic Organic COmpounds in the ENvironment).

The second type are more complex projects where monitoring of POPs in ambient air is one part of the whole program. Examples of this type are the common project of Czech institutions with the US EPA-Project TEPLICE (concerning the state of the environment in Nothern Bohemia) and the Project SILESIA (Silesia region around the town of Ostrava in the northeastern part of the Czech Republic).

The third type consists of local pilot studies concerning the state of air contamination by POPs in various regions of the Czech Republic (capital Praha, region Zljn).

Project SYMOS

The measurements in the Project SYMOS have been carried out during the period from July to December 1994 every week, from Wednesday 08.00 to Thursday 08.00 yielding 26 samples (Holoubek et al. 1995b). PCDDs/Fs have been sampled every four weeks from Thursday 08.00 to Friday 08.00 yielding 6 samples during this sampling campaign. POPs were collected using a high volume sampler equipped with an adsorbent polyurethane foam. The sampling flow rate was about 50 m^3 per hour.

Sample preparation and analysis involved chromatographic separation analysis by HPLC (PAHs) and HRGC-MSD (Cl-PEST, PCBs). HRGC-HRMS has been used to quantify the full range of 2,3,7,8-substituted PCDDs/Fs.

The median and range values of the PAHs in ambient air samples are summarised in Table 1, selected chlorinated pesticides and selected congeners of PCBs in Table 2 and PCDDs/Fs in Table 3. Some of the results from the SYMOS pilot study, from TOCOEN measurements, from the Projects TEPLICE and SILESIA and from some other programs are compared in Table 4.

The Czech Republic is a very industrialized country with a high density of towns, villages and various types of industry. Some parts of country, especially the very industrialized nothern part of Bohemia and Moravia and the capital Praha are very polluted, not only by "classical" types of pollutants, but also by persistent organic pollutants.

Table 1: Contents of PAHs in ambient air, observatory Košetice, southern Bohemia

Observed values [pg.m^{-3}]	NAP	ACE+ FLR	PHE	ANT	F LU	PYR
Average	12 258	7 674,1	9 229 6	219,1	2 054,3	1 060,9
Stand. deviation	12 517	6 450,4	7 727,4	152,9	1 361,8	907,6
Minimum	250	200	120	10	120	70
Maximum	39 100	30 600	31 200	640	5 970	4 720
Median	6 585	4 020	6 250	1 510	760	760

Observed values [pg.m^{-3}]	BaA+ CHR	BbF	BkF	BaP	Σ PAHs
Average	224,7	245	55	71,7	29 573,9
Stand. deviation	156,2	118,4	39,3	38,9	26 614,4
Minimum	20	40	20	10	540
Maximum	700	480	160	130	101 000
Median	210	235	40	75	19 100

Abbreviations used: NAP = naphthalene, ACE = acenaphthylene, FLR = fluorene, PHE = phenanthrene, ANT = anthracene, FLU = fluoranthene, PYR = pyrene, BaA = benzo[a]anthracene, CHR = chrysene, BbF, BkF = benzo[b or k]fluoranthene, BaP = benzo[a]pyrene, Σ PAHs = sum of polycyclic aromatic hydrocarbons

Table 2: Contents of chlorinated pesticides and selected congeners and selected congeners in ambient air, observatory Košetice, southern Bohemia

Observed values [pg.m^{-3}]	α-HCH	HCB	β-HCH	γ-HCH	PCB 52	PCB 101
Average	193	207.6	130.6	416.2	305.4	102
Standard deviation	126.5	96.5	54.2	190.2	272.3	45.5
Minimum	100	100	100	100	100	48
Maximum	670	340	245	720	1200	200
Median	150	240	100	400	200	100

Abbreviations used: HCH = hexychlorocyclohexane, HCB = hexachlorobenzene, PCBs = polychlorinated biphenyls

Table 3: Contents of PCDDs/Fs in ambient air, observatory GEMS Košetice, southern Bohemia

Sample	Observed values [fg.m^{-3}]			
	Σ PCDDs	Σ PCDFs	Σ PCDDs/Fs	I-TEQ
28/07/94	11 369	4 712,9	16 081,9	156,3
25/08/94	421	1 269,4	1 690,4	47,6
22/09/94	1 922,9	423,9	2 346,8	19,7
20/10/94	353,6	36,6	390,2	2
17/11/94	414,4	34,2	448,6	2,5
15/12/94	438,9	42,3	481,2	2,3
Average	2 486,6	1 086,6	3 573,2	38,4
Stand. dev.	4 010,8	1 679,4	5 641,7	55,2
Minimum	353,6	34,2	390,2	2
Maximum	11 369	4 712,9	16 081,9	156,3
Median	430	233,1	1 085,8	11,1

Table 4: Range and observed values of POPs in some monitoring sites in the Czech Republic

Sites	Range of observed values				
	B[a]P [pg.m^{-3}]	Σ PAHs [pg.m^{-3}]	Σ PCBs [pg.m^{-3}]	Σ PCDDs/Fs [pg.m^{-3}]	I-TEQ [fg.m^{-3}]
Košetice (SYMOS)	10 - 130	540 - 101 000	100 - 1 400	390,2 - 16 081,9	2- 156,3
Teplice (Project Teplice)	4 780				
Prachatice (Project Teplice)	3 490				
Ostrava (Project Silesia)	700 - 24 400	33 200 - 479 300			
Praha (MEGA)	1 000- 7 200	85 500 - 399 900	412,2- 3018,5		16,1- 161,1
Zljn (TOCOEN)	< 100 - 42 460		< 5 - 9 612	< 100 - 54 500	<0,2- 299
Výesová (TOCOEN)	100 - 13 500	55 000 - 1 158 000			
Valašské, Mezirici (TOCOEN)	1 200 - 62 700	144 900 - 99 334 900			
Mokrá (TOCOEN)			147 - 2 294	1 100	1,1

Acknowledgement

This work has been supported and granted by the Ministry for the Environment, Czech Republic, the US EPA and the CEC (PHARE Program EC-HEA-18C2).

References

Barrie L, Gregor D, Hargrave B, Lake R, Muir D, Shrearer R, Tracey B, Bidelman T (1992) Arctic contaminants: sources, occurence, parthways. Sci. Total Environ. 122, 75-134

Han S-L (1994) An overview of persistent organic pollutants in the environment. A report for the Task Force on Persistent Organic Pollutants under the Convention on Long-range Transboundary Air Pollution. Department of Indian Affairs and Nothern Development, Canada, Ottawa.

Holoubek I, Cáslavský J, Nondek L, Kocan A, Pokorný B, Lenícek J, Hajšlová J, Kocourek V, Matoušek M, Pacyna J, Thomas D (1995a) The Compilation of Emission Factors of POPs in Czech and Slovak Republics. AXYS Canada/Masaryk University.

Holoubek I, Cáslavský J, Kocan A, Petrik J, Chovancová J, Drobná B, Korínek P, Lenícek J (1995b) Monitoring of POPs (PAHs, PCBs, PCDDs/Fs) in ambient air, Czech Republic. Proceedings of 10th World Clean Air Congress, Espoo, Finland, May 28 - June 02, Vol 2, paper no. 271.

Muir DCG, Ford CA, Grifft NP, Stewart REA, Bidleman T (1992) Organochlorine contaminants in narwhal (Monodon monoceros) from the Canadian Arctic. Environ. Pollut. 75, 307-316.

Paasivirta J (1992) The nature, properties and fate of organochlorine compounds in the environment. In: Organochlorines from pulp mills and other sources. Research methodology studies 1988-1991, University Jyvaskyla, Research Report No. 38,Jyvaskyla, Finland.

Pacyna JM, Brorstrom-Lunden E, Cleerman M, Jansson B, Korhonen M, Oehme M, Paasivirta J, Erecius Poulsen M, Sinkkonen S, Zuber A (1995) Regional and global cycling of persistent organic pollutants: An assessment and feasibility study in a Nordic perspective. NILU Report, Kjeller, Norway.

Scholz-Voldner E (1992) Estimation of pesticide emissions to the air resulting from agricultural applications. Presented to the 9th World Clean Air Congress and Exhibition. August 30-September 04.

UN ECE (1994): State of knowledge report of the UN ECE Task Force on Persistent Organic Pollutants, UN, Genova.

Wania F, Mackay D (1993) Global fractionation and cold condensation of low volatility organochlorine compounds in polar region. Ambio 22, 10-18.

Interdisciplinary "Case Studies" and Training Experts for Primary Prevention of Environmentally - Born Health Hazard

Jan W. Dobrowolski
W. Goetel School of Environmental Management and Engineering AGH (Tech. Univ.) Open University AGH "Technology and the Human Environment", Kraków Com. for Protection Human Health, Polish Academy of Sciences, Cracow Branch

Some recommendations related to long-term interdisciplinary "case studies" on environmental deterioration, related health hazard and primery prevention.

Estimation of spatial differentiation of the degree of chemical contamination is necessary to detect both direct as well as indirect health hazards in industrial regions. The purpose of determining the direct hazard serve analyses of bioaccumulation of toxic elements and substances in samples taken from people inhabiting areas of various pollution of the natural environment. Of special value for studies of present state of organism contamination may be blood samples. Therefore they were treated as a basic evidence to evaluate the influence on the content of lead, cadmium zinc and iron in the regions of three metallurgical centres in Poland of the Lenin and "Katowice" Steelworks and "Boleslaw". In order to establish the indirect hazard to human health, bioindication can be successfully used. The relevant elements are accumulated for instance in liver and other parenchymatous organs as well as in animal hair. Hair samples of cattle and dogs were recognised as the basic index of spatial differentiation of intoxication in the vicinity of the Mining-Metallurgical combine "Boleslaw", in the protection zone of the former Lenin Steelworks (now called Sendzimir) and of the "Katowice" Steelworks. Investigation of the content of lead, cadmium and zinc in hair made it possible to establish in retrospect the exposure of animals staying for very long in a given area. Bioindication is part of complex ecotoxicological studies linked with environmental monitoring. They are especially important for proper shaping of the areas of protection zones around big industrial plants. This can then be related to a reasonable location of both dwelling houses and other objects as well as of arable land or of breeding centre in regions of industrial imission effects (Dobrowolski, Fertig 1990, Dobrowolski, Vohora, 1989).

It also seems advisable to apply new research methods of high sensitivity which enable to estimate early disturbances of vital functions due to an impact of various contamination of the environment. These methods include among other things cellular monitoring of elements in the single cell under SEM associated with investigation of these elements in the food chain. The method can be applied both in rural areas with a raised level of morbidity due to the chemical composition of the environment as well as in areas of farms in the vicinity of industrial complexes. Excessive concentration of heavy metals in blood cells is linked with disturbances of their action, also with a decrease in resistance. There is also the possibility of wide application of a simple method of evaluation of spatial differentiation by means of mean daily levels of NO_2, SO_2, and S in red cells; to this purpose samples of a new type were used which makes it possible to carry out exposure at many sites at the same time. One can estimate in the same way both the degree of hazard at various places of job as well as isolines showing the spatial distribution of mean daily concentration of the cited gases (Dobrowolski Amaya et al. 1990, Fertig, Dobrowolski, 1986 Vohora, Dobrowolski, 1990, Dobrowolski et al. 1990, Dobrowolski, 1993).

The evaluation of the influence of different environmental carcinogenic factors requires interdisciplinary co-operation. Related studies include epidemiological surveys and air, water and soil, chemical, toxicological, and microbiological analyses, supplemented by experimental verification of suspected ecological pathogens and cofactors. A balance of carcinogens and protective agents in the external environment and in the human body is recommended for an ecologically oriented prevention ecotoxicological control of the food chain using modern technology (Proton-induced X-ray emission (PIXE), nuclear activation analysis, and induced coupled plasma) should be integrated with microanalyses at the cellular level (by X-ray microanalysis under scanning electron microscopy, nuclear magnetic resonance (NMR) and spontaneous and delayed chemical-luminiscence for balance of free-radicals and their scavengers). A pilot cross-disciplinary study conducted in the area of a "cluster" of human neoplasms and cattle leukaemia, in comparison with control villages in Poland, showed an excess in Pb, Hg, Ni, Rb, K, Mn, Cr, and Zn, accompanied by a nutritional deficiency in Mg, Ca, Fe, Co, and Se in the food chain of the "cluster". The living and breeding houses in this area were significantly more contaminated with the toxicogenic moulds *Aspergillus flavus* and *Penicillum meleagrinum* and by nitrate and nitrite in the drinking water. Our experiments showed that selenium deficiency stimulated the growth of fungi and some bacteria and increased the immunosupressive and teratogenic effects of aflatoxin B_1. New methods of protection of the indoor environment against microbiological contamination and laser-related biotechnology for the nutritional prevention of selenium deficiency and associated risk of neoplasms have been introduced. Primary prevention requires a large scale application of highly sensitive methods for early detection of risk factors in the environment, food, water and at the personal level as well as education of the society and an integrated common corrective action. (Dobrowolski, 1976, Dobrowolski and Vohora 1989, Dobrowolski and Smyk 1993).

A systematic approach to the relation between the quality of the total human environment and pattern of incidence of selected diseases seems to be necessary for assessing the influence of complex external factors (with special reference to excess or deficiency of particular elements in different regions) on the public health (.Dobrowolski, 1994).

Interdisciplinary "case studies" should be supplemented with experimental research on various species of mammals. These should include investigations about synergistic or antagonistic relation of elements.

The risk of health hazards related to the deterioration of the environment is increased by a deficiency of essential elements. Higher incidence of cancer has been associated with concurrent pollution of living houses with toxinogenic moulds and deficiency of selenium in food and water with precursors of nitrosamines (Aleksandrowicz and Smyk, 1973, Aleksandrowic et al 1975, Dobrowolski and Vohora, 1989).

Studies on the interdependency of different elements (e.g. As, Cd, Hg, Se, Zn, Mg, Ca, Pb) and other factors affecting pathogenesis of common diseases have received considerable attention in recent years. Deficiency of selenium in the body is associated with not only higher risk of cancer, but also of Keshan cardiomyopathy, as well as with higher sensitivity to mycotoxins, benzene, ionising radiation and heavy metal toxicity (Aleksandrowicz and Wazewska-Czyzewska, 1976, Aleksandrowicz et al. 1977).

Application of newer methods of analysis enables the measurement of trace amounts of both necessary and harmful elements in health and disease (even in single cells by use of X-ray microanalysis or PIXE together with SEM observations) and also early detection of health hazard due to deterioration of the human environment (Vohora and Dobrowolski, 1990).

Nutritional prevention of diseases (including diseases of civilisation) require ecotoxicological control of the elemental composition of the trophic chain and a modification of the human diet with the large-scale application of natural food additives (eg essential elements contained in rock salt, mineral water, plants, bee-products, pollen, herb-honey etc.) as well as the application of new biotechnology like fortification of soil or water with deficient and necessary trace elements (including hydroponic culture of vegetables), laser stimulation of seeds, bulbs, etc. (Dobrowolski et al. in press).

Joint pilot projects related to ecotoxicological surveys, followed by a systematic approach to solving typical problems under typical conditions for Central and Eastern Europe are recommended (including application of new methods for early detection of environmental risk factors, alternative technologies for prevention against these risk factors at the sources and postgraduate courses in environmental management and policy, environmental medicine and biology focused on primary prevention of irreversible bio-medical effects as well as deterioration nature and culture heritage).

Regional centers for common action for sustainable development and primary prevention could be recognised as basic elements of a network of the International University of Bio-Environment (IUBE) and other institutions interested in cooperation in this field (Dobrowolski, in press).

Prague as the European Culture Heritage particularly stimulates reflections about the transmission of the ideas of democracy into the reality of everyday life, regarding continuous participation of local communities in decision-making where this concerns the state of the environment for all in the future. We could attempt to follow some models of sociotechnical actions that were taken in ancient Greece on a small scale and which later became a pattern followed by the whole World, as fundamental elements of contemporary civilisation.

The principle to adopt is that of constantly improving into practice the concept of interdisciplinary research together with the *training of experts to solve typical problems,* as well as ways of a more *integrated cooperation* between professionals and the whole society. This is related not only to the rigorous fulfillment of the *heuristic model,* but also to creating the basis in order to permit the exchange of experiences, related to the methods of getting different groups of people involved in environmentally-oriented activities. We should discuss the concept of a network for international co-operation for Common Action for Global Life Support System and Sustainable Development (including exchange of comments from different points of view). The immediate purpose of this concept is the creation of a system which would guarantee that every one can participate, if only he or she can bring up convincing and reasonable concepts for building and constant improvement in practice of the *model of environmentally friendly civilization,* which will be beneficial both to people alive today as well as to future generations of all creatures and to the whole Biosphere.

By this mean, for instance, system approach to the elaboration of ways of monitoring and predicting the state of the natural resources and the environment on local and global scale in feed back with effective primary prevention of irreversible loses, changes and deterioration. The participation is open to every inhabitant of the Central and Eastern Europe. Contribution of all professional groups in this common action, would be a real guarantee of higher efficiency of management for better environment and primary prevention in particular.

We have to focus on some practical implication of this point of view:

There is a common international consensus for the top priorities for sustainable development. The crucial question is: How should we transfer these general principles into daily activities of every politician, local decision-maker, and finally representative of every professional group in all countries all over the world ?

It is essential to understand that to meet the basic needs of every individual, preserving health is necessary. This should be (according to WHO definition) trough active management of the optimal state of our out- and in-door environment, regarding quality of our working places, living houses, areas for recreation and permanent control of food quality.

We are in a position to offer adequate tools for monitoring personal exposure in the total human environment (Dobrowolski et all, 1990) followed by primary prevention of related health hazard (Vohora, Dobrowolski, 1990).

Feed back mechanisms between common knowledge over conditions of environmentally-related preventive actions and optimisation of decision-making processes seems to be of special importance under democratic conditions. Let me give a few examples from this part of Europe, the inhabitants of which are under higher risks of civilisation diseases as the result of the ignorance by decision makers on the bioassesment of technologies and management of resources. The process of democratisation of social life in Poland and later on in other Central and Eastern European Countries, has favoured wider contribution of local communities in cutting-down at the sources rate of the most dangerous for human health pollutants e.g. carcinogens and teratogens.

Looking for models of co-operation between experts from different disciplines, inhabitants, authorities, managers, NGOs, we could refer to methodological experiences of over 25 years out of university. National and International Schools on sustainable development and primary prevention in different national parks and areas for recreation (including the oldest in Europe border parks). In this way the traditional narrowly-oriented system of training and education, was supplemented by interdisciplinary *"case studies"* focused on solving real problems of common interest.

The author used experiences related to the above-mentioned model of training in helping Russian scholars when introducing in 1976 a new subject of cross-disciplinary environmental studies at the Lomonosov University in Moscow and well-known Japanese experts in environmental ethiology of incurable civilisation diseases. They had a keen interest in a systematic approach to environmental and nutritional prevention of congenital malformations, cancer incidence, etc., including both basic, research-developing studies, as well as education for active contribution of consumers in prevention of environmentally-born diseases. Distance education may be really helpful in making large number of people active partners in common action in this field. Perspectives of international co-operation in this field were recognised as one of the most promising topics during a meeting of the European Distance Education Network at our University in 1991. We promoted the understanding of different aspects of the contributions of local people in environmental management, sustainable development and concepts of primary prevention (focused on pan-European integration) at the AGH Open University's cycle on Technology and the Human Environment since 6 years.

During lectures and discussions experts from different bio-medical, social, technical and other disciplines introduced up-to day knowledge about environmental risk factors on local and international scale, followed by practical recommendations and advice useful for participants of the cycle.

Our Open University has been founded in connection with the International Symposium on New Trends in Environmental Education on Tertiary Level in 1989.

At the same time we have started with a new subject of diploma study "System of Environmental Management" focused not only on integration by interdisciplinary, problem-oriented studies, but also common action of experts and local communities for improvement of environmental quality. Our multi-disciplinary Committee for Protection Human Health of the Polish Academy of Sciences in Kraków city organised during several years open lectures of well known foreign and Polish experts related to progress in primary prevention of environmentally-related diseases (leukaemia and cancer in particular) including partnership with producers and consumers, focused on promotion of pollutant-free food and research-developing studies on natural food-additives rich in deficient trace elements, vitamins, especially antioxidants, free-radical scavengers etc. We have also good experiences in long-term co-operation with mass media in this field. The key problem in common action seems to be more active contribution of non-experts, namely local inhabitants and authority as well as businessmen, as one society facing with common problems of crucial importance.

The progress of decentralisation of decision in Eastern Europe, favours more active participation of communities in action for improvement of environmental quality on local and international scales, but at the same time contemporary model of mass-media dominated culture results in creating passive behavioural patterns. To stimulate active participation of the whole society in Our Common Action for a Better Environment it is necessary to create a well working international system for permanent co-operation under equivalent, democratic partnership of all elements of this system. Let us exchange our concepts and experiences useful for spreading on large scale knowledge about relationship between local and global scales action for better future for all.

Adequate common knowledge should be supplemented by higher environmental sensitivity, moral maturity and long-term imagination, as well as by a sense of common responsibility. The relationship among people and between man and other living organisms (or relationship between diversity of cultures and biodiversity) is also a very important problem from the biocentric point of view. This assessment is reasonable not only in its ethical aspect, but also in relation to sustainable management of the natural resources and consciousness of *Homo sapiens* for all living organisms and proper for their environmental conditions. The criteria of real adoption of the biocentric approach versus the egoistic one may be our solidarity with both those who are in a worse than our situation (e.g. victims of overexploitation of the resources and ecological catastrophe) people, as well as ecosystems under risk of destruction in different regions of the world. There are many examples of long-distance co-operation. Our multi-disciplinary team took also part in co-operation with Japanese experts in primary prevention of some diseases of civilisation (by scientific and practical experiences useful for inhabitants of both countries).

Our experiences in environmentally-friendly and reasonable (in terms of cost-benefit analysis) biotechnologies have been recognised as useful for sustainable development and nutritional prevention of some diseases in rural areas in India and Latin America (Vohora and Dobrowolski, 1990). We do not intend to develop gen-tech and contribute to progress by the introduction of new mutants or hybrids versus biodiversity of strains better adopted to the local environment. We are trying to stimulate biomass production of local strains of cultivated plants e.g. by low energy laser stimulation of the seeds and optimisation of the amounts of essential trace elements, etc. Some organisational achievements of the "green revolution" related to training of experts and education of local communities seems to be very helpful also for introduction of such find biotechnology on large scale. This kind of large-scale application of new research-developing studies would be hopefully useful for the prevention of climate change by stimulation of growth of energy forests and CO_2 assimilation as well as reclamation of many polluted areas. Better communication and co-operation among teams of experts involved in similar field of activity for life support system and social contribution in primary prevention has to be the starting point for putting ideas of biocentric approaches into real life. Available knowledge about the relation of the BIOS concept (Vlavianos-Arvanitis, 1991) and common human expectations on the background of sustainable development seems to be necessary not only for consensus but also for active participation in the elaboration of models of environmentally-friendly action for each group of contemporary society. Let us try to integrate a review of basic tasks of European environmental policy (Jankowska et al. 1994) and theoretical proposals of bio-politics (Vlavianos-Arvanitis, 1991) with past practical experiences and international pilot projects.

We would like suggest also the application of telematics systems as very promising for more efficient environmental and resources management (including transfer of know-how, creation of new jobs connected with promotion of low-energy, environmentally-friendly technologies) as well as for easy access for the whole society to information about methods of solving similar environmental problems within pilot projects and to continuous distance training and education for bioasessment of decision of managers and authority by local communities in relation to environmental impact. New tools in relative heuristic processes could be offered by leading scientific centres offering supercomputer simulation using neural networks for solving common problems. Permanent co-operation in this field was recommended by the experts contributing to the first meeting (International Seminar) on environment and health in cities in Central and Eastern Europe held in Poland in 1992 and the International Conference on Training of Experts for Environmental Protection and Sustainable Development in Europe, organised at our University in December 1993.

We intend to contribute to disseminate these ideas by the development ofour International Schools on Primary Prevention and Sustainable Development in Kraków city and border parks during summer vacations, including next year. The main principle of our activity is problem-oriented training and education related to model interdisciplinary "case studies" (including nutritional prevention of health hazard, ecotourism, reclamation and restructuring of industrial regions, waste management, application of biotechnology, reduction of side effects of transport, etc.). We have long-term good experiences in integration of pilot NGO's projects in environmental education with modernisation of formal education in this field (Dobrowolski 1993).

At present we focus our attention on different aspects of the relation between Environment and Culture. I am convinced that this relationship as well as knowledge about environmental reasons of health hazard and principles of system of primary prevention are the most important for strong motivation and support of proposed permanent activity both by decision-makers and international societies, local communities.

The aims of the International Conference "Training of Experts for European Co-operation on Protection of the Environment and Promotion of Sustainable Development" in Kraków 1993 were the following:

- Exchange of experience in conducting programmes and methods of professional training (mainly at graduate and postgraduate levels), based on a system approach to management of the human environment;
- Discussion of co-operation to make university diplomas from different countries equivalent;
- Exchange of experience in
  -general environmental education for students and the whole society,
  -co-operation of academics scholars with school teachers, journalists, engineers, decision makers, NGOs and local communities for common pro-environmental action (e.g. restructuring and modernisation of industry, land reclamation, waste management, reduction of traffic pollution, sustainable development of protected areas, eco-tourism).

The main topics of the Conference were:

- Integrated action towards reducing trans-national pollution at source;

- Improvement of quality of the human environment and health - with particular attention to co-operation with Central and Eastern European countries;
- Experiences in training students and graduates about interdisciplinary approach to environmental management (including the promotion of sustainable development, primary prevention of health hazard and conservation of the cultural heritage);
- Effects of economic and political changes in Central and Eastern Europe and their impact on environmental quality;
- Conditions for effective co-operation of experts, decision-makers and local community under democratic conditions (including environmental education and contribution of NGO's);
- Other topics of interest (including distance education in this field).

There are following proposals for activities of the International Centre Of Sustainable Development And Primary Prevention (focused on pan-European Co-operation and Environmental Problems in Central and Eastern Europe) in Kraków city.

- Problem-solving training (co-ordination of diploma and Ph.D. works in different universities) based on interdisciplinary "case study" and followed by common action including:

1.a. team of scholars and scientific (experts in natural, social sciences and technology), university students, practitioners from local administration, public and private companies (including experts in environmental control, technology and ecotoxicology).

2.a. team of teachers, journalists, delegates of NGOs' involved in environmental education and contribution of local society in environmentally-oriented action, in links with the AGH Open University "Technology and the Human Environment",

3. Postgraduate Courses for:

a. experts in environmental problems in administrative bodies,

b. experts in environmental engineering from industry, etc.

c. physicians interested in human ecology and primary prevention (environmental medicine),

d. decision-makers (interested in sustainable development and optimisation of decision-making process in links with local action),

4. International Summer Schools (including interdisciplinary field study following 27 years of tradition of interdisciplinary research and problem-solving training initiated and conducted under the scientific leadership of Professor Dr. J.W. Dobrowolski),

5. International study-visits focused on exchange of methodological experiences in situ (with top priority for co-operation university cities in Europe) for promotion of both bilateral and multilateral co-operation, related to pilot projects and common action for reducing at the sources of long-distance (e.g. transboundary) pollution in this field.

We would like to propose interdisciplinary cooperation about new European Textbook for university students (stressing the crucial importance of ecotoxicological surveys followed by common action for the elimination of environmental risk factors at the sources).

Remarks about the suggested topics of the International Textbook On Sustainable Development And Primary Prevention :

Following these recommendations we would like to invite you to co-operate by contributing to an international European Textbook on Sustainable Development. The textbook should be problem-oriented and based on interdisciplinary "case studies". The textbook would first of all be aimed at training so-called environmental generalists for full integration of pro-environmental action on local scale, combined with regional, pan-European and global co-operation. The following aspects should be included.

1. Historical survey of environmental degradation and wasteful management of natural resources. Ecological, ethical, philosophical, economical and other aspects of pro-environmental civilisation (including methodological recomendation and "case studies").
2. The analysis of consequences resulting from technocratic approach. Irreversible effects of environmental contamination (referring to human health, unique monuments of nature and culture.
3. The principles of methodical interdisciplinary co-operation and system approach, taking primary prevention against civilisation diseases as an example.
4. The principles of sustainable development and pioneer traditions in this area (the concept of sozology and sozotechnics, experience of international schools "Man and Environment"). The implications from the Agenda 21 accepted at the Earth Summit: Environment and Development in Rio de Janeiro.
5. The promotion of technological development for more effective action to limit the most harmful pollution in eco-toxicological and radio-ecological aspects regarding the broader use of so-called clean energy sources, low energy and low material consuming technologies, biotechnologies etc. Economical, legal and other instruments in environmental protection. Environmental and technological awareness in the society and the limitation in the application of environmentally sound technologies as well as the perspectives of removing these obstacles by European integration.
6. Environmentally friendly town and country planning on local, regional, national and international scale (e.g. problems of transport).
7. Other topics according to recommendations of authors.
8. We would like to invite interested exsperts at the International Conference. European Action for Sustainable Development and Primary Prevention that is to be held at the University of Mining and Metallurgy (1st -2nd December 1995). The organisers are able to sponsor the authors of selected papers.
9. Would you please send your proposals (the title and conspectus of the chapter of the International Textbook) and the title and the summary of your report for the International Conference in this field to:

Prof. dr. Jan W. Dobrowolski
Goetel's School of Environmental Management and Engineering Dept. of Environmental Management and Protection c/o Wydz. Geodezji Gór. i Inz. Srodowiska AGH, 30-059 Kraków, al. Mickiewicza 30, Poland

Conclusions

1. An individual approach to personal monitoring is complemetary to interdisciplinary "case studies" including epidemiological and ecotoxicological surveys focused on primary prevention of health hazard associated with deterioration of the natural environment and contamination of the human food chain.

2. A systematic approach is recommended for both fields of reseach.

3. Linkeage between control of levels of some ecotoxins and metals in particular in the trophic chain (using such methods as AAS, PIXE, NA) and at a cellular level (i.e. in the single blood cells under SEM) has been proved in relation to the environment and individual contamination in selected industrial region (e.g. the biggest in Poland metallurgical centers and their inhabitants), as well as in cancer (leukemic clusters).

4. Amaya's method (modificated by Krochmal) was applied both for mesurement of the total individual exposure to NO_2 (in in-door and out-door environment) and SO_2 as well as for drawing isolines useful for environmental management according to spatial differentiation of irritating gases (including their synergistic effects).

5. Multifactoral assesment of environmetal quality and health status of populations under higher risk has to be followed by experimental recommendations related both to the elimination of risk factor at the sources, reclamation of contaminated areas (including supplementation of soil with protective additives and replacement of highly intoxicated vegetables by laser stimulated seeds, cutting, bulbs etc. of industrial plants of stimulation.) and also by fortification of human food with protective factors (e.g. antioxidants, „free radicals scavengers“ etc.) This is a way of sustainable development with high priority of common action for primary preventation.

6. The results of our study support development of submolecular methods in application for early detection of synergistic effects of different environmetal risk factors, including embryological and other highly sensitive biological criteria (related to simultaneous influence of chemical, physical and biological pollutants and changes not only in structure but also in biological function related to thermodynamic and informative aspects). Interdisciplinary field and experimental studies (including simulation of real situation in the natural environment) seem to be area of interdisciplinary cooperation in feed-back with future studies and preventive action of irreversible effects (like diseases of civilization, deterioration of nature and culture heritage) followed by integrated pro-ecological policy in Europe and problem-solving training at universities.

References

Aleksandrowicz J, Dobrowolski J W, and Lisiewicz J, (1973) Effect of selenium on immunosupressive and carcinogenic properties of aflatoxin B_1. Rev Espanola Oncol 22 (2): 239

Aleksandrowicz J, and Smyk B, (1973) The association of neoplastic diseases and mycotoxins in the environment. Texas Rep Biol Med 31 (4): 715

Aleksandrowicz J, Starek B and Moszczyäski P, Wplyw selenu na obraz krwi obwodowej u szczurów przewlekle eksponowanych na benzen. (Polish) Medycyna Pracy 28 (6), 454

Aleksandrowicz J and Wazewska-Czyzewska M, (1976): Selenium as radioprotective substance. Haematologica 2, 49

Dobrowolski J W, (1976) New aspects of environmental protection against developmental malformations and the cancer incidence.: Science for Better Environment, HESC Secretariat, Asahi Evning News, the Science council of Japan, Tokyo

Dobrowolski J W, (1993) Ekotoksykologiczna ocena czynników ryzyka zwiazanych ze skazeniem Pb, Cd, Zn a ekologiczna profilaktyka.:Prace Mineralogiczne 84 PAN Oddzial w Krakowie, Komisja Nauk Mineralogicznych

Dobrowolski J W, (1994) The concept of training and education for international cooperation in solving common problems of sustainable development, In: Training of experts for european cooperation on protection of the environment and promotion sustainable development. Proceedings of the International Conference, University of Mining and Metallurgy, Faculty of Management, Faculty of Mining Surveying and Environmental Engineering Kraków pp 89-98

Dobrowolski J W, Amaya K, Fertig S, Janas A, Skoczeä M, Marsza^ek M. (1990) Personal monitoring and bioindicators of exposure to some chemical pollutants in industry and residential areas. Ecotoxicology of protective zones. Environment Protection Engineering 16 (2): 105-112

Dobrowolski J W, Fertig S, (1990) Bioindication of the contamination by heavy metals of people and animals in the regions of some metallurgical works, 99-117 pod red. Jachimskiego J. Polish Academy of Sciences - The Kraków Section

Dobrowolski J W, and Smyk B, (1993) Environmental Risk Factors of Cancer and Their Primary Prevention. : Journal of Environmental Pathology, Toxicology and Oncology, 12 (1): 55-57

Dobrowolski J W, Vohora S B, eds team book (1989) Ekologizm w ochronie zdrowia, Komisja Ochrony Zdrowia Spolecznego PAN, oddzial w Krakowie, Ossolineum, Wroclaw, Kraków, Warszawa

Fertig S, Dobrowolski J W, (1986) Die Umweltbelastung mit Blei, Kadmium und Zinc in drei Industrieballungsgebieten in Polen: Spurenelementsymposium, Anke M. et al, eds., KMU, Leipzig, FSU Jena, Leipzig

Jankowska-Klapkowska A, Dobrowolski J W, Preisner L, Wagner A, Magnus G, eds team book (1994) Trainig of experts for european cooperation on protection of the environment and promotion of sustainable development, UUM (AGH), Kraków

Vlavianos-Arvanitis A, eds team book (1991) Biopolitics the Bio-Environment Vol III, Biopolitics International Organisation, Athens

Vohora S B, Dobrowolski J W, eds team book (1990) New horizons of health aspects of elements, Hamdard University, New Delhi

Annex 1

Scheme of possible international cooperation

Regional Approach :

Motivation: Each country in Central and Eastern Europe has its own specific environmental problems, which reflects historical patterns of development localization and use of the natural resources, but a have more environmental problems in common, than other regions, like ecotoxical and health implication of use of out - of- time high energy and highly polluted technologies.

Conclusion: Experts interdisciplinary "case studies" followed by problem-oriented training have to develop international and interdisciplinary pilot projects (including research, training of environmentalists - menagers, projects leaders, etc. at graduate and postgraduate levels, education of university students, integrated with whole system of education, including distance education, Open University, out-of-school environmental education in cooperation with Non-Governmental Organizations for Common Action for Better Environment for all) focused on system approach to solving typical problems under typical conditions for Central and Easter European Countries.

Recommendations: To establish as soon as possible Regional Centres for promotion joint pilot project for promotion of interdisciplinary "case studies", data-base and system of communication for distribution methodological experiences and proposals for development of international cooperation in related areas, beginning from Center of Sustainable Development and Primary Prevention of Eenvironmental Risk Factors (for Human Health, Nature and Culture Heritage) in Kraków city (refering to old tradition of related interdisciplinary research projects and problem-oriented training of environmental generalists and environmental education of the whole society), the World Culture Heritage under risk of environmental deterioration.

Annex 2

Scheme of interdisciplinary "case studies" on ecotoxicological survey followed by a systematic approach to environmentally-oriented primary prevention as scientific contribution to sustainable development.

1. **Environmental monitoring** (including measurement of local emission, imission - dry and wet input) and linkage to

a) Personal monitoring (e.g. measurement of individual exposure to as much as possible harmful environmental factors like mutagens, carcinogens, co-carcinogens immunosupresors, promotors of carcinogenesis, including synergistic effects and balance of risk factors and protecting factors e.g. oxidants and antioxidance, free radicals and free radical scavengers) see Dobrowolski et al. papers related to application of samplers for measurement of personal exposure in out-of-door and in-door-environment i.e. total exposure to some toxic gasses, as well as description of spatial and temporat differentiation in 24 h. levels of NO_2 in the air.

See Dobrowolski et al. on studies about levels of some metals in bioindicators eg human hair, whole blood, placenta, nails etc. samples and in the single blood cells (using so called cell monitoring in correlation with standards analyses by AAS, NA, PIXE and morphological study under SEM).

See Dobrowolski et al. on the investigation of synergistic effects of different environmental pathogens eg. carcinogens and co-carcinogens in relation to selected fractions of blood cells eg lymphocytes or blood serum samples at submolecular level (using measurement of photon emission, both by single counting technique and delay luminescence, or electrochemical potential by platine powder electrodes) under physiological and pathological conditions (eg of blood samples of healthy and leukemic subjects, etc.)

b) Corelation between results of personal monitoring (including cellular level) measurement of populational exposure eg to heavy metals, sulfur and imput of the some pollutants on ecosystems and human or animal food chain.

2. Ecotoxicological survey (multifactoral) in different industrial regions (eg. metallurgical centers in Poland the Sendzimir - former Lenin - Steel Works in Kraków, "Katowice", "Boleslaw"), in "cancer/leukemia clusters" in the villages, regions for recreation and eco-tourism like national (including border) parks.

3. Problem-solving training of environmentalists, decision-makers, NGO-leaders, teachers, journalists for Common Action.

Plants and Animals as Biomonitors of Heavy Metal Level in the Aquatic Ecosystem of the River Danube

N. Oertel
Hungarian Danube Research Station of H.A.S., Jávorka S. u. 14., 2131 Göd, Hungary

Introduction

The goal of biological monitoring is to give reliable and proper information on possible effects of chemicals, present in the water due to human activities, to enable the protection of the aquatic ecosystems and particularly to provide for the required scientific quidance in legislation and enforcement.

Most of the monitoring efforts starting in the 60s, concerned chemical and physical parameters. As awareness grew, it turned out that though most monitoring programmes can provide a sensitive indication of the concentration of substances that have been selected for study, but this information lacks essential biological meaning. Pollution implies deleterious effects and these must always be assessed in relation to biological systems. The biological approach - with the help of the time and effect integrating capability of aquatic plants and animals - can overcome the shortages of 'point-like' chemical analyses (Oertel 1991a). To mention a few:

- chemical analysis provides numbers which require translation into possible biological effects on the basis of available toxicity data,
- the chemical approach can not account for any additive, synergetic or antagonistic effect, that might occur,
- chemical data may lack some constituents, particularly trace metabolites and reaction products,
- toxicity is not only an expression of compound interest qualities, but also of the chemical, physical characteristic of the medium, in a wider sense, of the aquatic environment.

Biological monitoring techniques can be inexpensive and require less sophisticated instrumentation, but of course they also have limitations in precision and reproducibility. Most of these shortages could easily be overcome by the cautious evaluation and interpretation of the results. The question of time scaling demands the greatest care, because biological time scales are quite different from those customarily used in chemical monitoring. It is essential to realise, that with a few exceptions (like the prospective Biological Early Warning Systems), most of the biological responses will only provide usable signals after some time, by a retrospective way (De Zwart 1995). Biological monitoring does not replace chemical and physical monitoring, they all provide converging lines of information, that supplement each other, but are not mutually exclusive.

The bioaccumulative capability of aquatic organisms during bioaccumulation monitoring - as one of the most frequent techniques in aquatic toxicology - was used to detect heavy metal levels in the ecosystem of the River Danube in the 80s and 90s. Bioaccumulation methods have the following advantages:

- bioavailability is measured at the actual site of pollution,
- bioaccumulating organisms are to a certain extent capable of time-integrating ambient concentration of pollutants in water:
  * by giving a better picture of the severity of the situation than the actual, often erratic changes in outside concentrations,
  * the factors controlling this uptake are integrated in this measure,
- there is preconcentration of pollutants before chemical analysis.

One of the greatest difficulties is the incorporation of the still rather weak knowledge of the ecological aspects into the toxicological consideration. Therefore, during the last one and half decades a lot of efforts were made to enlighten the ecological side and background of ecotoxicological questions.

Materials and Methods

This paper reviews passive and active biomonitoring (PBM and ABM) techniques which were carried out in different sections of the main arm of the River Danube and in different side arms of the Szigetköz area between 1980 and 1994 (Figure 1.).

The sampling places, organisms and methods used in particular, separate experiments are listed below:

ABM with translocated mussels:

site: "control area" (Schisler backwater, Szigetköz); "translocation area" (main arm of the River Danube, Göd/1669 rkm)

organism: *Anodonta cygnea, Dreissena polymorpha, Unio pictorum*

method: 6-week of translocation in plastic baskets

PBM with macrophytes in a side-arm system:

site: Dunaremete side-arm system (Szigetköz area)

organism: macrophytes (*Ceratophyllum demersum, Nuphar lutea, Nymphaea alba, Potamogeton pectinatus, Potamogeton perfoliatus, Polygonum amphibium f. natans*)

method: collection of the native stands of submerse and floating-leaves rooted macrophytes

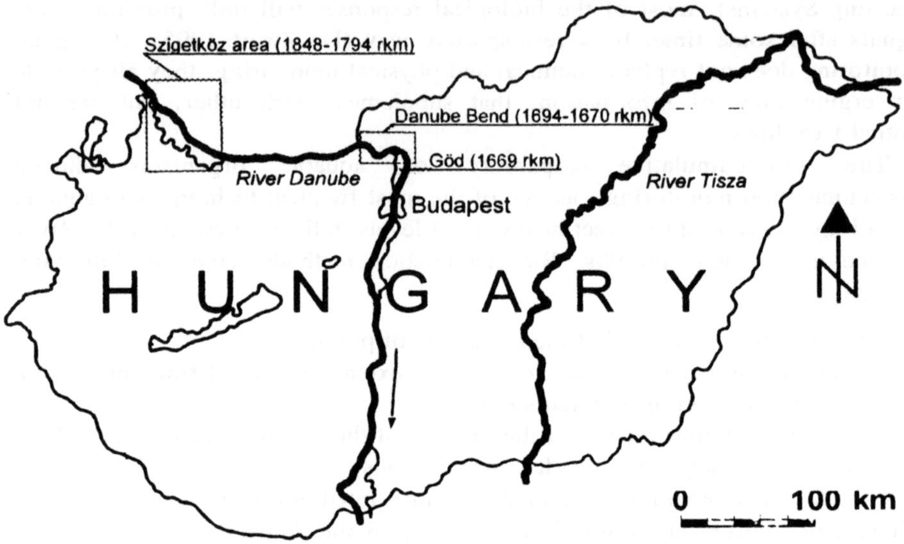

Fig. 1. Map of Hungary with the locations of biomonitoring activities in the water ecosystem of the River Danube between 1980 and 1994.

PBM with macrophyte in the main arm:
site: five sampling places in the main arm of the River Danube (1694-1670 rkm)
organism: *Potamogeton pectinatus*
method: collection of native stands of *Potamogeton pectinatus*

PBM with bivalves and gastropods in the Szigetköz area:
site: main channel (Old Danube); active alluvial plain (Schisler backwater); protected area (Zátonyi Danube, Mosoni Danube, Lipót ox-bow lake)
organism: bivalves (*Anodonta cygnea, Unio pictorum*), gastropods (*Lymnea auricularia, Lymnea stagnalis, Planorbarius corneus)*
method: collection of indigenous species

Combined PBM - ABM with *Cladophora*:
site: main arm of the River Danube at Göd (1669 rkm)
organism: *Cladophora glomerata* (indigenous, attached, filamentous green alga, most abundant periphyton organism of the littoral zone)
method: translocation of artificial floating substrate into the stream line

Combined PBM - ABM with macroinvertebrates:
site: main arm of the River Danube at Göd (1669 rkm)
organism: *Chironomidae, Ephemeroptera, Gammaridae, Trichoptera* (indigenous species in the *Cladophora* web of the littoral zone)
method: artificial floating substrate, translocated into the stream line

Filamentous green alga, submerged and floating-leaves rooted macrophytes species, molluscs (gastropods and bivalves) and macroinvertebrates were used as biomonitors. Generally, indigenous species collected from their natural habitats were used in passive biomonitoring (PBM). Active biomonitoring (ABM) technique was used, when the bivalves in plastic baskets, were translocated from the Szigetköz area to the main arm of the River Danube, for a 6-week exposure. In the case of our periphyton experiments the two methods (PBM and ABM) were combined, when the indigenous attached filamentous green alga were exposed, on artificial floating substrates and were translocated from the littoral zone to the stream line (Oertel 1989, Oertel 1992/a, Oertel 1994).

After sampling, selection and thorough cleaning both the plant material and the animal tissues were dried at 105 °C. After weighing the biological material was digested with a mixture of concentrated HNO_3 and H_2O_2 in a sealed teflon bomb. Heavy metals (Ag, Cd, Co, Cr, Cu, Fe, Mn, Ni, Pb and Zn) were determined by using a Varian Techtron AA 275 ABD type atomic absorption spectrophotometer (Praker 1972, Varian 1978, Salánki et al. 1982, Burrows and Whitton 1983, Duzzin et al. 1988). Usually the whole body - for the large mussels (*Anodonta*, *Unio*) only the gills - were used and the metal concentration was expressed as $\mu g\ g^{-1}$ dry weight.

Results and Discussion

Environmental monitoring systems are descriptive in terms of concentration and emission and regulatory in terms of control, in other words, monitoring must have a purpose and a function in the process of risk assessment and pollution control. During the performance of monitoring four objectives should be discussed, these are: signal, control, trend (recognition) and instrument function (De Zwart 1995).

During the presentation the results concerning aquatic organisms as biomonitors in the River Danube will be grouped and illustrated according to these four main objectives of biomonitoring.

Signal function

Mussels (*Anodonta cygnea, Dreissena polymorpha, Unio pictorum*) were used in active biomonitoring (ABM) for signal function to detect adverse changes in heavy metal level and localise the source of pollution. The bivalves were translocated from a backwater of the Szigetköz area to the main arm of the River Danube, to the up-stream section of the capital. After a six-week period the heavy metal concentration significantly decreased to about the level, which is characteristic for a non or moderately polluted section of a large river (Figure 2.). The relatively high concentration level in the 'control' site, in the Schisler backwater - previously believed as non-polluted - needs some explanation.

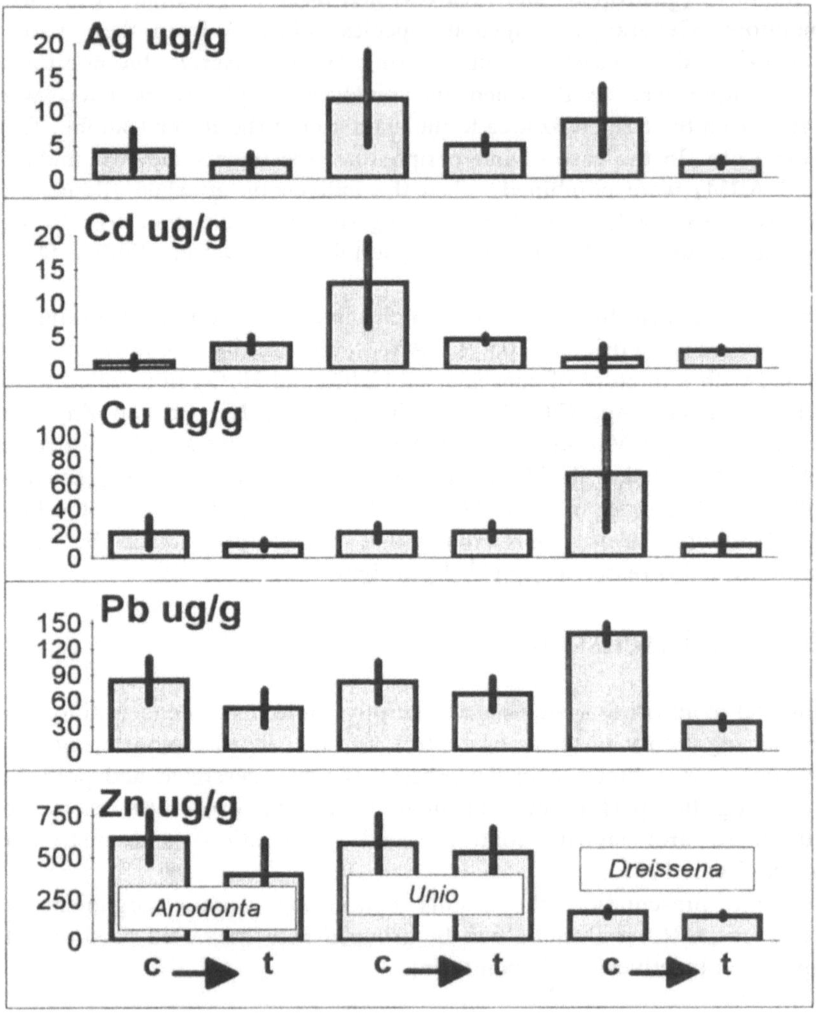

Fig. 2. Effect of six-week translocation on the heavy metal concentration of mussels. Control site (*c*) : Schisler backwater in the Szigetköz area; translocation site (*t*) : main arm of the River Danube (Göd, 1669 rkm). Average±S.D. are indicated (n=10).

Although the heavy metal concentration in the sediment was similar - except for Cd, which was higher - to those which were found in other sections of the river (VITUKI 1989a, VITUKI 1989b, Dinka 1994, Wachs 1994). These values are very close to the threshold concentration accepted for soils. In autumn 1992, after a very long lasting dry period, a very poor water quality - characterised by low dissolved oxygen, pH and redoxpotential (0.5-3.9 mg l^{-1},

6.68-7.01, 40-127 mV, respectively) - was traced in the fully separated backwater. This reductive condition was favourable to the release of heavy metals from the sediment which was followed by their increased uptake in mussels.

Finally we concluded, that the sediment that was polluted previously - by high streamflow or floods, during the periods of direct connection with the main channel of the River Danube - acted as an 'internal' source of heavy metal pollution in consequence of the changed environmental conditions. This situation could become more frequent after the damming of the Danube, in order to operate the Gabcikovo Barrage System. Both the direct surface and the indirect ground water connection, between the side-arm network and the River Danube, failed in consequence of decreasing water level in the main channel and were therefore, followed by the sinking of ground water table.

The signal function of biomonitoring was also used to analyse the cause of elevated heavy metal concentration after dredging (Figure 3.). In order to compensate the loss of water, because of the diversion the Danube a supplementary water supply system (SWS) was constructed in the side-arm system of the Szigetköz area. This is why a lot of natural side arms were dredged, canalised and unified into a network. In consequence of dredging the previously sedimentated heavy metals released from the sediment increased the particulated:dissolved ratio in the water phase. An increasing uptake of heavy metals was followed by a modified heavy metal accumulation capability of the different species of submerged (*Ceratophyllum demersum*, *Potamogeton pectinatus*, *Potamogeton perfoliatus*) and floating-leaf rooted (*Nuphar lutea*, *Nymphaea alba*, *Polygonum amphibium f. natans*) macrophytes (Ráth and Oertel 1994).

Trend (prediction) function

Based on a time series analysis the trend or prediction function of the monitoring contributes to the prediction of human activities and their effects and their future developments.

In the Danube Bend, up-stream of the capital, Budapest, a lot of spur-dikes and parallel dikes were constructed in order to operate a navigation channel at normal or low stream flow. Because of this river training activity an increased siltation, sedimentation occurred in the littoral zone of the main channel, mainly in the spur-dike bays. In this changed condition places became suitable for the colonization of macrophytes, even in the most canalised section of the River Danube. This recognition of the appearance of species in the newly formed biotopes serves also as an example of the signal function of biomonitoring (Ráth 1980/81, Ráth 1984).

Potamogeton pectinatus - excellent indicator of bioavailable pollutants - colonised the silted areas in the 70s and after that the records of their heavy metal content were used for long term observation of increasing heavy metal

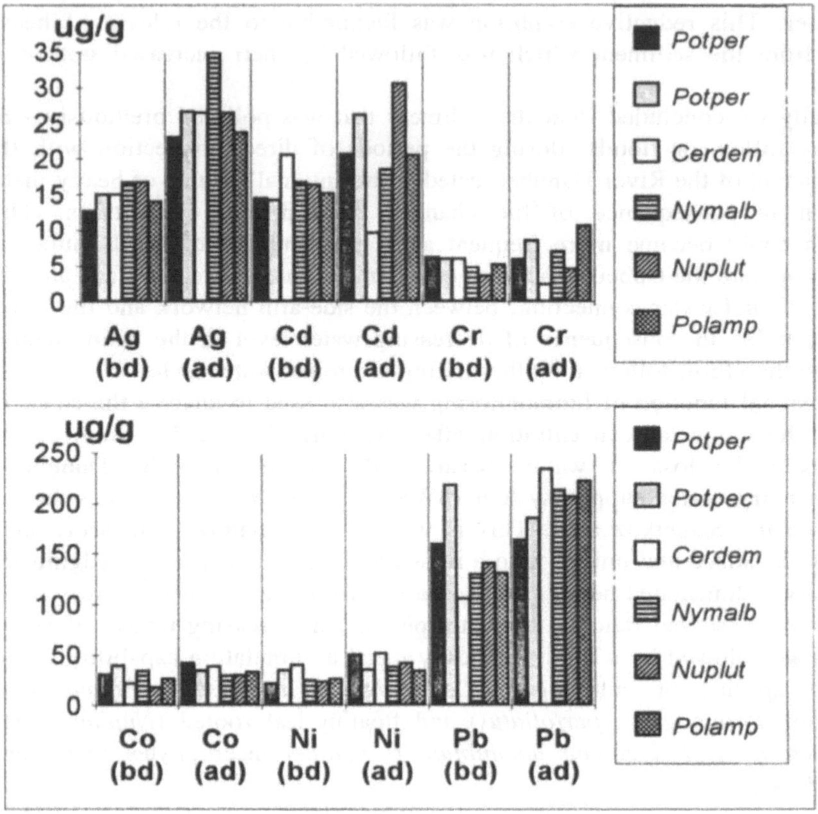

Fig. 3. Heavy metal concentrations of the macrophytes in the Szigetköz side-arm system before (*bd*) and after (*ad*) dredging. *Potper= Potamogeton perfoliatus*; *Potpec= Potamogeton pectinatus*; *Cerdem= Ceratophyllum demersum; Nymalb= Nymphaea alba; Nuplut= Nuphar lutea*; *Polamp= Polygonum amphibium*.

emission during the 80s (Figure 4.). The level of heavy metals shows a decreasing trend at the end of 80s and in the 90s in accordance with the measures taken by the authorities reducing the acceptable limit values of the emitted heavy metals (Ráth and Oertel 1991). Simultaneously, this dataset can also be used for control function.

The same tendencies could be observed in the suspended matter. Most of the metals take part in the sedimentation by binding to suspended particulated matter. The rooted submerse macrophytes can indicate this fraction of heavy metals, while they can also uptake these from the sediment (Oertel 1991b, Oertel 1992b). The fact, that it was not possible to detect the same phenomenon by analysing the water, gives further evidence of the advantage of biological approach over chemical analysis.

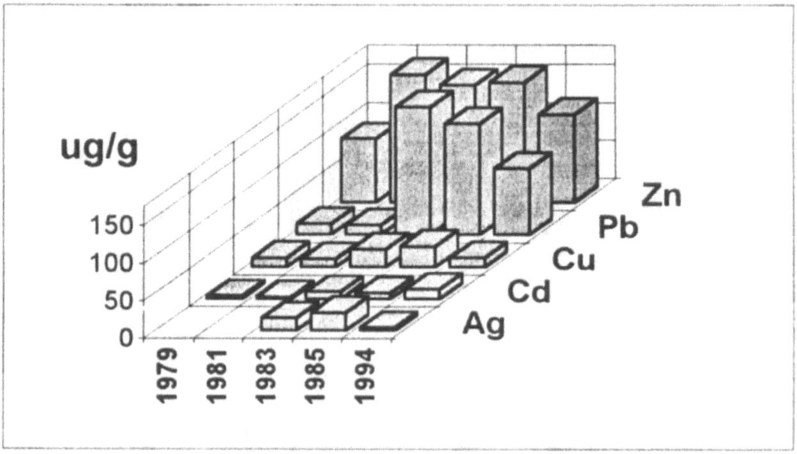

Fig. 4. Long-term monitoring of the level of heavy metals in the River Danube (Danube Bend, 1694-1670 rkm) by *Potamogeton pectinatus* between 1979 and 1994.

Control function

Due to increasing water engineering activity in order to construct the Gabcikovo Barrage System a lot of disturbances have occurred in the vulnerable water ecosystem of the Szigetköz area both in the hydrological and ecological conditions. Lacking the direct surface and indirect ground water connection with the main channel, the side arms, backwaters and ox-bow lakes have been fed by the supplementary water supply (SWS) system. Since the diversion of the Danube in 1992, the different types of water on the active alluvial plain and protected area have been differently supplied with reservoir water, seepage water or water from the main channel, either pumped or gravitationally gained.

According to the geographical location of the studied water bodies and their way of supply, an extremely diversified and changeable system developed both in structure and function. Our biomonitoring activities using gastropods as biomonitors for detecting heavy metals in the 90s could also control - indirectly - the effectiveness of the SWS system. The more the water body is separated from the other parts of streaming, lotic branch system, the more it is possible that it would develop poor, reductive water quality. In these kinds of water bodies (e.g. Schisler backwater in autumn 1993) the previously accumulated, sedimentated pollutants can be released in consequence of the changed hydrochemical conditions and the molluscs can easily uptake and accumulate the heavy metals (Figure 5.).

On the other side, the side arms and the branch sub-systems - which were previously more or less independent and of course uniquely different from each other - are connected by the SWS system and directly fed from the main

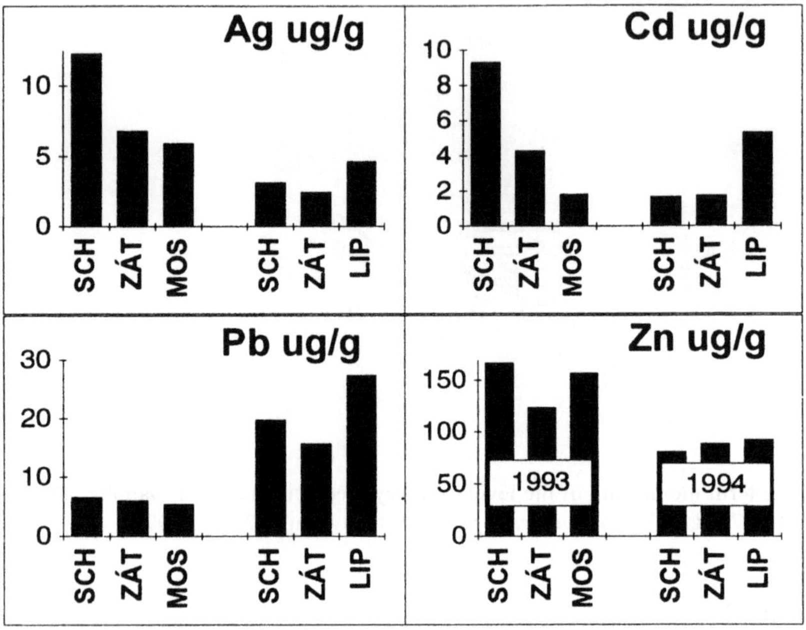

Fig. 5. The effect of the supplementary water supply system on the heavy metal distribution in the Szigetköz area, monitored by *Lymnea stagnalis,* in autumn 1993 and in spring 1994. SCH= Schisler backwater (indirectly supplied by ground water); ZÁT= Zátonyi Danube (directly supplied by seepage and reservoir water); MOS= Mosoni Danube (directly supplied by reservoir water); LIP= Lipót ox-bow lake (directly supplied by pumped water from the main channel of the River Danube).

channel or the reservoir. Therefore external pollution can spread through the common watercourses quickly and without difficulty (e.g. Zn in Mosoni Danube in autumn 1993 or Ag, Cd, Pb, Zn in Lipót ox-bow lake in spring 1994) (Oertel 1995).

Instrument function

Scientific research in as an activity-effect or cause-effect study fulfils the instrument function of biomonitoring by helping the recognition and clarification of the underlying processes. Using a biomonitor organism for indication, the first question is its applicability and the conditions of the usability of the species.

In the early 80s *Cladophora glomerata*, a filamentous green alga was used for control of the heavy metal levels in the River Danube. This attached, periphytic species was well known as good indicator of heavy metals, but we had to check this ability and the conditions in which it is valid in a large eutrophicated and industrially polluted river. Exposing *Cladophora* on an artificial floating substrate (AFS) we could define that the heavy metal concentration in alga depends on its developmental stages and its vertical position in the littoral zone. These changes in time and space strongly depend on and correlated with the dry weight of *Cladophora*. Besides the 'weight dilution' effect, as the most important internal factor, the ambient concentration of heavy metals proved to be the most dominant external factor, which influence directly the heavy metal level in the alga (Oertel 1994).

Other physico-chemical factors (flood, light, temperature, conductivity, pH, ORP) indirectly influence - in most cases through the changes of dry weight - the process of uptake and accumulation (Whitton 1970, Oertel 1994). After clarifying all these fundamental questions, and comparing the actual heavy metal concentration in *Cladophora* to the base level, which is characteristic for the alga on the studied river section, we could follow the pollution events. This could serve as a signal function, in long-term monitoring as trend, moreover as control function. After evaluating our results and comparing them to the literature, we could classify, that the studied reach of the Danube was industrially moderately polluted at the beginning of the 80s (Whitton et al. 1989).

As an example for the source orientation, the attached biomonitor, the *Cladophora* proved to be a good indicator of the dissolved form of heavy metals (e.g. Ag, Cd, Co) which can be easily uptaken and originated from close pollution sources. While attached alga proved to be very sensitive and source orientated indicator of the dissolved phase of heavy metals, *Trichopteras* living in the *Cladophora* web are - in some cases 1 or 2 magnitudes - higher concentrators of some heavy metals (e.g. Ag, Cd, Cu, Pb), than the alga itself (Figure 6.). This group of macroinvertebrates as collector organisms prefer suspended particulated matter (SPM), which acts as the preconcentrator of heavy metals, therefore they often have higher metal concentration, than, different types of feeding groups of macroinvertebrates (Oertel 1992b). Otherwise in the case of animals their mobility could challenge their validity to indicate and delimit the pollution sources.

Conclusions

As the results of the above demonstrated projects showed, aquatic plants and animals proved to be very efficient indicators for heavy metals in different biomonitoring strategies. But prior to planning and operating a monitoring network, the applicability of the species in question and the conditions in which the results of bioindication are valid and interpretable must be carefully investigated and a typology of water types and ambient physico-chemical conditions must be properly described.

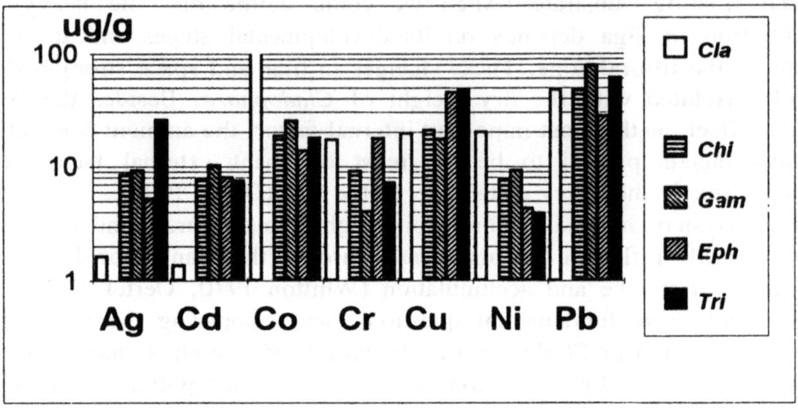

Fig. 6. Concentration of heavy metals in *Cladophora glomerata* (*Cla*) and in macroinvertebrates (*Chi=Chironomidae; Gam=Gammaridae*; *Eph=Ephemeroptera*; *Tri=Trichoptera*) in the River Danube (Göd, 1669 rkm) in the 80s.

Biological monitoring will never replace chemical analysis, but they both have specific advantages in the detection of various signals of the quality of the aquatic environment. And what is more, in environmental monitoring it is not permissible to separate the two from each other when considering the cause-effect relationship arising from the fundamental ecological principles. Their alternating dominance in the different phases of monitoring would change. In the development of monitoring strategies both technique should be considered and their advantages used at best, in order to obtain a monitoring program - integrated monitoring - that is suitable for its task. It must always be kept in mind, that the comparability and interpretability of their responses in time are quite different.

Moreover, we have to take into consideration the shift from biological to ecological assessment methods which development has occurred in the last decades. The surplus of ecological assessment with respect to biological valuation methods lies in its greater emphasis on the relationship between different, mostly simultaneously working abiotic steering factors, but also biotic interactions. Furthermore, ecological assessment provides the water manager with special tools to guide an aquatic ecosystem in a desired direction.

The biomonitoring activities that have been carried out can contribute to the better understanding and clarifying the ecological aspects of ecotoxicology. Those indigenous species, living in the benthic and littoral zone, and those communities (periphyton) were studied, along with those techniques - like artificial floating substrate, translocation -, which were promising to give complex and ecologically sound responses in ecotoxicological questions.

The biomonitor techniques that are summarised in this paper may provide a scientifically improved basis of an applied water quality assessment and control for the everyday management.

Acknowledgements

Thank to my colleague, to Mrs B. Ráth for her collaboration in studying heavy metals in macrophytes. The investigations in 1994 were financially supported by the Hungarian National Scientific Foundation under contract No. OTKA/T 012725.

References

Burrows IG, Whitton BA, (1983): Heavy Metals in Water, Sediments and Invertebrates from a Metal-Contaminated River free of Organic Pollution. Hydrobiologia 106:263-273

De Zwart D, (1995) Biomonitoring. In: Monitoring Water Quality in the future. Vol. 3., Ministry of Housing, Spatial Planning and the Environment, The Hague, The Netherlands, pp. 1-11

Dinka M, (1994) Vorstudie über die Schwermetallbelastung des Sediments in einem Altwasser der Kleinen Schüttinsel. 30. Arbeitstagung der IAD, Zuoz/Schweiz, Wissenschaftliche Kurzreferate pp. 331-335

Duzzin B, Pavoni B, Donazzolo R, (1988) Macroinvertebrate Communities and Sediments as Pollution Indicators for Heavy Metals in the River Adige (Italy). Wat. Res. 22:1353-1363

Oertel N, (1989) Heavy-metal Accumulation in *Cladophora glomerata* (L.) Kütz in the River Danube. AMBIO 20:264-268

Oertel N, (1991a) Biomonitoring for water quality determination. With particular reference to (semi)continuous monitoring systems. Individual Study Final Report. Individual Studies. AES III.2. IHE. Delft/The Netherlands, pp. 1-21

Oertel N, (1991b) Heavy metals in the River Danube: Water, suspended matter, periphyton. Verh. Internat. Verein. Limnol. 24:1961-1964

Oertel N, (1992a) Heavy metal investigations with mussels in active biomonitoring. In: Hydrobiological surveillance of the side-arm system of the Szigetköz area. 758 AKA/D. MTA ÖBKI/MDÁ. Project Report pp. 1-7 (in Hungarian)

Oertel N, (1992b) Heavy metals in the water, suspended matter and periphyton organisms of the River Danube. Thesis of Ph.D. MTA ÖBKI Magyar Dunakutató Állomás, Göd. MTA Könyvtára pp. 1-148 (in Hungarian)

Oertel N, (1994) The applicability of *Cladophora glomerata* (L.) Kütz in an active biomonitoring technique to monitor heavy metals in the river Danube. The Science of the Total Environment, Supplement 1993. 2:1293-1304

Oertel N, (1995) Molluscs as bioindicators of heavy metals in a side-arm system of the River Danube disturbed by water engineering activity. XXVI Congress of International Association of Theoretical and Applied Limnology, Verh. Internat. Verein. Limnol. (under publication)

Parker CR, (1972) Water Analysis by Atomic Absorption Spectroscopy. Varian Techtron Pty. Ltd., Springvale, Australia, pp. 1-78

Ráth B, (1980/81) Untersuchung der Laichkrautvegetation im Donauarm bei Vác und in seinen Nebengewässern (Stromkm 1669-1690). Ann. Univ. Sci. Budapest, Sect. Biol. 22-23:109-124

Ráth B, (1984) Die Chemische Zusammensetzung des Kammlaichkrautes (*Potamogeton pectinatus*) im Donauarm von Vác. Ann. Univ. Sci. Budapest, Sect. Biol. 24-26:81-89

Ráth B, Oertel N, (1991) Schwermetallakkumulation der *Potamogeton pectinatus* - Bestande in der Donau Oberhalb von Budapest (Stromkm 1640-1694). 29. Arbeitstagung der IAD. Kiew, September 1991., Wissenschaftliche Kurzreferate pp. 170-173

Ráth B, Oertel N, (1994) Die Auswirkung der Wasserbaulichen Massnahmen auf die Schwermetallakkumulation der Makrophyten im Nebenarmsystem von Dunaremete (Kleine Schüttinsel, Stromkm 1826, Ungarn). 30. Arbeitstagung der IAD, Zuoz/Schweiz, Wissenschaftliche Kurzreferate pp. 336-340

Salánki J, Balogh VK, Berta E, (1982) Heavy Metals in Animals of Lake Balaton. Water. Res. 16:1147-1152

VARIAN (1978) Analytical Methods for Flame Spectroscopy. Varian Techtron Pty. Ltd., Springvale, Australia, pp. 1-123

VITUKI (1989a) Water, sediment and hydrobiological investigations of the Danube, its tributaries and side arms. VITUKI, Budapest, Research Report, 7612/3/970. pp. 1-159 (in Hungarian)

VITUKI (1989b) Some water quality questions of GBS: Models and their applicability. VITUKI, Vízminőségvédelmi Intézet, Budapest, pp. 1-94 (in Hungarian)

Wachs B, (1994) Schwermetallgehalt von Sedimenten aus der Donau. 30. Arbeitstagung der IAD, Zuoz/Schweiz, Wissenschaftliche Kurzreferate, pp. 310-314

Whitton BA, (1970) Biology of *Cladophora* in Freshwaters. Water Research 4:457-476

Whitton BA, Burrows IG, Kelly MG, (1989) Use of *Cladophora glomerata* to Monitor Heavy Metals in Rivers. Journal of Applied Phycology 1:293-299

Ataxonomic Assessment of Phytoplankton Integrity by Means of Flow Cytometry

C.E.W. Steinberg[1], H. Schäfer[2], M. Siedler[2], W. Beisker[2]
[1]Institute of Freshwater and Fish Ecology, Müggelseedamm 310, D-12587 Berlin, FRG
[2]GSF Research Center of Environment and Health, P.O. Box 1129 Neuherberg, D-85758 Oberschleissheim, FRG

Abstract

Flow cytometry, a method well established in medicine and biotechnology, can also make an important contribution to (applied) limnological as well as ecotoxicological studies on phytoplankton. Flow cytometry can, for instance, contribute to the ataxonomic structural and functional assessment of phytoplankton. This approach may serve as a supplement to the well-established taxonomic evaluation by means of various microscope techniques.

We present some examples for such ataxonomic phytoplankton evaluation. These examples include phytoplankton of eutrophicated and acidified water bodies as well as slowly flowing rivers. Phytoplankters may be differentiated by their pigment contents into carotinoid-rich ones (such as Chrysophyceae, Bacillariophyceae, and Dinophyceae) and carotinoid-poor ones (such as Euglenophyceae and Chlorophyceae).

As a useful biomass parameter of phytoplankton algae we tested successfully protein staining by fluorescein isothiocyanate. We discuss the advantaged of this approach as compared with results obtained by Coulter counter or by biomass calculations from microscope analyses.

Up to now, evaluation of the biological quality of pelagic water bodies is still laborious and time consuming because of the microscopical examination of planktic communities usually practiced. As a possible improvement we present a structural ataxonomic approach for assessing the integrity of individual phytoplankters (on the basis of physiological parameters) as well as of the phytoplankton communities that is based on annual means of biomass spectra. Flow cytometry can provide considerable relief.

Introduction

In most aquatic systems, phytoplankton organisms are the most important primary producers. Thus, any alterations in the phytoplankton community can have strong effects on the whole system. Phytoplankton examination with respect to productivity estimations or evaluation of human impacts such as eutrophication, acidification or introducing xenobiotica into aquatic systems is laborious and time consuming if done by microscope.

On the other hand, quite often no distinct taxonomic record of the organisms is necessary. In many instances, summarizing parameters, such as cell density and volume or other biomass parameters, i.e. protein content, may be sufficient. This approach is the ataxonomic description of an organismic community. Structure as well as several functions of a community can be described ataxonomically. Furthermore, the ataxonomic approach can provide additional information as compared to traditional taxonomic evaluations. This applies especially to evaluations of phytoplankton integrity on various levels of ecological aggregations (individual to biocenosis levels). Our contribution will focus on integrity evaluations on both the individual level and the biocenosis level.

In the case of ataxonomic evaluations, flow cytometry can facilitate and accelerate sample evaluation, especially for water samples with low cell density, high detritus concentration or with a high amount of very small organisms like the picoplanktic cyanophytes (Campbell and Yentsch 1989 a, b; Campbell et al. 1989, Legendre and Yentsch 1989, Li 1994, Troussellier et al. 1993, Phinney and Cucci 1989). Flow cytometry can be the method of choice also for routine applications like monitoring of drinking water reservoirs or controlling eutrophication processes due to release of nutrients from wastewater plants or input by diffuse sources caused by excessive fertilization in agriculture.

Photosynthesis pigment autofluorescence is a practical way not only to quantify phytoplankton cell density, but also to discriminate the pigment groups of a phytoplankton community giving information about the trophic state of a water body or the course of seasonal processes (Beeler SooHoo et al. 1986, McMurter and Pick 1994, Yentsch and Phinney 1985).

Instead of measuring cell volume by microscope or a Coulter counter as described in many papers, FITC protein staining can be used as biomass descriptor for flow cytometry (Darzynkiewicz et al. 1985, Freeman and Crissman 1975). CHLa fluorescence of the phytoplankton cells can be recorded simultaneously and employed as identification parameter to discriminate between detritus and bacterio-, phyto- and microzooplankton.

Among the numerous dyes some exist that specifically indicate physiological features of cells, for instance, intracellular pH, esterase activity or membrane potential for the assessment of cellular functions. Sieracki et al. (1993) have demonstrated the potential application of these probes to marine phytoplankton by preliminary experiments. Several attempts to find a sensitive test for cell vitality have been carried out. Fluorescein diacetate (FDA) and various other probes are frequently used with vitality tests of mammalian cells. Flow cytometrical measurements with these fluorogenic substrates allow investigations of enzyme activity and enzyme kinetics of *viable* cells on an individual cellular basis. In contrast to bulk measurements, flow cytometry provides information about the cell-to-cell variability of metabolic activity. The assessment of phytoplankton individuals is based on, for instance, the intracellular accumulation of fluorescein after FDA hydrolysis (Dorsey et al. 1989). Here we report on an application of this approach to freshwater algae using the kinetics of FDA uptake and hydrolysis. In the long-term, our results appear to encourage the establishment of a quantitative method for assessing cellular responses of freshwater organisms to natural and anthropogenic stresses and stressors.

For detecting and quantifying anthropogenic influence on plankton organisms chemical analyses can supply information about nature and concentration of an introduced and potentially harmful substance, but this is a weak base for drawing any conclusions on the real effects on the organism community. Laboratory single species tests are also of limited use since for the need of practicability they have to be designed for standard conditions not considering the influence of nutrient supply, light, temperature, interactions between species, long time exposition to low concentrations or/and mixtures of xenobiotica and the presence of geochemical matrices which all can severely alter the sensitivity of affected organisms. Therefore, field surveys and observations using ecologically relevant endpoints such as changes in species (genetic) diversity, spatial and temporal distribution of species, and alterations in energy flow are indispensable (Steinberg et al. 1994).

Biomass spectra are an approach for determining plankton system integrity based on allometric relationships, i.e. cell/body volume/mass or organic carbon content of plankton organisms (Gaedke 1992, Platt 1985, Sprules and Munawar 1986).

Material and Methods

Algae: For protein content analysis and CFR-examinations the following algal species were used including Chlorophytes (*Chlamydomonas reinhardii*, *Chlorogonium elongatum*, *Haematococcus lacustris*, *Scenedesmus subspicatus*, *Ankistrodesmus bibraianus*, *Chlorella vulgaris*), Chrysophytes (*Ochromonas danica*, *Poteriochromonas malhamensis*), Cryptophytes (*Cryptomonas ovata*, *Chroomonas* spec., a *Rhodomonas* isolate from Lake Constance), a Euglenophyte (*Euglena gracilis*), Bacillariophytes (*Cyclotella cryptica*, *Stephanodiscus subsalsus*, *Phaeodactylum tricornutum*, *Navicula pelliculosa*) and picoplanktic Cyanophytes (*Aphanocapsa salina*, *Synechocystis pevalekii*, *Synechococcus rubescens*, two *Synechococcus* isolates from Lake Constance). A ciliate, *Tetrahymena pyriformis*, was used in addition to the algae for protein content analysis. For further information see Schäfer et al. (in press).

Flow Cytometry: A dual laser flow cytometer (FACStar^{Plus}, Becton Dickinson) equipped with two argon ion lasers was used for all flow cytometric measurements. The construction of a typical commercially available flow cytometer is shown in Fig. 1. To simplify matters, only one laser and one fluorescence detection channel are depicted. The suspended cells in the sample vessel are transported to a fine nozzle. Particles up to a diameter of about 40 µm were measured. In the nozzle they are hydrodynamically focused and separated from each other by the sheath flow thus they are flowing one after the other in the fine water jet formed by the nozzle. The water jet is passing the laser focus and each cell is illuminated individually by the laser beam and causes light scattering and/or fluorescence light pulse. A light detector [photodiode or photomultiplier (PMT)] converts the light pulse to an electrical pulse. Side scatter and fluorescence light pulses are collected by a microscope lens at right angle. Having passed the lens, the fluorescence and scatter light are divided by a beam splitter into the side scatter and the fluorescence emission pathways. In front of the side scatter detector a fluorescence blocking optical filter is mounted analogous to the forward scatter channel. The fluorescence light passes an optical filter that only transmits the fluorescence wavelength range and blocks the excitation wavelength. Fluorescence light is registered by another PMT. The electrical pulses from the PMTs are fed to a computerized data acquisition unit that creates a data set for each measured particle containing the pulse intensity data of all photo detectors.

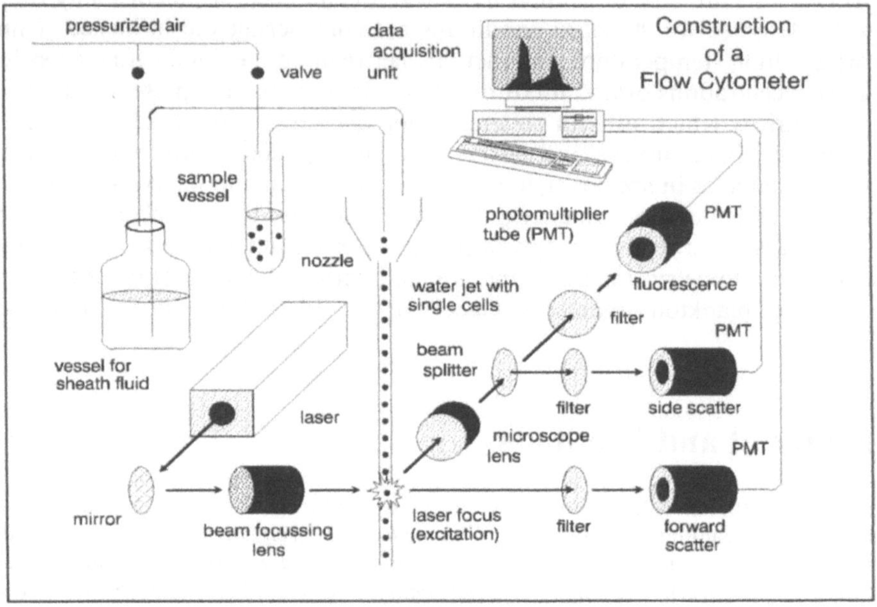

Fig. 1. Schematic view of a flow cytometer

Data were stored in listmode files and transferred for further analysis using the FAST-FILE program (Becton Dickinson) to an IBM compatible computer. The ***D****ata* ***A****nalysis* ***S****ystem* DAS (version 4.29) was used for all data analysis and graphics display as well (Beisker 1994). All parameters were measured logarithmically.

Pigment composition analyses: Excitation wavelengths were set to 458 nm and 528 nm with a laser power of 100 mW each. The fluorescence emission of chlorophyll was measured at wavelengths longer than 665 nm (Filter RG665, Schott). The amplifier gains of both chlorophyll fluorescence channels were adjusted to get a chlorophyll fluorescence ratio (CFR) of about 1 for chlorophytes. Phycoerythrin was excited at 528 nm and measured using a 575 ± 13 nm bandpass filter. FITC was excited at 488 nm (500 mW) and measured with a 530 ± 15 nm bandpass filter in separate runs.

Protein Staining with FITC for Flow Cytometry: All cells were fixed using 0.2% glutaraldehyde, spun down, washed with 0.5 M $NaHCO_3$ solution (pH 8) and stained with 200 µg/ml FITC (Fluorescein isothiocyanate, isomer I, Sigma F-7250, dissolved in DMSO, final DMSO concentration 0.1%) in the $NaHCO_3$ solution (4°C, 36 hours). Before flow analysis the cells were washed in the $NaHCO_3$ solution to remove unbound FITC (see also Schäfer et al. in press).

Photometrical Quantification of Cellular Protein: Cells in exponential growth phase were spun down after taking a subsample for cell titer determination by microscopical counting. The pellets were resuspended in 0.1 M NaOH containing 0.025 % (w/vol) SDS, and frozen in liquid nitrogen until French press treatment to disrupt the cells. After TCA-precipitation protein was measured photometrically using the SMITH bicinchonic acid assay (Berges 1993, Stoschek 1990).

Fluorescein Diacetate Studies: Fluorescein diacetate (FDA) is a nonfluorescent compound readily taken up by cells. Nonspecific esterases break the ester bonds producing the fluorescent compound fluorescein. The intensity of fluorescein fluorescence depends mainly on esterase activity and the integrity of cell membranes. After the assay conditions had been optimized with respect to temperature and algal cell density as well as FCD concentrations, a standard procedure was developed for flow cytometric analysis of FDA hydrolysis by intracellular esterases in *Chlamydomonas rheinhardii*: Aliquots of log phase cells were centrifuged at 1000g for 3 min, washed twice and resuspended. After the cell suspension (20,000 cells/ml) had been cooled down to 15°C, the reaction was started by adding 30 µM FDA (stock solution 20 mg/ml in acetone). Data were immediately collected for an interval of 15 min.

For FDA reaction kinetics, a single laser at an excitation wavelength of 488 nm (500 mW) was applied. Fluorescence emission intensity of fluorescein was recorded with a bandpath filter around 530 nm and the fluorescence intensity of chlorophyll - as stated above - at wavelengths longer than 665 nm (RG 665 nm, Schott).

Water samples. All water samples were fixed using 0.2 % glutaraldehyde immediately after sampling (Ruttner sampler for subsurface samples of Lake Tegernsee) and were kept in dark at 4°C for a maximum of two weeks until analysis.

Results and Discussion

Biomass Evaluations: Since flow cytometry is based on optical detection parameters such as light scattering, fluorescence or, scarcely practised, absorption of light there are principally two ways for measuring biomass of cells: Registering cell shape/volume by light scattering or using fluorescent dyes for staining cellular components such as DNA, RNA, lipids or proteins. Protein content is thought to be the best fit for biomass for there are some well known and highly specific dyes, for example FITC (Darzynkiewicz et al. 1985, Freemann and Crissmann 1975). The problem is the standardisation of the fluorescence units: What is the equivalence factor between one FITC fluorescence and one biomass (protein) unit?

Cell volume determination based on light scattering can be biased by vacuoles causing an overestimation of cellular material. The same applies to microscopic measurements and calculations. An exemplification is presented in Fig. 2. Here the FITC fluorescence of 12 algal species is plotted *vs.* their cell volume determined by microscopic measurements. The FITC fluorescence of most algal species fits well with the cell volume except for one diatom (*Cyclotella cryptica*) and the two chrysophytes *Ochromonas* and *Poteriochromonas*, three highly vacuolated algal species.

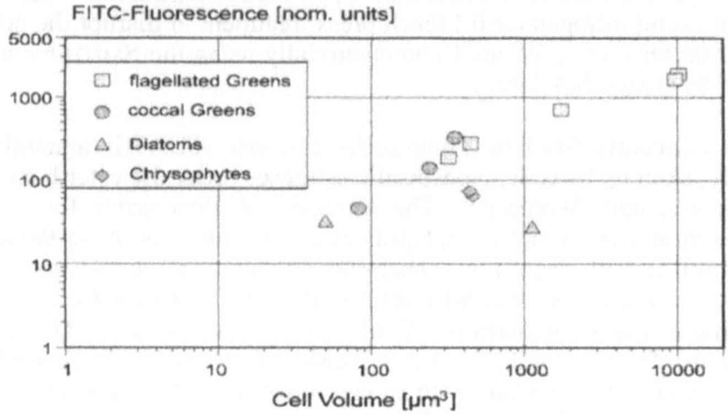

Fig. 2. FITC fluorescence *vs.* cell volume of 12 algal species

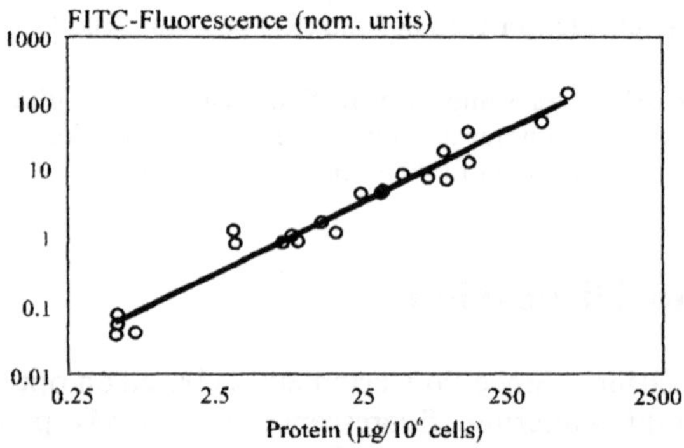

Fig. 3. FITC protein fluorescence *vs.* cellular protein content

As next protein reference we tested the quantification with photometrical procedures. Fig. 3 shows the high correlation (r=0.979, determined in a logarithmic scale) of mean protein FITC fluorescence and cellular protein content measured photometrically by the Smith bicinchonic acid method with laboratory grown algae and the ciliate. The range of mean cell volume determined by image analysis spreads over 5 decades from 0.8 (*Synechococcus*) to about 10,000 µm^3 (*Tetrahymena*). Though extraction of pigments would slightly improve the staining results in some species, extraction is not appropriate for evaluating natural samples since chlorophyll fluorescence is strictly needed as identification parameter.

Integrity of Phytoplankton Individuals: Several fluorescent probes have been tested on a range of different systematic phytoplankton species. These intracellular indicators of physiological processes can become fluorescent or change the intensity or spectral quality of their fluorescence. Prior to flow cytometry it had to be examined microscopically whether or not the stains were able to penetrate the cell walls of the vital phytoplankton organisms (Schäfer et al. in press).

Results of the FDA series are displayed in Figs. 4a-b. The typical reaction kinetic of the selected cells is shown in Fig. 4a. Only active cells, that are cells with increasing fluorescein fluorescence intensity with time, were selected for further calculations.

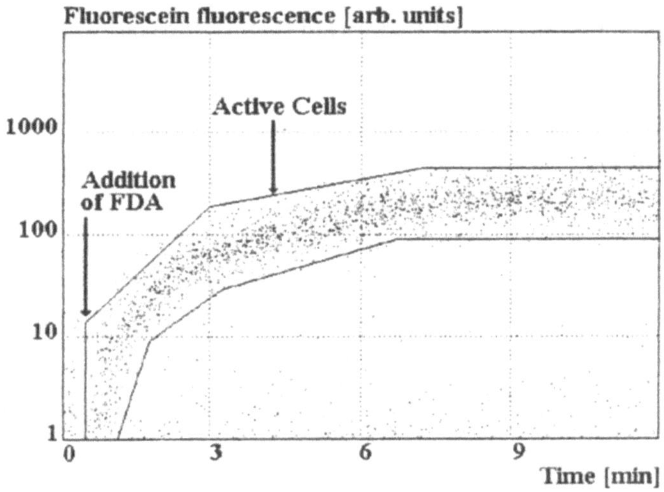

Fig. 4a. Hydrolysis of FDA in *Chlamydomonas rheinhardii*: FDA reaction kinetics. arb. = arbitrary

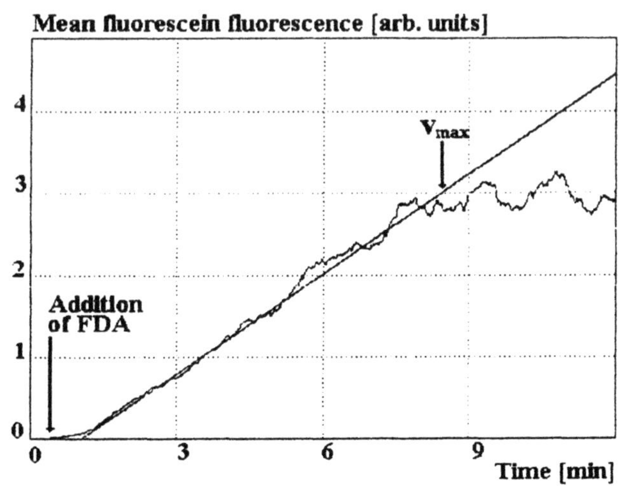

Fig. 4b. Hydrolysis of FDA in *Chlamydomonas rheinhardii*: Determination of v_{max}

For time dependent analysis of FDA hydrolysis for each time point t the means of the fluorescence emission intensity of cells measured in an interval ΔTs around t were calculated and transferred to a one dimensional data set for further regression analysis (Fig. 4b). From this data set the steepest slope of the curve was determined. This is equal to the maximum reaction velocity (v_{max}) expressed in arbitrary units of the increase of fluorescein fluorescence per second.

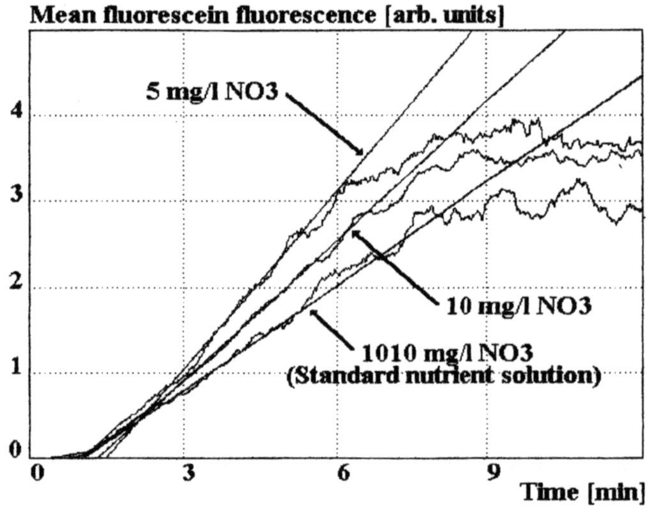

Fig. 4c. Hydrolysis of FDA in *Chlamydomonas rheinhardii*: Impact of various nitrate concentrations

To demonstrate the feasibility of this assay, one example will be given. Besides the effects of toxic compounds (see Schäfer et al. in press.) on cellular activity, physiological changes in nutrient supply can be determined. In Fig. 4c the reaction kinetics of log phase cells grown on standard nitrate concentration (1010 mg/l KNO_3) and reduced nitrate concentrations (5 and 10 mg/l KNO_3) are shown with their calculated maximum reaction velocity v_{max}. The reaction velocity of intracellular FDA hydrolysis under lowered nitrate conditions increased as compared with the control. This can be explained by the stimulating effect of low nitrate availability on the intracellular enzyme system of *C. rheinhardii*. At the same time the growth rate of *C. rheinhardii*, grown under low nitrate conditions, was reduced significantly (not shown).

Though the presented integrity assessment of phytoplankton on the basis of individuals is not of very sensitivity, it clearly expresses the feasibility of this approach. In combination with the assessment of ataxonomic structure evaluation (see below) it will improve our knowledge of biocenotic structures on a relatively abstract level.

Photosynthesis Pigment Composition: Fig. 5 illustrates the CFR analysis as one example for an ataxonomic structural evaluation of freshwater samples with regard to photosynthesis pigment autofluorescence. This analysis differentiates between algal groups with low carotenoid content as accessory pigments and groups with high carotenoid content.

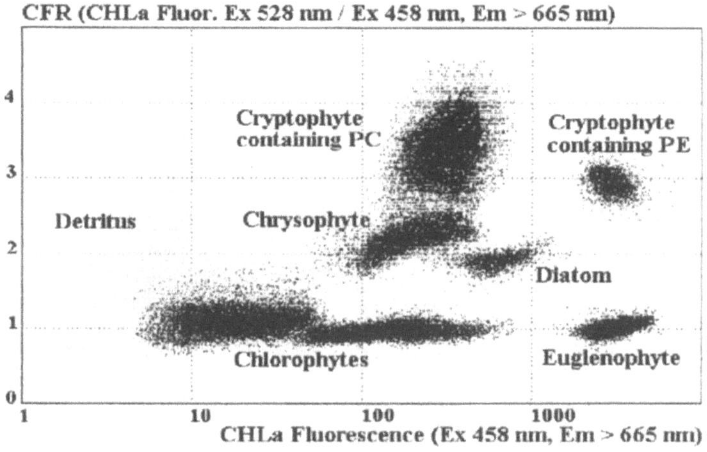

Fig. 5. CFR of Algae of Different Systematic Groups; CFR: CHL fluorescence (>665nm) Ex 528nm / CHLa fluorescence Ex 458; PC = phycocyanin, PE = phycoerythrin

Carotenoids are effectively excited around 530 nm and pass the absorbed energy immediately to CHLa which then releases excessive energy as dark red fluorescence light emission (>660 nm). Since CHLa has one excitation maximum around 450 nm, and shows only weak absorption in the green (around 530 nm), algae with minor carotenoid content exhibit strong CHLa fluorescence if illuminated in the blue (458 nm) and show only weak fluorescence if excited with green light. However, cells rich in carotenoids fluoresce strongly both if illuminated with blue or green light.

If the ratio of CHLa-fluorescence excited with 528 nm and of 458 nm excitation (CFR) is calculated, algae containing minor amounts of carotenoids show CFR values around 1 (the chlorophytes and the euglenophyte in Fig. 5), whereas cells rich in carotenoids exhibit higher CFR values (>1.5) (the chrysophyte and the diatom in Fig. 5). Cryptophytes also show high CFR values (>>2).

Fig. 6 depicts another combination of pigment autofluorescence evaluation especially suitable for species containing PE (many cyanophytes and cryptophytes). Algae without PE exhibit much less fluorescence intensity around 575 nm with respect to cell volume or pigment content. The small PE containing cyanophyte in Fig. 6 shows much higher fluorescence intensity than the chlorophyte and the diatom even though the latter are much larger. The fluorescence of the PE containing cryptophyte is also quite strong, whereas the two phycocyanin rich cyanophytes exhibit only weak fluorescence emission.

Another accessory pigment of cyanophytes and cryptophytes, phycocyanin (Ex around 630 nm, Em 650 nm) cannot be measured with our equipment right now, and even if a HeNe laser (632 nm) is available, discrimination of phycocyanin fluorescence emission from chlorophyll fluorescence is almost impossible since CHLa is excited also very effectively at 632 nm.

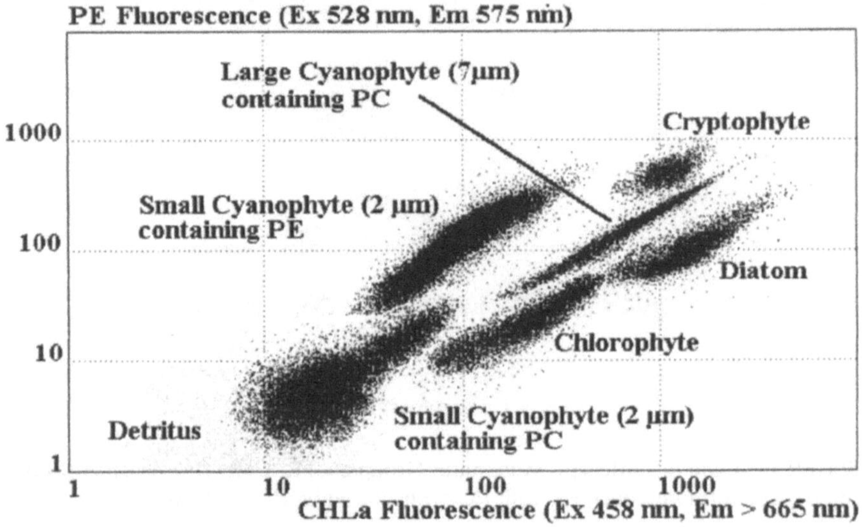

Fig. 6. PE Fluorescence of Algae of Different Systematic Groups; PC = phycocyanin PE = phycoerythrin

Structural Evaluation of Phytoplankton Samples: The vertical distribution of the **Lake Tegernsee** phytoplankton (sampling date 20.8.94) correlates well with the temperature gradient (Fig. 7). In the upper mixing layer (0 to ca 9 m) the phytoplankton concentration is high and rather constant, but drops to less than 25% from 10 to 20 m. In the deeper layer (>20 m) the cell concentration is again quite constant, although the eukaryotic algae are decreasing with increasing depth. The small picoplanktic cyanophytes (SLPs, „*Synechococcus/Synechocystis*-like particles") are dominating in number from surface to bottom. In classical evaluation, they have been excluded, as simulated in the above part. Centric diatoms (together with chrysophytes indicated as Carot.+) represent the largest part of the eukaryotic algae (around 2 % of the total number of phytoplankton cells), the cell density of chlorophytes (Carot.-) and cryptophytes (Crypt.) is even less (about 1, respectively 0.5 %).

Since CHLa fluorescence is decreasing quite slowly after cell death, the particles showing CHLa fluorescence in the deep water samples can be vital cells, but also dead or dying algae sedimenting from the upper layers. Vital staining, for example an enzyme activity assay using fluorescein diacetate (FDA) could be applied for answering this question (Schäfer et al. in press).

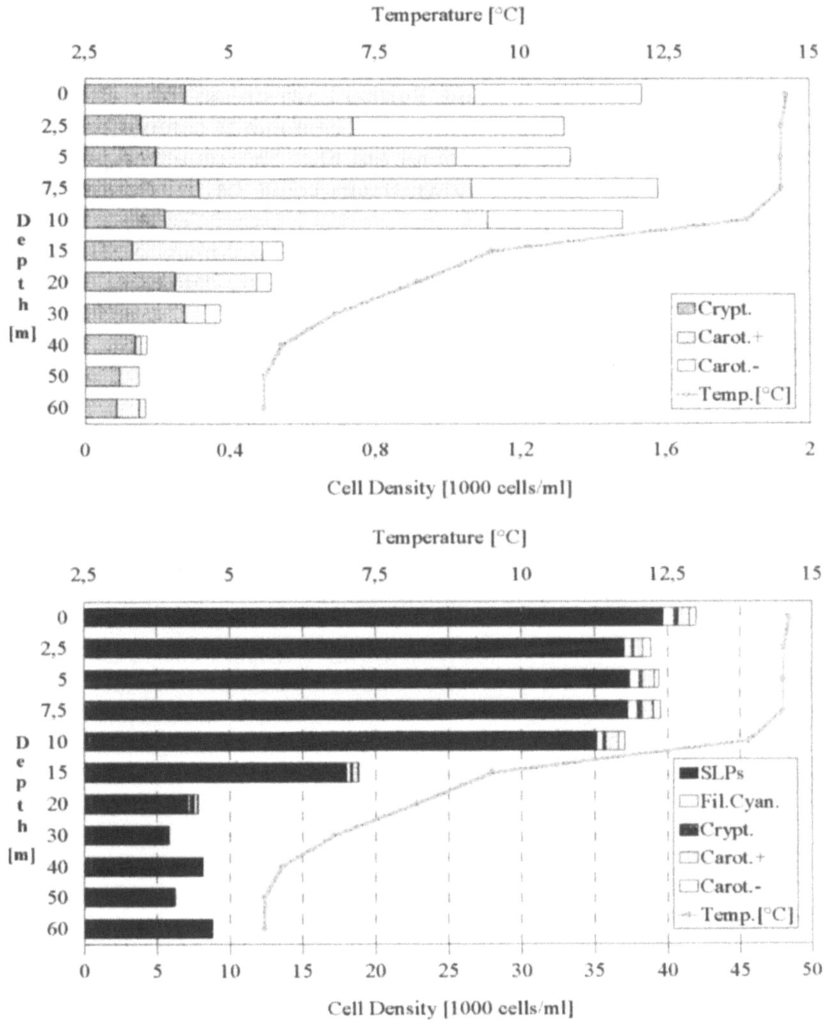

Fig. 7. Phytoplankton Community Structure of Lake Tegernsee which is clearly dominated by picoplanktic cyanobacteria. Above: "classical" structure evaluation of eucaryotics only. Below: Complete structural evaluation including picoplanktic procayots as well.
Car-: chlorophytes and euglenophytes; Car+: diatoms and chrysophytes; Cry: cryptophytes; fCyan: small filamentous cyanophytes; SLPs small picoplanktic cyanophytes.

River Wörnitz is a typical lowland river of northern Bavaria with much agricultural activities within the catchment area. From its source south of Rothenburg ob der Tauber it flows over a distance of about 100 km into the Danube nearby Donauwörth. Except of its upper course (up to km 75) it is flowing very sluggishly and it is dammed up several times. There are also several bayous in the lower course and therefore a rich phytoplankton community can be found (Fig. 8). Up to km 75 chlorophytes are dominating, mostly epilithic and epiphytic species ripped off by the water current. Further downstream (from km 68) centric diatoms and chrysophytes are dominating, and community composition as well as cell density remain quite stable. Around km 68 a maximum of phytoplankton density is attained, the decrease downstream could be caused by increasing populations of zooplanktic grazers.

In contrast to the mentioned Lake Tegernsee, no specimens of SLPs could be found in the nutrient-rich River Wörnitz at that sampling date.

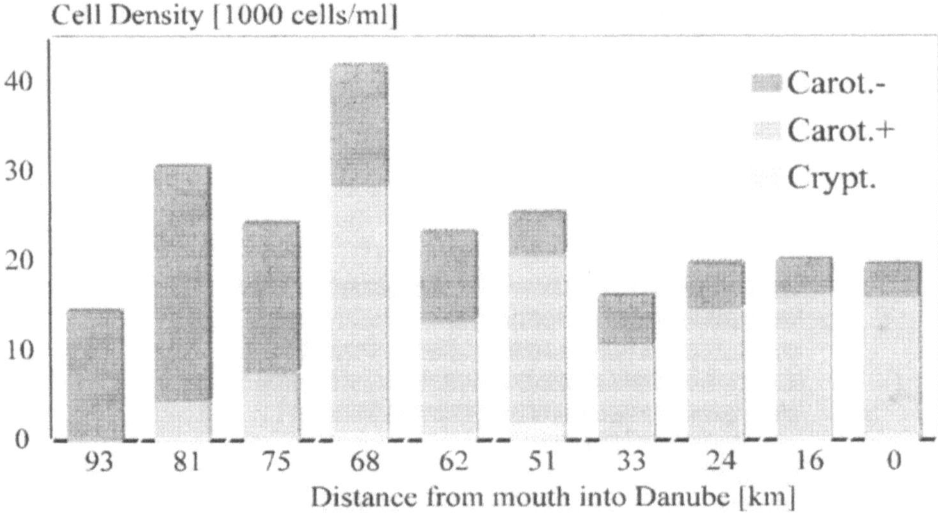

Fig. 8. Phytoplankton Community Structure in the River Wörnitz
For abbreviations, see Fig. 7

These two examples of phytoplankton structure evaluations clearly confirm the feasibility of the ataxonomic approach. The identification and quantification of major phytoplankton classes (according to their pigment content) combined with biomass determinations (as outlined above) will allow the determination of biomass spectra of the bacterio-, phyto- and microzooplankton. Biomass spectra itselves can serve as an indicator of the energy flow within the biocenosis under study. From energy flow structure, a measure of integrity of the biocenosis in question can be derived. This will be outlined below.

Theoretical Consideration[1]: Sheldon et al. (1967, 1972) used electronic counters to automatically count size-classified planktic particles in the ocean. Unexpectedly, they found near-equal biomass of differently sized 'particles', meaning

---

[1] modified from Steinberg and Geller (1993)

organisms, when subsequent classes of size were ordered on a logarithmic scale. It is known that physiological rates of respiration, of growth, or generation lengths are first order dependents of body size (e.g. Peters 1983). These allometric relationships hold over about 20 orders of magnitude of body weights from unicells to big mammals, including the planktic organisms. Fenchel (1974) gave the first comprehensive overview in which both rates of respiration and of growth were treated as a common theme. With these relationships quantified in operational form, all the necessary pieces were available to establish a theory on a continuous size distribution in the pelagic ecosystem.

Progress towards development of a continuous model depended on leaving behind the notion of discrete trophic levels (Platt 1985), demanding organisms to be assigned to one trophic level, and that only a small number of mutually-exclusive trophic categories exist. Quantitative conclusions about overall-food-chain efficiencies furthermore require simplified assumptions about the ratio between the body sizes of predators and their prey, say a factor of 10 in body length or 10^3 by weight (Kerr 1974). Adverse to this assumption is the fact that the autotrophs in the sea are covering a broad range of cell size of at least 10^7 by weight, and to assign a given predator, probably feeding on *all* particles of suitable size, to only one trophic level appears as an unjustified simplification of the pelagic system (Platt 1985). Furthermore, in nutrient-rich systems with high concentrations of food particles, new predators appear and dominate, which use filtering apparatuses to acquire their food which is much smaller relative to their own size than in the above mentioned predator type, and may show a relation of body sizes of 10^2 to 10^3 by length, or 10^6 to 10^8 by weight (Geller 1991). The existence of different types of predators feeding on very differently sized food leads to more complex food webs, making obsolete the idea of simple food 'chains'.

Within the context of the size distribution in the sea, Azam et al. (1983) have analyzed the ecological role of heterotrophic microorganisms. They conclude that about 25% of the net primary production passes via dissolved organic matter through bacteria, which are grazed by microzooplankton (principally heterotrophic flagellates in the size range of 3 to 10 µm). This pathway, called the 'microbial loop', may be seen as an example of the material flux in the trophic continuum against the general direction of small too large. Microzooplankton, through its excreted metabolites, thus becomes implicated as playing a major part in the remineralization of nutrients in the sea, a role traditionally assigned to bacteria.

In the literature considering the relationship between the microbial loop and the known 'classical' components of the planktic food web, some kind of artificial separation emerged. In his concise review, Fenchel (1988) pointed to the unsuitability of this separation and showed the microbial loop to be a 'normal' constituent of planktic food chains consisting of multiple, parallel strains of energy flow, the basic organisational trait being the size-specific feeding, as well of 'herbivorous' grazers as of 'carnivorous' predators both feeding on all organic particles that are about one order of magnitude smaller than the consumer organism. All consumers, therefore, might be considered as 'omnivorous' since they feed on equally sized green and 'non-green' food as well. This basic structure of the planktic food web appears as a sufficient description of nutrient-poor planktic systems. Other systems with higher levels of nutrients and denser suspensions of food seemingly are amended by the appearance of additional filter-feeders as mentioned above.

Both the hereby given view of an organismic food web structure and the finding of an 'ataxonomic' continuous distribution of planktic biomass over all size classes appear as consistent and may be seen also from the physiological view. Body size is known (Peters 1983) as the first-order determinant of all physiological rates, e.g. of respiration and of growth (R_W), which decline with increasing individual body weight (W), $R_w \sim W^{-0.25}$, when measured on a weight-specific base. The respective time constants, for example generation length (T), increase with body weight in a comparable way: $T \sim W^{0.25}$, resulting by a total cumulative energy demand per life time (E_T) of $E_T \sim (T * R_W) * W^0$. The transfer efficiency of energy from food to consumer was also found to be independent of body size (Platt 1985), as well the efficiency of food utilization (Kleiber 1975), as the net growth efficiency (Banse 1976) and, resulting from both, also the production efficiency.

In the meantime, several investigations of the structure of planktic communities indicate a homogeneous distribution of biomass over all size classes from bacteria up to large zooplanktic predators in undisturbed stable water bodies like the open ocean and large lakes. Abundance is decreasing with increasing biomass, and a linear relationship between biomass and abundance is resulting as shown for Lake Constance in Fig. 9 (from Gaedke 1992).

As an example of irregularities in a biomass spectrum caused by waste water and xenobiotica pollution Fig. 9 (from Sprules and Munawar 1986) shows a biomass spectrum of Lake Superior. Apart from a large gap resulting from the absence of organisms in these size classes the biomass distribution is less regular if compared with Lake Constance.

It appears likely that the ataxonomic approach may serve as a measure of the completeness of the ecosystem, assuming that all possible niches determined by the first-order determinant body size, are filled. From extreme or polluted ecosystems, it is known that certain size classes are missing, leaving gaps in the size spectrum. Such an incomplete spectrum was found in an obviously non-balanced plankton system of a bay of Lake Superior by Sprules and Munawar (1986) (Fig. 9). This incomplete or disturbed system (gap in the size spectrum, greater standard errors, smaller r^2, see Table 1) appears as less mature than the described full spectrum. Intuitively, we expect the full systems to be more stable against perturbations, that is, diversity (of size) might produce stability (of the biomass pool and its size distribution).

The assessment procedure can involve two approaches:

(1) from the statistical parameters of the fitted straight line (see Table 1)

and

(2) from the time for recovery the system needs after the disturbance has been stopped.

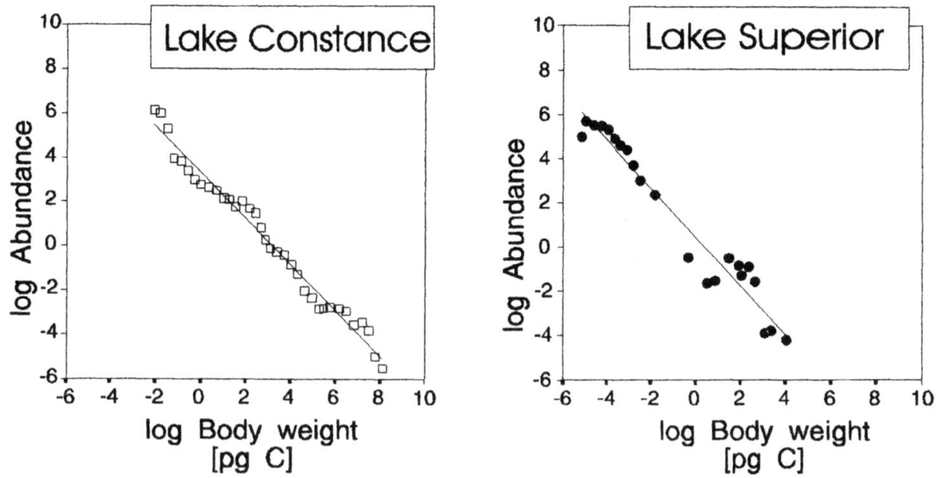

Fig. 9. Normalized biomass spectrum and fitted straight line of the entire plankton community in Lake Constance (left) and Lake Superior (right), respectively, averaged over the season and water column. (After Gaedke 1992, Sprules and Munawar 1986)

Table 1. Statistical parameters of the fitted lines for Lake Constance and Lake Superior in Fig. 9 according to: log Abundance = A - B•log Body weight

	Lake Constance	Lake Superior
A	3.277 $\pm$ 0.109	0.472 $\pm$ 0.176
B	-1.020 $\pm$ 0.026	-1.135 $\pm$ 0.057
r^2	0.98	0.95

These assumptions, however, to date are based more on intuitive feeling than on hard facts, and should be proven in future studies. Furthermore, such holistic approaches must be expanded to other ecosystems, and time as well as severity of both the impact and recovery has to be evaluated in case studies. Meanwhile, some studies have empirically proved this allometric relationship between body size and population abundance with biocenoses other than planktic ones. This applies to benthic animals of lakes (Strayer 1994) as well as animals of soils (Blackburn et al. 1993, Cotgreave 1993). Some of the studies show only rather weak relationship that is probably due to fact that the studies were restricted to animals only, excluding primary producers and microbes. Nevertheless, these studies indicate the feasibility to identify major natural and anthropogenic stressors for the biocenoses under study.

For routinely evaluations of the planktic system integrity, both approaches mentioned above (Lake Superior as well as Lake Constance) have some severe limitations: The biomass spectrum of Lake Superior has been analyzed by means of a Coulter counter that means semi-quantitatively. However, this application is of limited value, since the apparatus only counts particles and is not able to distinguish between living (bacteria, algae, microzooplankters) and dead (detritus) ones. With the Lake Constance biomass spectrum, this limitation has been overcome by means of individual microscopic countings. However, this approach is time- and labor-consuming and can lead to overestimations of biomass, as shown above.

The outlined application of flow cytometry will overcome the limitations of both previous approaches. It is able to

- distinguish between living and dead particles due to the individual and specific pigment content of the particles
- count automatically
- evaluate biomass (protein) content of single particles.

The combination of integrity evaluation on different levels of ecological aggregation (individual, population or even biocenosis level) will provide a tool of identifying short-term as well as (relatively) long-term impacts on planktic systems and help to perceive major stressors.

Conclusion

Some examples encouraging a more intense application of flow cytometry in limnology and aquatic ecotoxicology have been presented in this contribution. CFA analysis and FITC protein staining for creating biomass spectra of bacterio-, phyto- and microzooplankton are assumed to be powerful structural-ataxonomic approaches for describing plankton communities and identifying alterations due to internal and external stressors. In addition, physiological parameters such as for instance membrane integrity and esterases activity evaluate the integrity of individual plankton organisms on a short-term basis.

Acknowledgement

The authors wish to express their gratitude to Dr. Andreas Schrimpf, Wasserwirtschaftsamt Rosenheim, for kindly supply of the Tegernsee samples and the physical data and to Mr. Ulrich Kentner from the University of Constance for the *Synechococcus* isolates from Lake Constance.

References

Azam F, Fenchel T, Field JG, Gray JS., Meyer-Reil LA and Thingstad F (1983): The ecological role of water-column microbes in the sea. Mar. Ecol. Prog. Ser. 10:257-263

Banse K (1976) Rates of growth, respiration and photosynthesis of unicellular algae as related to cell size - a review. J Phycol 12:135-140

Beeler SooHoo JB, Kiefer DA, Collins DJ, McDermid IS (1986) In vivo fluorescence excitation and absorption spectra of marina phytoplankton: I. Taxonomic characteristics and responses to photoadaptation. J Plankt Res 8:197-214

Beisker W (1994) A new combined integral-light and slit-scan data analysis system (DAS) for flow cytometry. Comp Meth Progr Biomed 42:15-26

Berges JA, Fisher AE, Harrison PJ (1993) A comparison of Lowry, Bradford and Smith protein assays using different standards and protein isolated from the marine diatom *Thalassiosira pseudonana*. Mar Biol 115:187-193

Blackburn TM, Brown VK, Doube BM, Greenwood JJD, Lawton JH and Stork NE (1993) The relationship between abundance and body size in natural animal assemblages. J Animal Ecol 62:519-528

Campbell JW, Yentsch CM (1989a) Variance within homogeneous phytoplankton populations, I.: Theoretical framework for interpreting histograms. Cytometry 10:587-595

Campbell JW, Yentsch CM (1989b) Variance within homogeneous phytoplankton populations, II.: Analysis of clonal cultures. Cytometry 10:596-604

Campbell JW, Yentsch CM, Cucci TL (1989) Variance within homogeneous phytoplankton populations, III.: Analysis of natural populations. Cytometry 10:605-611

Cotgreave P .: The relationship between body size and population abundance in animals. Trends Ecol. Evol. 8, 244-248.

Darzynkiewicz HA, Crissman Z, Tobey RA, Steinkamp JA (1985) Correlated measurements of DNA, RNA, and protein in individual cells by flow cytometry. Science 228:1321-1324

Dorsey J, Yentsch CM, Mayo S, McKenna (1989) Rapid analytical technique for the assessment of cell metabolic activity in marine microalgae. Cytometry 10:622-631

Fenchel T (1974) Intrinsic rate of natural increase: The relationship with body size. Oecologia 14:317-316

Fenchel T (1988): Marine plankton food chains. Ann Rev Ecol Syst 19:19-38

Freeman DA, Crissman HA (1975) Evaluation of six fluorescent protein stains for use in flow microfluorometry. Stain Technology 50:279-284

Gaedke U (1992) The size distribution of plankton biomass in a large lake and its seasonal variability. Limnol Oceanogr 37:1202-1220

Geller W (1991) Die Planktongemeinschaft als dynamisches System. In '*Ökologie der oberbayerischen Seen*'. Komm Ökol Bayer Akad Wiss, Pfeil Verl, Munich, pp31-47

Kerr SR (1974) Theory of size distribution in ecological communities. J Fish Res Board Can 31:1859-1862

Kleiber M (1975) *The fire of life*. 2nd edn, Krieger, New York.

Legendre L, Yentsch CM (1989) Overview of flow cytometry and image analysis in biological oceanography and limnology. Cytometry 10:501-510

Li WKW (1994) Primary production of prochlorophytes, cyanobacteria, and eukaryotic ultraplankton: Measurements from flow cytometric sorting. Limnol Oceanogr 39:169-174

McMurter HJG, Pick FR (1994) Fluorescence characteristics of a natural assemblage of freshwater picocyanobacteria. J Plankt Res 16:911-925

Peters RH (1983) *The ecological implications of body size*. Cambridge Univ Press, 329pp

Phinney DA, Cucci TL (1989) Flow cytometry and phytoplankton. Cytometry 10:511-521

Platt T. (1985) Structure of the marine ecosystem: Its allometric basis. Can Bull Fish Aquat Sci 213:55-64

Schäfer H, Siedler M, Beisker W Müller K, Steinberg C (1995) Die Verwendung der Durchflußzytometrie zur ataxonomischen Charakterisierung von Phytoplankton. Vom Wasser 84:379-389

Schäfer H, Siedler M, Beisker W, Steinberg C: Assessment of functional and structural parameters of unicellular planktic algae by use of flow cytometry. Inter Rev ges Hydrobiol (in press)

Sieracki G, Valet G, Cucci T (1993) Report on the advanced workshop on fluorescent probes for marine flow cytometry. Use of fluorescent probes in the study of phytoplankton physiology and cellular biochemistry. Signal Noise 6

Sheldon RW and Parsons TR (1967) A continuous size spectrum for particulate matter in the sea. J Fish Res Board Can 24:909-915

Sheldon RW, Prakash A and Sutcliffe WH (1972) The size distribution of particles in the ocean. Limnol Oceanogr 17:327-340

Sprules WG, Munawar M (1986) Plankton size spectra in relation to ecosystem productivity, size, and perturbation. Can J Fish Aquat Sci 43:1789-1794

Steinberg CEW and Geller W (1993) Biodiversity and interaction within pelagic nutrient cycling and productivity. In: ***Biodiversity and Ecosystem Function.*** Eds.: Schulze ED and Mooney HA, Ecological Studies 99, Springer-Verlag, Heidelberg: pp43-64.

Steinberg CEW, Geyer HJ, Kettrup AF (1994) Evaluation of xenobiotic effects by ecological techniques. Chemosphere 28:357-374

Stoscheck CM (1990) Quantitation of protein. Methods in Enzymology 182:50-68

Strayer DL (1994).: Body size and abundance of benthic animals in Mirror Lake, New Hampshire. Freshwat Biol 32:83-90.

Troussellier M, Courties C, Vaquer A (1993) Recent applications of flow cytometry in aquatic microbial ecology. Bio Cell 78:111-121

Yentsch CS, Phinney DA (1985) Spectral fluorescence: An ataxonomic tool for studying the structure of phytoplankton populations. J Plankt Res 7:617-632

The Young Scientist Poster Award

The Young Scientist Poster Award is given to the first author of the best poster, who must be younger than 35 years.

The winner is announced at the Closing Ceremony of the EUROTOX-Congress, and the abstract is published in the Congress Proceedings.

The winner 1995 was P. Soucek, Prague.

Identification of the Highly Polymorphic S-Mephenytoin Hydroxylase in Humans

P. Soucek[1], Z. Guo[2], P. Sandhu[2], M.V. Martin[2], J. Cosme[3], P.H. Beaune[3], and F.P. Guengerich[2]
[1]National Institute of Public Health, Center of Occupational Diseases, Šrobárova 48, Prague 10, 100 42, Czech Republic
[2]Department of Biochemistry and Center in Molecular Toxicology, Vanderbilt University, Nashville, TN, USA
[3]INSERM, U75, Université René Descartes, Paris, France

The metabolism of the anticonvulsant drug mephenytoin is highly polymorphic in Oriental populations (about 20 % of slow metabolizers) in comparison with Caucasians, where 2 - 5 % of slow metabolizers are found. It was reported that 4-hydroxylation of the S-enantiomer of mephenytoin is catalyzed by a cytochrome P450 enzyme from the 2C-family. A large number of efforts have been made to characterize this enzyme, but so far its real identity remained unknown.

S-Mephenytoin hydroxylation activity was assayed in microsomes from genetically engineered yeast, expressing cytochromes P450 2C8, 2C9, and 2C18, resp. Activity was also assayed in a panel of human liver microsomes, and the most extensively metabolizing microsomes were used for affinity purification of P450 proteins crossreacting with anti-2C10 antibodies.

The first experiment showed that microsomes from genetically engineered yeast, overexpressing P450 isoforms 2C8, 2C9/10, 2C18, and 3A4, do not metabolize S-mephenytoin at measurable rates. All these microsomes, however, were metabolically active, as judged by tolbutamide hydroxylation. When S-mephenytoin was assayed in a panel of human liver microsomes, reasonably high activities were observed. Immunoblots, using these microsomes, were stained with anti-human-2C10-antiserum, and they showed four bands. The staining intensity of band 3 correlated with S-mephenytoin hydroxylation activity in human liver microsomes. The 2C10 IgG fraction was then used for immunoaffinity chromatography of the most extensively metabolizing human liver microsome sample. The preparation was electrophoretically separated, blotted onto nitrocellulose and stained for total protein. This blot again showed four bands, which were then excised and subjected to protein sequencing. The direct comparison of the first 15 aminoacids from the N-terminus demonstrated that band 1 corresponded to the human CYP 2C9/10, and that band 3 matched the sequence of the first 9 aminoacids of CYP 2C19. The conclusion of this study, that cytochrome P450 2C19 is responsible for the 4-hydroxylation of S-mephenytoin in humans is further supported by recent findings of Goldstein et al. (Biochemistry **33:** 1743-1752, 1994), who expressed CYP 2C19 in yeast and proved that this protein is the principal S-mephenytoin hydroxylase in humans.

Subject Index

MIX
Papier aus verantwortungsvollen Quellen
Paper from responsible sources
FSC® C105338

If you have any concerns about our products,
you can contact us on
ProductSafety@springernature.com

In case Publisher is established outside the EU,
the EU authorized representative is:
Springer Nature Customer Service Center GmbH
Europaplatz 3, 69115 Heidelberg, Germany

Printed by Libri Plureos GmbH
in Hamburg, Germany